A. J. E. van Bel W. J. P. van Kesteren (Eds.)

Plasmodesmata

Structure, Function, Role in Cell Communication

Springer
Berlin
Heidelberg
New York
Barcelona
Hong Kong
London
Milan
Paris
Singapore
Tokyo

Aart J. E. van Bel Wilhelmus J. P. van Kesteren (Eds.)

Plasmodesmata

Structure, Function, Role in Cell Communication

With 105 Figures

Springer

Professor Dr. A. J. E. van Bel
Dr. W. J. P. van Kesteren
Institut für Allgemeine Botanik
und Pflanzenphysiologie
Justus-Liebig-Universität Giessen
Senckenbergstr. 17
D-35390 Giessen
Germany

ISBN-13: 978-3-642-64229-6 e-ISBN-13: 978-3-642-60035-7
DOI: 10.1007/978-3-642-60035-7

Library of Congress Cataloging-in-Publication Data
Plasmodesmata : structure, function, role in cell communication / Aart J. E. van Bel, Wilhelmus J. P. van Kesteren (eds.). p. cm. Includes bibliographical references and index. ISBN-13: 978-3-642-64229-6
1. Plasmodesmata. I. Bel, Aart Jan Eeuwe van, 1943- . II. Kesteren, Wilhelmus J. P. van (Wilhelmus Johannes Petrus), 1960- . QK725.P596 1999 571.6'42--dc21

Softcover reprint of the hardcover 1st edition 1999

Production: PRO EDIT GmbH, D-69126 Heidelberg

Typesetting: Zechnersche Buchdruckerei, Speyer
Cover Design by design & production, D-69121 Heidelberg

SPIN: 10567541 31/3137-5 4 3 2 1 0 –

Preface

Plasmodesmata have fascinated generations of plant biologists whose initial attempts to unravel the secrets of the plasmodesmata were frustrated by the lack of the right tools. This changed during the 1960s. With the advent of electron microscopy, the basics of the plasmodesmal substructure were disclosed. Subsequently, physiological research was boosted by the availability of specialized fluorescent dyes. Since then, integrated application of molecular-biological approaches has potentiated the capacity of the researchers. It is no surprise, therefore, that insights into plasmodesmal substructure and functioning have been gained at an increasing rate over the past years. A large part of the book deals with acquired knowledge on the substructure and physiology of plasmodesmata and the tasks they fulfil. Account will be given of current concepts and attention be paid to exciting prospects concerning the significance of plasmodesmata for the plant as a whole.

It seems obvious that we are only beginning to understand function and functioning of the plasmodesmata despite all progress. One reaches this conclusion when overlooking the amazing discoveries over the past decade, but also the numerous claims and speculations with regard to plasmodesmal properties. These range from the various ways in which plasmodesmal gating is induced by viruses, their ability to modulate in response to certain stimuli, the potential to transport macromolecules such as dextrans and proteins, and the capacity of some plasmodesmata to traffic mRNA.

Admittedly, some of the knowledge need not to be entirely solid, as we are not quite sure to what extent experimental manipulation of the cells provokes artefactual reactions. Interpretation of electron microscope images strongly depends on the fixation and contrasting methods and suffers from the minute size of the plasmodesmata. Moreover, intracellular injection of fluorochromes, either by pressure or by electrical current, may effect the opening status of the plasmodesmata. Several chapters highlight the experimental pitfalls and a few methodological limitations.

The urge to develop intercellular connections has probably emerged more than once during evolution. Given the diversity of intercellular connections, there is more than one solution to cope with the demands of multicellularity as a number of chapters testify.

Comprehensive contributions on some emerging themes were envisaged, but time needed for inclusion of these matters would have outdated the other chapters. Editing a survey book is a race against the clock that one is bound to lose if time constraints are ignored. Nevertheless, the information collected here is a prelude to future research and indicates the way research must take such as identifying genes and molecules involved in the construction of and trafficking through plasmodesmata.

Finally, we are very grateful to the authors who joined this project. We admire their commitment and dedication and have enjoyed the vivid exchange of ideas on various subjects. Their patience to comply with the ever-lasting flow of corrections and remarks was immense. Hopefully, their efforts will further stimulate the research on plasmodesmata as the outstanding book of Brian Gunning and Tony Robards did more than 20 years ago.

January 1999 Aart van Bel and Pim van Kesteren

Contents

List of Contributors

Dr. A. J. E. van Bel
Institut für Allgemeine Botanik und Pflanzenphysiologie,
Justus-Liebig-Universität Giessen, Senckenbergstr. 17, 35390 Giessen, Germany

Dr. Pim van Kesteren
Institut für Allgemeine Botanik und Pflanzenphysiologie,
Justus-Liebig-Universität Giessen, Senckenbergstr. 17, 35390 Giessen, Germany

Dr. C. E. J. Botha
Department of Botany and Electron Microscopy Unit, Rhodes University, P. O. Box 94,
Grahamstown 6140, South Africa

Dr. B. Ding
Department of Botany, Oklahoma State University, Stillwater, Oklahoma 74078, USA

Dr. H. V. M. van Rijen
Dept. of Medical Physiology and Sports Medicine, Faculty of Medicine,
Utrecht University, Universiteitsweg 100, 3584 CG Utrecht, The Netherlands

Dr. P. B. Goodwin
Department of Crop Sciences, University of Sydney, New South Wales 2006, Australia

Dr. K. J. Oparka
Unit of Cell Biology, Department of Virology, Scottish Crop Research Institute,
Invergowrie, Dundee DD2 5DA, UK

Dr. M. E. Cook
Department of Biological Sciences, Illinois State University, Campus Box 4120,
Normal, Illinois 61790-4120 , USA

Dr. W. H. Müller
Utrecht University, Molecular Cell Biology, EMSA-Centraal Bureau voor Schimmel-
cultures, Padualaan 8, NL-3584 CH Utrecht, The Netherlands

Dr. R. L. Overall
School of Biological Sciences, Macleay Bldg A12, The University of Sydney,
New South Wales 2006, Australia

Dr. R. Kollmann
Botanisches Institut der Christian-Albrechts-Universität, 24098 Kiel, Germany

Dr. A. Schulz
Dept. of Plant Biology, The Royal Veterinary and Agricultural University, DK-1871 Frederiksberg C, Copenhagen, Denmark

Dr. Maria Kwiatkowska
Department of Cytophysiology, University of Lódz, Ul. Pilarskiego 14, 90-231 Lódz, Poland

Dr. C. van der Schoot
Agrotechnical Research Institute, ATO-DLO, P. O. Box 17, 6700 AA Wageningen, The Netherlands

Dr. Katrin Ehlers
Institut für Allgemeine Botanik und Pflanzenphysiologie, Justus-Liebig-Universität Giessen, Senckenbergstr. 17, 35390 Giessen, Germany

Dr. D. U. Beebe
Institut de recherche en biologie végétale, Université de Montréal, 4101 est, rue Sherbrooke, Montréal (Québec) H1X 2B2, Canada

Dr. G. A. Thompson
Department of Plant Sciences, 303 Forbes Building, University of Arizona, Tucson, Arizona 85721, USA

Dr. R. S. Nelson
Samuel Roberts Noble Foundation, 2510 Sam Noble Parkway, P. O. Box 2180, Ardmore, Oklahoma 73402, USA

Abbreviations

AER	appressed endoplasmic reticulum
AM	apical meristem
AMV	alfalfa mosaic virus
ATP	adenosine triphosphate
ATPase	adenosine triphosphatase
BDMV	bean dwarf mosaic virus
BMV	brome mosaic virus
BNYVV	beet necrotic yellow vein virus
BS	bundle sheath
BSC	bundle sheath cell
BSMV	barley stripe mosaic virus
BWYV	beet western yellow virus
CaMV	cauliflower mosaic virus
CB	cascade blue hydrazide
CC	companion cell
CCCP	carbonylcyanide-m-chlorophenylhydrazone
CCD	charge-coupled device
CCMV	cowpea chlorotic mottle virus
CD	cytochalasin D
cDNA	complementary DNA
CF	5(6)-carboxyfluorescein
CLSM	confocal laser scanning microscope
CMotV	carrot mottle virus
CMV	cucumber mosaic virus
CP	coat protein
CPMV	cowpea mosaic virus
cx	connexin
Da	dalton
DCB	dichlobenil
DCN	dye-coupling number
DDG	2-deoxy-D-glucose
def	deficiens (gene)
DEF	deficiens (protein)
DGD	diethylene glycol distearate
DNA	deoxyribonucleic acid

dpi	days post inoculation
dsDNA	double-stranded DNA
EM	electron microscopy
ER	endoplasmic reticulum
F	fluorescein
FDA	fluorescein diacetate
FGlu	fluorescein glutamate
FITC	fluorescein isothiocyanate
flo	floricaula (gene)
FLO	floricaula (protein)
FM	floral meristem
FRAP	fluorescence recovery after photobleaching
GA	gibberelic acid
GFP	green fluorescent protein
Gln	glutamine
glo	globosa (gene)
GLO	globosa (protein)
Glu	glutamate
GRV	groundnut rosette virus
GUS	β-glucuronidase
HC-Pro	helper component proteinase
His	histidine
HPTS	8-hydroxypyrene-1,3,6-trisulphonic acid
IAA	indole-3-acetic acid
IC	intermediary cell
IgG	immunoglobulin G
IM	inflorescence meristem
IP_3	inositol triphosphate
kn	knotted (gene)
KN	knotted (protein)
kDa	kilodalton
kPa	kilopascal
LRB	lissamin rhodamine B
LY	lucifer yellow
LYCH	lucifer yellow CH
Mb	megabyte
MC	mesophyll cell
MEL	molecular exclusion limit
MP	movement protein
MR	middle region
MS	mestome sheath
MSC	mestome sheath cell
MSV	maize streak virus
MW	molecular weight
NA	nucleic acid
nam	no apical meristem
ORF	open reading frame

PAGE	polyacrylamide gel electrophoresis
PAP	plasmodesma-associated protein
PAS procedure	periodic acid Schiff procedure
PBS	parenchymatous bundle sheath
Pd	plasmodesma
PepMoV	pepper mottle virus
Phe	phenylalanine
PI-model	positional information model
PLRV	potato leafroll virus
PM	plasma membrane
PP	phloem parenchyma
PP 1, 2	phloem protein 1, 2
PPC	phloem parenchyma cell
ppi	pixels per inch
PPP	plasmodesma-permeant protein
P-protein	phloem protein
PPU	pore-plasmodesma unit
PR	pathogenesis-related
PVM	paraveinal mesophyll
PVX	potato virus X
PVY	potato virus Y
RCNMV	red clover necrotic mosaic virus
RNA	ribonucleic acid
SCP	supracellular control protein
SCSI	standard computer system interface
SDS	sodium dodecyl sulphate
SE	sieve element
SEL	size exclusion limit
SER	sieve element reticulum
SHMV	sunn-hemp mosaic virus
SPC	septal pore cap
STEP	sieve tube exudate protein
sxd	sucrose export defective
TAV	tomato aspermy virus
TBRV	tomato black ring virus
TBSV	tomato bushy stunt virus
TC	transfer cell
TCV	turnip crinkle virus
TEM	transmission electron microscopy
TEV	tobacco etch virus
TGB	triple gene block
ThWST	thin-walled sieve tube
TIFF	tagged information file format
TkWST	thick-walled sieve tube
TMV	tobacco mosaic virus
TOTO	benzothiazolium-4-quinolinium dimer
TPN-ADP	2'-0-(trinitrophenyl)adenosine-5"-diphosphate

TR	Texas red
TRITC	tetramethyl-rhodamine isothiocyanate
TRV	tobacco rattle virus
TYMV	turnip yellow mosaic virus
Tyr	tyrosine
UV	ultraviolet
vDNA	viral DNA
VP	vascular parenchyma
VPC	vascular parenchyma cell
vRNA	viral RNA
wt	wild-type
ZIO	zinc iodide osmiumtetroxide

Chapter 1

Plasmodesmata, a Maze of Questions

A. J. E. van Bel · S. Günther · W. J. P. van Kesteren

Institut für Allgemeine Botanik und Pflanzenphysiologie,
Justus-Liebig-Universität, Senckenbergstrasse 17, 35390 Giessen, Germany

This book bears witness to quantum leaps in the knowledge of plasmodesmal structure, functioning and function. At the same time, much is still obscure. Any overview of the literature on plasmodesmata reveals a multitude of untouched areas, uncertainties, uncertified assumptions and controversies. The number of those pertaining to basic issues is striking. One is tempted to believe, therefore, that we are only beginning to explore the amazing properties of plasmodesmata. In this chapter, an attempt is made to list some of the unresolved questions. Several of them are discussed more extensively in the following chapters, while others are only touched upon or hardly taken into consideration. Our goal here is to identify a few challenging areas and sketch some prospects on future plasmodesmal research.

1 What Was the Selective Pressure to Produce Plasmodesmata?

It is a widespread opinion that the emergence of plasmodesmata was an unavoidable event in the evolution to multicellular bodies of Viridiplantae. The communication corridors are supposed to be developed in order to sustain trafficking of metabolites, coordinate and integrate the activities of individual cells, and enable the differentiation between the cells (see Lucas et al. 1993). Despite the obvious logic of this idea, the need for plasmodesmata for the maintenance of multicellular Viridiplantae is not absolute. Some trichoid or thalloid representatives of Dinophyta, Haptophyta and most Xanthophyceae, Cladophorophyceae and Zygnematophyceae do not possess intercellular cytoplasmic bridges (Raven 1984; Cole and Sheath 1990; Sandgren et al. 1994; van den Hoek et al. 1995). Moreover, several syncytial algae (e.g. *Botrydium*, *Caulerpa*, *Vaucheria*) possess clearly differentiated bodies (Raven 1997), which renders the necessity of plasmodesmata for cellular differentiation questionable. These examples may simply be the exceptions to the rule. Too many good arguments support the view that plasmodesmata are essential for the functioning of multicellular plants. A compelling one is that plasmodesm(-like) structures have evolved in apparently unrelated groups of the Viridiplantae (Raven 1997). This functional convergence appears to show the success of the plasmodesmal concept.

What was the evolutionary stress to induce the emergence of plasmodesmata? Trying to provide the plausible answers, one cannot escape from the likelihood that the increase in plant size or cell number has triggered the development of cellular communication. Intercellular connections exist in algal taxa containing large and complex representatives, such as the Floridophyceae (multicellular Rhodophyta), the Phaeophyceae, and several groups of multicellular Chlorophyta (Raven 1997). In these taxa, there is a clear division of labour between the cells. The evolutionary strain for building intercellular connections may therefore have been cell heterogeneity rather than being multicellular above a critical size. Several investigators hypothesize a relationship between the presence of plasmodesmata and the mechanism of cell division (see Marchant 1976; Lucas et al. 1993). As Marchant noted (1976), however, "the ability to form plasmodesmata is not so much a consequence of a particular cytokinetic mechanism but rather depends upon whether an organism has the genetic capability for the expression of plasmodesmata".

That division of labour was the decisive factor to produce symplasmic contacts may be established by a future systematic study of the distribution of symplasmic connections among the Chlorophyta. The postulate predicts that symplasmic communication is absent as long as the cells are autotrophic and dispose of identical metabolic machineries. It must also be investigated if symplasmic bridges exist between vegetative and generative cells. When symplasmic connections occur only at the interface between generative and vegetative cells, it would support the view that division of labour was the evolutionary strain to produce intercellular connections.

Admittedly, the above comments oversimplify the factors having led to the evolution of symplasmic connections as the structural-physiological solution for integration of the cell activities. The structural responses to the evolutionary stress are intriguingly diverse and a source of numerous questions. Why do the pit connections in Rhodophyta (Pueschel 1990), for instance, differ so strongly from plasmodesmata? Why do ER-equipped plasmodesmata occur in a very limited number of fungi (Hashimoto et al. 1964; Hawker and Gooday 1967; Powell 1974; Marchant 1976), whereas the cell walls of the other fungi contain septal pores or dolipores (see chapt. 8, Müller et al.)? Why are ER corridors absent in the plasmodesmata of the major part of Phaeophyta and Chlorophyta (Fraser and Gunning 1969; Franceschi and Lucas 1982: Kwiatkowska and Maszewski 1986)? Only in Charophyceae, ER strands run through plasmodesmata with certainty as was shown by Cook et al. (1997) for *Chara zeylanica*. In view of the polymorphic ultrastructure, it is hard to believe that the evolution of symplasmic connections was a monophyletic event. For more detailed information the reader is referred to three excellent reviews (Lucas et al. 1993; Cook et al. 1997; Raven 1997) and to the Chapters 7, 8 and 12.

The ubiquitous presence of plasmodesmata in the Embryophyta suggests that it is the division of labour that requires communication between cells. Hence, the presence of plasmodesmata has probably been an important factor in the successful conquest of the land area. The evolution of plasmodesmata is outlined by Cook and Graham (Chapt. 7), with special emphasis on the structural description of plasmodesmata in the various plant groups. The functional role of plasmodesmata in Charophyceae is discussed by Kwiatkowska (Chapt. 12). As an alternative to plasmodesmata, the intercellular apparatus in Basidiomycetes is described by Müller et al. (Chapt. 8).

2 Why Did Plants and Animals Develop Different Evolutionary Solutions for Intercellular Communication?

Symplasmic connections have evolved along remarkably different lines in plants and animals. In some boundary regions of animal cells, apposed plasma membranes lie as close together as 2 to 4 nm. This gap is spanned by the protruding ends of many identical transmembrane proteins (connexins) forming a field of connexons which is named gap junction. Animal cells communicate via these connexons, six transmembrane connexin units arranged in a circle and enclosing a pore (Meiners et al. 1988).

Plant cells are linked more intimately by plasmodesmata. Plasma membrane and ERmembrane are continuous through the plasmodesm, the cytoplasmic solutes are in direct contact, and even cytoskeleton elements may interconnect through the plasmodesmata (White et al. 1994; see Overall, Chapt. 9). Incidentally, it should be noted that the cyanobacterial microplasmodesmata (Marchant 1976; Giddings and Staehelin 1981) may show a resemblance to the connexons (Lucas et al. 1993) and also that fungal septal pores may contain protein complexes with a connexon-like exclusion limit of 800 Da (Raven 1984).

The evolutionary strain to produce such intimate intercellular contacts in plants may have been the appreciable distance between the protoplasts and the nature of the extraprotoplasmic materials. In contrast to most animal tissues, plant cells possess a solid and thick cell wall barrier, and unlike animal cells, plants cells are hardly motile as these are encased in this skeleton. Nature, growth pattern and thickness of the cell wall have possibly prevented plant cells from communicating in the same way as their animal counterparts. Extracellular tools for intercellular sensing and messaging, such as matrix proteins like cadherins and integrins present in adhesion belts, desmosomes, and focal contacts in animals, have not been found in plants so far. To what extent intercellular communication occurs by proteins via the plant cell walls is unclear. Investigators are just beginning to discover a variety of wall proteins, some of which may be involved in intercellular messaging.

Thickness and composition of the cell walls could have urged the plants to develop cytosolic corridors when division of labour was practised. The resulting intimate cellular network allows more direct intercellular sensing and orchestration than those in animal tissues, and a commensurate degree of intercellular traffic. The higher potential to cooperation may have resulted in the "supracellular organization" of plants; plants are organized on a tissue basis rather than on a cell basis (Lucas et al. 1993). Plants are presumably subdivided in clusters – symplasm domains (Section 7; see Ehlers and van Bel, Chapt. 14) – in which the cells are connected via plasmodesmata (Erwee and Goodwin 1985; van der Schoot and van Bel 1990). This leaves us with the speculation that the supracellular organization is indirectly due to the presence of cell walls. Such an explanation looks more acceptable than the postulate of an organization master design being developed by evolutionary quests.

3 Are There Several Types of Plasmodesmata?

The majority of literature gives the impression that higher plants possess one single type of plasmodesm which modulates in response to the intra- and extracellular conditions. Some authors, alluding to more than one type of plasmodesmata, are proponents of a more differentiated opinion (e.g. Robinson-Beers and Evert 1991; Waigmann et al. 1997). Remarkably enough, any touch of differentiation is lost when it comes to the description of the plasmodesmal substructure; the attempt is made to squeeze all observations into one single concept. The question arises as to whether this strive for uniformity is not useless, since there are three lines of evidence in favour of functionally disparate plasmodesmata:

1. The multimorphology of plasmodesmata within a single species. The architecture of plasmodesmata within one species is diverse. Along the path from the mesophyll to the sieve elements in the small veins of *Saccharum officinale* leaves, covering a distance of about 100 mm, five types of plasmodesmata were identified that are morphologically different (Robinson-Beers and Evert 1991). The diversity may be associated with the respective tasks of the cells along this trajectory. An important feature of the multimorphology is the degree of branching, which is supposed to be correlated with the direction and bulk of intercellular transport (Gamalei 1990).

For instance, plasmodesmal branching at the interface between bundle sheath cells and intermediary cells intensifies during the sink-source transition in leaves (Gamalei 1990; Gagnon and Beebe 1996; Volk et al. 1996; see Beebe and Russin, Chapt. 15).

2. The origin of plasmodesmata. A second line of evidence in favour of a distinction between plasmodesmata concerns their meristematic origin. Plasmodesmata in tissues of vascular origin may deviate functionally from those in tissues of parenchymatic origin. In leaves, parenchymatic and provascular tissue develops independently (Esau 1965; Nelson and Dengler 1997) and are linked at a later stage by secondary plasmodesmata at the interface between mesophyll and vascular tissue (Ding and Lucas 1996; see Kollmann and Glockmann, Chapt. 10). The disparity of the plasmodesmata from the respective meristems is exemplified by the differential maturation of plasmodesmata in yeast invertase-transgenic tobacco plants (Ding et al. 1993)
3. Virus movement through plasmodesmata. The differential, functional behaviour of plasmodesmata becomes manifest through the traffic of viruses. Transport of virus from the bundle-sheath cells to vascular parenchyma or bundle-sheath cells in the minor veins is restricted for some viruses (Ding et al. 1995). Evidence that the limitation is due to a different plasmodesmal passageway comes from various studies. The movement protein of tobacco mosaic virus only increases the size exclusion limit of the plasmodesmata between mesophyll cells; plasmodesmata beyond the bundle-sheath demarcation are not modified (Ding et al. 1992a). Further, expression of the 2a protein of cucumber mosaic virus in transgenic tobacco plants prevents the virus from entering at the interface between bundle-sheath and phloem cells (Wintermantel et al. 1997). In contrast, luteoviruses are only capable of crossing the pore/plasmodesm units between sieve elements and companion cells (Sanger et al. 1994; Derrick and Barker 1997; Schmitz et al. 1997). Their movement proteins do not even seem to be capable of opening up plasmodesmata of other phloem cells. When combining these observations, we must conclude that at least three functional types of plasmodesmata occur in higher plants. Most likely, other types of plasmodesmata do exist; for instance, plasmodesmata between leaf hairs deviate from other plasmodesmata in that the ER-structures are able to widen (Waigmann et al. 1997).

4 The Ultrastructure of Plasmodesmata

The interpretation of plasmodesmal pictures has been the subject of fervent debates over the past years (Tilney et al. 1991; Ding et al. 1992b; Botha et al. 1993; Badelt et al. 1994; Turner et al. 1994; White et al. 1994; Overall and Blackman 1996; Ding 1997; Radford et al. 1998; see Botha and Cross, Chapt. 2; Ding, Chapt. 3; Overall, Chapt. 9). The discussion was mainly focused on the visible impact of different fixation and contrast procedures on the macromolecular appearance and spatial arrangement in the cytoplasmic sleeve. For further information we refer to the chapters pertinent to this matter (Chapt. 2, 3 and 9).

The dissimilar interpretation of the electron microscope pictures (Ding et al. 1992b; Overall and Blackman 1996) and the chemical dissection of plasmodesmata (Tilney et

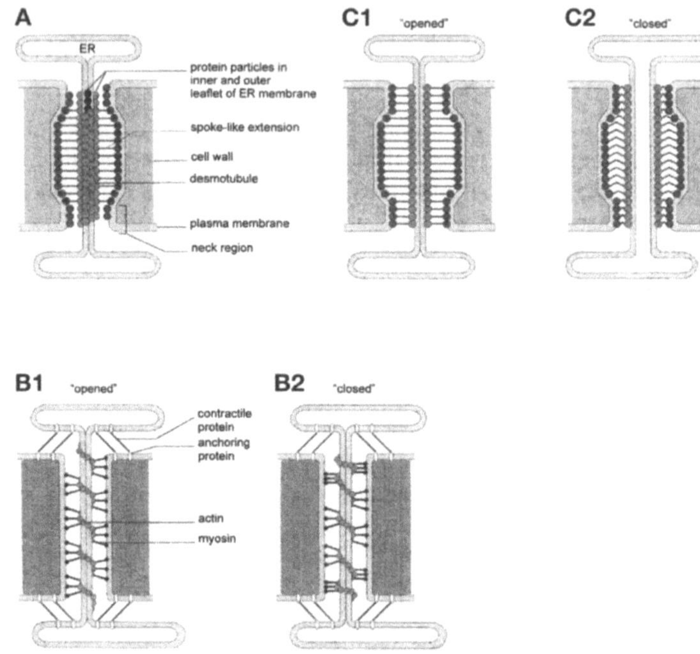

Fig. 1 A–C. Models of plasmodesmal structure and gating. A The leaflets of the appressed ER membrane and the inner leaflet of the plasma membrane are transformed into, or covered with probably helically arranged proteins within the cell wall area. The proteins at the inner leaflet of the plasma membrane and the outer leaflet of the ER membrane are connected by "spokes" (Ding et al. 1992a; Botha et al. 1993). Whether these spokes occur in the neck region is unclear. Regulation of the effective pore size is brought about by the position of the respective protein particles. B Protein particles (most likely actin; White et al. 1994) are arranged helically along the central ER rod (or appressed ER). Other particles (possibly myosin; Radford and White 1998) link the actin helices to the plasma membrane (Overall and Blackman 1996; Radford and White 1998). The ER at the plasmodesmal orifice is proposed to be anchored to the plasma membrane by contractile proteins (possibly centrins; Overall and Blackman 1996). B1, B2 Changes in the arrangement of the actin and myosin particles in the neck region regulates molecular trafficking (cf. Fig. 2C in McLean et al. 1997). C The cytoplasmic sleeve as well as the desmotubule are gatable (C1, C2). Conformational changes are brought about by the interaction between actin and myosin (Gamalei et al. 1994) or by contraction of the spokes and associated proteins (McLean et al. 1997). Opening of the desmotubule correlates inversely with closure of the cytoplasmic sleeve and vice versa. The contractile complexes have a higher density in the neck region (cf. Fig. 2B in McLean et al. 1997). See A for an explanation of the respective plasmodesmal components

al. 1991; Turner et al. 1994) has left us with several structural interpretations (Fig. 1). The question is whether the observed variations result from differences in the preparation procedures, and we are only asked to pick the right universal model. Or are the visible variations due to fundamental differences in plasmodesmal structure reflecting a functional variance? As yet, we believe that the macromolecular network in the cytoplasmic corridor displays an essentially uniform arrangement with some differences in functional domains.

Apart from the structural assessment of the cytoplasmic sleeve, the existence of collars (Turner et al. 1994), the presence of external (Badelt et al. 1994) and internal (Evert et al. 1977) sphincters, and the configuration of the ER strand traversing the plasmodesmata (Gamalei et al. 1994; Waigmann et al. 1997) are still matters of debate. Collars have been observed only once in the few studies using chemical dissection of the plasmodesmal apparatus (Turner et al. 1994). By contrast, external sphincters have been shown frequently (Olesen 1979, 1980; Thomson and Platt-Aloia 1985; Badelt et al. 1994). The prevailing opinion that the sphincters are composed of callose (Robards and Lucas 1990; Lucas et al. 1993) is not consistent with the claim that callose formation is an artefact (Radford et al. 1998). As for the ER strands through the plasmodesmata, the current opinion is that these are massive rods (Ding et al. 1992b; Botha et al. 1993; Lucas et al. 1993; Overall and Blackman 1996; Ding 1997). On the other hand, some plasmodesmata may have the potential to transport via the ER channels as was claimed for the plasmodesmata between bundle-sheath cells and intermediary cells in symplasmically loading species (Gamalei et al. 1994) and between trichome cells in chickpea (Lazzaro and Thomson 1996) and tobacco (Waigmann et al. 1997).

5 What Are the Protein Building Blocks of Plasmodesmata?

In 1989, a workshop was devoted to the biology of gap junctions and plasmodesmata (Robards et al. 1990). The basis for this combined effort of botanists and zoologists was the speculation that plasmodesmata and connexons had strong functional similarities (Meiners et al. 1988). This concept turned out to be fundamentally wrong: structural and functional similarities are quite modest. Yet, the presumptive resemblance played a major role in initial efforts (Meiners and Schindler 1989; Yahalom et al. 1991) to identify plasmodesm-associated proteins (PAPs).

A major difficulty in characterizing PAPs is the purification procedure. As PAPs are anticipated to be firmly attached to cell walls, attempts were undertaken to produce cell wall material containing plasmodesmal remnants (Monzer and Kloth 1991; Yahalom et al. 1991). The proteinaceous material from the wall was separated by standard protein electrophoresis techniques (SDS-PAGE). Cross-reaction with anti-connexin antibodies corroborated the idea that PAPs are related to the connexins (Yahalom et al. 1991; Hunte et al. 1992; Schulz et al. 1992). The amino acid sequence, however, hints at a relationship with protein kinases rather than with connexins (Mushegian and Koonin 1993).

In further studies with improved cleaning techniques (Kotlizky et al. 1992; Epel et al. 1995), a variety of other possible PAPs was discovered (Table 1). One could argue that these findings make the vision behind the PAP identification as well as the isolation procedures suspect. Firstly, the mere supposition that PAPs are homologous to the connexins (Yahalom et al. 1991) may be erroneous since the plasmodesmata and connexons are very different in evolutionary, structural and functional respect. Secondly, part of the PAPs may be aggregated as a plug in the plasmodesmal neck region (Turner et al. 1994) and be missed out by some cell wall isolation procedures. At least a few identical proteins are to be detected irrespective of the cleaning procedure anyway. This does seem to be the case with cell wall-bound proteins which are often found in the 26-28 kDa and the 40–43 kDa regions (Table 1). These overlap zones may be indicative of the existence of two major PAP classes.

New attempts in which viruses labelled with green fluorescent protein (GFP) are targeted to plasmodesmata are in progress (Epel et al. 1998). Hopefully, the fluororoch-

Table 1. The molecular weights of plasmodesma-associated proteins (PAPs) determined by SDS-PAGE after various cell wall isolation procedures

Proteins detected	Remarks	Species/Tissue	Reference
29 kDa[a]	Detected with antibodies raised against rat-liver connexin	*Glycine max*	Meiners and Schindler (1989)
28, 43 kDa	Enrichment of two bands in cell wall fractions	*Solanum nigrum* Cell wall fractions of tissue homogenates	Monzer and Kloth (1991)
26 kDa (PAP26[b]), 27 kDa (PAP27[c])	Homologous domains of cx32 (PAP27) and cx43 (PAP26)	*Zea mays* Mesocotyls of etiolated maize seedlings	Yahalom et al. (1991)
Molecular weight not determined	cx26-like protein related to the mouse liver connexin 26	*Vicia faba* and *Helianthus annuus* Mesophyll protoplasts	Schulz et al. (1992)
PAP26, PAP27 also 80, 21, 18 kDa	Probably also proteins of 60, 51, 41, and 32 kDa	*Zea mays* Mesocotyls of etiolated maize seedlings	Kotlizky et al. (1992)
29, 40[d], 21 kDa	Detected with antibodies against cx32 (29 kDa) or cx26 (40 kDa), respectively	*Vicia faba* Plasma membranes, cell wall fractions	Hunte et al. (1992, 1993)
43[e], 80, 100, 130 kDa	Different treatments removed material from exposed PD ends	*Zea mays* Root tips	Turner et al. (1994)
41 kDa[f]	In immunolocalization, antibodies against the 41-kDa protein (S-41) also cross-reacted with PAP26 (Yahalom et al. 1991)	*Zea mays* Mesocotyls of etiolated maize seedlings	Epel et al. (1996)
95, 45, 44, 33 kDa	Detected with monoclonal antibodies (MAB 45/22[g])	*Chara corallina* Nodal walls	Blackman et al. (1998)

[a] A cx27-like protein.
[b] PAP26 appeared to be distributed along the entire length of the plasmodesmata.
[c] PAP27 appeared to be localized more at the neck region.
[d] Probably a dimeric form of a 21-kDa, cx26-like protein.
[e] Although not reproducibley.
[f] Localized to PDs and cytoplasmic structures that are apparently Golgi membranes.
[g] MAB45/22 also labelled higher plant plasmodesmata.

rome tags the proteins which are part of the plasmodesmal apparatus. Successful identification of PAPs may eventually enable the production of plants which are transgenic for plasmodesmal functioning.

6 What Is the Functional Molecular Exclusion Limit of Plasmodesmata?

A feature frequently investigated is the size exclusion limit (SEL) as an indicator of the maximal molecular size of the compounds transported through the plasmodesmal corridor. As the potential to pass a plasmodesm does not only depend on the size of the molecule, we would prefer the term molecular exclusion limit (MEL) for reasons outlined elsewhere (Kempers and van Bel 1997). The MEL is determined by microinjection of fluorochromes of various sizes using pressure or current pulses. Apart from the size, the exclusion limit depends on the shape and charge of the fluorochrome (Erwee and Goodwin 1984; Terry and Robards 1987; Tucker and Tucker 1993; van Bel and Kempers 1997; see Goodwin and Cantrill, Chapt. 5). The packing of the molecule is also vital for exclusion assessment; proteins of about 40 kDa have about the same Stokes radius as dextrans of 15 kDa (Böckenhoff 1995). Circumstantial factors such as fluorochrome bleaching, compartmentation and enzymatic breakdown strongly affect the fluorescence and, hence, can influence the molecular exclusion limit measured - the apparent exclusion limit (Goodwin et al. 1990; see Goodwin and Cantrill, Chapt. 5). It should be underlined that the MELs present only a basal value which can vary with the physiological condition of the tissue (see Schulz, Chapt. 11).

Besides the nature of the fluorescent probe, the manner of fluorochrome application may affect the exclusion limit. Storms et al. (1998) reported that the apparent molecular exclusion limit in virus-infected plants was affected by the injection method. Fluorescent dextrans injected by pressure evidenced a much larger diameter than those introduced by iontophoresis in virus-infected plants. Consequently, the observed increase in plasmodesmal diameter was ascribed to the impact of the pressure device. The absence of this pressure effect in control plants was explained by weakening of the plasmodesmal construction by the virus infection (Storms et al. 1998).

A survey of the literature gives some credibility to the view that the apparent MEL is related to the way in which fluorochromes are being introduced (Table 2). A major handicap in tackling this issue is the fact that macromolecules like fluorescent dextran conjugates can hardly be introduced by iontophoresis (Kempers and van Bel 1997), so any comparison seems to lack an adequate base. It is striking, however, that the basal exclusion limit of plasmodesmata between parenchymatic cells is in the order of less than 1 kDa when dyes are introduced by iontophoresis (Erwee and Goodwin 1983, 1985; Terry and Robards 1987; Tucker 1993; Table 2) or by hydraulic injection (Derrick et al. 1990, 1992). When pneumatic injection (i.e. Wolf et al. 1989; Ding et al. 1996; Table 2) is used, molecules far larger than those observed with iontophoresis slip through the plasmodesmata. The same trend is visible when the exclusion limits of plasmodesmata between leaf trichome cells are compared (Derrick et al. 1992 vs. Waigmann and Zambryski 1995).

That pneumatic injection tends to overinflate plant cells is not unlikely. According to the specifications, pneumatic injectors allow injections in the "microlitre to femtolitre range" (Narashige IM-200), or the minimal dose to be injected in a controlled

Table 2. Exclusion limits of various plasmodesmal connections determined by fluorochromes introduced by iontophoresis, pneumatic or hydraulic microinjection. In most studies on the movement protein (MP)-induced enlargement of the plasmodesmal channel, the exclusion limits of the non-infected control tissues have not been measured systematically. However, the size of the macromolecules that are able to pass the control plasmodesmata has been given occasionally

Method of injection	Plant / tissue, cell type	Treatment	Fluorescent dyes (kDa)	PD opening measured (kDa)		Reference
				Control	Test tissue	
Diffusion	*Setcreasea purpurea*/ staminal hairs	—	F-conjugates (0.376 to 0.996)	0.700–0.800 depending on side groups	—	Tucker (1982)
Injection of liposomes containing the fluorochromes	*Nicotiana tabacum*/ spongy mesophyll cells	TMV MP transgenic plants	LYCH (0.457) and F-dextrans (9.4)	Not determined	9.4[a]	Deom et al. (1990)
Iontophoresis	*Egeria densa*/ leaf	Group II ions	CF, F-conjugates (0.376 to 0.874)	0.665	Movement depends on ions applied	Erwee and Goodwin (1983)
Iontophoresis	*Egeria densa*[b]/ extrastelar tissues (cortical symplasts in root stem and leaf)	—	CF, F-conjugates (0.376 to 0.749)	0.665[c] 0.674[d] 0.376[e]	—	Erwee and Goodwin (1985)
Iontophoresis	*Abutilon stratum*/ trichomes from open flowers or closed buds	—	F, LRB, LYCH, F-conjugates (0.376 to 19)	Depending on Stokes radius	—	Terry and Robards (1987)
Iontophoresis	*Setcreasea purpurea*/ staminal hairs	—	CF, F-conjugates (0.376 to 0.926)	Diffusion depending on structure of the molecule	—	Tucker and Tucker (1993)
Iontophoresis	*Setcreasea purpurea*/ staminal hairs	Azide	CF, F-conjugates (0.376 to 1.386)	ca. 0.9	PDs become enlarged[f]	Tucker (1993)
Pressure injection and iontophoresis	*Elodea canadensis*/ leaf	—	CF, LRB, FITC, F-conjugates, LRB-insulin A chain (0.376 to 19.4)	0.874	—	Goodwin (1983)

Table 2. Continued

Method of injection	Plant / tissue, cell type	Treatment	Fluorescent dyes (kDa)	PD opening measured (kDa)		Reference
				Control	Test tissue	
Pressure injection of liposomes containing the dyes	*Nicotiana tabacum*/ spongy mesophyll cells	TMV MP transgenic plants	LYCH, F-Gly_6, F-dextrans (0.457 to 17.2)	0.7 to 0.8[g]	9.4	Wolf et al. (1989)
Pressure injection, Narishige IM-5B microinjector	*Nicotiana clevelandii*/ epidermal leaf cells of fully expanded young leaves	Virus infections (TRV, TBRV, PVY, CMotV, GRV)	LYCH, CF, F-$(AA)_n$ (0.376 to 0.839)	Between 0.666 and 0.839	No modification found[h]	Derrick et al. (1990)
Pressure injection	*Nicotiana clevelandii*/ trichome cells	TRV infection by microinjection of virus particles	LYCH, F-$(AA)_n$, F-insulin A chain, F-dextran, LYCH-dextran (0.457 to 10)	Between 0.537 and 0.850	4.4	Derrick et al. (1992)
Pressure injection, Pneumatic PicoPump WPI PV830	*Nicotiana tabacum*/ MCs, BSCs and PPCs in leaves of intact plants or cut sections	TMV MP transgenic plants	LYCH (0.457), F-dextrans (3.9 to 17.2)	Not determined	9.4[i]	Ding et al. (1992b)
Pressure injection	*Cucurbita maxima*/ extrafascicular stem phloem	—	LYCH, F-dextran (0.457 to 3)	3	—	Kempers et al. (1993)
Pressure injection PLI–II, Medical Systems	*Nicotiana tabacum* and *N. benthamiana*/ leaf epidermal cells	AMV MP transgenic plants	F-dextrans (3 and 4.4)	Not determined[j]	4.4	Poirson et al. (1993)
Pressure injection, Pneumatic PicoPump WPI PV830	*Vigna unguiculata*/ mesophyll cells of a leaf attached to the plant	RCNMV MP injection	F-dextrans (3.9 and 9.4), F-MP (35)	Not determined	35	Fujiwara et al. (1993)
Pressure injection	*Triticum aestivum*/ epidermal cells or cortical cells in the root hair zone of seedlings	Azide or N_2 gas	LYCH (0.453), TPN-ADP (0.681) and F-dextrans (1 to 10)	< 1	5–10	Cleland et al. (1994)
Pressure injection PLI-II, Medical Systems	*Nicotiana tabacum*/ leaf trichome cells	CMV 3a[k] transgenic plants	F-dextrans (4.4 and 10)	Not determined[l]	10	Vaquero et al. (1994)

Table 2. Continued

Method of injection	Plant / tissue, cell type	Treatment	Fluorescent dyes (kDa)	PD opening measured (kDa)		Reference
				Control	Test tissue	
Pressure injection (as described in Wolf et al., 1989, except that liposomes were not employed)	*Nicotiana benthamiana* and *Phaseolus vulgaris*/ mesophyll cells of mature attached leaves	BL1 and BR1[m] injection	F-BL1, F-BR1, F-dextrans (10 and 20), TOTO-ssDNA, TOTO-dsDNA	Not determined[n]	BL1: >10 and <20[o] BR1: no increase	Noueiry et al. (1994)
Pressure injection, Pneumatic PicoPump WPI PV830	*Nicotiana tabacum*/ spongy mesophyll cells	TMV MP injection	LY-dextran, F-dextran (10, 20, 40)	Not determined	>20 <40[p]	Waigmann et al. (1994)
Pressure injection	*Nicotiana clevelandii*/ trichome cells	TMV MP injection	F-dextrans (4 to 13)	>7 <9	No change although MP itself moves	Waigmann and Zambryski (1995)
Pressure injection, Pneumatic PicoPump WPI PV820	*Nicotiana tabacum*/ mesophyll cells of a mature leaf attached to the plant	CD, profilin	F-dextrans (10 and 20)	10	20	Ding et al. (1996)
Pressure injection	*Nicotiana tabacum*/ epidermal cells of leaves	TMV infection	TR-dextran (>10)	Not	10[q]	Oparka et al. (1997)
Pressure injection and iontophoresis	*Vicia faba*/ stem phloem	—	LYCH, LYCH-dextrans, F-dextrans (0.457 to 40)	10	—	Kempers and van Bel (1997)
Pressure injection	*Nicotiana tabacum*/ leaf mesophyll cells	RPP13-1[r] injection	LYCH, F-RPP13-1, F-TMV MP, F-dextran (9.4 and 20)	Not determined	>9.4 <20	Ishiwatari et al. (1998)

[a] Depending on the developmental stage of the leaf. [b] Barrier to dye movement is found between the epidermis and the cortex in stem and root. [c] Shoot-apex symplast. [d] Leaf epidermal symplast. [e] Stem and root epidermal cells. [f] Coefficients of diffusion increase after azide treatment. [g] Also movement of 3.9-kDa F-dextran in some control plants (14%)! [h] Except reduction of SEL in CMotV-infected and, to a smaller extent, in GRV-infected tissue. [i] Modification of SEL is restricted to MCs and BSCs and depends on the developmental stage of the leaf. [j] Movement of 3-kDa dextrans in control plants (11 to 22%)! [k] The putative MP. [l] Movement of 4.4-kDa dextrans in control plants (12%)! [m] Proteins of bean dwarf mosaic geminivirus (BDMV) involved in systemic infection of the plant. [n] In 9 out of 76 (11.8%, bean) and 7 out of 82 (8.5%, tobacco) microinjections of controls, 10-kDa dextrans moved through plasmodesmata. [o] BL1 moves through plasmodesmata and also mediates movement of dsDNA. [p] Differences in MEL depending on the way of treatment (wildtype with coinjection of MP or MP-transgenic plant). [q] Restricted to the leading edge of expanding infection sites. [r] A thioredoxin h protein from rice (*Oryza sativa*).

manner is 1 nanolitre (Eppendorf). The latter volume is about 100 times higher than that of an average mesophyll cell. It should be noted, however, that the factory specifications pertain to oocytes without any notable osmotic pressure. The volumes injected in highly turgescent plant cells will be appreciably lower (but always too large?) than the specifications given. Moreover, visual surveillance during the injection also helps to prevent overinflation.

Obviously, one cannot generalize the effects of pressure injections (Table 2). Due to a finer control, hydraulic injection by a micropressure probe certainly introduces smaller quantities than the pneumatic injection. The former manner of application seemingly produces exclusion limits similar to those obtained with iontophoresis, and smaller than those obtained with pneumatic injection (footnotes j, l, n, Table 2). The claim that iontophoresis is the most reliable injection method for determination of the exclusion limit (Storms et al. 1998) has not been proven experimentally and may therefore be premature for reasons outlined in Section 8. In conclusion, we need a thorough reevaluation of the direct impact of the injection methods on the exclusion limits through comparative experiments.

7 What About the Autonomy of Cells in a Symplasmic Domain?

Experiments with low-molecular fluorochromes have led to the belief that molecules of <1 kDa are able to move freely within a symplasmic domain. Small metabolites such as sugars, amino acids and organic acids are envisioned to move within a symplasmic domain down a concentration gradient (Tyree 1970). For instance, it is a long-standing, but unproven assumption that intercellular transport of photosynthates takes place by diffusion through the plasmodesmata (for reviews, see Giaquinta 1983; van Bel 1993). As a result, almost every textbook claims that photoassimilates are transported in the mesophyll toward the minor veins down a concentration gradient, without experimental verification of the concentrations in the successive mesophyll cells (Gunning 1976; for criticism, see van Bel 1996). Similarly, it is also postulated that carbohydrate production in C_4plants cannot possibly function without diffusional exchange of metabolites via the plasmodesmata between Kranz mesophyll and bundle-sheath cells (Osmond and Smith 1976).

In the above mechanisms, symplasmic domains are conceived as syncytia with very narrow and selective bridges. Compounds of low molecular weight are thought to diffuse through the plasmodesmata, whereas only specific macromolecules are allowed to pass by active transfer. Recent publications on the trafficking of macromolecules and mRNA (e.g. Lucas et al. 1995; Mezitt and Lucas 1996; Perbal et al. 1996; Balanchandran et al. 1997; Ghoshroy et al. 1997; Ishiwatari et al. 1998) support the view that plant cells combine individual integrity with a high degree of interaction. The balance between autonomy and interaction may shift with the developmental or physiological state of the tissue (see van der Schoot and Rinne, Chapt. 13; Ehlers and van Bel, Chapt. 14). It is obvious that such a type of organization demands a tight control on the balance between cell autonomy and intercellular cooperation.

However attractive and logical this concept may appear, the consequences of unlimited diffusional exchange of low-molecular compounds have hardly been taken into consideration. Metabolite concentrations would tend to level off between the cells, pH will be equalized, and the membrane potentials may become identical due to electrical coupling. Such a system may work for tissues with relatively homogeneous cells executing the same tasks. However, cell autonomy solely at the macromolecular level will cause problems for any tissue in which cells execute distinct tasks. This issue is also discussed by Schulz (Chapt. 11).

The question therefore emerges if cell specialization within a tissue can be consistent with the potential for unlimited diffusional exchange via plasmodesmata. At first sight, radical solutions such as a complete symplasmic isolation of cells seem to indicate that symplasmic autonomy is an absolute prerequisite for the functioning of differently specialized neighbouring cells. As for guard cells, only an unequivocal lack of symplasmic communication (Wille and Lucas 1984; Erwee et al. 1985; Palevitz and Hepler 1985) appears to guarantee a perfect functioning.

Other examples, however, suggest the opposite. A clear physiological difference also exists between epidermis and mesophyll, and one anticipates a strict symplasmic barrier at their borderline in keeping with the observations on guard cells. Plasmodesmata were indeed reported to be rare at the ab- and adaxial interface between epidermis and mesophyll in *Populus* leaves (Russin and Evert 1985). In *Commelina* leaves, by contrast, the absolute amount of plasmodesmata was high at the interface between epidermis and palisade parenchyma, but low at the interface between epidermis and spongy mesophyll (P. van Kesteren and A.van Bel, unpubl. results). Thus, the symplasmic segregation of mesophyll and epidermis is far from absolute despite their distinct tasks. The incomplete isolation between mesophyll and epidermis questions the way in which the cellular integrity of spezialized neighbours is maintained.

Even more puzzling is the situation with regard to the symplasmic coupling of sieve element and companion cell. They have an intimate and compulsory interrelationship and probably share specialized plasmodesmata with high exclusion limits (Kempers et al. 1993; Kempers and van Bel 1997; see Schulz, Chapt. 11; Thompson, Chapt. 16). Yet the metabolic machineries of sieve element and companion cell are extremely different. Symplasmic communication is thus required whilst a highly divergent physiology must be maintained. The contrasting demands make the plasmodesmata between sieve element and companion cell likely to be corridors that are selective to metabolites, minerals (calcium!?) and phytohormones. This may also hold true for cells with less conspicuous physiological dissimilarities; the conflict between autonomy and cooperation is then met by a certain selectivity for low-molecular compounds. The striking disability of aromatic amino acids to pass plasmodesmata indicates such a potential to select small molecules (Erwee and Goodwin 1984; Tucker and Tucker 1993).

8 Phytohormonal and Electrical Signalling Through Plasmodesmata?

Some years ago, Oparka (1993) advanced the view that plasmodesmata may provide a pathway for electrical and, in particular, hormonal signals. Given their relatively low molecular weight, phytohormones are candidates for plasmodesmal transfer (while keeping reservations concerning free symplasmic movement in mind, Sect. 7). Plasmodesmal transport of phytohormones may be of paramount importance for developmental processes (Kwiatkowska 1991; see Kwiatkowska, Chapt. 12; van der Schoot and Rinne, Chapt. 13). Unfortunately, the "tyranny of the free space" in the literature on phytohormones has left its mark on the research on phytohormone transport. To date, very little work on symplasmic movement of phytohormones has been done (Drake and Carr 1979; Kwiatkowska 1991).

Another way of signalling that is largely neglected is the electrical transmission between plant cells. In early experiments, electrical conductivity between cells was taken as evidence for plasmodesmal operation (e.g. Spanswick and Costerton 1967; Brinckmann and Lüttge 1974; Overall and Gunning 1982). Recent experiments showed that plasmodesmata can operate as regulatable conductors for electricity (Reid and Overall 1992; Lew 1994, 1996; Holdaway-Clarke et al. 1996). In the intact plant, electrical currents are therefore expected to flow to adjacent cells (Williams and Pickard 1972a,b, 1974). In this manner, electrical currents could convey short and long-distance messages. The electrical transmission through plasmodesmata could resemble that through connexons (see van Rijen et al., Chapt. 4).

Electrical signals may function as a trigger for metabolic cascades in plants. Self-generated or triggered electrical currents may also regulate the gating of plasmodesmata and, hence, influence intercellular communication. If that were true, dye injection by iontophoresis affects the symplasmic properties of the system under investigation and the results obtained should be treated with great care (see Sect. 6; Storms et al. 1998).

9 What Is the Mechanism of Trafficking Macromolecules Through Plasmodesmata?

In contrast to connexons, plasmodesmata are capable of trafficking macromolecules, such as proteins and mRNA. The very first indications for protein passage through plasmodesmata were inferred from studies on the composition of phloem sap. Phloem sap contains about 150 proteins (Fisher et al. 1992; Nakamura et al. 1993; Sakuth et al. 1993). The molecular weights of these compounds (called P-proteins) mostly lie in the range of 10 to 25 kDa (Fisher et al. 1992; Nakamura et al. 1993; Sakuth et al. 1993; Schobert et al. 1995, 1998) and are remarkably lower than those of proteins collected from hypocotyl cells (Sakuth et al. 1993). Most of the P-proteins are water-soluble (Fisher et al. 1992) and can be obtained from sieve-tube exudates (sieve tube-exudate proteins or STEPs; Schobert et al. 1995). The enucleate sieve elements manage to survive over months and years, and in some species over decades (Raven 1991). Over such a long period, turnover of P-proteins has to be accomplished, but cannot possibly take place in enucleate cells. Therefore, production of P-proteins in the companion cells and subsequent transport of these proteins through the pore-plasmodesma units (PPUs) was postulated (for a more comprehensive survey, Thompson, see Chapt. 16).

Turnover of P-proteins was actually demonstrated by use of ^{35}S-labelled methionine (Fisher et al. 1992; Sakuth et al. 1993). Moreover, the genes of the PP2 lectin present in the sieve elements of pumpkin (Smith et al. 1987) are expressed exclusively in the companion cells (Bostwick et al. 1992). The findings indeed suggest collectively that P-proteins are produced in the companion cells and then transported to the sieve elements (for more detailed information, Chapt. 16 by Thompson). The exceptionally large exclusion limits (about 20 kDa; Kempers et al. 1993; Kempers and van Bel 1997) found for the PPUs conform with a transit of low-molecular proteins. The large exclusion limits may be induced by the P-proteins themselves. Some STEPs are able to enlarge the molecular exclusion limits and enable cell-to-cell trafficking of 10–to 20-kDa dextran conjugates (Balanchandran et al. 1997; Ishiwatari et al. 1998). As they have a chaperone-like nature (Schobert et al. 1995), some of the P-proteins are thought to be involved in the passage of macromolecules through the PPUs.

The transit of P-proteins through PPUs may reflect a universal mechanism for intercellular protein trafficking. The absence of RNA encoding the plant transcription factor knotted 1 (KN1) in the L1 layer of maize seedlings (Smith et al. 1992; Jackson et al. 1994) raised the suspicion that the KN1 protein could move through plasmodesmata. Lucas et al. (1995) demonstrated that KN1 not only moves through plasmodesmata, but also facilitates the movement of other proteins and lucifer yellow-dextran conjugates. Surprisingly, KN1 seems to permit the trafficking of its own kn1 sense mRNA. Recently, transfer of mRNA encoding the StSUT1 transporter through PPUs was postulated (Kühn et al. 1997).

Numerous plant virus species invade the neighbouring cells via the plasmodesmata (for the use of plant viruses for studies on plasmodesmata, see Oparka et al., Chapt. 6; for short- and long-distance transport of viruses, see Nelson, Chapt. 17). Most of the viruses produce movement proteins (MPs) that are able to cross plasmodesmata, as evidenced by labelling with fluorescein (Fujiwara et al. 1993; Noueiry et al. 1994; Ding et al. 1995), immunocytochemistry (Waigmann and Zambryski 1995), and GFP fusion (Canto et al. 1997; see also Chapt. 6 by Oparka et al.). The MPs also induce plasmodesmal gating for passing their genetic material (Wolf et al. 1989; Derrick et al. 1992; Ding et al. 1992b; Fujiwara et al. 1993; Poirson et al. 1993; Noueiry et al. 1994; Vaquero et al. 1994; Waigmann et al. 1994; Oparka et al. 1997). These viruses are likely to exploit the same trafficking mechanism as the one for host nucleic acids developed earlier in evolution.

Pathogen-induced proteins have also been reported to move through plasmodesmata. For instance, the pathogenesis-related PRms of maize moves through plasmodesmata between the parenchyma cells of the central pith (Murillo et al. 1997).

Recently, several attempts have been undertaken to integrate the observations on the plasmodesmal transport of macromolecules (Fig. 2). MPs, STEPs, PRms, and KN1 (plasmodesma-permeant proteins or PPPs) may share some features that permit their passage through the plasmodesmata. The cytoskeleton is probably involved in tracking the PPPs towards the plasmodesma. MPs colocalize with the microtubules and, to a lesser extent, to actin filaments (Heinlein et al. 1995; McLean et al. 1995). The other PPPs may also be attached to the cytoskeleton either directly or by linking complexes (Gilbertson and Lucas 1996). Whether the PPPs are subject to unwinding or alignment during attachment is obscure. Having arrived at the plasmodesmal orifice, a PPP is presumed to attach to a docking protein. Protein kinases may be engaged in the docking event (Citovsky et al. 1993; Mushegian and Koonin 1993; Tacke et al. 1993; Sokolova et al. 1997) which may trigger a conformational change of the plasmodesmal corridor that allows the passage of the PPP. The unselective increase of the exclusion limit is puzzling. Despite their structural dissimilarity to proteins, high-molecular fluorochrome-dextran conjugates are able to pass plasmodesmata in the "gated" configuration (Deom et al. 1990; Derrick et al. 1992; Poirson et al. 1993; Waigmann and Zambryski 1995; Oparka et al. 1997).

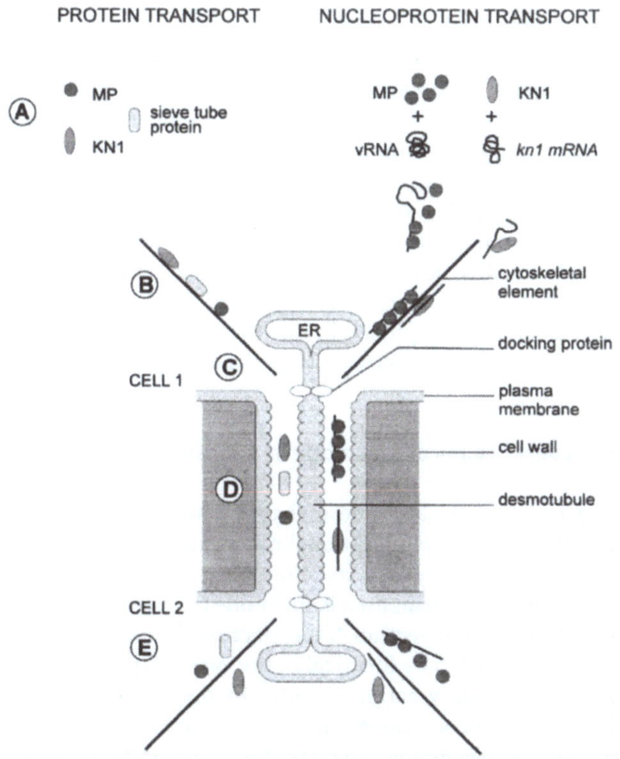

Fig. 2. Hypothetical trafficking of macromolecules through plasmodesmata (modified after Ghoshroy et al. 1997). Protein transport (*left-hand side*): Plasmodesma-permeant proteins (*MP*, *KN1*, several sieve tube proteins; *A*) attach to the cytoskeleton (*B*) and are tracked toward the plasmodesmal orifice, where the proteins interact with a docking protein (*C*). After transfer through the plasmodesma (*D*), the proteins are released at the other side of the plasmodesma (*E*). Nucleoprotein transport (*right-hand side*): Some plasmodesma-permeant proteins have binding sites for nucleic acids (viral RNA, *vRNA*; messenger RNA, *mRNA*; knotted 1 RNA, *kn 1*) that enable the formation of nucleoprotein complexes (*A*). The nucleoprotein complexes are trafficked through the plasmodesma processed in the same manner as depicted for proteins only (*B*, *C*, *D*, *E*)

What happens after the docking is unclear. There is almost general agreement on the fact that the trafficking through the plasmodesma is a guided process, but opinions on the mechanism diverge. Some authors (Overall and Blackman 1996) prefer an activated transport by interaction of actin and myosin molecules traversing the plasmodesma (see Fig. 1B). Others advocate channelling mediated by proteins that are attached to the membranes lining the plasmodesmal sleeve (see Fig. 1A).

Provided that PPP trafficking takes place along these lines, all PPPs should have at least two common domains or signal peptide sequences, one for interaction with the cytoskeleton or cytoskeleton ligand, the second for docking at the plasmodesmal orifice. Trafficking of mRNA would require a supplementary binding domain at the PPP for nucleic acid (NA) attachment. Yet identical sequences in the PPPs have not been identified to date.

Trafficking of viral NAs includes the formation of NA/MP complexes which may be processed in the manner described above (Fig. 2; Ghoshroy et al. 1997). The fate of the MP is a matter of dispute; it may be left behind within the plasmodesmal corridor or may traverse the plasmodesma as part of the NA/MP complex. Trafficking of complete NA/MP complexes through plasmodesmata, however, is the most likely option (McLean et al. 1993; Waigmann and Zambryski 1995; Carrington et al. 1996; Ghoshroy et al. 1997; Blackman et al. 1998). The prevailing view is that viral NAs pass the plasmodesmata in a linear form accompanied by several MPs (Ghoshroy et al. 1997; Blackman et al. 1998) after which the MPs may be uncoupled. There are indications that the conformational change of the plasmodesma is redressed after the NA/MP complex has passed (Oparka et al. 1997). Mesophyll cells lose the capacity to traffic macromolecules some time after the passage of the viruses (Oparka et al. 1997).

10 Are There Several Types of Plasmodesmal Gating?

The potential mechanisms of plasmodesmal gating are being discussed comprehensively by Schulz (Chapt. 11). A few additional notes may be appropriate in view of the possible differences in plasmodesmal structure (Fig. 1; Sect. 4). If the plasmodesmal structure is variable, how universal then is the mode of operation including the gating mechanism? As put forward in Section 4, the macromolecular organization of the cytoplasmic sleeve is supposed to be identical in the various types of plasmodesmata, at least structurally. Consequently, gating executed by the macromolecular apparatus in the cytoplasmic sleeve would be relatively similar between various plasmodesmal types.

In numerous papers, plasmodesmata are described as being constricted by so-called sphincters, ring-like bodies around the orifices of the plasmodesmata (Olesen 1979, 1980; Thomson and Platt-Aloia 1985; Badelt et al. 1994). The external sphincters presumed to be located in the apoplasm would act as a complement to, or as an alternative for, intracellular gating. Several authors argue that the callose depositions found around the plasmodesmata are involved in plasmodesmal gating and may reside in the sphincters (Delmer et al. 1993; Brown et al. 1997; Roy et al. 1997). However, a universal occurrence of sphincters is disputable as sphincters are not visible around the plasmodesmata of many species. Additionally, callose deposition was reported to be an artefact (see Overall, Chapt. 9; Radford et al. 1998). As usual, this finding raises more questions than it solves. Is callose synthesis an artefact emerging in all types of plasmodesmata or in all fixation procedures? Do sphincters occur anyway or only in specific plasmodesmata? Or are sphincters composed of materials other than callose?

The concept of plasmodesmal gating was strongly modified over the past years (for a comprehensive treatment, see Schulz, Chapt. 11). The discovery that metabolic poisons and anaerobic conditions enlarge the molecular exclusion limit was striking (Schenk 1974; Tucker 1993; Cleland et al. 1994; Zhang and Tyerman 1997; Wright and Oparka 1997). This newly-acquired notion leads us through a few conflicting thoughts, as our mind is conditioned to see opening of plasmodesmata as an energy-requiring process.

The crucial discrepancy between these observations and those on the passage of macromolecules lies in the fact that the deactivated state allows fluorescent dextrans to pass the plasmodesmata (e.g. Cleland et al. 1994). Thus far, passage of dextrans was observed in a state when also other macromolecules are being trafficked (Fig. 2). A complete energy shortage is difficult to reconcile with an activated actin-myosin interaction engaged in the passage of these macromolecules (Fig. 2; White et al. 1994; Ding et al. 1996; Overall and Blackman 1996; Overall, Chapt. 9).

Furthermore, we must assume that closure of plasmodesmata is a strongly energy-consuming process. This is acceptable if we regard the sealing of plasmodesmata as a defence reaction in the case of wounding, in order to prevent leakage of cytoplasmic compounds.

We speculate that plasmodesmata can adopt different conformation states. The transition from one state to the other may occur in a gradual or abrupt fashion and may be dependent on the energy supply. In this frame of thinking, we could conceive three plasmodesmal configurations: (1) A maximally open one, without any energy investment, e.g. anaerobiosis. (2) The "normal" inactivated state with a low-energy input. (3) The sealed state with a high-energy input. The question is whether the degree of opening of the plasmodesmata is commensurate with the degree of energy input. We expect that the energy channelling toward the plasmodesmata in the case of sealing is different from that under "normal" conditions. In Chapter 11, Schulz makes several intriguing proposals on this issue.

11 How Important is the Versatility of Symplasmic Domains and the Turnover of Plasmodesmata for Whole-Plant Physiology?

One of the central issues in developmental physiology is how intercellular interaction directs the ontogeny of cells, tissues and organs (van der Schoot and Rinne, Chapt. 13). Plasmodesmata appear to play a pivotal role in the processing of communication and orientation signals. The discovery that cells in plant meristems communicate by exchange of genetic information highlights the performance of plasmodesmata (Jackson et al. 1994; Lucas et al. 1995; Perbal et. 1996; Jackson and Hake 1997). Impressive examples of the power of plasmodesmal trafficking are the events during the development of vegetative shoots (Jackson et al. 1994) and root tips (Duckett et al. 1994; van den Berg et al. 1995), the patterning of root hairs (Masucci and Schiefelbein 1996; Schiefelbein et al. 1997), trichome patterning (Kragler et al. 1998) and organ differentiation in floral meristems (Becraft 1995; van der Schoot and Rinne, Chapt. 13). Thus, ontogeny seems to be controlled by a permanently changing pattern of gated plasmodesmata, among others (van der Schoot and Rinne, Chapt. 13; Ehlers and van Bel, Chapt 14).

There are several indications that not only symplasmic communication is indispensable for cell development and differentiation. Symplasmic isolation may be equally important. The necessity of plasmodesmal plugging for antheridial development in *Chara* has been well documented (Kwiatkowska and Maszewski 1986; Kwiatkowska 1988). Generally speaking, symplasmic isolation seems to be compulsory for cell differentiation in *Chara* (Kwiatkowska, chapter 12). Full symplasmic isolation also precedes division and differentiation of the cambial precursors of sieve/element companion cell complexes (van Bel and van Rijen 1994). Similarly, differentiation in calluses is associated with symplasmic isolation. The plasmodesmata close at the interface between dividing and non-dividing cells (Ehlers and van Bel, Chapt. 14).

An important question pertains to the dynamics of symplasmic domains during the mature state of the tissues. Are the domains static or are they dynamic, giving rise to varying alliances between cells? Symplasmic rearrangement would be part of a flexible response to injuries or unusual environmental stresses. The shifting communication or transport orientation patterns allow the plants to react in an appropriate fashion.

It is certain that plasmodesmata can be produced de novo. An impressive list of examples of secondary plasmodesmata formation is given by Kollmann and Glockmann (Chapt. 10). It is questionable if the potential to form secondary plasmodesmata also implies permanent plasmodesmal turnover during maturity. Breakdown of plasmodesmal constructions has been documented and some indication for the breakdown mechanism has been given (Ehlers et al. 1996). However, no evidence has been produced thus far for the rate of turnover of plasmodesmata, for the occurrence of plasmodesmal turnover in specific cell types, or for the signals that induce the formation or degradation of plasmodesmata.

12 Concluding Remarks

A central issue to be solved is the elucidation of the molecular structure and composition of the plasmodesmal gating apparatus. As soon as this goal is achieved, the mechanisms of intercellular transport and communication can be studied by manipulating plasmodesmal functioning at the molecular level.

Acknowledgements. We are much indebted to Dr. Katrin Ehlers for extensive discussions and helpful comments. The expert advice of Dr. Martha Cook is gratefully acknowledged. We thank Dr. Chris van der Schoot for supplying data on injection methods.

References

Badelt K, White RG, Overall RL, Vesk M (1994) Ultrastructural specializations of the cell wall sleeve around plasmodesmata. Am J Bot 81:1422–1427

Balanchandran S, Xiang Y, Schobert C, Thompson GA, Lucas WJ (1997) Phloem sap proteins from *Cucurbita maxima* and *Ricinus communis* have the capacity to traffic cell to cell through plasmodesmata. Proc Natl Acad Sci USA 94:14150–14155

Becraft PW (1995) Intercellular induction of homeotic gene expression in flower development. Trends Genet 11:253–255

Blackman LM, Gunning BES, Overall RL (1998) A 45 kDa protein isolated from the nodal walls of *Chara corallina* is localised to plasmodesmata. Plant J 10, 401–411

Blackman LM, Boevink P, Santa Cruz S, Palukaitis P, Oparka KJ (1998) The movement protein of cucumber mosaic virus traffics into sieve elements in minor veins of *Nicotiana clevelandii*. Plant Cell 10:525–537

Böckenhoff A (1995) Untersuchungen zur Physiologie der Nährstoffversorgung des Rübenzystennematoden *Heterodora schachtii* und der von ihm induzierten Nährzellen in Wurzeln von *Arabidopsis thaliana* unter Verwendung einer speziell adaptierten in situ Mikroinjektionstechnik. PhD Thesis, University of Kiel, Kiel

Bostwick DE, Dannenhoffer JM, Skaggs MI, Lister RM, Larkins BA, Thompson GA (1992) Pumpkin phloem lectin genes are specifically expressed in companion cells. Plant Cell 4:1539–1548

Botha CEJ, Hartley BJ, Cross RHM (1993) The ultrastructure and computer-enhanced digital image analysis of plasmodesmata at the Kranz mesophyll-bundle sheath interface of *Themeda triandra* var. imberbis (Retz) A. Camus in conventionally-fixed leaf blades. Ann Bot 27:255–261

Brinckmann E, Lüttge U (1974) Lichtabhängige Membranpotentialschwankungen und deren interzelluläre Weiterleitung bei panaschierten Photosynthesemutanten von *Oenothera*. Planta 119: 47–57

Brown RC, Lemmon BE, Stone BA, Olsen OA (1997) Cell wall (1→3)- and (1→3,1→4)-β-glucans during early grain development in rice (*Oryza sativa* L.) Planta 202:414–426

Canto T, Prior DAM, Hellwald KH, Oparka KJ, Palukaitis P (1997) Characterization of cucumber mosaic virus. IV. Movement protein and coat protein are both essential for cell-to-cell movement of cucumber mosaic virus. Virology 237:237–248

Carrington JC, Kasschau KD, Mahajan SK, Schaad MC (1996) Cell-to-cell and long-distance transport of viruses in plants. Plant Cell 8:1669–1681

Citovsky V, McLean BG, Zupan JR, Zambryski P (1993) Phosphorylation of tobacco mosaic virus cell-to-cell movement protein by a developmentally regulated plant cell wall-associated protein kinase. Genes Dev 7:904–910

Cleland RF, Fujiwara T, Lucas WJ (1994) Plasmodesmal-mediated cell-to-cell transport in wheat roots is modulated by anaerobic stress. Protoplasma 178:81–85

Cole KM, Sheath RG (1990) Biology of the red algae. Cambridge University Press, Cambridge

Cook ME, Graham LE, Botha CEJ, Lavin CA (1997) Comparative ultrastructure of plasmodesmata of *Chara* and selected bryophytes: toward an elucidation of the evolutionary origin of plant plasmodesmata. Am J Bot 84:1169–1178

Delmer DP, Volokita M, Solomon M, Fritz U, Delphendal W, Herth W (1993) A monoclonal antibody recognizes a 65 kDa higher plant membrane polypeptide which undergoes cation-dependent association with callose synthase in vitro and co-localizes with sites of high callose deposition in vivo. Protoplasma 176:33–42

Deom CM, Lapidot M, Beachy RN (1990) Plant virus movement proteins. Cell 69:211–224

Derrick PM, Barker H (1997) Short and long distance spread of potato leaf roll virus: effect of host genes and transgenes conferring resistance to virus accumulation in potato. J Gen Virol 78: 243–251

Derrick PM, Barker H, Oparka KJ (1990) Effect of virus infection on symplastic transport of fluorescent tracers in *Nicotiana clevelandii* leaf epidermis. Planta 181:555–559

Derrick PM, Barker H, Oparka KJ (1992) Increase in plasmodesmatal permaeability during cell-to-cell spread of tobacco rattle virus from individually inoculated cells. Plant Cell 4:1405–1412

Ding B (1997) Cell-to-cell transport of macromolecules through plasmodesmata: a novel signalling pathway in plants. Trends Cell Biol 7:5–9

Ding B, Lucas WJ (1996) Secondary plasmodesmata: biogenesis, special functions and evolution. In: Smallwood M, Knox JP, Bowles DJ, (eds) Membranes: specialized functions in plants. Bios, Oxford, pp 489–506

Ding B, Turgeon R, Parthasarathy MV (1992a) Substructure of freeze-substituted plasmodesmata. Protoplasma 169:28–41

Ding B, Haudenshield JS, Hull RJ, Wolf S, Beachy RN, Lucas WJ (1992b) Secondary plasmodesmata are specific sites of localization of the tobacco mosaic virus movement protein in transgenic tobacco plants. Plant Cell 4:915–928

Ding B, Haudenshield JS, Willmitzer L, Lucas WJ (1993) Correlation between arrested secondary plasmodesmal development and onset of accelerated leaf senescence in yeast invertase transgenic tobacco plants. Plant J 4:179–189

Ding B, Kwon M-O, Warnberg L (1996) Evidence that actin filaments are involved in controlling the permeability of plasmodesmata in tobacco mesophyll. Plant J 10:157–164

Ding XS, Shintaku MH, Arnold SA, Nelson RS (1995) Accumulation of mild and severe strains of tobacco mosaic virus in minor veins of tobacco. Mol Plant-Microbe Interact 8:32–40

Drake GA, Carr DJ (1979) Symplastic transport of gibberellins: evidence from flux and inhibitor studies. J Exp Bot 30:439–447

Duckett CM, Oparka KJ, Prior DAM, Dolan LAM, Roberts K (1994) Dye-coupling in the root epidermis of *Arabidopsis* is progressively reduced during development. Development 120:3247–3255

Ehlers K, Kollmann R (1996) Formation of branched plasmodesmata in regenerating *Solanum nigrum* protoplasts. Planta 199:126–138

Ehlers K, Schulz M, Kollmann R (1996) Subcellular localization of ubiquitin in plant protoplasts and the function of ubiquitin in selective degradation of outer-wall plasmodesmata in regenerating protoplasts. Planta 199:139–151

Epel BL, Kuckuck B, Kotlizky G, Shurtz S, Erlanger M, Yahalom A (1995) Isolation and characterisation of plasmodesmata. Methods Cell Biol 50:237–253

Epel BL, van Lent JWM, Cohen L, Kotlizky G, Katz A, Yahalom A (1996) A 41 kDa protein from maize mesocotyl walls immunolocalizes to plasmodesmata. Protoplasma 191:70–78

Epel BL, Katz A, Kotlizky G, Erlanger M, Heinlein M, Padgett H, Beachy R, Boulton M (1998) GFP lights up the tunnel: studies on plasmodesmal composition and function as probed with viral

movement proteins labeled with green fluorescent protein. Workshop on Plasmodesmata and transport of plant viruses and plant macromolecules, Madrid. Instituto Juan March 80, p25
Erwee MG, Goodwin PB (1983) Characterisation of the *Egeria densa* Planch. leaf symplast: inhibition of the intercellular movement of fluorescent probes by group II ions. Planta 158:320–328
Erwee MG, Goodwin PB (1984) Characterization of the *Egeria densa* leaf symplast: response to plasmolysis, deplasmolysis and to aromatic amino acids. Protoplasma 112:162–168
Erwee MG, Goodwin PB (1985) Symplast domains in extrastelar tissues of *Egeria densa* Planch. Planta 163:9–19
Erwee MG, Goodwin PB, van Bel AJE (1985) Cell-cell communication in the leaves of *Commelina cyanea* and other plants. Plant Cell Environ 8:173–178
Esau K (1965) Plant anatomy. John Wiley, New York.
Evert RF, Eschrich W, Heyser W (1977) Distribution and structure of plasmodesmata in mesophyll and bundle-sheath cells of *Zea mays* L. Planta 136:77–89
Fisher DB, Wu Y, Ku MSB (1992) Turnover of soluble proteins in the wheat sieve tube. Plant Physiol 100:1433–1441
Franceschi VR, Lucas WJ (1982) The relationship of the charasome to chloride uptake in *Chara corallina*: physiological and histochemical investigations. Planta 154:525–537
Fraser TW, Gunning BES (1969) The ultrastructure of plasmodesmata of the filamentous green algae, *Bulbochaete hiloensis* (Nordst.) Tiffany. Planta 88:244–254
Fujiwara T, Giesman-Cookmeyer D, Ding B, Lommel SA, Lucas WJ (1993) Cell-to-cell trafficking of macromolecules through plasmodesmata potentiated by the red clover necrotic mosaic virus movement protein. Plant Cell 5:1783–1794
Gagnon MJ, Beebe DU (1996) Minor vein differentiation and the development of specialized plasmodesmata between companion cells and contiguous cells in expanding leaves of *Moricandia arvensis* (L) DC (Brassicaceae). Int J Plant Sci 157:685–697
Gamalei YV (1990) Leaf phloem. Nauka, Leningrad (in Russian)
Gamalei YV, van Bel AJE, Pakhomova MV, Sjutkina AV (1994) Effects of temperature on the conformation of the endoplasmic reticulum and on starch accumulation in leaves with the symplasmic minor-vein configuration. Planta 194:443–453
Ghoshroy S, Lartey R, Sheng J, Citovsky V (1997) Transport of proteins and nucleic acids through plasmodesmata. Annu Rev Plant Physiol Plant Mol Biol 48:27–50
Giaquinta RT (1983) Phloem loading of sucrose. Annu Rev Plant Physiol 34:347–387
Giddings TH, Staehelin LA (1981) Observation of microplasmodesmata in both heterocyst-forming and non-heterocyst forming filamentous cyanobacteria by freeze-fracture electron microscopy. Arch Microbiol 129:295–298
Gilbertson RL, Lucas WJ (1996) How do viruses traffic on the vascular highway? Trends Plant Sci 1: 260–268
Goodwin PB (1983) Molecular size limit for movement in the symplast of *Elodea* leaf. Planta 157:124–130
Goodwin PB, Shepherd V, Erwee MG (1990) Compartmentation of fluorescent tracers injected into the epidermal cells of *Egeria densa* leaves. Planta 181:129–136
Gunning BES (1976) The role of plasmodesmata in short distance transport from and to the phloem. In: Robards AW, Gunning BES (eds) Intercellular communication in plants: studies on plasmodesmata. Springer, Berlin Heidelberg New York, pp 203–227
Hashimoto T, Kishi T, Yoshida N (1964) Demonstration of micropores in fungal crosswall. Nature 202:1353
Hawker LE, Gooday MA (1967) Delimitation of the gametangia of *Rhizopus sexualis* (Smith) Callen: an electron microscope study of septum formation. J Gen Microbiol 49:371–376
Heinlein M, Epel BL, Padgett HS, Beachy RN (1995) Interaction of tobamovirus movement proteins within the plant cytoskeleton. Science 270:1983–1985
Holdaway-Clarke TL, Walker NA, Overall RL (1996) Measurement of the electrical resistance of plasmodesmata and membranes of corn suspension culture cells. Planta 199:537–544
Hunte C, Schnabl H, Traub O, Willecke K, Schulz M (1992) Immunological evidence of connexin-like proteins in the plasma membrane of *Vicia faba* L. Bot Acta 105:104–110
Hunte C, Janßen M, Schulz M, Traub O, Willecke K, Schnabl H (1993) Age-dependent modifications and further localization of the CX 26-like protein from *Vicia faba* L. Acta Bot 106:207–212
Ishiwatari Y, Fujiwara T, McFarland KC, Nemoto K, Hayashi H, Chino M, Lucas WJ (1998) Rice phloem thioredoxin h has the capacity to mediate its own cell-to-cell transport through plasmodesmata. Planta 205:12–22
Jackson D, Hake S (1997) Morphogenesis on the move: cell-to-cell trafficking of plant regulatory proteins. Curr Opin Gen Dev 7:495–500
Jackson D, Veit B, Hake S (1994) Expression of maize KNOTTED1 related homeobox genes in the shoot apical meristem predicts patterns of morphogenesis in the vegetative shoot. Development 120:405–413

Kempers R, van Bel AJE (1997) Symplasmic connections between sieve element and companion cell in the stem phloem of *Vicia* have a size exclusion limit of at least 10 kDa. Planta 201:195–201
Kempers R, Prior DAM, van Bel AJE, Oparka KJ (1993) Plasmodesmata between sieve element and companion cell of extrafascicular phloem of *Cucurbita maxima* permit passage of 3 kDa fluorescent probes. Plant J 4:567–575
Kotlizky G, Shurtz S, Yahalom A, Malik Z, Traub O, Epel BL (1992) An improved procedure for the isolation of plasmodesmata embedded in clean maize cell walls. Plant J 2:623–630
Kragler F, Lucas WJ, Monzer J (1998) Plasmodesmata: dynamics, domains and patterning. Ann Bot 81:1–10
Kühn C, Franceschi VR, Schulz A, Lemoine R, Frommer WB (1997) Macromolecular trafficking indicated by localization and turnover of sucrose transporters in enucleate sieve elements. Science 275:1298–1300
Kwiatkowska M (1988) Symplasmic isolation of *Chara vulgaris* antheridium and mechanisms regulating the process of spermatogenesis. Protoplasma 142:137–146
Kwiatkowska M (1991) Autoradiographic studies on the role of plasmodesmata in the transport of gibberellin. Planta 183:294–299
Kwiatkowska M, Maszewski J (1986) Changes in the occurrence and ultrastructure of plasmodesmata in antheridia of *Chara vulgaris* L. during different stages of spermatogenesis. Protoplasma 132:179–188
Lazzaro MD, Thomson WW (1996) The vacuolar-tubular continuum in living trichomes of chickpea (*Cicer arietinum*) provides a rapid means of solute delivery from base to tip. Protoplasma 193: 181–190
Lew RR (1994) Regulation of the electric coupling between *Arabidopsis* root hairs. Planta 193:67–73
Lew RR (1996) Pressure regulation of the electrical properties of growing *Arabidopsis thaliana* L. root hairs. Plant Physiol 112:1089–1100
Lucas WJ, Ding B, van der Schoot C (1993) Plasmodesmata and the supracellular nature of plants. New Phytol 125:435–476
Lucas WJ, Bouché-Pillon S, Jackson DP, Nguyen L, Baker L, Ding B, Hake S (1995) Selective trafficking of KNOTTED1 homeodomain protein and its mRNA through plasmodesmata. Science 270: 1980–1985
Marchant HJ (1976) Plasmodesmata in algae and fungi. In: Robards AW, Gunning BES (eds) Intercellular communication in plants: studies on plasmodesmata. Springer, Berlin Heidelberg New York, pp 59–80
Masucci JD, Schiefelbein JW (1996) Hormones act downstream of TTG and GL2 to promote root hair outgrowth during epidermis development in the *Arabidopsis* root. Plant Cell 8:1505–1517
McLean BG, Waigmann E, Citovsky V, Zambryski P (1993) Cell-to-cell movement of plant viruses. Trends Microbiol 1:105–109
McLean BG, Zupan J, Zambryski P (1995) Tobacco mosaic virus movement protein associates with the cytoskeleton in tobacco cells. Plant Cell 7:2101–2114
McLean BG, Hempel FD, Zambryski PC (1997) Plant intercellular communication via plasmodesmata. Plant Cell 9:1043–1054
Meiners S, Schindler M (1989) Characterization of a connexin homologue in cultured soybean cells and diverse plant organs. Planta 179:148–155
Meiners S, Baron-Epel O, Schindler M (1988) Intercellular communication – filling in the gaps. Plant Physiol 88:791–793
Mezitt LA, Lucas WJ (1996) Plasmodesmal cell-to-cell transport of proteins and nucleic acids. Plant Mol Biol 32:251–273
Monzer J, Kloth S (1991) The preparation of plasmodesmata from plant tissue homogenates: access to the biochemical characterization of plasmodesmata-related polypeptides. Bot Acta 104:82–84
Murillo I, Cavallarin L, San Segundo B (1997) The maize pathogenesis-related PRms protein localizes to plasmodesmata in maize radicles. Plant Cell 9:145–156
Mushegian AR, Koonin EV (1993) The proposed plant connexin is a protein kinase-like protein. Plant Cell 5:998–999
Nakamura S, Hayashi H, Mori S, Chino M (1993) Protein phosphorylation in the sieve tubes of rice plants. Plant Cell Physiol 34:927–933
Nelson T, Dengler N (1997) Leaf vascular pattern formation. Plant Cell 9:1121–1135
Noueiry AO, Lucas WJ, Gilbertson RL (1994) Two proteins of a plant DNA virus coordinate nuclear and plasmodesmal transport. Cell 76:925–932
Olesen P (1979) The neck constriction in plasmodesmata. Evidence for a peripheral sphincter-like structure revealed by fixation with tannic acid. Planta 144:349–358
Olesen P (1980) A model of a possible sphincter associated with plasmodesmatal neck region. Eur J Cell Biol 22:250
Olesen P, Robards AW (1990) The neck region of plasmodesmata. General ultrastructure and some functional aspects. In: Robards AW, Lucas WJ, Pitts JD, Jongsma HJ, Spray DC (eds) Parallels in cell

to cell junctions in plants and animals. NATO ASI Series, vol H46, Springer, Berlin Heidelberg New York, pp 145–170
Oparka KJ (1993) Signalling via plasmodesmata – the neglected pathway. Semin Cell Biol 4:131–138
Oparka KJ, Prior DAM, Santa Cruz S, Padgett HS, Beachy RN (1997) Gating of epidermal plasmodesmata is restricted to the leading edge of expanding infection sites of tobacco mosaic virus (TMV). Plant J 12:781–789
Osmond CB, Smith FA (1976) Symplastic transport of metabolites during C_4-photosynthesis. In: Robards AW, Gunning BES (eds) Intercellular communication in plants: studies on plasmodesmata. Springer, Berlin Heidelberg New York, pp 229–241
Overall RL, Blackman L (1996) A model of the macromolecular structure of plasmodesmata. Trends Plant Sci 1:307–311
Overall RL, Gunning BES (1982) Intercellular communication in *Azolla* roots. I Electrical coupling. Protoplasma 111:151–160
Palevitz BA, Hepler PK (1985) Changes in dye coupling of stomatal cells of *Allium* and *Commelina* demonstrated by microinjection of lucifer yellow. Planta 164:473–479
Perbal MC, Haughn G, Saedler H, Schwarz-Sommer S (1996) Non-autonomous function of the *Antirrhinum* floral homeotic proteins DEFICIENS and GLOBOSA is exerted by their polar cell-to-cell trafficking. Development 122:3433–3441
Poirson A, Turner AP, Giovane C, Berna A, Roberts K, Godefroy-Colburn T (1993) Effect of the alfalfa mosaic virus movement protein expressed in transgenic plants on the permeability of plasmodesmata. J Gen Virol 74:2459–2461
Powell MJ (1974) Fine structure of plasmodesmata in a chytrid. Mycologia 66:606–614
Pueschel CM (1990) Cell structure. In: Cole KM, Sheath RG (eds) Biology of red algae. Cambridge University Press, Cambridge, pp 7–41
Radford JE, White RG (1998) Localisation of a myosin-like protein to plasmodesmata. Plant J 14:743–750
Radford JE, Vesk M, Overall RL (1998) Callose deposition at plasmodesmata. Protoplasma 201: 30–37
Raven JA (1984) Energetics and transport in aquatic plants. AR Liss, New York
Raven JA (1991) Long-term functioning of enucleate sieve elements: possible mechanisms of damage avoidance and damage repair. Plant Cell Environ 14:139–146
Raven JA (1997) Multiple origins of plasmodesmata. Eur J Phycol 32:95–101
Reid RJ, Overall RL (1992) Intercellular communication in *Chara*: factors affecting transnodal electrical resistance and solute fluxes. Plant Cell Environ 15:507–517
Robards AW, Lucas WJ (1990) Plasmodesmata. Annu Rev Plant Physiol Plant Mol Biol 41:369–419
Robards AW, Lucas WJ, Pitts JD, Jongsma HJ, Spray DC (1990) Parallels in cell to cell junctions in plants and animals. NATO ASI, series vol H46, Springer, Berlin Heidelberg New York
Robinson-Beers K, Evert RF (1991) Fine structure of plasmodesmata in mature leaves of sugarcane. Planta 184:307–318
Roy S, Watada AE, Wergin WP (1997) Characterisation of the cell-wall microdomain surrounding plasmodesmata in apple fruit. Plant Physiol 144:539–547
Russin WA, Evert RF (1985) Studies on the leaf of *Populus deltoides* (Salicaceae): ultrastructure, plasmodesmal frequency, and solute concentrations. Am J Bot 72:1232–1247
Sakuth T, Schobert C, Pecsvaradi A, Eichholz A, Komor E, Orlich G (1993) Specific proteins in the sieve-tube exudate of *Ricinus communis* L. seedlings: separation, characterization and in-vivo labelling. Planta 191:207–213
Sandgren CD, Smol JP, Kristiansen J (1994) Chrysophyte algae: ecology, phylogeny and development. Cambridge University Press, Cambridge
Sanger M, Passmore B, Falk BW, Bruening G, Ding B, Lucas WJ (1994) Symptom severity of beet western yellows virus strain ST9 is conferred by the ST9-associated RNA and is not associated with virus release from the phloem. Virology 200:48–55
Schenk E (1974) Similar effect of cytokinins and carbonyl cyanide m-chlorophenylhydrazone (CCCP) on amino acid translocation in leaves of *Sagittaria graminea* Michx. Acta Bot Neerl 23:739–746
Schiefelbein JW, Masucci JD, Wang H (1997) Building a root: the control of patterning and morphogenesis during root development. Plant Cell 9:1089–1098
Schmitz J, Stüssi-Garaud C, Tacke E, Prüfer D, Rohde W, Rotfritsch O (1997) In situ localization of the putative movement protein (pr17) from potato leaf roll luteovirus (PLRV) in infected and transgenic potato plants. Virology 235:311–322
Schobert C, Großmann P, Gottschalk M, Komor E, Pecsvaradi A, zur Nieden U (1995) Sieve-tube exudate from *Ricinus communis* L. seedlings contains ubiquitin and chaperones. Planta 196:205–210
Schobert C, Baker L, Szederkenyi J, Großmann P, Komor E, Hayashi H, Chino M, Lucas WJ (1998) Identification of immunologically related proteins in sieve-tube exudate collected from monocotyledonous and dicotyledonous plants. Planta 206:245–252

Schulz M, Traub O, Knop M, Willecke K, Schnabl H (1992) Immunofluorescent localization of a connexin 26-like protein at the surface of mesophyll protoplasts of *Vicia faba* and *Helianthus annuus* L. Bot Acta 105:111–115
Smith LM, Sabnis DD, Johnson RPC (1987) Immunocytochemical localisation of phloem lectin from *Cucurbita maxima* using peroxidase and colloidal-gold labels. Planta 170:461–470
Smith LG, Breene B, Hake S (1992) A dominant mutation in the maize homeobox gene *knotted-1* causes its ectopic expression in leaf cells with altered fates. Development 116:21–30
Sokolova M, Prüfer D, Tacke E, Rohde W (1997) The potato leafroll virus 17K movement protein is phosphorylated by a membrane-associated protein kinase from potato with biochemical features of protein kinase C. FEBS Lett 400:201–205
Spanswick RM, Costerton JWF (1967) Plasmodesmata in *Nitella translucens*: structure and electrical resistance. J Cell Sci 2:451–464
Stewart KD, Mattox KR, Floyd GL (1973) Mitosis, cytokinesis, the distribution of plasmodesmata and other characteristics in the Ulotrichales, Ulvales and Chaetophorales: phylogenetic and taxonomic considerations. J Phycol 9:128–141
Storms MMH, van der Schoot C, Prins M, Kormelink R, van Lent JWM, Goldbach RW (1998) A comparison of two methods of microinjection for assessing altered plasmodesmal gating in tissues expressing viral movement proteins. Plant J 13:131–140
Tacke E, Schmitz J, Prüfer D, Rohde W (1993) Mutational analysis of the nucleic acid-binding 17 kDa phosphoprotein of potato leafroll luteovirus identifies an amphipathic α-helix as the domain for protein/protein interactions. Virology 197:274–282
Terry BR, Robards AW (1987) Hydrodynamic radius alone governs the mobility of molecules through plasmodesmata. Planta 171:145–157
Thomson WW, Platt-Aloia K (1985) The ultrastructure of the plasmodesmata of the salt glands of *Tamarix* as revealed by transmission and freeze-fracture electron microscopy. Protoplasma 125: 13–23
Tilney LG, Cooke TJ, Connelly PS, Tilney MS (1991) The structure of plasmodesmata as revealed by plasmolysis, detergent extraction, and protease digestion. J Cell Biol 112:739–747
Tucker EB (1982) Translocation in the staminal hairs of *Setcreasea purpurea.* I A study of cell ultrastructure and cell-to-cell passage of molecular probes. Protoplasma 113:193–201
Tucker EB (1993) Azide treatment enhances cell-to-cell diffusion in staminal hairs of *Setcreasea purpurea.* Protoplasma 174:45–49
Tucker EB, Tucker JE (1993) Cell-to-cell diffusion selectivity in staminal hairs of *Setcreasea purpurea.* Protoplasma 174:36–44
Turner A, Wells B, Roberts K (1994) Plasmodesmata of maize root tips: structure and composition. J Cell Sci 107:3351–3361
Tyree MT (1970) The symplast concept. A general theory of symplastic transport according to the thermodynamics of irreversible processes. J Theor Biol 26:181–214
van Bel AJE (1993) The transport phloem. Specifics of its functioning. Prog Bot 54:134–150
van Bel AJE (1996) Interaction between sieve element and companion cell and the consequences for photoassimilate distribution. Two structural hardware frames with associated physiological software packages. J Exp Bot 47:1129–1140
van Bel AJE, Kempers R (1997) The pore/plasmodesm unit; key element in the interplay between sieve element and companion cell. Prog Bot 58:278–291
van Bel AJE, van Rijen HVM (1994) Microelectrode-recorded development of the symplasmic autonomy of the sieve element/companion cell complex in the stem phloem of *Lupinus luteus* L. Planta 192:165–175
van den Berg C, Willemsen V, Hage W, Weisbeek P, Scheres B (1995) Cell fate in the *Arabidopsis* root meristem determined by directional signalling. Nature 378:62–65
van den Hoek C, Mann DG, Jahns HM (1995) Algae: an introduction to phycology. Cambridge University Press, Cambridge
van der Schoot C, van Bel AJE (1990) Mapping membrane potential differences and dye coupling in internodal tissues of tomato (*Solanum lycopersicum* L.). Planta 182:9–21
Vaquero C, Turner AP, Demangeat G, Sanz A, Serra MT, Roberts K, Garcia-Luque I (1994) The 3a protein from cucumber mosaic virus increases the gating capacity of plasmodesmata in transgenic tobacco plants. J Gen Virol 75:3193–3197
Volk G, Turgeon R, Beebe D (1996) Secondary plasmodesmata formation in the minor-vein phloem of *Cucumis melo* and *Cucurbita pepo* L. Planta 199:425–432
Waigmann E, Zambryski P (1995) Tobacco mosaic virus movement protein mediated protein transport between trichome cells. Plant Cell 7:2069–2079
Waigmann E, Lucas WJ, Citovsky V, Zambryski P (1994) Direct functional assay for tobacco mosaic virus cell-to-cell movement protein and identification of a domain involved in increasing plasmodesmatal permeability. Proc Natl Acad Sci USA 91:1433–1437

Waigmann E, Turner A, Peart J, Roberts K, Zambryski P (1997) Ultrastructural analysis of leaf trichome plasmodesmata reveals major differences from mesophyll plasmodesmata. Planta 203: 75–84

White RG, Badelt K, Overall RL, Vesk M (1994) Actin associated with plasmodesmata. Protoplasma 180:169–184

Wille AC, Lucas WJ (1984) Ultrastructural and histochemical studies on guard cells. Planta 160: 129–142

Williams SE, Pickard BG (1972a) Receptor potentials and action potentials in *Drosera* tentacles. Planta 103:193–221

Williams SE, Pickard BG (1972b) Properties of action potentials in *Drosera* tentacles. Planta 103: 222–240

Williams SE, Pickard BG (1974) Connections and barriers between cells of *Drosera* tentacles in relation to their electrophysiology. Planta 116:1–16

Wintermantel W, Banerjee N, Oliver JC, Paolillo DJ, Zaitlin M (1997) Cucumber mosaic virus is restricted from entering minor veins in transgenic tobacco exhibiting replicase-mediated resistance. Virology 231:248–257

Wolf S, Deom CM, Beachy RN, Lucas WJ (1989) Movement protein of tobacco mosaic virus modifies plasmodesmatal size exclusion limit. Science 246:377–379

Wright KM, Oparka KJ (1997) Metabolic inhibitors induce symplastic movement of solutes from the transport phloem of *Arabidopsis* roots. J Exp Bot 48:1807–1814

Yahalom A, Warmbrodt RD, Laird DW, Traub O, Revel JP, Willecke K, Epel BL (1991) Maize mesocotyl plasmodesmata proteins cross-react with connexin gap junction protein antibodies. Plant Cell 3:407–417

Zhang WH, Tyerman SD (1997) Effect of low oxygen concentration on the electrical properties of cortical cells of wheat roots. J Plant Physiol 150:567–572

Notes added in proof:

1. Meanwhile B. Ding published an outstanding review (Intercellular protein trafficking through plasmodesmata, Plant Mol. Biol. 38, 279–310, 1998) dealing with several themes discussed in this chapter. Among others, the mechanism of intercellular protein trafficking, the role of protein trafficking in plant growth and development and the existence of different functional types of plasmodesmata are reviewed in detail.
2. Evidence was provided that plasmodesmata at the mature stage have protein-trafficking properties different from those at the young stage (A. Itaya et al.; Developmental regulation of intercellular protein trafficking through plasmodesmata in tobacco leaf epidermis, Plant Physiol. 118, 373–385, 1998).
3. A strong involvement of the endoplasmic reticulum in virus guidance was highlighted by M. Heinlein et al.; Changing patterns of localization of tobacco mosaic virus movement proteins and replicase to endoplasmic reticulum and microtubules during infection, Plant Cell 10, 1107–1120, 1998.

CHAPTER 2

Plasmodesmal Imaging – Towards Understanding Structure

C. E. J. BOTHA · R. H. M. CROSS

Department of Botany and Electron Microscopy Unit, Rhodes University, Grahamstown 6140, South Africa

1 Introduction

Robards' paper in Nature, (1968a) entitled *Desmotubule – a plasmodesmatal substructure* started a debate which, some 29 years later, remains largely unresolved. The concept of the desmotubule, its substructure, and its relationship to the other components of higher plant plasmodesma remains challenged with respect to its substructure, its presence in most higher plant plasmodesma and, therefore, its role in cell-to-cell communication. Almost as many interpretations of structure exist in the literature (see Gunning and Robards 1976; Robards 1971, 1976; Robards and Lucas 1990 and literature cited; Ding et al. 1992, and literature cited; Botha et al. 1993 and literature cited, Overall and Blackman 1996) as there are references.

In their review Robards and Lucas (1990) alluded to the difficulties associated with electron imaging of plasmodesma. Much has been gained by using alternative methods of fixation such as cryofixation and freeze substitution (see Ding et al. 1992). However, the problem remains one of interpreting the electron image as seen in transverse view and the reconciliation of this with longitudinal views of plasmodesma from the same tissues.

Botha et al. (1993) observed that, despite the difference in time and advances made in fixation technique, the views of plasmodesmal ultrastructure presented by Robards (1968a, b), for example, in *Salix fragilis* cambial cells (see Figs. 2–4, Robards 1968b) differ little from chemically fixed plasmodesma in, for example, grass leaf vascular bundles in *Saccharum officinarum* (Robinson-Beers and Evert 1991) or those in *Themeda triandra* (Botha et al. 1993) or *Eragrostis plana* (Botha 1992).

The use of freeze substitution techniques has allowed resolution of particulate material within plasmodesma more convincingly than has been the case with chemically fixed material. However, close examination of Fig. 8 in Robards' 1968b paper clearly shows granularity and what appears to be particulate substructural detail. The problem thus does not lie primarily with interpretation of structure in transections of plasmodesma, but in longitudinal sections of plasmodesma. Here there remain enormous difficulties, associated for the most part with noise generated within the electron image and the plane of the section. In addition, average section thickness exceeds that of the diameter of the desmotubule (15–25 nm). Ding et al. (1992) and subsequently Botha et al. (1993) have demonstrated that there is correlation between freeze substitution and careful chemical fixation with respect to the structure of the desmotubule in *Nicotiana tabacum* and *Themeda triandra*, respectively, when viewed in transverse as well as in longitudinal section.

Given the technological advances that have been made in electron microscopy and the great strides made in preparative equipment, the one most singularly frustrating issue which remains is that there has been little significant increase in the resolution of these important molecular trafficking channels. Thus, despite the many years which have elapsed since Robards' images appeared in the literature (1968a, b), these still remain classic transectional images of higher plant plasmodesma.

2 Digital Imagery

Botha et al. (1993) have demonstrated that digital false colour images can be used to interpret the cross-sectional as well as longitudinal structure of complex plasmodesma such as those that occur in C_3 and C_4 grasses. The figures produced using this approach were based upon relatively low- resolution (400 dots per inch) digital images of otherwise high-resolution electronmicrographs of plasmodesma in transection as well as in longitudinal section. Clearly, the high agreement between the interpretation made by Ding et al. (1992) and that of Botha et al. (1993) suggested that digital imaging would be a valuable tool in determining possible plasmodesmal ultrastructure. The principal aim of this chapter is to explore the potential of digital imaging with respect to the structure of plasmodesma.

3 Problems and Pitfalls

Creating a "good" image is a difficult process. Photographically, image processing involves three important steps – the first being exposure of the negative, the second, developing the negative, and the third, processing of a positive image. In effect what is achieved in the darkroom is adjustment of contrast and brightness to produce an "acceptable" image which reveals the details that the author requires (or likes). Unfortunately, in producing the acceptable image by changing the brightness and contrast level of the image, this may result in the loss of information from the image itself. Thus the only totally acceptable source of the full spectrum of information is the negative and not the final darkroom print. However, it is important to note that the characteristics of the negative are also dependent upon exposure time and choice of film, as well as processing technique.

Images produced on the viewing screen of a transmission electron microscope (TEM) are analogue and must be digitized before any of the processing or analytical procedures discussed here are carried out. Analogue to digital conversion is normally undertaken using a close-coupled device (CCD) TV camera, or a slow-scan CCD camera in the electron beam path. Whilst the former is relatively inexpensive, it offers only TV resolution (512 × 512 pixels). Slow-scan cameras, on the other hand, yield higher resolution (up to 2048 × 2048 pixels), but at much greater cost and with images of 4 megabytes (Mb) or more in size. The third alternative is to acquire the image by conventional (analogue) photographic means and to digitize the image by scanning either the negative or a high resolution photographic print (given the qualification in the previous paragraph).

Digital imaging via either negative or positive require that the analogue image be converted to a digital image, ensuring thus that the smallest possible components of the image (ideally each silver grain present on the photograph) is interpreted as a number between 0 and 255. These values make up the grey levels in a computer-generated grey scale, and represent the information that we are interested in storing to disc. Each number reflects information and should thus be considered important to the composition of the image. Ideally, digitized images should contain at least 200 grey levels of information, without eliminating true black (level = 0) and true white (level = 255). Each image will thus contain a very large number of bits of information, which are stored as pixels. What we interpret as resolution in a photograph, therefore, is related to the number of pixels of information stored with our digitized image. The more pixels there are, the higher the image resolution, and concomitantly, the larger the physical image file size becomes. Image enhancement of electronmicrographs of small structures such as plasmodesma thus requires careful selection of the base information to ensure that all relevant information is present (assuming that the negative is properly exposed and sharply focused as well). This will ensure that no additional noise-related artefacts are introduced into the digital image. It is imperative that images are digitized at different resolutions, in order to determine which shows the most useful information.

4 Gamma Correction Using Lookup Tables

The most fundamental mistake made during digitization processes is making the assumption that brightness and contrast adjustment is the most obvious adjustment to make. It is important that the electronmicrograph negative (the analogue image) is carefully examined before any adjustment of brightness or contrast is contemplated either in the darkroom (analogue print) or with the scanner (the digital image). Enhancing brightness (lightening the image) may well result in a loss of information (pixels) in the darker region of the image (shaded area, Fig. 1A). Conversely, enhancing the contrast of the image may result in loss of information primarily in the lightest

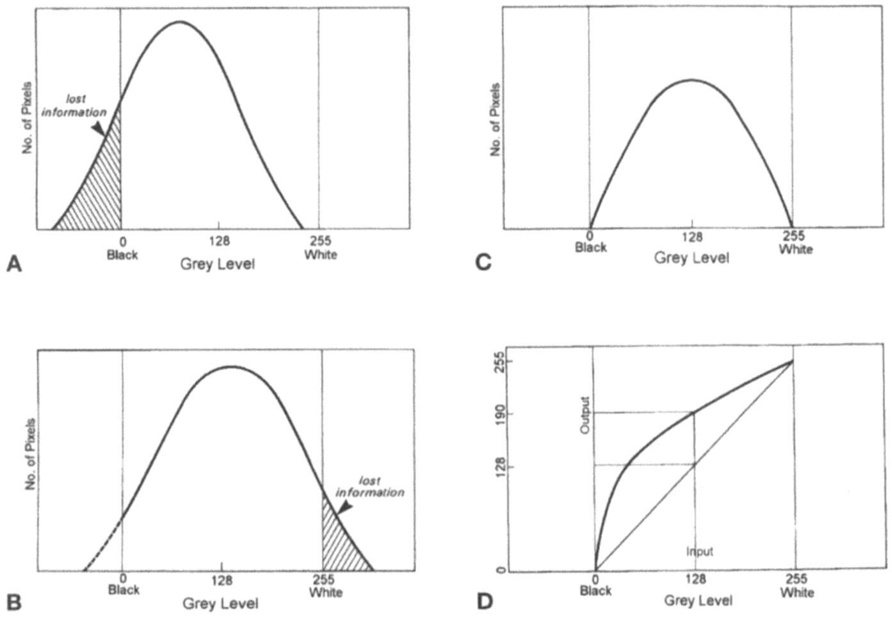

Fig. 1A–D. A An idealized histogram showing the effect of adjusting the brightness of an image. Decreasing brightness (*left*) results in a loss of information in the darker regions of the image and vice versa. **B.** The effect of adjusting the contrast (gain) of a digital image. Increasing it beyond acceptable limits results in a loss of information in the lighter regions of the image and vice versa. **C** An "ideal" distribution of information. Here grey scale pixels 0–255 are represented and the image should thus show maximal information in the dark as well as in the light pixel range. **D.** The effect of changing Gamma settings from 1 (linear) to 2 (logarithmic). Grey level 128 becomes approximately 190, thus boosting and enhancing detail in the darker areas of the image, without loss of detail in the lighter regions

regions of the image (shaded area, Fig. 1B) but also from the darker (0–50 on the grey scale, dashed line, on left). The operator may think that the images look "good", but overzealous adjustments must cause loss of important information. Care should be taken also in selecting the image processing package – our personal choice is Adobe Photoshop (version 2.5 and above, Adobe Systems Inc., Seattle, WA, USA) and Paint Shop Pro (version 4.X and above, Jasc Inc., Eden Prairie, MN, USA). These packages (and we acknowledge that there are others which may be equally well suited to the task) allow the user to make corrections which enhance the image without concomitant data loss. The ideal manipulated image will end up with a distribution of pixels through the full range (0 to 255 pixels), and the manipulation that is applied must remove as few pixels as possible, to yield an image which effectively "stretches" the histogram to include all available data-containing pixels (Fig. 1C).

Gamma (γ) measures the contrast that affects the midlevel greys (midtones) of an image. Adjusting the Gamma allows changes in the brightness values of the middle range of grey tones without dramatically changing the shadows (very dark pixels) and highlights (very light pixels).

Gamma correction thus affects contrast and brightness settings, and must therefore be adjusted first. Many packages, monitors, and printers have varying Gamma correction built in, so it becomes important to know what these default settings are, and how easily they may be altered, or indeed are "adjusted" by the software. Figure 1D shows the effect of adjusting the pixel values for the midrange (128) by applying a Gamma correction of approximately 2 to the original linear output (the straight line). The midpoint value has been increased on the Y axis to approximately 190, thus enhancing the information in the black to mid grey (0 to 128) regions of the image by approximately 68% but without any loss of image information across the full 0–255 pixel range. Gamma correction thus becomes a very powerful tool in the arsenal available in grey scale image processing.

5 Digital Scanners and Image Storage

Irrespective of the image source (i.e. negative or positive), converting analogue to digital images requires the use of a scanner which is capable of creating accurate halftone images (interpreted as levels of grey) to at least a 16 × 16-pixel cell size. The initial image size becomes a major factor and has to be considered very carefully – for example, a postcard-sized black and white photograph, scanned at 400 dpi will be approximately 4.4 Mb in size and the same image scanned at 600 dpi will create a 9.8 Mb file. Clearly, a fast standard computer system interface-based (SCSI) personal computer is necessary in order to handle and process these images. What must be avoided at all costs is any form of software image compression, which effectively reduces the image's physical size, but which achieves this at the expense of pixel information, by a process of similar pixel averaging. Pixels with digital values such as 241, 242 and 243 may be stored as an average value of 242. The end result is a loss of digital (grey scale) information and aggregated clumping of similar pixels. Tagged information file format (TIFF) is the format of choice for image storage, as most imaging programmes recognize TIFF properly – other formats may be proprietary and non-transferable from package to package.

6 Tissue Preparation and Digital Imaging Procedures

All the transmission electron micrographs presented in this chapter, were made using sections taken from the mesocarp of 226-day-old avocado tissue (*Persea americana* Mill. cv. Hass) . The embedding, staining and sectioning procedures are essentially similar to those discussed in detail elsewhere (Botha 1992; Botha et al. 1993).

Sections were viewed and photographed using a JEOL JEM 1210 (Jeol, Tokyo) transmission electron microscope at 100 kV. Selected electron micrographs were scanned at either 600 or 1200 pixels per inch, using a Hewlett-Packard HP-5P scanner (Hewlett-Packard, Palo Alto CA USA) and stored as uncompressed TIFF images. Scanned images were manipulated using Adobe PhotoShop 4.0, and Paint Shop-Pro version 4.12.

7 Examples of Digital Imaging

The micrographs that have been selected to illustrate the effects of resolution and Gamma control, we believe, show sufficient detail and are taken from a representative series of collections made from near-ripe avocado mesocarp tissue. We deliberately did not choose images that were perfect, as we felt that very little would be achieved by this – as such, fairly average images were chosen to illustrate the points that are made in this chapter about image enhancement.

The effect of scanning resolution and Gamma correction is clearly demonstrated in the high magnification images shown in Fig. 2. Higher resolution scanning enhances image sharpness, but both series show that enhancing Gamma from 1.0 to 3.5 reveals potentially important substructural details, especially those associated with the desmotubule. Comparison of the analogue (Fig. 2J) with the digital images scanned at 600 (Fig. 2A–E) and 1200 pixels per inch (ppi) (Fig. 2F–L), demonstrate no loss of resolution in the scanned images, compared with the analogue image. However, the effect of Gamma correction becomes evident in Fig. 2D, ($\gamma = 2.5$), where the granular nature of the desmotubule becomes more easy to see. In contrast, the higher resolution (1200 ppi) scanned images (Fig. 2F–I) show enhanced desmotubular detail in images with a Gamma correction of 2.50, and show improving resolution of the desmotubule at $\gamma = 3$ (Fig. 2L). Comparison of the 600 and 1200 ppi images demonstrates that the higher resolution images enhance differences and enhance the detail (at the same γ value) with respect to the cell wall, the matrix structure of the desmotubule and, most importantly, digital imaging reveals that the cytoplasmic sleeve is electron-lucent. It further reveals details of the interconnected matrix, which makes up the expanded extra-plasmodesmal collar outside of this, which is common in cytokinin-treated avocado mesocarp.

Figure 3 shows (non-transporting) plasmodesma from abscisic acid (ABA)-treated mesocarp tissue. The analogue micrograph (Fig. 3E) was printed as a normal high-contrast image, and was subsequently scanned at 1200 ppi and digitally enhanced (Figs. 3B–D). Of interest was that plasmodesma in mesocarp of ABA-treated tissue, were constricted by what appears to be an electron-dense collar-like structure, composed of a number of closely spaced but overlapping spherical subunits approximately 2.5 to 3.5 nm in diameter; compare these subunits in the analogue image to those in the digitally enhanced images (Fig. 3A–D, unlabelled paired arrows). These figures il-

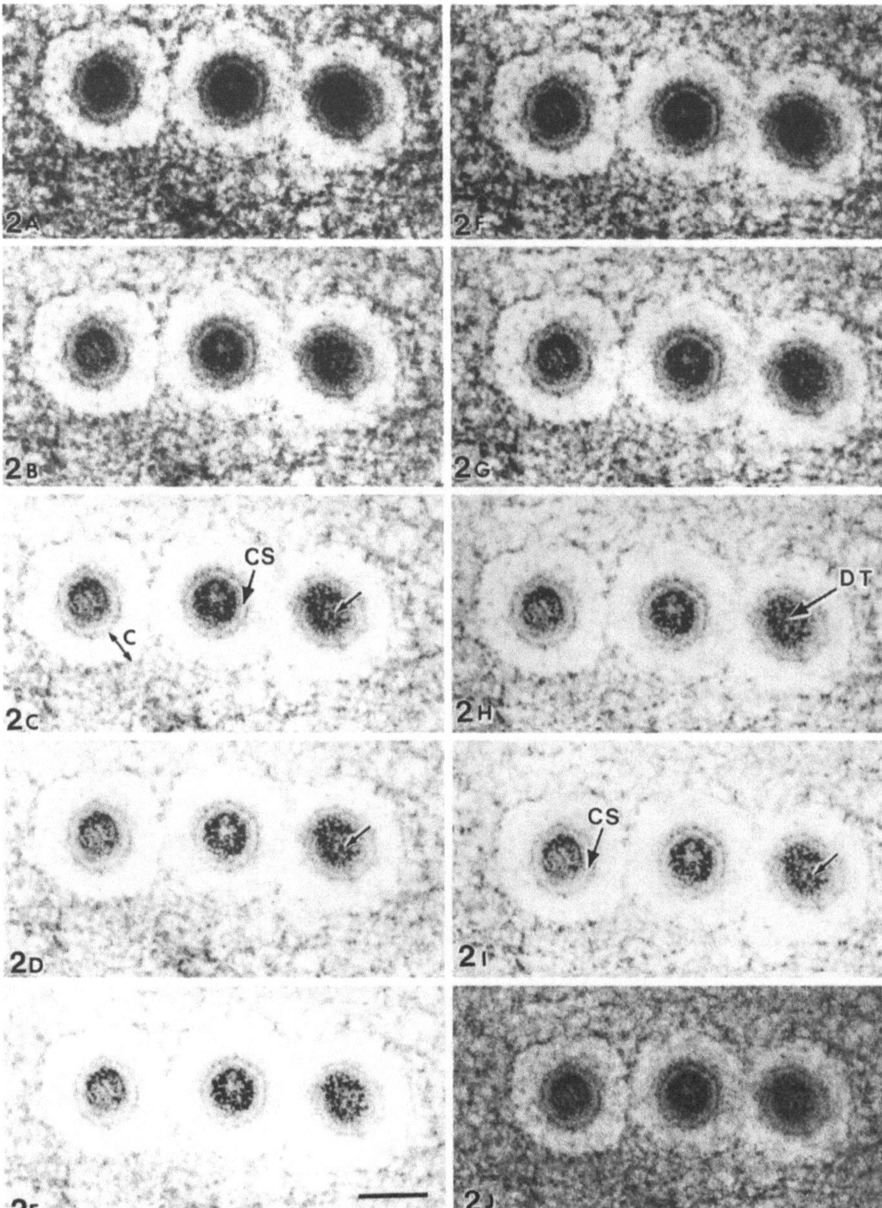

Fig. 2A–J. Composite plate, showing a comparison of analogue (J) with digital images made at 600 (A–E) and 1200 (F–L) pixels per inch, of a triplet of plasmodesmata, in a radial wall of 226-day-old mesocarp tissue in avocado which had been treated with 10 mol m^{-3} cytokinin. The effect of increasing Gamma correction from $\gamma = 1.0$ to 2.5 (A–C) and above clearly illustrates a subtle but positive effect on substructural detail enhancement. Note especially the desmotubule which becomes progressively more granular in appearance. The higher resolution images reveal more detail at $\gamma = 3$ than the corresponding 600-ppi image, where some detail (loss of granularity of the plasmalemma leaflets) has been lost. *Arrows* point to granulate components of the desmotubule; *C* plasmodesmal collar; *CS* electron lucent cytoplasmic sleeve; *DT* desmotubule. *Bar* 50 nm

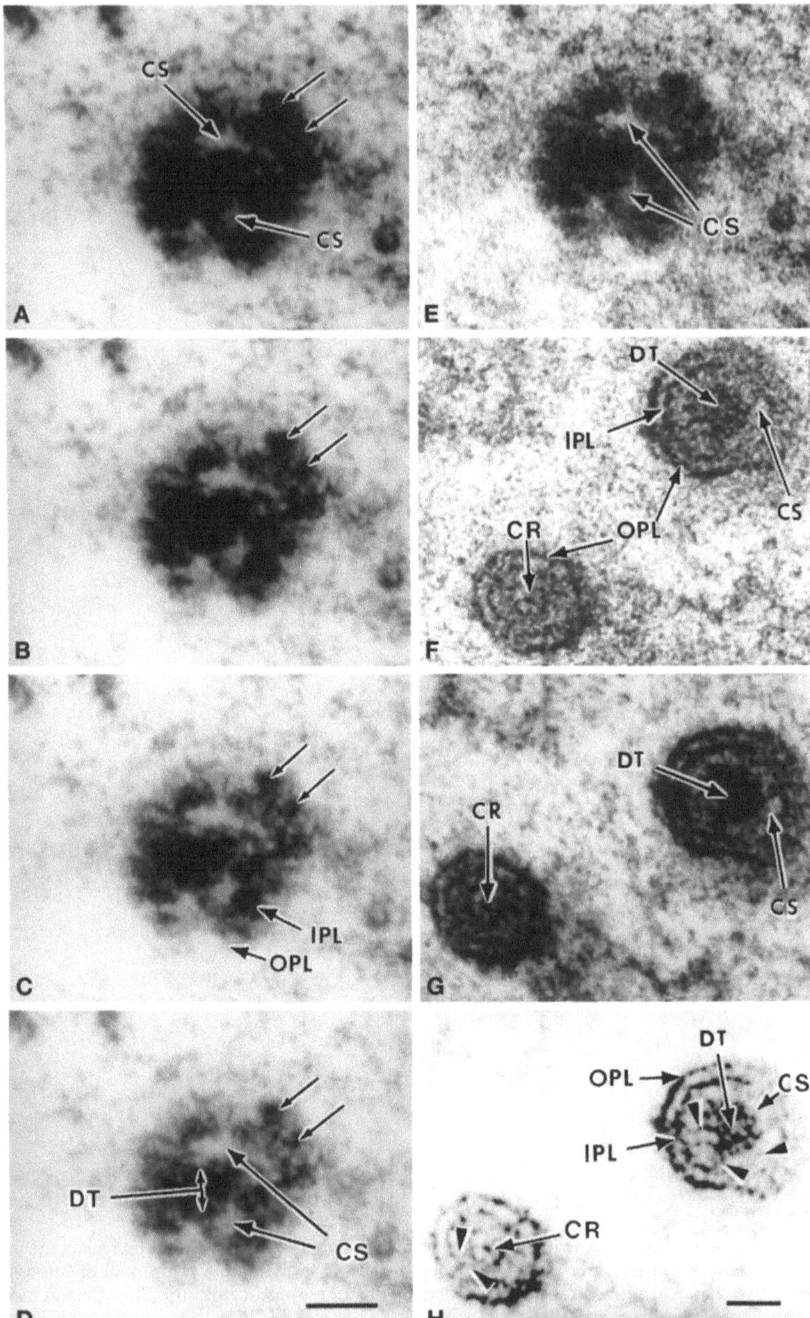

Fig 3A–H. Plasmodesma, from 226-day-old mesocarp tissue of avocado, treated with 10 mol m^{-3} abscisic acid (**A–E**) and control tissues (**F–H**). The analogue image of plasmodesma from ABA-treated tissue (**E**) shows this to be surrounded by a dense matrix of granulate material, but little detail can be lustrate quite clearly the effect of Gamma correction and its remarkable effect on image resolution. A Gamma correction of 2.0 to 2.5 emphasizes and enhances resolution of the individual subunits within the substructure (compare paired arrows, Fig. 3A–E). Some of these units seem to be associated with the desmotubule whilst others (apparently interconnected) form the tightly packed inner and outer plasmalemma leaflets, which are clearly enhanced through digital imaging and Gamma correction. Of note is the digital enhancement achieved, to reveal the electron-lucent regions of the cytoplasmic sleeve in Fig. 3D, – a feature which is not very clear in the analogue image (Fig. 3E).

Figure 3F–H has been included to illustrate the acceptable extreme to which digital images of plasmodesma could be subjected (at least in this study) without major or significant loss of image information. The (slightly oblique) plasmodesma illustrated in the analogue (Fig. 3F) and digitally enhanced (Fig. 3G–H) image was taken from 226-day-old mesocarp control tissues. The analogue image clearly demonstrates the presence of the characteristic double-layered plasmamembrane, which surrounds a granular cytoplasmic sleeve, which, in turn, surrounds the desmotubule. The central rod is visible in the analogue and digital images. Digital enhancement and correcting Gamma to $\gamma = 2.5$ effectively removes much of the background speckled (noise) effect, revealing more information, particularly in relation to the two plasmamembrane leaflets which are some 3 nm apart, and the spoke-like structures (arrowheads) connecting the desmotubule to the inner plasmamembrane leaflet. Comparison of Fig. 3F with Fig. 3G reveals a more electron lucent, open cytoplasmic sleeve, which is, in this instance, about 6.5–8 nm across.

8 Concluding Remarks

Researchers interested in plasmodesmal ultrastructure and more specifically "functional" structure, will remain fascinated by them for a long time – simply because electronmicrographs remain but two-dimensional snapshots in time (Robards and Lucas, 1990). Unravelling their structure is an extremely difficult task, made more so by difficulties associated with our understanding and interpretation of longitudinal sections. However, transections of average plasmodesma such as those illustrated here in this chapter show unequivocally that prudent use of digital imaging techniques such as Gamma correction can be enormously beneficial. Furthermore, we believe that these

made out in the normally printed image. In contrast, enhancement of γ from 1.0 to 2.5 (**B–D**) shows enhanced detail. Note particularly the granular, possibly spherical structures (*arrows*, **A–E**) and the close association of these spherical units with the desmotubule (*DT*) or, for apparently interconnected ones, with the inner (*IPL*) and outer (*OPL*) plasma membrane area; *CS* cytoplasmic sleeve. In **F–H** two slightly oblique plasmodesma are shown, as analogue (**F**) and digitally enhanced ($\gamma = 1.0$, **G**; $\gamma = 2.5$ **H**) images from 226-day mesocarp control tissues. Note a classical and characteristic double-layered (*IPL* and *OPL*) plasma membrane, which surrounds a typically granular, electron-dense cytoplasmic sleeve (*CS*), encasing the desmotubule, and a central rod (*CR*) in the analogue image. After digital enhancement to $\gamma = 2.5$, much of the background speckled effect is removed and a great deal more information is obtained, particularly in relation to the two plasma membrane leaflets. Of particular interest are the spoke-like structures (*unlabeled arrowheads*) which here seem to connect the desmotubule to the IPL. *Bars* 20 nm techniques will lead to revelations of more substructural detail than the currently accepted use of contrast or brightness control and manipulation in conventional photographic procedures. Digital manipulation is a tool – but, we argue, one of the most powerful available in the image preparation arsenal. We have shown that the substructural details in plasmodesma present in near-ripe avocado are enhanced through digital manipulation. The enormous advantage offered by PC-based digital manipulation results from its being a relatively inexpensive technique. All that is needed is a reasonably good flatbed scanner, an adequately powered computer and good software packages (which need not be the most expensive available on the market). Anyone, after a little practice, may start to produce images which offer exciting new information relating to the structure of intriguingly frustrating objects such as plasmodesma. Hopefully, further development of this technique will ultimately improve our understanding of the structure of this intriguing intercellular transport channel.

References

Botha CEJ (1992) Plasmodesmatal distribution, structure and frequency in relation to assimilation in C_3 and C_4 grasses in Southern Africa. Planta 187:343–358

Botha CEJ, Hartley BJ, Cross RHM (1993) The ultrastructure and computer-enhanced digital image analysis of plasmodesmata at the Kranz mesophyll-bundle sheath interface of *Themeda triandra* var. *imberbis* (Retz) Camus, A. in conventionally fixed leaf blades. Annals of Botany 72:255–261

Ding B, Turgeon R, Parthasarathy MV (1992) Substructure of freeze-substituted plasmodesmata. Protoplasma 169:28–41

Gunning, BES, Robards, AW (1976) Intercellular communication in plants: Studies on plasmodesmata. Gunning BES, Robards AW (eds). Springer, Berlin Heidelberg

Overall RL, Blackman LM (1996) A model for the macromolecular structure of plasmodesmata. Elsevier Trends in Plant Sciences 1:307–311

Robards AW (1968a) Desmotubule – a plasmodesmatal substructure. Nature 218:784

Robards AW (1968b) A new interpretation of plasmodesmatal ultrastructure. Planta 82:200–210

Robards AW (1971) The ultrastructure of plasmodesmata. Protoplasma 72:315–323

Robards AW (1976) Plasmodesmata in higher plants. In: Gunning BES, Robards AW (eds) Intercellular communication in plants: Studies on plasmodesmata. Springer, Berlin Heidelberg, pp 15–57

Robards AW, Lucas WJ (1990) Plasmodesmata. Annu Rev Plant Physiol Plant Mol Biol 41:369–419

Robinson-Beers K, Evert RF (1991) Ultrastructure of and plasmodesmatal frequency in mature leaves of sugarcane. Planta 184:291–306

CHAPTER 3

Tissue Preparation and Substructure of Plasmodesmata

B. DING

Department of Botany, Oklahoma State University, Stillwater, Oklahoma 74078, USA

1 Introduction

Research on the plasmodesma seeks to address two basic issues: what kinds of molecules are transported through this intercellular organelle, and how are these molecules transported? Elucidation of the transport mechanisms requires that the substructure of the plasmodesma be understood. This understanding may include at least two aspects: ultrastructural features and molecular compositions.

Views of the plasmodesmal substructure have mainly been obtained through the use of transmission electron microscopy (TEM). This chapter attempts to discuss in general terms the prevailing methods used to prepare tissue samples for TEM studies of plasmodesmata, focusing on the limitations of these methods and on the interpretation of data obtained with these methods. I will also discuss briefly the use of integrated approaches to study the plasmodesmal structure at the molecular level.

2 Structure of the Plasmodesma

The structure of the plasmodesma is reviewed extensively in Chapter 9 and only some general features are outlined here to facilitate subsequent technical discussions. Through the exploration of many workers over the past 20 years (e.g. López-Sáez et al. 1966; Robards 1968, 1971; Zee 1969; Olesen 1979; Hepler 1982; Overall et al. 1982;

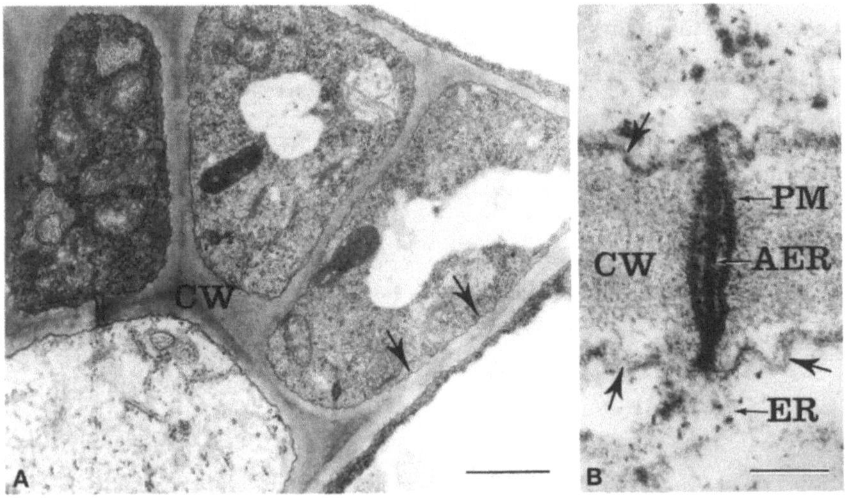

Fig. 1 A, B. A Vascular cells from a tobacco leaf fixed first with 4% glutaraldehyde/0.1% tannic acid and then with 2% osmium tetroxide. Note that the plasma membrane appears wavy and pulled away from the cell wall (*CW*), as indicated by arrows. *Bar* 1 μm. **B** A plasmodesma between bundle-sheath cells of a tobacco leaf fixed with the same fixation protocol as in **A**, showing major structural components of the organelle at this resolution: the plasma membrane (*PM*) and the appressed endoplasmic reticulum (*AER*). The AER is continuous with the endoplasmic reticulum (*ER*) in the cytoplasm. *Unlabelled arrows* indicate the plasma membrane that is pulled away from the cell wall (*CW*). This is a typical artifact of chemical fixation. *Bar* 100 nm (Ding et al. 1992a)

Thomson and Platt-Aloia 1985; Olesen and Robards 1990; Tilney et al. 1991; Ding et al. 1992b; Botha et al. 1993; Badelt et al. 1994; Ehlers and Kollmann 1996; Glockmann and Kollmann 1996), it is now established that a plasmodesma consists of the plasma membrane (PM) which is continuous between adjacent cells and forms a porous structure across the cell walls, and that the appressed endoplasmic reticulum (AER) cylinder (formerly desmotubule; see Lucas et al. 1993 for terminology) is positioned longitudinally in the center of the pore (Fig. 1A, B). The AER is structurally continuous with the rest of the cytoplasmic endoplasmic reticulum (ER) system. Within the plasmodesma, the space between the plasma membrane and the AER cylinder are thought to form microchannels for intercellular transport (Ding et al. 1992b). The AER itself may also play a role in intercellular transport of lipids (Grabski et al. 1993) and photoassimilates (Gamalei et al. 1994; Glockmann and Kollmann 1996). In particular, an open ER lumen is postulated to be the pathway for intercellular transport of at least photoassimilates (Gamalei et al. 1994).

Despite so many intensive investigations, some key issues of plasmodesmal substructure still remain to be resolved. Different views have been expressed concerning various aspects of the substructure of the plasmodesma. It is possible that variations in the plasmodesmal structure as proposed by different workers are dependent on plant species, tissue or cell types, and developmental stages in some cases. However, these variations may have also arisen when different methods were used to prepare tissues for microscopy. The latter is the focus of discussion of this chapter.

3 Methodologies of Tissue Preparation for Studying the Plasmodesmal Structure

Two methods have been used to fix plant tissues for studying the plasmodesmal structure: chemical fixation and cryofixation. Chemically fixed tissues were usually processed to obtain thin sections for TEM examination. Cryofixed samples have either been freeze-fractured to obtain replicas or freeze-substituted to obtain thin sections for TEM examination.

3.1 Chemical Fixation

Generally speaking, a typical chemical fixation protocol to prepare samples for structural studies at the TEM level consists of fixing small pieces of samples first in a primary fixative such as glutaraldehyde and/or paraformaldehyde, and then in a secondary fixative which is usually osmium tetroxide. Glutaraldehyde/paraformaldehyde cross-links proteins and osmium tetroxide fixes lipids (Glauert 1975; Bozzola and Russel 1992). Chemical fixation was the primary method used by many workers to prepare plant materials for plasmodesmal structure studies (e.g. López-Sáez et al. 1966; Robards 1968, 1971; Zee 1969; Olesen 1979; Overall et al. 1982; Olesen and Robards 1990; Tilney et al. 1991; Botha et al. 1993; Badelt et al. 1994; Ehlers and Kollmann 1996; Glockmann and Kollmann 1996). These studies have provided a general and fundamental understanding of the structure of the plasmodesma. In particular, that the plasmodesma is basically composed of the plasma membrane and the AER which is continuous with the cytoplasmic ER is elegantly demonstrated (e.g. López-Sáez et al. 1966; Hepler 1982; Overall et al. 1982; Ehlers and Kollmann 1996; Glockmann and Kollmann 1996).

In a number of studies, tannic acid of 1–2% was included in the primary fixative (e.g. Olesen 1979; Overall et al. 1982; Thomson and Platt-Aloia 1985; Tilney et al. 1991; Badelt et al. 1994). Since tannic acid stains microtubules negatively (Fujiwara and Linck 1982), it was suggested that tannic acid could also stain plasmodesmal structures negatively. In particular, "electron-lucent" (negatively stained) particles were proposed to exist between the AER and the plasma membrane (Overall et al. 1982; Thomson and Platt-Aloia 1985; Olesen and Robards 1990).

However, Ding et al. (1992b) suggest that tannic acid does not stain plasmodesmal structures negatively. First, in tissues processed with and without the use of tannic acid, the staining patterns of plasmodesmal structures are the same. Second, overstaining by high concentrations of tannic acid could produce spurious images that appear negatively stained. As discussed by Fujiwara and Linck (1982), it is extremely important to use tannic acid of low-molecular weight and at low concentrations. When used at a concentration of more than 1%, tannic acid can cause extensive precipitation inside and outside a cell, thereby obscuring fine cell structures (Fujiwara and Linck 1982). For example, the diameter of microfilaments is increased from 6–7 nm to 13 nm by the use of 1% tannic acid (Seagull and Heath 1979). Ding et al. (199la, 1992b) used 0.1% tannic acid and found that the 6–7 nm diameter of microfilaments was not increased by this treatment. Therefore, use of low concentrations of tannic acid results in minimal distortion of the native dimension of cell structures. On the basis of these considerations, Ding et al. (1992b) suggest that the electron-lucent areas between the AER and the plasma membrane represent spaces, rather than particles. Their study further indicates that these spaces are created by electron-dense particles embedded in the plasma membrane and the AER cylinder of the plasmodesma (see below).

A number of workers have reported that "sphincter"-like structures are present in the cell wall region surrounding the orifice of the plasmodesma. These structures are particularly evident when 1–2% tannic acids were used in the primary fixative (Robards 1976; Olesen 1979; Mollenhauer and Morré 1987; Badelt et al. 1994), and when Driselase was used to digest the cell walls (Badelt et al. 1994). It has been suggested that these sphincter structures may function to regulate the opening or closing of the plasmodesmal transport channels between the plasma membrane and AER by some sort of contraction and expansion mechanism. However, a close examination of some of the published micrographs (Overall et al. 1982; Badelt et al. 1994) reveals that similar globular structures are also present at the interface between the plasma membrane and the cell wall, when the former is pulled away from the latter, in regions where plasmodesmata are absent. It is important to keep in mind that the effect of Driselase treatment on cell structures other than the cell walls is unknown. As discussed above, use of high concentrations of tannic acid can also be problematic. Thus, it is possible that these sphincter structures are pure artefacts induced by a combination of poor chemical fixation (evidenced by the distortion of membrane structures), Driselase treatment and tannic acid precipitation. Even though Badelt et al. (1994) used Freon to cryofix some samples for freeze substitution, the images presented (Fig. 2 in their paper) showed gross distortion of the cell structure - the plasma membrane is pulled far away from the cell wall. Thus, the freezing technique used in that particular study

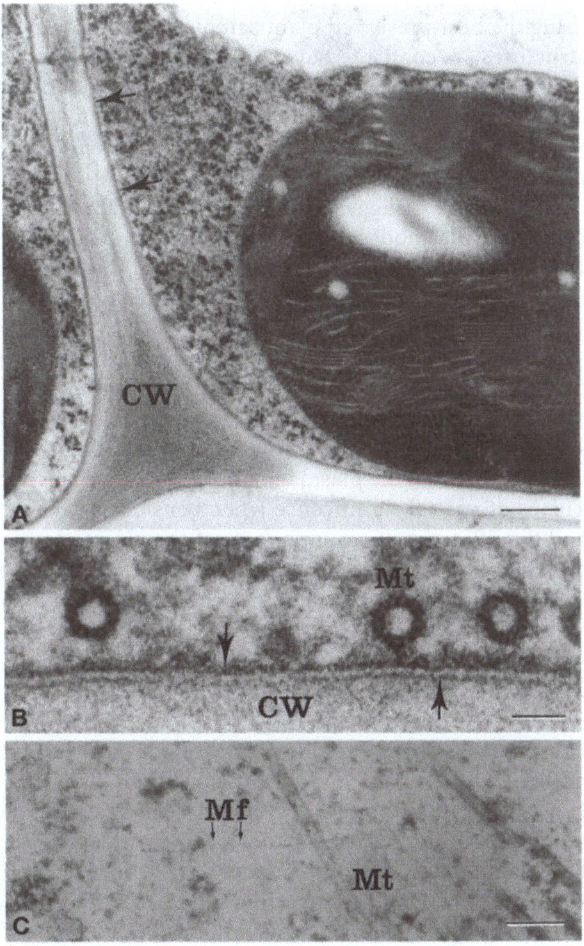

Fig. 2 A–C. Tobacco cells prepared by cryofixation and freeze substitution. The leaf samples were cryofixed by propane jet freezing (**A**) and (**C**) or high pressure freezing (**B**), freeze substituted in acetone containing 0.1 % tannic acid and then in acetone containing 2% uranyl acetate/2% osmium tetroxide. **A** Tobacco mesophyll cells. There is no visible ice damage at this resolution level. Note that the plasma membrane (*unlabelled arrows*) *is* smooth and tightly appressed to the cell walls (*CW*). *Bar* 0.2 µm (Ding et al. 1991a). **B** A tobacco mesophyll cell, showing good preservation of the subunit structures of the microtubules (*Mt*). *Unlabelled arrows* indicate the plasma membrane that is smooth and tightly appressed to the cell wall (*CW*). *Bar* 30 nm (Ding et al. 1991b). **C** A tobacco root cortical cell showing good preservation of a single microfilament (*Mf*), in addition to microtubules (*Mt*). *Bar* 80 nm (Ding et al. 1991a)

apparently did not preserve cell structures as well as expected. In conclusion, the images published so far can hardly be considered convincing evidence for the presence of sphincters, due to inferior fixation quality. This conclusion, however, does not exclude the possibility that some special structures are present. It is possible that the presence of such structures is species-dependent. Tannic acid may indeed stain some important and unique structures. The nature of these structures needs to be reevaluated using alternative fixation protocols that cause minimal distortion of cell structures. Furthermore, if such structures indeed exist, their functions need to be carefully studied.

In addition to the sphincter structures, spiral structures have also been suggested to exist in the cell wall and encircle the whole length of the plasmodesma (Badelt et al. 1994; Overall and Blackman 1996). These structures were proposed to function together with sphincters to control opening and closing of the plasmodesma. As for sphincters, whether such spiral structures indeed exist in nature requires reevaluation using the best-preserved plant samples and innovative TEM techniques. Even if such structures are confirmed to exist by various means possible, one has to wonder about their real functions. If they were involved in regulating plasmodesmal transport, then how does a cytoplasmically localized molecule for transport efficiently communicate with complicated cell wall-localized sphincter and/or spiral structures to act upon the plasmodesma?

A recent study suggests that an open tubular structure is present in place of an AER cylinder in plasmodesmata of trichome cells of *Nicotiana clevelandii* (Waigmann et al. 1997). Waigmann et al. (1997) suspect that this tubular structure may be composed entirely of proteins, as suggested earlier by Tilney et al. (1991). Direct evidence for this assumption remains outstanding.

When it comes to substructural studies, it is important to keep in mind that chemical fixatives penetrate cells slowly, take several minutes to immobilize cellular structures, and induce many artefacts in cellular structures (Mersey and McCully 1978; Gilkey and Staehelin 1986). A notable example of such artefacts is illustrated in Fig. 1, which shows that the plasma membrane is wavy and is pulled away from the cell wall. This is of special significance when considering the fact that the plasma membrane and the ER are major components of the plasmodesma and that plasmodesmata are dynamic entities. Therefore, caution should be exercised in interpreting the details of the plasmodesmal substructure in chemically fixed materials.

3.2 Cryofixation

Cryofixation, or ultrarapid freezing, can physically stabilize cellular structures in a few milliseconds (Plattner and Bachmann 1982; Gilkey and Staehelin 1986; Menco 1986). In simple terms, a sample is cryofixed by bringing it rapidly into contact with cryogens such as Freon or liquid propane or with a cooled metal (usually copper) surface that has a temperature of approximately $-180°C$ or lower. There are a variety of freezing techniques, all developed to achieve the fastest freezing speed and the greatest depth of good freezing possible (Plattner and Bachmann 1982; Gilkey and Staehelin 1986; Menco 1986). The depth of good freezing is the sample thickness that can be frozen without visible ice damage at the TEM level. The simplest freezing method is to plunge the samples into a cryogen manually. At the best, this method can yield good freezing of a sample thickness of 10–15 μm (e.g. Tiwari et al. 1984). A propane jet freezer is able to freeze up to 80 μm in the presence of appropriate cryoprotectants (Ding et al. 1991a). A high-pressure freezer can freeze a sample well up to 600 μm (Gilkey and Staehelin 1986; Dahl and Staehelin 1989).

Because of its ultrarapid and purely physical action, cryofixation is superior to chemical fixation in preserving cell structures close to their native state (Plattner and Bachmann 1982; Fernandez and Staehelin 1985; Gilkey and Staehelin 1986; Menco 1986; Staehelin and Chapman 1987). As shown in Fig. 2A and B, the plasma membrane in cryofixed and freeze-substituted plant materials is smooth and tightly appressed to the cell wall. Cryofixation is also the best approach to preserve the cytoskeleton, in particular the actin filaments which are labile and difficult to preserve by chemical fixation in plant cells (Fig. 2C; Tiwari et al. 1984; Lancelle et al. 1986, 1987; Tiwari and Polito 1988; Lichtscheidl et al. 1990; Ding et al. 1991a). Because of its distinct advantages over chemical fixation in preserving dynamic cell structures, cryofixation has also been used in combination with other methods to process tissue samples to study the plasmodesmal substructure.

3.2.1 Freeze Fracture

In freeze fracture, a very sharp and cooled knife is used to cut the frozen tissue at −100°C. The frozen tissue is so brittle that the knife passage does not usually yield a clean-cut surface of the tissue; rather, the tissue fractures during knife passage. Because the hydrophobic membrane interior requires less energy to fracture than the cytoplasm, the membrane lipid bilayer is often split open during fracturing. Platinum/carbon is then evaporated onto the fractured tissue surface to form a replica, which essentially copies the topography of the surface. The replica, made free of cell debris, is examined directly in the TEM. Readers interested in the technique are referred to Bozzola and Russell (1992) for a general description of the technique and references cited therein for detailed information.

The freeze-fracture technique was used by Willison (1976) and Thomson and Platt-Aloia (1985) to study the plasmodesmal structure. Robards and Clarkson (1984) also made observations of plasmodesmata in freeze-fractured maize root cells. These studies revealed distinct particles as the plasmodesmal components. The main advantage of the technique is that the replica offers a three-dimensional view of cell structures. Such a view cannot be gained directly from thin sections, which produce only a two-dimensional view of any structures. In terms of gaining insight into the substructure of the plasmodesma, the freeze-fracture method has a number of drawbacks. First, the fracture plane occurs randomly and rarely exposes structures of interest. In particular, it is difficult to expose and visualize the internal structure of the plasmodesma. Although Thomson and Platt-Aloia (1985) presented some longitudinal views of freeze-fractured plasmodesmata, the resolution is very low. Second, the platinum particles evaporated onto the fractured tissue surfaces are approximately 2 nm in diameter, which is similar to the diameter of plasmodesmal particles (see Fig. 3 and discussion below), as revealed by TEM examination of thin sections. Thus, the details of the plasmodesmal structure may well be buried in a replica. Furthermore, when the evaporated platinum/carbon coats the fractured cell membranes, the cell structures are inevitably augmented in dimension in the replicas. For instance, the smallest cell structure that can be clearly identified in a replica above platinum particle background, such as an intramembrane protein particle, has a replica diameter of approximately 8–10 nm (Robards and Clarkson 1984; Bozolla and Russell 1992). Thus, when one observes a particle of 10 nm in a replica of a plasmodesmal structure, it is uncertain whether such a particle represents a 2-nm plasmodesmal particle augmented in the replica, or two or three closely spaced 2-nm plasmodesmal particles showing up as one big particle when coated with platinum/carbon. Despite these drawbacks, freeze fracture could be a useful technique in studying the plasmodesmal structure when data generated with this technique are corroborated by data obtained with other techniques.

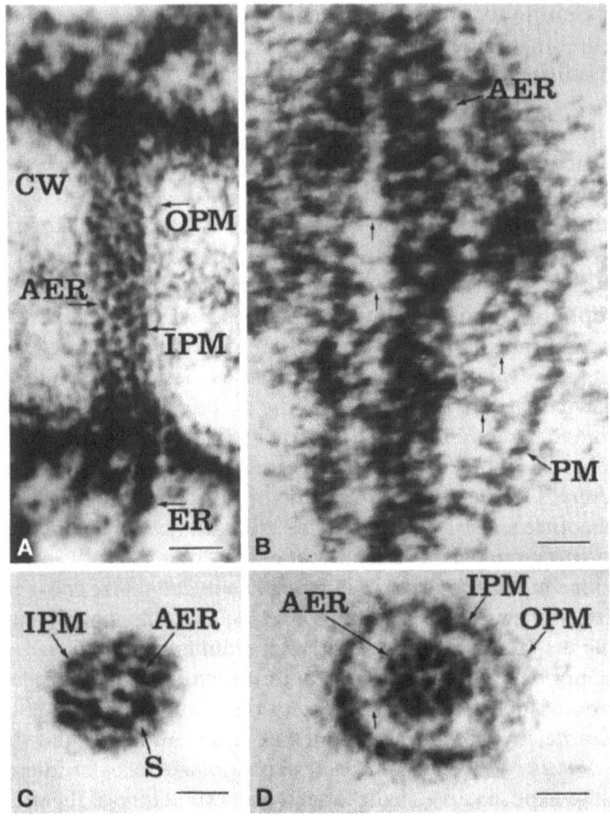

Fig. 3 A–D. Computer-enhanced high-resolution images of plasmodesmata in cryofixed/freeze-substituted tobacco leaves. **A** Longitudinal view of a newly formed primary plasmodesma between mesophyll cells. Electron-dense particles are embedded in the appressed endoplasmic reticulum (*AER*) and the inner and outer leaflets of the plasma membrane (*IPM* and *OPM*, respectively); *CW* cell wall; *ER* endoplasmic reticulum. *Bar* 18 nm. **B** A developing primary plasmodesma between phloem parenchyma cells, showing enlarged space between the plasma membrane (*PM*) and the AER that contains electron-dense "spoke-like" structures (*unlabelled arrows*). *Bar 13* nm. **C** Transverse view of a plasmodesma between phloem parenchyma cells. The spaces (S) between the electron dense particles of the AER and the IPM are presumably microchannels for intercellular transport. *Bar* 10 nm. **D** Oblique-transverse view of the middle portion of a developing primary plasmodesma between phloem parenchyma cells as shown in **B**. The *unlabelled arrows* indicate "spoke-like" structures interconnecting the *IPM* and *AER*. Because the section plane is oblique, several layers of electron-dense particles of the AER can be seen. *Bar* 12 nm (Ding et al. 1992b)

3.2.2 Freeze Substitution

An alternative approach to freeze fracture is freeze substitution, which has been shown to yield superior preservation of even the very labile plant cytoskeletal elements, microfilaments, in addition to other cellular components (Fig. 2; Tiwari et al. 1984; Lancelle et al. 1987; Tiwari and Polito 1988; Lichtscheidl et al. 1990; Ding et al.

1991a). In freeze substitution, the frozen samples are usually placed in acetone or ethanol at −90°C for a period of 1 to several days to allow substitution of ice by the dehydrant. The dehydrated samples are gradually brought to room temperature, infiltrated with and then embedded in plastic for obtaining thin sections. Because the cell structures are already physically stabilized by freezing, chemical fixatives such as glutaraldehyde and osmium tetroxide can be used in the substitution fluid to chemically stabilize the structures. This chemical fixation step presumably alters very little the cell structures.

Ding et al. (1992b) obtained high-resolution electron microscopic images of freeze-substituted tobacco plasmodesmata, which revealed that distinct particles (presumably proteinaceous) of 2–3 nm are embedded in both the plasma membrane and the AER membranes of the plasmodesma (Fig. 3). Spoke-like structures are found to extend from the AER to the plasma membrane in the central cavity region of a plasmodesma. The biochemical and functional nature of these extensions remains to be determined. The spaces between the plasma membrane and AER particles are thought to form microchannels for intercellular transport (Ding et al. 1992b; Botha et al. 1993). These microchannels may be tortuous (Ding et al. 1992b).

Thus far, thin sections of cryofixed tissues are probably the most suitable samples for high-resolution TEM studies of the plasmodesmal substructure. The main limitations are the same as with any thin sections. The images are two-dimensional and a section thickness of 60–70 nm imposes severe limitations on the resolution of the precise spatial arrangement of overlapping structures, such as the 2–3 nm particles, within the plasmodesma. The resolution can be partially improved by examining a section at various tilt angles in the microscope so that overlapping structures can be optically separated (Ding et al. 1992b), but it works only to a limited section depth.

In previous studies using the cryofixation method (Willison 1976; Thomson and Platt-Aloia 1985; Ding et al. 1992b), plant materials were cut into small pieces necessary to accommodate cryofixation. Thus, one cannot exclude the possibility that some artefacts may have already been introduced during the cutting procedure. Nevertheless, the superior structural preservation by cryofixation as compared to chemical fixation is widely recognized. The ultimate goal of preserving the plasmodesma in a completely native and undisturbed state is to find a plant system which can be cryofixed without cutting or pretreating the materials.

4 Elucidating the Molecular Structure of the Plasmodesma by Integrative Approaches

Recent studies have indicated that the cytoskeleton may interact with the plasmodesma. White et al. (1994) localized actin to the plasmodesma via immunolabelling. However, the labelling intensity is disappointingly low. If actin filaments are indeed present in the plasmodesma, then the low intensity of labelling could be explained in a number of ways. First, actin filaments, especially single filaments, are notoriously labile in plants and only the best cryofixation protocols can preserve them well (e.g. Tiwari et al. 1984; Lancelle et al. 1987; Lichtscheidl et al. 1990; Ding et al. 1991a, 1992c). Second, an actin filament has to be exposed, longitudinally, on the very surface of a tissue section for maximal access by antibodies. Any oblique orientation of an actin filament with respect to the section surface will allow only partial labelling of the filament and thus make the labelling signal very low. Clearly, other methods need to be developed to improve the accessibility of actin filaments to antibodies in tissue sections. For example, samples may be embedded in diethylene glycol distearate (DGD; Capco et al. 1984; Nickerson et al. 1990) or polyethylene glycol (PEG; Wolosewick 1984). The advantage of this technique is that DGD and PEG can be removed from sections so that all cellular structures, including the cytoskeleton and plasmodesmata, are fully exposed to antibodies. Indeed, such sections have been shown to give superb immunolabelling of the cytoskeleton (Wolosewick et al. 1983; Wolosewick 1984; Nickerson et al. 1990). If necessary, soluble proteins may be extracted from sections to improve visualization of the cytoskeleton (Nickerson et al. 1990).

White et al. (1994) found that treatment of cells with cytochalasin B to disrupt actin filaments led to an enlargement of the diameter of the plasmodesma. Ding et al. (1996) showed that disruption of actin filaments with cytochalasin D or maize profilin (Staiger et al. 1994) increased the size exclusion limit (SEL) of plasmodesmata in tobacco mesophyll cells from a basal value of 1 kDa to at least 20 kDa, whereas treatment of these cells with phalloidin, a drug that stabilizes actin filaments, had no such effect. These results, taken together, suggest that actin filaments may control the permeability of plasmodesmata in tobacco mesophyll in some manner. It will be of great interest to determine whether or not the spoke-like structures as shown in Fig. 3B represent any cytoskeletal elements (i.e. actin or myosin?). The identity of those electron-dense particles is also of special interest.

Isolation and functional characterization of the biochemical components of plasmodesmata will be a major focus in elucidating the molecular structures of these intercellular organelles. There has been some progress made in this direction (e.g. Epel 1994; Epel et al. 1996; Waigmann et al. 1997; Ding 1998). In order to make further progress, molecular, biochemical, genetic and electron microscopic approaches need to be integrated. Immunolabelling will become a critical tool in this effort. The minute amount of some of the plasmodesmal proteins may render immunolabelling ineffective, especially when using thin sections, which limit immunodetection of antigens exposed on the section surface. When the labelling density is low, uncertainty arises as to whether the labelling truly represents the presence of antigen or is merely background labelling. Furthermore, indirect immunolabelling, where a gold-conjugated secondary antibody is used to detect the location of the primary antibody, may not be the ideal method to localize plasmodesmal proteins in some cases. The problem is that the antibodies are a few nanometers long (e.g. a 150 000 kDa-IgG molecule is approximately 8 nm long) and the observed location of an antibody does not necessarily correspond exactly to the location of the antigen. Such position shift is further amplified when a secondary antibody is used. Thus, the observed localization of a gold particle conjugated to the secondary antibody may be many nanometers away from the antigen location. While this may not be a serious problem in localizing antigens in most cases, it is a critical factor in evaluating the localization pattern of putative plasmodesmal proteins by indirect immunolabelling due to the minute size of the plasmodesma. When gold particles are observed in the cell wall areas immediately surrounding the plasmodesma, it becomes uncertain whether the antibody indeed decorates the plasmodesma or the cell walls. Thus, it may be desirable to use gold-conjugated primary antibodies alone so as to improve the spatial resolution of localization in such situations. Clearly, innovative methods need to be developed for the detection or localization of plasmodesmal proteins.

5 Concluding Remarks

Our ultimate goal of elucidating the structure of the plasmodesma is to facilitate an understanding of its function in mediating cell-to-cell transport. Future studies in this direction will benefit from a wealth of new information that has been accumulated in recent years about the function of the plasmodesma. In particular, the discovery that the plasmodesma has the capacity to transport macromolecules (Mezitt and Lucas 1996; Ding 1998; Ding et al. 1999) provides a new incentive for research on the molecular architecture of this intercellular organelle. Pure TEM investigation, especially using chemically fixed samples, will probably yield limited new information on the plasmodesmal substructure. Cryofixation is clearly the choice of method at present to preserve samples for high resolution structural and immunolabelling studies. When integrated with molecular, biochemical and genetic approaches, TEM will be more powerful than ever.

It will be ideal to establish a plant system where multidisciplinary approaches are possible. Such a system should allow easy investigation of transport functions by such cell biological means as microinjection, structural analysis by cryofixation with minimal pretreatment, and biochemical isolation of plasmodesmal components. Furthermore, it should be amenable to genetic analysis of mutants defective in plasmodesmal transport functions. Once such a system is established, research on the molecular structure and function of plasmodesmata will be greatly facilitated.

References

Badelt K, White RG, Overall RL, Vesk M (1994) Ultrastructural specializations of the cell wall sleeve around plasmodesmata. Am J Bot 81:1422–1427

Botha CEJ, Hartley BJ, Cross RHM (1993) The ultrastructure and computer-enhanced digital image analysis of plasmodesmata at the Kranz mesophyll-bundle-sheath interface of *Themeda triandra* var. imberbis (Retz) A. Camus in conventionally fixed leaf blades. Ann Bot 72:255–261

Bozzola JJ, Russell LD (1992) Electron microscopy. Jones and Bartlett, Boston

Capco D, Krochmalnic G, Penman S (1984) A new method of preparing embeddment-free sections for transmission electron microscopy: applications to the cytoskeletal framework and other three-dimensional networks. J Cell Biol 98:1878–1885

Dahl R, Staehelin LA (1989) High pressure freezing for the preservation of biological structure: theory and practice. J Electron Microsc Tech 13:165–174

Ding B (1998) Intercellular protein trafficking through plasmodesmata. Plant Mol Biol 38:279–310

Ding B, Turgeon R, Parthasarathy MV (1991a) Routine cryofixation of plant tissue by propane jet freezing for freeze substitution. J Electron Microsc Tech 19:107–117

Ding B, Turgeon R, Parthasarathy MV (1991b) Plasmodesmatal substructure in cryoflxed developing tobacco leaf tissue. In: Bonnemain JL, Delrot S, Lucas WJ, Dainty J (eds) Recent advances in phloem transport and assimilate compartmentation. Ouest Editions, Nantes

Ding B, Haudenshield JS, Hull RJ, Wolf S, Beachy RN, Lucas WJ (1992a) Secondary plasmodesmata are specific sites of localization of the tobacco mosaic virus movement protein in transgenic tobacco plants. Plant Cell 4:915–928

Ding B, Turgeon R, Parthasarathy MV (1992b) Substructure of freeze-substituted plasmodesmata. Protoplasma 169:28–41

Ding B, Turgeon R, Parthasarathy MV (1992c) Effect of high-pressure freezing on plant microfilament bundles. J Microsc 165:367–376

Ding B, Kwon MO, Warnberg L (1996) Evidence that actin filaments are involved in controlling the permeability of plasmodesmata in tobacco mesophyll. Plant J 10:157–164
Ding B, Itaya A, Woo YM (1999) Plasmodesmata and cell-to-cell communication in plants. Internatl Rev Cytol 190:251–316
Ehlers K, Kollmann R (1996) Formation of branched plasmodesmata in regenerating *Solanum nigrum* protoplasts. Planta 199:126–138
Epel BL (1994) Plasmodesmata: composition, structure, and trafficking. Plant Mol Biol 26:1343–1356
Epel BL, Van Lent JWM, Cohen L, Kotlizky G, Katz A, Yahalom A (1996) A 41-kDa protein isolated from maize mesocotyl cell walls immunolocalizes to plasmodesmata. Protoplasma 191:70–78
Fernandez DE, Staehelin LA (1985) Structural organization of ultrarapidly frozen barley aleurone cells actively involved in protein secretion. Planta 165:455–468
Fujiwara R, Linck RW (1982) The use of tannic acid in microtubule research. In: Wilson L (ed) Methods in cell biology, vol 24. Academic Press, San Diego, pp 217–233
Gamalei YV, Van Bel AJE, Pakhomova MV, Sjutkina AV (1994) Effects of low temperature on the conformation of the endoplasmic reticulum and on starch accumulation in leaves with the symplasmic minor-vein configuration. Planta 194:443–453
Gilkey JC, Staehelin LA (1986) Advances in ultrarapid freezing for the preservation of cellular ultrastructure. J Electron Microsc Tech 3:177–210
Glauert AM (1975) Fixation, dehydration and embedding of biological specimens. In: Glauert AM (ed) Practical methods in electron microscopy, vol 3. North-Holland Publishing, Amsterdam, pp 1–207
Glockmann C, Kollmann R (1996) Structure and development of cell connections in the phloem of *Metasequoia glyptostroboides* needles. I. Ultrastructural aspects of modified primary plasmodesmata in Strasburger cells. Protoplasma 193:191–203
Grabski S, de Feijter AW, Schindler M (1993) Endoplasmic reticulum forms a dynamic continuum for lipid diffusion between contiguous soybean root cells. Plant Cell 5:25–38.
Hepler P (1982) Endoplasmic reticulum in the formation of the cell plate and plasmodesmata. Protoplasma 111:121–133
Lancelle SA, Callaham DA, Hepler PK (1986) A method for rapid freeze fixation of plant cells. Protoplasma 131:153–165
Lancelle SA, Cresti M, Hepler PK (1987) Ultrastructure of the cytoskeleton in freeze-substituted pollen tubes of *Nicotiana alata*. Protoplasma 140:141–150
Lichtscheidl IK, Lancelle SA, Hepler PK (1990) Actin-endoplasmic reticulum complexes in *Drosera*. Their structural relationship with the plasmalemma, nucleus, and organelles in cells prepared by high pressure freezing. Protoplasma 155:116–126
López-Sáez JF, Giménez-Martín G, Risueno MC (1966) Fine structure of the plasmodesm. Protoplasma 61:81–84
Lucas WJ, Ding B, van der Schoot C (1993) Plasmodesmata and the supracellular nature of plants. New Phytol 125:435–476
Menco BPM (1986) A survey of ultra-rapid cryofixation methods with particular emphasis on applications to freeze-fracturing, freeze-etching, and freeze-substitution. J Electron Microsc Tech 4: 177–240
Mersey B, McCully ME (1978) Monitoring of the course of fixation of plant cells. J Microsc 114:49–76
Mezitt LA, Lucas WJ (1996) Plasmodesmal cell-to-cell transport of proteins and nucleic acids. Plant Mol Biol 32:251–273
Mollenhauer HH, Morré DJ (1987) Some unusual staining properties of tannic acid in plants. Histochemistry 88:17–22
Nickerson JA, Krockmalnic G, He D, Penman S (1990) Immunolocalization in three dimensions: immunogold staining of cytoskeletal and nuclear matrix proteins in resinless electron microscopy sections. Proc Natl Acad Sci USA 87:2259–2263
Olesen P (1979) The neck constriction in plasmodesmata – evidence for a peripheral sphincter-like structure revealed by fixation with tannic acid. Planta 144:349–358
Olesen P, Robards AW (1990) The neck region of plasmodesmata: general architecture and some functional aspects. In: Robards AW, Jongsma H, Lucas WJ, Pitts J, Spray D (eds) Parallels in cell to cell junctions in plants and animals. Springer, Berlin Heidelberg New York, pp 145–170
Overall RL, Blackman LM (1996) A model for the macromolecular structure of plasmodesmata. Trends Plant Sci 1:307–311
Overall RL, Wolfe J, Gunning BES (1982) Intercellular communication in *Azolla* roots: I. Ultrastructure of plasmodesmata. Protoplasma 111:134–150
Plattner H, Bachmann L (1982) Cryofixation: a tool in biological ultrastructural research. Int Rev Cytol 79:237–304
Robards AW (1968) A new interpretation of plasmodesmatal ultrastructure. Planta 82:200–210
Robards AW (1971) The ultrastructure of plasmodesmata. Protoplasma 72:315–323

Robards AW (1976) Plasmodesmata in higher plants. In: Gunning BES, Robards AW (eds) Intercellular communication in plants: studies on plasmodesmata. Springer, Berlin Heidelberg New York, pp 15–57

Robards AW, Clarkson DT (1984) Effects of chilling temperatures on root cell membranes as viewed by freeze-fractured electron microscopy. Protoplasma 122:75–85

Seagull RW, Heath IB (1979) The effect of tannic acid on the in vivo preservation of microfilaments. Eur J Cell Biol 20:184–188

Staehelin LA, Chapman RL (1987) Secretion and membrane recycling in plant cells: novel intermediary structures visualized in ultrarapidly frozen syvamore and ċarrot suspension-culture cells. Planta 171:43–57

Staiger CJ, Yuan M, Valenta R, Shaw PJ, Warn RM, Lloyd CW (1994) Microinjected profilin affects cytoplasmic streaming in plant cells by rapidly depolymerizing actin microfilaments. Curr Biol 4: 215–219

Thomson WW, Platt-Aloia K (1985) The ultrastructure of the plasmodesmata of the salt glands of *Tamarix* as revealed by transmission and freeze-fracture electron microscopy. Protoplasma 125: 13–23

Tilney LG, Cooke TJ, Connelly PS, Tilney MS (1991) The structure of plasmodesmata as revealed by plasmolysis, detergent extraction, and protease digestion. J Cell Biol 122:739–747

Tiwari SC, Polito VS (1988) Organization of the cytoskeleton in pollen tubes of *Pyrus communis:* a study employing conventional and freeze-substitution electron microscopy, immunofluorescence, and rhodamine-phalloidin. Protoplasma 147:100–112

Tiwari SC, Wick SM, Williamson RE, Gunning BES (1984) Cytoskeleton and integration of cellular function in cells of higher plants. J Cell Biol 99:63s–69s

Waigmann E, Turner A, Peart J, Roberts K, Zambryski P (1997) Ultrastructural analysis of leaf trichome plasmodesmata reveals major differences from mesophyll plasmodesmata. Planta 203: 75–84

White RG, Badelt K Overall RL, Vesk M (1994) Actin associated with plasmodesmata. Protoplasma 180:169–184

Willison JHM (1976) Plasmodesmata: a freeze fracture view. Can J Bot 54:2842–2847

Wolosewick JJ (1984) Cell fine structure and protein antigenicity after polyethylene glycol processing. In: Revel JP, Barnard T, Haggins GH (eds) Science of biological specimen preparation. Scanning Electron Microscopy, O'Hare, Illinois, pp 83–96

Wolosewick JJ, Demey J, Meininger V (1983) Ultrastructural localization of tubulin and actin in polyethylene glycol embedded rat seminiferous epithelium by immunogold staining. Biol Cell 49: 219–226

Zee SY (1969) The fine structure of differentiating sieve elements of *Vicia faba.* Aust J Bot 17:441–456

CHAPTER 4

Electrical Coupling

H. V. M. VAN RIJEN · R. WILDERS · H. J. JONGSMA

Dept. of Medical Physiology and Sports Medicine, Faculty of Medicine,
Utrecht University, Universiteitsweg 100, 3584 CG Utrecht,
The Netherlands

1 Introduction

Almost all cells, in either plant or animal tissues, are intercellularly coupled. In plants, intercellular coupling is established by plasmodesmata (for a review see Robards and Lucas 1990), while in animal tissues an intercellular pathway is provided by gap junctions (for a review see Bruzzone et al. 1996).

Electrical coupling, i.e. the intercellular diffusion of inorganic ions between cells, allow not only tissue homeostasis, but also provides an intercellular pathway for the transmission of electrical signals. For instance, in the heart, gap junctions provide a low resistance cell-to-cell pathway for intercellular currents, thereby facilitating fast action potential propagation throughout the heart (Gros and Jongsma 1996). Action potential-like activity is also found in excitable plants like *Mimosa pudica*, with a putative role for plasmodesmata in the transmission of electrical signals (Samejima and Sibaoka 1983). Since membrane potential changes and transmitted electrical activity were also detected in other plants in response to, for example electrical pulses (Paszewski and Zawadzki 1973, 1974, 1976a; Zawadzki 1980; Zawadzki and Trebacz 1982, 1985), wounding (Sambeek and Pickard 1976a, 1976b; Paszewski and Zawadzki 1976b; Roblin 1985; Roblin and Bonnemain 1985) or changes in leaf illumination (Eschrich et al. 1988), plasmodesmata most likely play a role in electrical intercellular coupling in plants.

Electrical coupling between animal cells and the biophysical properties of gap junctions has been studied extensively. Although gap junctions were first regarded as simple ohmic resistors coupling adjacent cells, multidisciplinary research has revealed that gap junctions are specified membrane structures with different properties, like electrical conductance, permeability and ion selectivity, in different tissues. Electrophysiological experiments have shown shown that gap junctions exhibit voltage- and time-dependent behaviour, which is strongly dependent on the type of gap junction protein of which the channels are composed. Furthermore, is has been shown that the conductance and permeability of gap junctions is highly regulated by means of phosphorylation and expression (for a review see Bruzzone et al. 1996).

Although the structure and physiology of gap junctions and plasmodesmata are fundamentally different, strong evidence exists that plasmodesmata are the pathway for electrical signalling in plants, like gap junctions in animal tissues. However, very little is known about the electrical properties of plasmodesmata and the regulation thereof.

This chapter will present the methods that are commonly used to measure eletrical coupling between cells, and discuss the limitations of these techniques. Furthermore, we will evaluate whether these techniques can be applied in higher plant tissues in order to investigate the electrical behaviour of plasmodesmata.

2 Determination of Electrical Coupling Between Cells

2.1 Determination of Electrical Coupling by Current Injection

A relatively easy way to measure electrical coupling between cells in a one-, two- or three-dimensional array is presented in Fig. 1A. Neighbouring cells that are electrical-

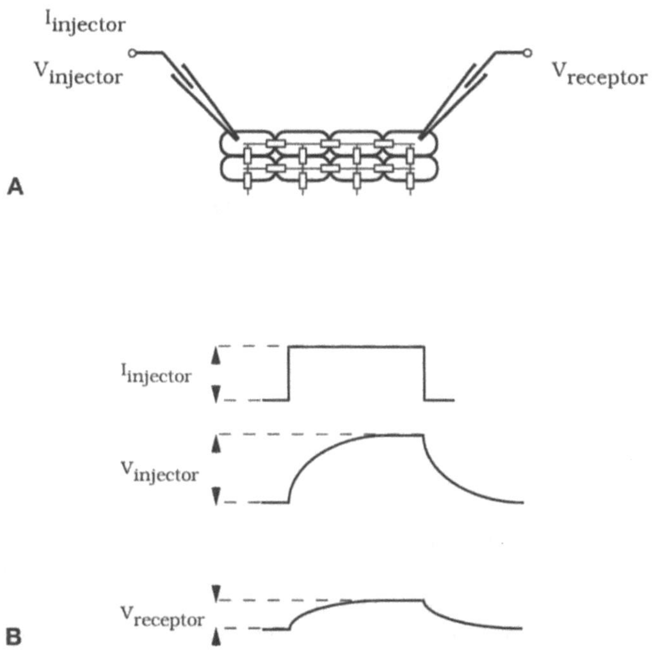

Fig. 1. A, B Experimental procedure for determination of electrical coupling by means of current injection and voltage measurement. **A** Two microelectrodes are impaled into a monolayer of cells or a tissue, with a certain distance between the electrodes. The intercellular conductances are depicted as ohmic resistors. **B** A current ($I_{injector}$) is intracellularly injected in one of the cells which causes voltage deflections in both cells ($V_{injector}$ and $V_{receptor}$). The slow rising phase of the voltage deflections is caused by capacitive properties of the plasma membranes. Electrical coupling between the cells can be expressed as the ratio $V_{receptor}/V_{injector}$ (coupling ratio), or more specifically calculated using two- or three-dimensional Bessel cable models (Siegenbeek van Heukelom et al. 1972; Jongsma and Van Rijn 1972)

ly coupled (as depicted by ohmic resistors) are both impaled by microelectrodes that are connected to microelectrode amplifiers with high input impedance (10^{12} MΩ). Via one of the electrodes, current is intracellularly injected, which will evoke a voltage deflection in both cells as a result of electrotonic current spread (Fig. 1B). The amplitude of the voltage deflection in the cells depends on the value and the ratio of the different resistors. This method was used by Jongsma and van Rijn (1972) and Larson et al. (1983), for example, to determine electrical coupling in monolayers of neonatal rat heart cells and bovine endothelial cells, respectively. This approach was also used in plant science to determine the electrical coupling via plasmodesmata between *Elodea* cells (Spanswick 1972), apical root cells of *Azolla* (Overall and Gunning 1982), *Chara* (Coté et al. 1987; Ding and Tazawa 1989; Reid and Overall 1992), corn suspension-culture cells (Holdaway-Clarke et al. 1996) and cells in the stem phloem of *Lupinus* (van Bel and van Rijen 1994).

The amount of electrical coupling can be expressed as the ratio of the voltage deflection in the receptor electrode, divided by the voltage deflection in the injector electrode, which is commonly referred to as coupling ratio.

The coupling ratio, however, does not directly provide for quantitative information on the electrical coupling between the cells. The coupling ratio is determined not only by the resistance of the plasmodesmata, but also by that of the plasma membrane. Provided that all cells have an equal number of neighbours and that all intercellular pathways have identical resistance values and all membranes possess identical resistance values, quantitative information about the magnitude of intercellular coupling can be achieved. In cell pairs, these calculations are relatively simple (Holdaway-Clarke et al. 1996), but become progressively more difficult when the model system is extended to a two- or three-dimensional system (Larson et al. 1983).

A modification on the current injection technique was introduced by Lew (1994, 1996). In these experiments on adjacent root hairs of *Arabidopsis thaliana* a combination of voltage clamp and conventional voltage recording was performed. The older root hair was impaled with a double-barreled microelectrode connected to a voltage clamp ampifier, while the adjacent younger root hair was impaled with a single-barreled micropipette connected to an electrometer. In the older cell, voltage steps were applied, resulting in voltage changes in the younger cell, due to (clamping) current passage via plasmodesmata from the older to the younger root hair. The voltages of both cells and the clamping current in the older root hair can be measured. Although this technique allows voltage control in one cell, it is not fundamentally different from the conventional current injection technique. The current measured in the voltage clamped cell still depends on both the membrane resistance of this cell and the intercellular resistance and the voltage deflection in the second cell is determined by the intercellular current and the membrane resistance of this cel. Again, calculation of the intercellular resistance cannot be performed without assumptions.

A more direct and less speculative way to gain biophysical information on intercellular conductance is by voltage clamping both cells: the dual voltage clamp technique.

2.2 Determination of Electrical Coupling Using Dual Voltage Clamp

The dual voltage clamp method was introduced by Spray et al. in 1979. Kinetics and conductance of gap junctions were accurately measured using two seperate electrodes for voltage control and current injection in each cell of a cell pair (Fig. 2A). To allow measurements on gap junctions between small mammalian cells, dual voltage clamp was later performed using tight seal whole cell recording (Hamill et al. 1981), with two patch pipettes, one on each cell (dual whole cell patch clamp, Fig 2B).

With both techniques, the experimental protocol to determine the electrical properties of the intercellular junction is identical (Fig. 2C). Both cells are held at common holding potential, as a result of which no current will flow between the cells. Then the holding potential of one of the cells is changed, e.g. of cell A. This will result in pipette currents in both cells. In the stepped cell A, the measured current (I_a) will be the sum of the membrane current I_{m1}, and the current which flows through the intercellular conductor from cell A to cell B, I_j. In the non-stepped cell B, a current (I_b) of opposite sign will be seen which is the intercellular current I_j ($I_b = -I_j$), when the four-electrode dual voltage clamp technique is applied, and the intercellular current (I_j) minus the membrane current (I_{m2}) of cell B ($I_b = I_{m2} - I_j$) when using the dual whole cell patch clamp technique.

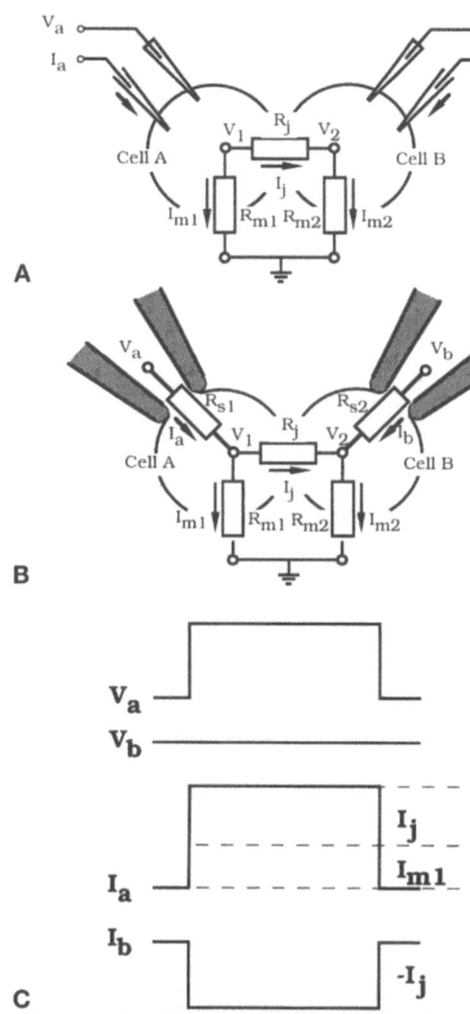

Fig. 2. A Equivalent resistive circuit of a cell pair under four electrode dual voltage clamp conditions. The intercellular current (I_j) flows across the intercellular resistance (R_j) as a result of the difference between the membrane potential of cell A (V_1) and cell B (V_2). **B** Equivalent resistive circuit of a cell pair under dual whole cell voltage clamp conditions. Again I_j flows across R_j as a result of the difference between the membrane potential of V_1 and V_2. The difference between the pipette potentials V_a and V_b and the membrane potentials V_1 and V_2, respectively, of the cells is determined by the series resistances and pipette currents of *cell A* and *B* (R_{s1} and I_a, and R_{s2} and I_b, respectively). In *cell A* the pipette current I_a equals the sum of the junctional current and the membrane current I_{m1} (flowing across the membrane resistance, R_{m1}). In *cell B* the pipette current I_b equals the intercellular current ($I_b = -I_j$) in the case of four electrode dual voltage clamp, and equals the membrane current I_{m2} (flowing across the membrane resistance R_{m2}), minus the intercellular current I_j ($I_b = I_{m2} - I_j$)when the dual whole-cell patch clamp technique is applied. C Experimental protocol of the determination of intercellular conductance. Both cells see panel B) are held at common holding potential, which will cause no current flow between the cells. In one of the cells the potential is changed stepwise, which will evoke currents in both cells, being the sum of the membrane and intercellular current in the stepped cell A, and the intercellular current in the non-stepped cell B. Intercellular conductance is calculated by dividing the intercellular current I_j as measured in cell B by the applied transcellular voltage $V_a - V_b$

The most commonly used technique of the two dual voltage clamp techniques is the dual whole-cell patch clamp technique (Fig. 2B). The dual whole cell patch clamp technique is easier to carry out, because only one (patch) pipette is required for each cell. However, the introduction of two additional series resistances (R_s), caused by the resistance of the patch pipette and residual resistance of the disrupted membrane under the patch electrode, can induce considerable errors in measuring the intercellular electrical conductance. Because the patch pipette is used for recording the membrane potential and injecting current simultaneously, a voltage drop will occur across each series resistance, as a result of which the actual membrane potentials V_1 and V_2 will always differ from the clamp potentials V_a and V_b, respectively. Strictly seen, the true intercellular conductance ($g_{j,t}$) is calculated using:

$$g_{j,t} = \frac{I_j}{V_1 - V_2} = \frac{-I_b + \dfrac{-I_b \cdot R_{s2}}{R_{m2}} + \dfrac{V_b}{R_{m2}}}{V_a - I_a \cdot R_{s1} - (V_b - I_b \cdot R_{s2})}, \tag{1}$$

from which R_{s1}, R_{s2}, the series resistances of cell A and cell B, respectively, and R_{m2}, the membrane resistance of cell B, are unknown parameters.

Therefore, the measured intercellular conductance ($g_{j,m}$) is usually calculated using:

$$g_{j,m} = -I_b/(V_a - V_b). \tag{2}$$

However, even under very favourable experimental conditions, the measured intercellular conductance $g_{j,m}$ will deviate from the true intercellular conductance $g_{j,t}$. This 'experimental' error is largely influenced by the ratio of the R_s, R_m, and R_j.

Intuitively, it can be deduced that the measured fraction of the true intercellular conductance,

$$F_j = g_{j,m}/g_{j,t}, \tag{3}$$

becomes smaller (i.e. the error becomes larger) when series resistance is high (membrane potentials V_1 and V_2 deviate largely from the applied holding potentials V_a and V_b, respectively), and membrane resistance is low (I_b=I_{m2}-I_j, I_b becomes a poor estimation of I_j if I_{m2} becomes larger as a result of low R_{m2}).

We have used a computer model approach to quantify the effect of series resistance alone and series resistance in combination with membrane resistance on measured fraction of $g_{j,t}$ (F_j). The equations used for these numerical experiments are given in the Appendix.

2.2.1 Effect of Series Resistance

The effect on the F_j of series resistance alone, i.e. in the case of infinitely high membrane resistances, is shown in Fig. 3. If in both cells non-junctional membrane currents are negligible due to very high membrane resistance ($\gg 1$ GΩ), the equivalent resistive circuit of Fig. 2B reduces to that of Fig. 3A, which is composed of three resistances in series, so that:

$$I_a = I_j = -I_b = (V_a - V_b)/(R_{s1} + R_j + R_{s2}). \tag{4}$$

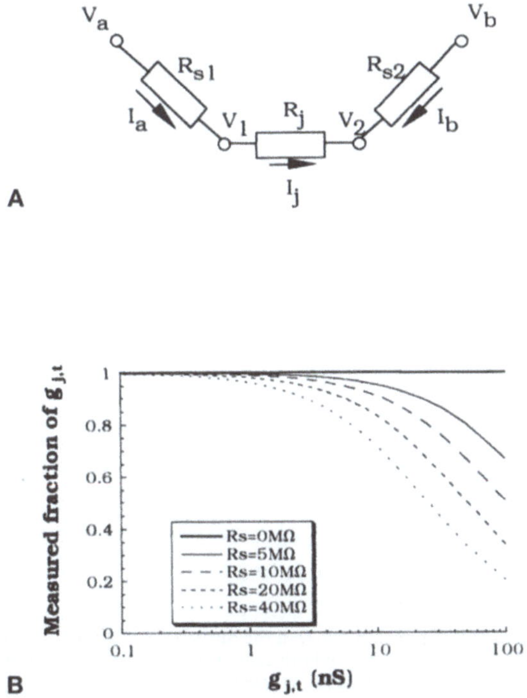

Fig. 3. A Equivalent resistive circuit of the dual whole cell patch clamp configuration with infinitely high membrane resistances. V_a and V_b clamp potentials; I_a and I_b pipette currents; V_1 and V_2 membrane potentials; R_{s1} and R_{s2} series resistances of cell A and cell B, respectively; R_j intercellular resistance; I_j intercellular current. **B** Relationship between the measured fraction of the true junctional conductance ($g_{j,t}$), and $g_{j,t}$, as calculated using Eq. (4), at total series resistance ($R_s=R_{s1}+R_{s2}$) of 0, 5, 10, 20 or 40 MΩ, in case of negligible membrane currents ($R_{m1}=R_{m2}=\infty$)

Figure 3B (in the legend of which the meaning of the symbols is given) demonstrates to what extent the measured value of intercellular conductance ($g_{j,m}$) deviates from the true value ($g_{j,t}$), as a function of series resistance (R_s, where $R_s=R_{s1}+R_{s2}$, and $R_{s1}=R_{s2}$) when Eq. (2) is used to calculate junctional conductance. The measured fraction of $g_{j,t}$ [F_j, see Eq. (3)] is shown as a function of $g_{j,t}$, with $g_{j,t}$ ranging between 0.1 and 100 nS. In the ideal case of zero series resistance ($R_s=0$), $g_{j,m}$ does not deviate from $g_{j,t}$, and $F_j=1$. At low intercellular conductance values of 0.1 nS, the series resistance-induced error is very close to zero, and up to a few percent at 1 nS ($F_j=0.996$ and $F_j=0.962$ at 40 MΩ series resistance, respectively). With increasing intercellular conductance, however, significant errors occur. Even at very favourable experimental conditions, with total series resistance as low as 10 MΩ, an intercellular conductance of 20 nS will be measured as 16.7 nS (17% error); at a total series resistance of 20 MΩ, only 14.2 nS wil be measured (29% error). Conversely, a measured conductance of 25 nS with 20 MΩ total series resistance results from a true intercellular conductance of 50 nS.

2.2.2 Effect of Series And Membrane Resistance

The effect on F_j of membrane resistance of both cells on (measured) intercellular conductance at different total series resistance (R_s), without correction for series resistance, i.e. using Eq. (2) is shown in Fig. 4. When R_s equals zero, the pipette potential is

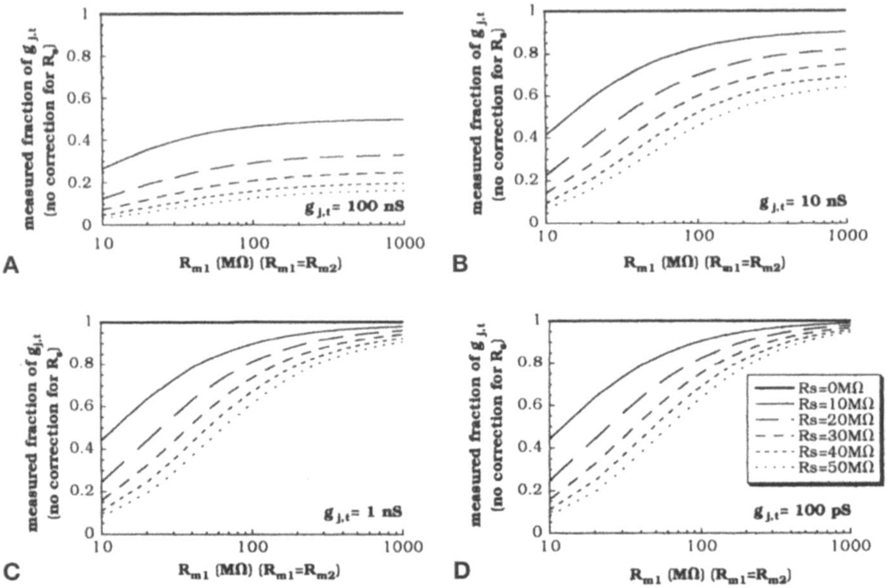

Fig. 4. A–D Relationship between the measured fraction of the true intercellular conductance $g_{j,t}$, and the membrane resistance of both cells, as determined by computer simulations. The membrane resistance of the stepped cell A (R_{m1}) was set equal to that of the non-stepped cell B (R_{m2}). Total series resistance (R_s=R_{s1}+R_{s2}) was set to 0-50 MΩ as indicated. No correction for R_s was made. **A** $g_{j,t}$=100 nS. **B** $g_{j,t}$=10 nS. **C** $g_{j,t}$=1 nS. **D** $g_{j,t}$=100 pS

identical to the membrane potential (V_a=V_1 and V_b=V_2). So the junctional current (I_j) is elicited by the command voltage. Also, V_2=0, resulting in zero current flow across R_{m2}, so that I_b equals $-I_j$. As a consequence, no errors occur [cf. Eq. (2)]: $g_{j,m}$=$g_{j,t}$ and F_j=1. When R_m is 1 GΩ, F_j has values very similar to those presented in Fig. 3, where R_{m1}=R_{m2}=∞. When R_m is lowered, a significant decrease in F_j is observed. With decreasing values of $g_{j,t}$, the errors induced by R_s become smaller, as can be seen from the values for F_j at R_m=1 GΩ. Similarly the errors induced by R_m become smaller with decreasing $g_{j,t}$. However, even at values for $g_{j,t}$ as low as 100 pS, considerable errors are still observed.

2.2.3 Correction for Errors Induced by Series and Membrane Resistance

When errors induced by series resistance alone or in combination with membrane resistance become unacceptably large, offline correction for these parameters can be applied. However, this requires that series and/or membrane resistances of both cells are determined during the experiment.

In single-cell voltage or current clamp, the commonly accepted procedures to determine the values of R_s are by applying a voltage or current step, respectively (see Fig. 5).

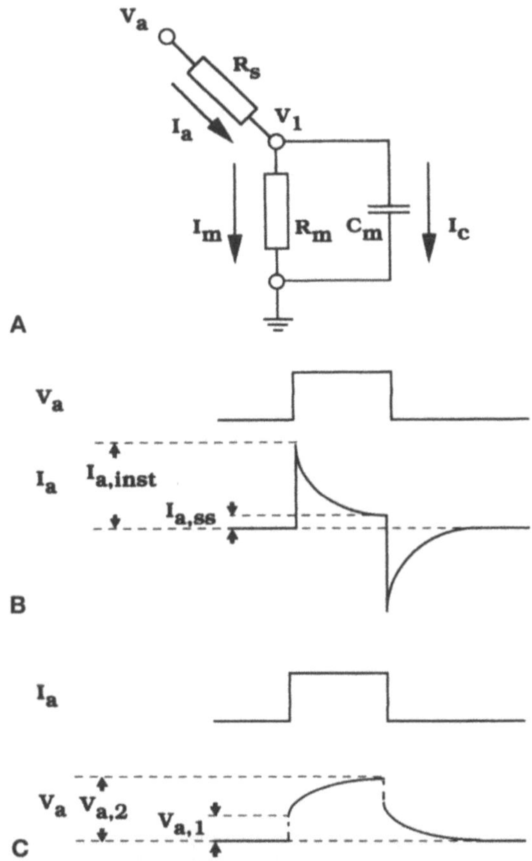

Fig. 5. A Equivalent circuit of a single cell under whole cell patch clamp conditions: a series resistance to the cell (R_s) in series with an RC circuit, representing the membrane resistance (R_m) and the membrane capacitance (C_m). I_a pipette current; I_m membrane current; I_c capacitive current **B** Determination of R_s and R_m under voltage clamp conditions. A stepwise change in clamp voltage V_a will result in a large instantaneous current ($I_{a,\,inst}$), that rapidly decreases to a steady-state value ($I_{a,\,ss}$). At the end of the voltage step a similar pattern of opposite sign is seen. The large instantaneous current at the beginning of the voltage step is solely determined by R_s, since the membrane capacitance effectively shortcircuits the membrane resistance: $R_s = V_a/I_{a,\,inst}$. The steady state current is determined by R_s+R_m, since no current will flow through C_m. R_m is determined by : $R_m = V_a/I_{a,\,ss} - R_s$. C Determination of R_s and R_m under current clamp conditions. A stepwise change in I_a will evoke a voltage deflection that consists of a fast rise of V_a to $V_{a,\,1}$, followed by a slower rise to a steady state value, $V_{a,\,2}$. The opposite is seen at the end of the current step. The fast rise to $V_{a,\,1}$, at the beginning of the current step is again caused by an effective shortcircuiting of the membrane resistance by C_m. Thus, the voltage deflection $V_{a,\,1}$ is representing the voltage drop across R_s, due to I_a. Therefore, R_s is determined by: $R_s = V_{a,\,1}/I_a$. $V_{a,\,2}$ is determined by the R_s+R_m, so that $R_m = V_{a,\,2}/I_a - R_s$

Both procedures are, however, also suitable under dual voltage or dual current clamp conditions, when an identical voltage or current step is applied to both cells simultaneously. At the beginning and end of the voltage or current step, when R_s is determined, the membrane capacitance (C_m) effectively shortcircuits the membrane resistance, which makes the membrane potential of both cells 0 mV. At this stage, there is no potential difference between the cells (V_j), and no junctional current will flow. Values for R_s can be determined independently in each cell.

The most commonly used method for determination of membrane resistances in the dual whole cell patch clamp configuration is to apply an identical command voltage step (ΔV) to both cells at the same time and to measure the change in steady-state currents ($\Delta I_{a, ss}$ and $\Delta I_{b, ss}$) in both cells (see Fig. 5B). Membrane resistances (R_{m1} and R_{m2}) are then calculated by:

$$R_{m1}=\Delta V/\Delta I_{a, ss} \tag{5}$$

$$R_{m2}=\Delta V/\Delta I_{b, ss} \tag{6}$$

Correction for series restance alone can be performed using Eqs. (A10) or (A11), and Eqs. (A8) or (A9) when correction for both series and membrane resistance is required. If membrane resistances are infinitely large, correction for series resistance alone, using Eq. (A10) or (A11), will bring F_j values close to 1. If errors larger than 10% ($F_j<0.9$) are regarded as unaccaptably large, correction for series resistance is neccessary when, e.g. total R_s amounts to 20 or 40 MΩ and $g_{j,t}$ becomes larger than 6 and 3 nS, respectively.

Figure 6 shows the relationship between F_j and membrane resistance, when, through Eq. (A11), offline correction for R_s, but not R_m, was applied. F_j is no longer dependent on $g_{j,t}$ if offline correction for series resistance is applied. It does, however, depend on both R_{s2} and R_{m2} through:

$$F_j=1/[1+(R_{s2}/R_{m2})], \tag{7}$$

which can be obtained from Eqs. (A3)-(A6) and Eq. (2).

If errors >10% are again regarded as unacceptable, correction for series resistance alone will suffice if both membrane resistances are larger than 300 MΩ. If membrane resistances become lower than 300 MΩ, correction for both membrane and series resistance will reduce the error to near zero values.

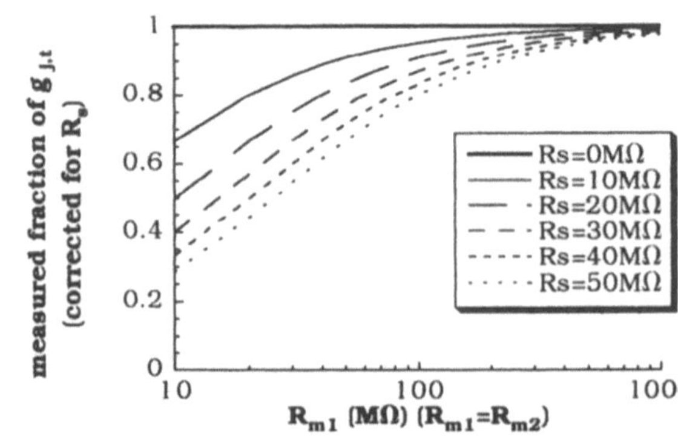

Fig. 6. Relationship between the measured fraction of true intercellular conductance $g_{j,t}$ and the membrane resistance of the stepped cell A (R_{m1}), and that of the non-stepped cell B (R_{m2}) after offline correction for series resistance according to Eq. (A11), as determined by computer simulation. Total series resistance ($R_s=R_{s1}+R_{s2}$) was set to 0–50 MΩ, as indicated

3 Dual Voltage Clamp Experiments on Plasmodesmata

To our knowledge, no dual voltage clamp experiments have been performed on plant cells to elucidate the biophysical properties of plasmodesmata. This is probably due to the fact that many of the requirements for reliable dual voltage clamp experiments in plant cells are difficult to meet.

3.1 Model System

Dual voltage clamp experiments require a pair of cells that are electrically coupled to each other, but not to other cells. Therefore, dual voltage clamp on animal cells is performed in vitro on cell pairs. A good candidate would be the pairs of adjacent internodal cells of *Chara*. Potential candidates for such an approach on higher plant cells would be: suspension callus of soybean (Parsons and Sanders 1989), suspension of corn cells (Holdaway-Clarke et al. 1996) or suspension of potato cells (C. van der Schoot, pers. comm.). Suspension callus of soybean and the suspension of corn cells form small clusters of cells of which the membrane resistance was about 700 and 132 MΩ for soybean (Parsons and Sanders 1989) and corn (Holdaway-Clarke et al. 1996), respectively.

3.2 Two- or Four-Electrode Dual Voltage Clamp?

All given candidates for a plant cell-pair system are models with cell walls. The use of patch pipettes is therefore virtually impossible, because patch-pipettes require protoplasts without cell walls, which is incompatible with the presence of plasmodesmata. Therefore, we are limited to the use of normal microelectrodes, with either one or two barrels.

When only two one-barreled electrodes are used, it is essential that the electrode resistances are as low as possible to avoid series resistance-induced problems. The membrane resistance of plant cells is between 16 MΩ and 4 GΩ (Overall and Gunning 1982; Blatt 1988; Blatt and Clint 1989; Holdaway-Clarke et al. 1996). Correction for current leakage through the plasmamembrane is therefore necessary under certain circumstances. As was shown in Fig. 3B, the measured fraction of the true intercellular conductance ($g_{j,t}$) strongly depends on the actual value of $g_{j,t}$ and on R_s. Several estimates for plasmodesmal conductance have been published in the literature. The conductance of plasmodesmata between nodal cells of *Chara* was very high, in the range of 5–10 μS (Coté et al. 1987; Ding and Tazawa 1989; Reid and Overall 1992). Of the higher plants, the plasmodesmal conductance was much lower, i.e. 48 nS between *Elodea* leaf cells (Spanswick 1972), 50–100 nS between *Äzolla* root cells (Overall and Gunning 1982), and 24 nS between corn suspension cells (Holdaway-Clarke et al. 1996). From the literature, it is known that the resistance of electrodes used to measure membrane potentials of different cell types in the plant can have resistances as low as 10–20 MΩ (Findlay and Hope 1976; Overall and Gunning 1982 ; van der Schoot and van Bel 1989, 1990).

With these electrode resistances in mind, we can read from Fig. 3B, that the minimal and maximal errors at an intercellular conductance of 25 to 100 nS amount to 33 and 80%, respectively. This shows that correction for series resistance in such experiments is inevitable. This approach cannot be used for experiments on *Chara*, because the electrode resistance is an order of magnitude higher than the plasmodesmal resistance.

The problem of series resistance-induced errors can be avoided by the four-electrode dual voltage clamp method, carried out by using two double-barreled electrodes, one in each cell (Fig. 2A). This technique might largely overcome the problems of series resistance, but is technically more complex. Some cross talk between the two channels in the electrode might occur. However, double-barreled electrode voltage clamp was successfully applied in guard cells of *Vicia faba* (Blatt 1988; Blatt and Clint 1989).

3.3 Other Considerations

Accurate measurement of intercellular currents is only possible if both microelectrodes are in the cytoplasmic compartment of the cells. Electrically, it will be hard to distinguish between a well-coupled cell pair with one or both electrodes placed in the vacuole, and a poorly coupled cell pair with both electrodes in the cytoplasm (see chapter 5). These measurements require the injection of a fluorescent tracer, present in the electrode tip. Intracellular injection of the dye identifies the cellular compartment impaled.

4 Concluding Remarks

Although dual voltage clamp measurements contain many pitfalls, it is a technique that allows for determination of biophysical properties of intercellular conductance with high resolution. In appropriate plant cell systems, it will certainly contribute to elucidating the electrical properties of plasmodesmata and their modulation by exogenic factors.

5 References

Blatt MR (1988) Potassium-dependent, bipolar gating of K^+ channels in guard cells. J Membr Biol 102:235–246

Blatt MR, Clint, GM (1989) Mechanisms of fucicoccin action: Kinetic modification and inactivation of K^+ channels in guard cells. Planta 178:509–523

Bruzzone R, White TW, Paul DL (1996) Connections with connexins, the molecular basis of direct intercellular signaling. Eur J Biochem 238:1–27

Coté R, Thain JF, Fensom DS (1987) Increase in electrical resistance of plasmodesmata of *Chara* induced by an applied pressure gradient across the node. Can J Bot 65:509–511

Ding DQ, Tazawa M (1989) Influence of cytoplasmic streaming and turgor pressure gradient on the transnodal transport of rubidium and electrical conductance of *Chara corallina*. Plant Cell Physiol 30:739–748

Eschrich W, Fromm J, Evert RF (1988) Transmission of electric signals in sieve tubes of zucchini plants. Bot Acta 101:327–331

Findlay GP, Hope AB (1976) Electrical properties of plant cells: Methods and findings. In: Lüttge M, Pittmann MG (eds) Transport in plants. Encyclopedia of plant physiology, vol 2A. Springer, Berlin Heidelberg New York, pp 53–92
Gros DB, Jongsma HJ (1996) Connexins in mammalian heart function. BioEssays 18:719–730
Hamill OP, Marty A, Neher E, Sakmann B, Sigworth FJ (1981) Improved patch clamp techniques for high resolution current recording from cells and cell-free membane patches. Pfluegers Arch – Eur J Physiol 391:85–100
Holdaway-Clarke TL, Walker NA, Overall RL (1996) Measurement of the electrical resistance of plasmodesmata and membranes of corn suspension-culture cells. Planta 199:537–544
Jongsma HJ, Van Rijn HE (1972) Electrotonic spread of current in monolayer cultures of neonatal rat heart cells. J Membr Biol 9:341–360
Larson DM, Kam EY, Sheridan JD (1983) Junctional transfer in cultured vascular endothelium: I. Electrical coupling. J Membr Biol 74:103–113
Lew RR (1994) Regulation of electrical coupling between *Arabidopsis* root hairs. Planta 193:67–73
Lew, RR (1996) Pressure regulation of the electrical properties of growing *Arabidopsis thaliana* L. root hairs. Plant Physiol. 112:1089–1100
Overall RL, Gunning BES (1982) Intercellular communication in *Azolla* roots: II. Electrical coupling. Protoplasma 111:151–160
Parsons A, Sanders D (1989) Electrical properties of soybean plasma membrane measured in heterotypic suspension callus. Planta 177:499–510
Paszewski A, Zawadzki T (1973) Action potentials in *Lupinus angustifolius* L. shoots. J Exp Bot 24: 804–809
Paszewski A, Zawadzki T (1974) Action potentials in *Lupinus angustifolius* L. shoots. II. Determination of the strength-duration relation and the all-or-nothing law. J Exp Bot 25:1097–1103
Paszewski A, Zawadzki T (1976a) Action potentials in *Lupinus angustifolius* L. shoots. III. Determination of the refractory periods. J Exp Bot 27:369–374
Paszewski A, Zawadzki T (1976b) Action potentials in *Lupinus angustifolius* L. shoots. IV. Application of thermal stimuli and investigations on the conduction pathways of the excitation. J Exp Bot 27:859–863
Reid RJ, Overall RL (1992) Intercellular communication in *Chara*: factors affecting transnodal electrical resistance and solute fluxes. Plant Cell Environ 15:507–517
Robards AW, Lucas WJ (1990) Plasmodesmata. Annu Rev Plant Physiol Plant Mol Biol 41:369–419
Roblin G (1985) Analysis of the variation potential induced by wounding in plants. Plant Cell Physiol 26:455–461
Roblin G, Bonnemain J (1985) Propagation in *Vicia faba* stem of a potential variation induced by wounding. Plant Cell Physiol 26:1273–1283
Sambeek JW, Pickard BG (1976a) Mediation of rapid electrical, metabolic, transpirational, and photosynthetic changes by factors released from wounds. I. Variation potentials and putative action potentials in intact plants. Can J Bot 54:2642–2650
Sambeek JW, Pickard BG (1976b) Mediation of rapid electrical, metabolic, transpirational, and photosynthetic changes by factors released from wounds. II. Mediation of the variation potential by Ricca's factor. Can J Bot 54:2651–2661
Samejima M, Sibaoka T (1983) Identification of the excitable cells in the petiole of *Mimosa pudica* by intracellular injection of Procion Yellow. Plant Cell Physiol 24:33–39
Siegenbeek van Heukelom J, Denier van der Gon JJ, Prop FJA (1972) Model approaches for evaluation of cell coupling in monolayers. J Membr Biol 7:88–110
Spanswick RM (1972) Electrical coupling between cells of higher plants: a direct demonstration of intercellular communication. Planta 102:215–227
Spray DC, Harris AL, Bennett MVL (1979) Voltage dependence of junctional conductance in early amphibian embryos. Science 204:432–434
van Bel AJE, van Rijen HVM (1994) Microelectrode-recorded development of the symplasmic autonomy of the sieve element/companion cell complex in the stem phloem of *Lupinus luteus* L. Planta 192:165–175
van der Schoot C, van Bel AJE (1989) Glass microelectrode measurements of sieve tube membrane poentials in internode discs and petiole strips of tomato (*Solanum Lycopersicum* L.). Protoplasma 149:144–154
van der Schoot C, van Bel AJE (1990) Mapping membrane potential differences and dye-coupling in internodal tissues of tomato (*Solanum Lycopersicum* L.). Planta 182:9–21
Zawadzki T (1980) Action potentials in *Lupinus angustifolius* L. shoots. V. Spread of excitation in the stem, leaves, and root. J Exp Bot 31:1371–1377
Zawadzki T, Trebacz K (1982) Action potentials in *Lupinus angustifolius* L. shoots. VI. Propagation of action potential in the stem after the application of mechanical block. J Exp Bot 33:100–110
Zawadzki T, Trebacz K (1985) Extra- and intracellular measurements of action potentials in the liverwort *Conocephalum conicum*. Physiol Plant 64:477–481

6 Appendix: Formulas

6.1 Basic Equations

A full set of basic equations from application of Ohm's law and Kirchhoff's law to the diagram of Fig. 2B. For abbreviations see next section)

Ohm's law:

$$V_a - V_1 = I_a \cdot R_{s1} \tag{A1}$$

$$V_1 = I_{m1} \cdot R_{m1} \tag{A2}$$

$$V_1 - V_2 = I_j \cdot R_j \tag{A3}$$

$$V_b - V_2 = I_b \cdot R_{s2} \tag{A4}$$

$$V_2 = I_{m2} \cdot R_{m2} \tag{A5}$$

Kirchhoff's law:

$$I_b = I_{m2} - I_j \tag{A6}$$

$$I_a = I_j + I_{m1} \tag{A7}$$

6.2 Calculation of Intercellular Conductance

Combining Eqs. (A1)–(A4) and (A7), and solving for g_j ($g_j = 1/R_j$), one arrives at:

$$g_j = \frac{I_a + \frac{I_a \cdot R_{s1}}{R_{m1}} + \frac{V_a}{R_{m1}}}{V_a - I_a \cdot R_{s1} - (V_b - I_b \cdot R_{s2})}. \tag{A8}$$

Similarly, using Eqs. (A1), and (A3)–(A6), one obtains:

$$g_j = \frac{-I_b + \frac{-I_b \cdot R_{s2}}{R_{m2}} + \frac{V_b}{R_{m2}}}{V_a - I_a \cdot R_{s1} - (V_b - I_b \cdot R_{s2})}. \tag{A9}$$

In the ideal case of infinitely high membrane resistances, Eqs. (A8) and (A9) reduce to

$$g_j = \frac{I_a}{V_a - I_a \cdot R_{s1} - (V_b - I_b \cdot R_{s2})} \tag{A10}$$

and

$$g_j = \frac{-I_b}{V_a - I_a \cdot R_{s1} - (V_b - I_b \cdot R_{s2})}. \tag{A11}$$

6.3 Modelling Currents of the Dual Whole-Cell Patch Clamp

For simulations of the dual whole cell voltage clamp determination of gap junctional conductance, amplifier currents were calculated. V_b was set to 0 mV. For calculation of the amplifier currents I_a and I_b of cells A and B, respectively, the substitute resistance of the network, and parts thereof, was calculated (see Fig. 2B), resulting in

$$I_a = \frac{V_a \cdot [R_j \cdot (R_{m2}+R_{s2})+R_{m2} \cdot R_{s2}+R_{m1} \cdot (R_{m2}+R_{s2})]}{R_{s1} \cdot [R_j \cdot (R_{m2}+R_{s2})+R_{m2} \cdot R_{s2}+R_{m1} \cdot (R_{m2}+R_{s2})+R_{m1} \cdot (R_j \cdot (R_{m2}+R_{s2})+R_{m2} \cdot R_{s2}]} \quad \text{(A12)}$$

and

$$I_b = \frac{R_{m1} \cdot R_{m2}}{(R_{m1}+R_j)\,(R_{m2}+R_{s2})+R_{m2} \cdot R_{s2}} \cdot I_a \,. \quad \text{(A13)}$$

6.4 Abbreviations Used

V_a voltage clamped pipette potential of cell A
V_b voltage clamped pipette potential of cell B
V_1 membrane potential of cell A
V_2 membrane potential of cell B
I_a pipette current (clamp current) of cell A
$I_{a,\,inst}$ instantaneous pipette current (clamp current) of cell A
$I_{a,\,ss}$ steady state pipette current (clamp current) of cell A
I_b pipette current (clamp current) of cell B
R_j intercellular resistance
I_j intercellular current
R_s series resistance
R_{s1} series resistance of cell A
R_{s2} series resistance of cell B
R_{m1} membrane resistance of cell A
I_{m1} membrane current of cell A
R_{m2} membrane resistance of cell B
I_{m2} membrane current of cell B
g_j intercellular conductance ($1/R_j$)
$g_{j,\,t}$ true intercellular conductance
$g_{j,\,m}$ measured intercellular conductance
F_j measured fraction of the true intercellular conductance ($g_{j,\,m}/g_{j,\,t}$)

CHAPTER 5

Use and Limitations of Fluorochromes for Plasmodesmal Research

P. B. GOODWIN · L. C. CANTRIL

Department of Crop Sciences, University of Sydney, New South Wales 2006, Australia

1 Introduction

Fluorochromes have been essential tools to our exploration of cell-to-cell communication in plants. They have provided us with a measure of cell coupling during development and differentiation, of the gating of plasmodesmata during changes in cell physiology, and of virus and macromolecule movement from cell to cell. Fluorochromes have demonstrated the dynamism within the intricate substructure that we see in EM images of plasmodesmata, and have forever dispelled the idea that plasmodesmata are slowly changing, passive channels restricted by a rigid plant cell wall. Techniques for the use of fluorochromes will be discussed. Like any measurement technique, the use of fluorochromes will only give valid results if care is taken to avoid errors. Some of the potential sources of error will be discussed, together with measures to avoid them.

2 Delivery of Fluorochromes

Various methods of delivery of these fluorescent compounds have been used. The loading of fluorochromes via cut or abraded surfaces is a technique that was used by pioneers in the field of cell-to-cell communication in plants (Tyree and Tammes 1975) and remains in common use for loading tracers into the phloem (McDonald et al. 1995). One of the simplest techniques is the passive loading of acetate forms of various fluorochromes including CF (carboxyfluorescein) and HPTS (8-hydroxypyrene-1,3,6-trisulphonic acid; Wright and Oparka 1996). The non-fluorescent acetate forms pass through the plasmalemma and in the cytoplasm they are cleaved to a non-plasmalemma permeable fluorescent form which is free to move from cell to cell via plasmodesmata (Wright and Oparka 1996). While this method of introducing fluorescent tracers is still useful for some studies of cell-to-cell communication due to its non-invasive approach (Duckett et al. 1994; Oparka et al. 1995; McLean et al. 1997), it is not possible to use it for the large majority of fluorescently labelled molecules. Electroporation or electropermeabilization has also been utilized to introduce membrane-impermeable molecules into plant systems (Oparka and Read 1994; Obermeyer and Weisenseel 1995) but is probably of value primarily to the investigator of long-range cell-to-cell communication, since the probe is likely to be introduced into most cells exposed to the medium, meaning that localized cellular movement is difficult to determine. The development of microinjection for the loading of individual cells with tracers has been the technique underlying many of the great successes in the exploration of the living plant symplasm. This technique relies on the insertion of a hollow glass micropipette containing the tracer or tracers of interest, and then delivery of them to the cell. Passive diffusion of the tracer from the micropipette tip was an early technique (Tucker 1982) but in most cases a force must be applied to drive the tracer into the cell. This force is either in the form of a voltage gradient (iontophoresis) or a pressure gradient (pressure injection) between the micropipette and the cell, so that the fluorochrome moves down this gradient into the cell.

3 Microinjection Techniques

There are numerous papers dealing with the techniques of iontophoresis, e.g. Goodwin (1983), van der Schoot and Lucas (1995) and pressure injection (Wolf et al. 1989; Oparka et al. 1990; Zhang et al. 1990). It is important to be aware that these techniques have their hazards and disadvantages. Iontophoresis is limited to the use of tracers with a net charge such as simple fluorochromes, and labelled peptides and dextrans. Current has also been reported to transiently enhance plasmodesmal conductance (van der Schoot and Lucas 1995). Pressure has been shown to close plasmodesmata and shut down cell-to-cell communication (Oparka and Prior 1992) and pressure changes within the cell can occur with both delivery techniques. Pressure injection, in addition to actively changing the turgor pressure during probe delivery, may require the use of wider bore needle tips to prevent clogging. There is thus an increased risk of "bursting" the cell and ruining the experiment. However, simultaneous monitoring of cell turgor pressure during microinjection is possible, reducing the pressure change problems with this delivery system (Oparka et al. 1990).

Insertion of the needle into the cell must be done so that the cell is not damaged. It is essential that the plant material is held rigid and the movement of the needle be coincident with the long axis of the needle, so that the cell is stabbed, not cut. Techniques for mounting plant tissues include pinning (Erwee and Goodwin 1983), taping (Ding et al. 1992a; van der Schoot and Lucas 1995), anchoring with low-temperature gelling agarose (Hepler and Callaham 1987; Holdaway-Clarke et al. 1996), or steadying with micropipettes (Derrick et al. 1992). Nevertheless, it is sometimes necessary to bring the needle up to the cell wall and to tap the microscope gently to puncture the cell wall and membrane. This ensures the smallest possible diameter of glass entering the cell and the smallest hole in the membrane. Sudden pressure changes from puncturing the cell can thus be avoided.

3.1 Cell Viability

It is essential to know that the cell has survived the insertion of the microelectrode, lived through delivery of the tracer, and remained intact after removal of the tip. In some larger cell systems the ability to observe cytoplasmic streaming before, during, and after microinjection has been exploited as a marker of cell integrity. For example, the leaves of *Egeria densa* (Goodwin 1983), the cells of *Chara corallina* (Shepherd and Goodwin 1992a), and the stamen hairs of *Setcreasea purpurea* (Tucker 1982, 1993; Tucker and Boss 1996) and *Tradescantia virginiana* (Hepler and Callaham 1987; L. C. Cantrill et al., unpub. data) have been particularly useful for microinjection studies of cell-to-cell communication and plasmodesmal physiology. However, it has been noted that the streaming observed within *Egeria densa* may be an artefact of leaf detachment (Gamalei et al. 1994) indicating a possible wounding response.

3.2 Long-Term Cell Survival

Cell survival for extended periods (several days) after microinjection is also another area which could provide good data on the damage caused by various methods of microinjection. Monitoring of injected cells up to 24 h after microinjection has shown cell survival and revealed low rates of macromolecular movement (Kikuyama et al. 1992; Plieth and Hansen 1996). Long-term studies have shown that dye-filled cells undergo cell division (Steinbiss and Stabel 1983; Palevitz and Hepler 1985; Kost et al. 1995). The latter authors followed injected cells for several days. More work could and should be done in this area.

3.3 Cell Membrane Potentials

The technique of monitoring the cell membrane potential offers the simplest and best indication of cell condition after microinjection. It has generally been a feature of the iontophoretic technique of microinjection but it is possible to combine the pressure injection of large and difficult molecules with monitoring of the membrane potential (Goodwin 1983; Plieth and Hansen 1996). The membrane potential should be stable, negative, and within a normal range, although what is normal can vary between species, season (Shepherd and Goodwin 1992a), and cell type (van der Schoot and Van Bel 1989, 1990). The major criterion, excepting reports of depolarising action potentials (Shepherd and Goodwin 1992a, b) is that the membrane potential during and after pressure microinjection (Kempers and van Bel 1997) or after electroporation be stable and close to the preinjection value. Some workers have used the membrane potential in conjunction with dye-coupling studies to determine levels of intercellular communication (van der Schoot and van Bel 1990; Shepherd and Goodwin 1992a, b; van Bel and van Rijen 1994), or as the chief means of determining cell coupling or plasmodesmal conductivity with dye injection relegated to locating the needle tip within the cell (van der Schoot and van Bel 1989; Holdaway-Clarke et al. 1996). Of course, much early work on cell-to-cell communication in plants (Spanswick 1972; Overall and Gunning 1982) and on gap junctions in animals (Loewenstein 1979) utilized electrical coupling to explore cell-to-cell communication.

4 Fluorescent Probes

The most popular probes for investigating cell coupling in the last decade have undoubtedly been the fluorescein derivatives [fluorescein isothiocyanate (FITC) and CF] and Lucifer Yellow carbohydrazide (LYCH). FITC has been conjugated to a wide variety of peptides, dextrans, and proteins for microinjection (Simpson 1978; Oparka and Read 1994). Conjugations with FITC have even been performed on proteins that are proposed to gate plasmodesmata such as KNOTTED 1 (Lucas et al. 1995) and viral movement proteins (Fujiwara et al. 1993; Noueiry et al. 1994) and have been achieved with retention of activity for subsequent microinjection. The fluorescent dye TOTO-1, which specifically labels DNA and RNA, has been used for labelling viral DNA and RNAs (Fujiwara et al. 1993; Noueiry et al. 1994) and plant RNA (Lucas et al. 1995) to demonstrate their trafficking from cell to cell after microinjection.

Despite some problems with sequestration, membrane permeability, and loss of fluorescence under acid conditions, the simple fluorochrome carboxyfluorescein has retained its usefulness as the smallest (lowest molecular weight) fluorescent symplasmic tracer available and has the advantage of high fluorescence at low concentrations (Warner and Bate 1987). LYCH and Cascade Blue hydrazide (CB) are relatively membrane-impermeant probes which are gaining great favour for microinjection studies (Bergmans et al. 1993; van der Schoot et al. 1995; Kempers and van Bel 1997). Both probes are aldehyde-fixable and have enabled live cell studies to be combined with histological localization work (Fisher 1988; Oparka and Prior 1988). Antibodies to these probes are also available (Haughland 1996).

In our experience, many plant tissues autofluoresce quite strongly under the UV illumination needed to reveal CB, (L. C. Cantrill, R. L. Overall and P. B. Goodwin, unpub. data), and this may pose a barrier to widespread use of Cascade Blue hydrazide or its conjugates as symplasmic tracers, just as the rhodamine dyes have been difficult to use inside chlorophyll-containing cells, since chlorophyll fluorescence is excited by the wavelengths needed to reveal rhodamine and its conjugates (Goodwin 1983). Careful selection of filter sets and the use of modern confocal or other digital detection equipment have the potential to overcome these problems. Different fluorochrome tracers can be coinjected or injected in the presence of other fluorescent molecules as in many studies of cell-to-cell communication between animal cells (Bryant and Fraser 1988; Nicolson et al. 1988). In plants, the full potential for multicolour microinjection and tracing of cell-to-cell communication is still to be realized, having only been reported by Bergmans et al. (1993).

5 Limitations of Technique

There are a number of potential sources of artefacts in examining cell-to-cell movement using injected fluorochromes. Probes could be contaminated with breakdown products or unbound fluorochrome and they could be subject to cytoplasmic binding, enzyme degradation, sequestration, bleaching or pH effects. Ways of testing for these problems are discussed now.

5.1 Impurities

All molecules labelled with one of the low molecular weight fluorochromes potentially contain quantities of the unbound fluorochrome. Thin layer chromatography can be used to determine if this is present (Goodwin 1983). Commercially available dextrans have also been shown to be contaminated with unbound fluorochrome, but this can be easily removed by reprecipitation in 80% ethanol (Derrick et al. 1992). Furthermore, commercial dextrans are polydisperse; that is, they contain a range of dextrans of different molecular sizes with the nominal value being the most common component of the mixture. These dextrans can be further separated using size-exclusion gel chromatography to gain greater precision in size-exclusion limit studies (Ding et al. 1993; Waigmann and Zambryski 1995).

5.2 Binding

This is a problem which is easily checked for, since fluorochromes which bind to large cell components, whether they be the cell walls, the cytoplasm, nucleus or starch grains will be visible in fluorescence microscopy in the tissue if the cells are killed, for example by a "delicate" squash, and soaked in a dilute solution of the dye, and then rinsed. Bound dyes will remain in the cells. Similarly, bound dye will remain in cells which have been injected if the cells are killed and rinsed with buffer. As an example, binding has been found with calcofluor M2R, but not with CF or the frequently used FITC conjugates (Goodwin 1983; Goodwin and Erwee 1985). Binding of LYCH to cell components is reported to occur after short periods and this is possibly caused by free cytoplasmic potassium ions (Warner and Bate 1987). Of course, these crude procedures will not reveal binding to readily dispersed proteins or other molecules, nor readily reversible associations with cell components.

5.3 Enzymatic Degradation

Fluorochromes are often conjugated with another molecule such as a peptide or dextran to obtain a specific molecular size for probing plasmodesmal channels. It is possible that cells might degrade the molecules, releasing the fluorochrome and giving false results. This is normally tested by incubating the conjugates with tissue homogenates, and looking for degradation products (Tucker 1982; Goodwin 1983) or by extracting the injected probes from the cells and separating them using chromatography (Goodwin and Erwee 1985; Terry and Robards 1987; Wolf et al. 1989). All reports of such studies have indicated stability of the conjugates.

5.4 The Injected Compartment

It is important that workers using microinjection to study cell-to-cell communication between plant cells be able to distinguish between injections into the apoplasm, the cytoplasm, and the vacuole. Figure 1A–C illustrates typical examples of these different regions microinjected with various fluorescent probes. With our current understanding of the location of the permeable channel through plasmodesmata, it is obvious that the microelectrode tip has to be located in the cytoplasm. The most striking feature of a cytoplasmic injection is the presence of a large fluorescent nucleus and peripheral location of the dye. Strands of cytoplasm may also be seen to traverse the region of the central vacuole.

Cell walls are extremely permeable to probes smaller than 10–20 kDa (O'Driscoll et al. 1993) and apoplasmic injections can give false positives for cell-to-cell movement. Injections into the vacuole alone give false negatives, as commonly used fluorochromes have been found to be unable to pass the tonoplast once inside the vacuole or to pass it only very slowly (Goodwin and Erwee 1990; Goodwin et al, 1990). The critical features to be wary of are dark cells finely outlined with fluorescence but lacking bright fluorescent nuclei, which indicates apoplasmic movement. The filling of the large central area of the cell, again without a bright nucleus or cytoplasmic strands be-

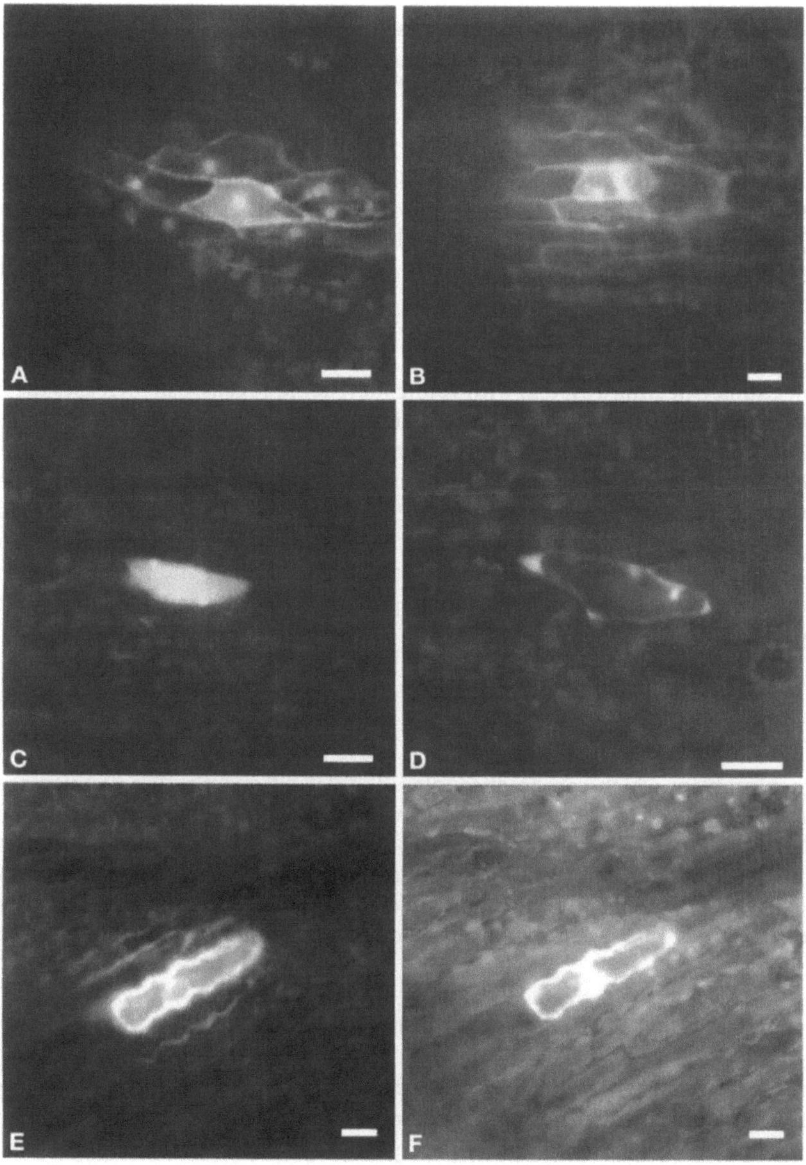

Fig. 1. A Carboxyfluorescein injected into the cytoplasm of a *Torenia* epidermal cell (*brightest cell*) showing characteristic location of fluorescence in nucleus and cell periphery. Similar probe location can be seen in neighbouring cells. **B** Fluorescent probe "injection" into the apoplasm. Note the absence of fluorescent nuclei in cells. **C** A vacuolar injection of an FITC-labelled peptide showing the cell filled with the probe and the absence of a well-defined fluorescent cytoplasm and nucleus. It is unclear whether failure of the probe to move from cell to cell is due to its vacuolar location or to plasmodesmata closure. **D** Cytoplasmic injection showing pooled cytoplasm in the corners of the cell. These regions are ideal targets for cytoplasmic injection. **E** Reflection of fluorescence on cell walls and air/water interfaces within the plant material can give falsely positive results for cell-to-cell movement. **F** Extended exposure of chlorophyll to blue illumination causes photobleaching or leads to yellow fluorescence by the tissue which masks any movement of the injected probe. *Bars* 20 μm

ing evident, indicates a vacuolar injection, assuming that one is dealing with a cell with a large central vacuole. In many cases, the nucleus will appear as a dark structure after a vacuolar injection. However, the practical reality is that it is difficult to inject only into the cytoplasm, as in most cells it occupies only a small fraction of the cell volume. Rupturing the vacuole mechanically or via a high current pulse or even inducing leakiness in the membrane with high current will transfer fluorochromes to the cytoplasm and then intercellularly (Fisher 1988; van der Schoot and van Bel 1990; Saito et al. 1996). However, the resulting fall in pH of the cytoplasm probably disrupts cell physiology (van der Schoot and van Bel 1990) and could affect plasmodesmal structure and function.

Targeting the needle tip to the nucleus or to the corners of cells (see Fig. 1D) where cytoplasm often pools is a better method of increasing the frequency of successful microinjections. Selecting a system with young meristematic cells also increases the likelihood of a cytoplasmic injection due to their greater relative cytoplasmic volume and small vacuoles that can be pushed away from the needle tip. Leaving the micropipette in the cell for a few minutes can also result in exclusion of the tip from the vacuole after movement of the tonoplast (Ding et al. 1992a). Another approach taken by Madore et al. (1986; see also Wolf et al. 1991) was to encapsulate the probes within liposomes. These fused with the tonoplast upon injection into the vacuole and released the probe into the cytoplasm. This exploited the high probability of entering the vacuole during microinjection.

5.5 Reflections

Another problem which can be encountered is flaring or reflection of fluorescence from a bright injected cell into neighbouring cells which can give a false positive (see Fig. 1E). It is easy to check on the presence of dye in any cell by closing down the aperture of the incident fluorescent light within the microscope so that the illuminated area is narrowed. In this way, the bright injected cell is excluded and it is then possible to look for fainter fluorescence in the usual cytoplasmic pattern in neighbouring cells.

5.6 Sequestration and Membrane Permeability

Under ideal circumstances, the fluorescent probes in common use for cell-to-cell communication studies would not pass membranes and would be retained in the cytoplasm. However, some fluorochromes are partly lipid-soluble. For example, CF will pass through the plasma membrane in its undissociated form if the apoplasmic pH is low (Oparka 1991). Despite this, for the purposes of observing cell-to-cell communication using epifluorescence microscopy, the amounts of this probe which would diffuse in either direction across the plasmalemma have been shown to be negligible (Erwee and Goodwin 1983; Goodwin 1983; Goodwin et al. 1990). Other fluorochromes such as LYCH (Stewart 1978) and CB (Haughland 1989) and the FITC conjugates (Hawes et al. 1995) are considered lipid-insoluble. Early work microinjecting LYCH demonstrated that this probe was retained within the injected symplasm and not sequestered to the vacuole (Erwee et al. 1985; Terry and Robards 1987; Oparka and Prior 1988), but the story is not so simple. LYCH and CB have been shown to be sequestered from the apoplasm to the vacuole (Oparka and Prior 1988; Oparka et al. 1991; Wright and Oparka 1994) from the apoplasm to the cytoplasm (O'Driscoll et al. 1991; Oparka et al. 1991) and from the cytoplasm to the vacuole (Fisher 1988; Goodwin et al. 1990; Cole et al. 1991; Oparka et al. 1991). The potential for sequestration of FITC-labelled dextrans over many hours has also been demonstrated (Cole et al. 1990). Undoubtedly, this area is very complicated and still requires much work before any conclusions can be drawn (Hawes et al. 1995). It is probably wise for users of fluorochromes to be aware that these events occur and that fluorochrome sequestration varies in rate and location between different species (Tucker and Spanswick 1985; Goodwin et al. 1990; Oparka and Hawes 1992).

As outlined by Goodwin et al. (1990) and Goodwin and Erwee (1990), the chief concern with microinjected CF and FITC conjugated peptides was that sequestration into one of the cellular compartments might favour some probes over others. This would greatly affect the distribution of the probes and thus affect interpretation of their cell-to-cell movement, particularly if sequestration was much faster with the larger tracers. Their apparent lack of intercellular movement could have been an artefact (Goodwin and Erwee 1990). The probes were found to move from the cytoplasm of the injected cell into three compartments: the cytoplasm of adjoining cells, the vacuole, and the nucleus. Neither movement into chloroplasts or mitochondria was visible, nor rapid loss across the plasmalemma. Movement into the nucleus was reversible, but the dye was sequestered in the vacuole in an irreversible fashion. The estimated permeability coefficients were as given in Table 1. Thus it was concluded that this family of probes were all sequestered similarly into the vacuole and that the difference in their cell-to-cell movement was due to the permeability of the plasmodesmata.

5.7 Bleaching Effects

Photobleaching can be induced by the high light intensities used to illuminate probes in epifluorescence microscopy. Bleaching can occur to the fluorochrome being used or to the cells being microinjected, the bleaching of cells being particularly obvious in chlorophyll-containing cells (Fisher 1988; see Fig. 1F). Various methods for reducing these effects have been discussed by Oparka and Read (1994). Photobleaching rapidly leads to a cessation of streaming in *Egeria*, and presumably cell damage in all species. For this reason, exposure to high light intensities needs to be minimized, and possible

Table 1. Estimated permeability coefficients of membranes of *Egeria* leaf cells ($cm \cdot s^{-1}$) (Goodwin et al. 1990)

Probe	MW	Tonoplast ($\times 10^{-6}$)	Nuclear envelope ($\times 10^{-5}$)	Adjoining cell walls ($\times 10^{-6}$)
6COOHF	376	2.4	9.4	112
FGlu	536	4.4	2.6	15.5
$F(Glu)_2$	666	2.6	3.7	11.4
$F(Gly)_6$	749	3.8	2.8	0.66
FLGGL	874	5.1	4.1	0.009

damaging affects of the light exposure be considered in designing and interpreting experiments.

Photobleaching, rather than being a problem, can also be harnessed for studying cell-to-cell communication in plants. Fluorescence redistribution after photobleaching (FRAP) has been utilized only a small number of times for investigations of cell-to-cell communication in plants and could offer further insights. It has been used to demonstrate cell coupling using bleaching of CF (Baron-Epel et al. 1988) and to observe the cell-to-cell diffusion of fluorescent analogues to ER-specific lipids (Grabski et al. 1993).

5.8 pH Effects

The fluorescence of the commonly used fluorescein derivatives is very sensitive to pH, being high at pH 8 and above, and slight below pH 4.5 (Haughland 1996). Given the relatively low pH of the vacuole, of the order of 6 (Beffagna and Romani 1988), there will be a substantial loss of probe fluorescence as it is sequestered into the vacuole. In contrast, LYCH and Cascade Blue have a stable fluorescence over a wide pH range (Stewart 1978, 1981) and the newer FITC-like fluorochromes such as Oregon Green are reported to have a broader pH stability (Haughland 1996). As already mentioned above, low pH can increase the lipid solubility of CF and it has also been shown to increase the uptake of a range of other probes across membranes (Oparka et al. 1991).

One of the early concerns with the fluorescein derivatives was whether CF and the different FITC-conjugated probes showed similar responses to pH. If the larger probes were less sensitive to pH, then their concentration would appear to be higher in the vacuole of the injected cell (where most of the probe comes to reside). This would lead to the impression of relatively less movement by such probes. However, the study by Goodwin et al. (1990) showed that there was relatively little difference between CF and the various FITC conjugates in their pH response, the differences not matching the differences in apparent mobility between cells.

6 Fluorochromes and Cell-to-Cell Communication

6.1 Assessing Symplasmic Permeability

Mapping tissues using fluorescent probes of different sizes has been a successful means of understanding the regulation of plasmodesmata during different developmental and physiological states. Determining the MEL (molecular exclusion limit) of a tissue by introducing probes of increasing molecular weight until a probe is identified which does not move from cell to cell is one means of quantifying the permeability of the symplasm. For example, domains within whole plant tissues that permit molecules of different sizes to move intercellularly have been identified in several species and found to be characteristic of a particular cell or organ type (Erwee and Goodwin 1985; Shepherd and Goodwin 1992a, b). Some of these domains are delineated by sharp boundaries that the fluorescent probe is unable to traverse, reflecting fine regulation of the MEL between different organs (Erwee and Goodwin 1985; Erwee et al.

1985; Bergmans et al. 1993; Duckett et al. 1994). More strikingly, microinjection of fluorescent molecules has demonstrated that MELs change during development (Erwee and Goodwin 1985; Palevitz and Hepler 1985; Shepherd and Goodwin 1992a, b) and that vastly different MELs can be part of the normal functioning of the plant (Kempers et al. 1993; Kempers and van Bel 1997).

Quantification of symplasmic permeability can also be achieved using a single probe in different cell types or under differing conditions. Erwee and Goodwin (1984) in a study of cell to cell communication and plasmolysis assessed the movement of a FITC conjugate [$F(Glu)_2$] by scoring the number of cells that contained the probe after microinjection. They showed a significant increase in $F(Glu)_2$ movement after recovery from plasmolysis. Likewise, van der Schoot and van Bel (1990) devised a dye-coupling number (dcn) for their investigation around vascular tissues in tomato internodes. This revealed increasing movement of LYCH from the central to the outer pith. Larger dcn's were obtained for CF (MW 376) than LYCH (MW 457) and this was suggested to be a result of the smaller relative molecular mass of CF and thus its greater permeability through plasmodesmata (van der Schoot and van Bel 1990).

6.2 Size of Probes

With the fluorochromes and their larger conjugates there is no simple relationship between molecular weight and size, although by convention, MELs have been expressed in terms of molecular weights. For the small peptide conjugates, molecular models have provided a useful guide to the "true" size of the molecules (Terry and Robards 1987). These workers found a relationship between the modelled hydrodynamic radius (Stokes radius) and cell-to-cell mobility for a range of peptide conjugates, with the molecular weight and the side groups on each probe influencing the hydrodynamic radius. Values for the Stokes radii of dextran probes are also available from calculations or empirical studies (Jøergensen and Møller 1979; Squire 1981; Poitevin and Wahl 1988). Dextrans are less dense than peptides, so that a pore able to pass a given molecular weight of dextran would pass a heavier protein (see Kempers and van Bel 1997).

In recent times, with the large body of work emerging on viral movement proteins and plant proteins such as Knotted 1, a problem has arisen of apparently larger proteins moving through plasmodesmata than the dextran MEL would predict (Wolf et al. 1989; Lucas et al. 1995). To illustrate, Waigmann and Zambryski (1995) observed slow cell-to-cell movement of a 7-kDa dextran, but much more rapid movement of a 30-kDa tobacco mosaic virus (TMV) motor protein (MP) and even the movement of a 90-kDa TMV MP-β-glucuronidase (GUS) fusion protein in tobacco leaf trichomes. This large discrepancy between MELs as measured by dextrans and proteins is probably not explicable by the differences in density. Molecular weight would thus appear to be virtually irrelevant as a measure of size exclusion for trafficked proteins. If unfolding and refolding is the key to the cell-to-cell movement of these relatively large molecules, then we obviously require modelling of the Stokes radii of unfolded nucleotide and peptide chains to determine the diameter of the plasmodesmal pores that accommodate them.

6.3 MEL of Tissues in the Absence of Viruses

The classical view of the MEL of around 800 Da for plasmodesmata under normal conditions has its origins in the dye-coupling studies on animal tissues (Loewenstein 1979) which sparked the early experiments of Tucker (1982), Goodwin (1983), and Terry and Robards (1987). Calculations of the pore sizes of fixed plasmodesmata (Overall et al. 1982) reinforced the idea of an 800-Da limit, but recently this view has begun to change. The level of the normal MEL appears to be very much dependent on both the species and the type of cells under investigation, as well as the physiological state of those cells.

The first demonstration of a MEL greater than 1 kDa in non-virus-infected tissues was made in the green alga *Nitella*, which over a period of 24 h permitted the cell-to-cell movement of a 45-kDa protein labelled with FITC (Kikuyama et al. 1992). This slow movement of macromolecules within characean algae has since been exploited for loading internode cells with Ca^{2+}- sensitive dyes conjugated to dextrans (Plieth and Hansen 1996). Recent work on leaf trichomes of tobacco (Waigmann and Zambryski 1995), stamen hairs of *Setcreasea purpurea* (Yang et al. 1995) and *Tradescantia* stamen hair cells (L.C. Cantrill, R.L. Overall and P.B.Goodwin, unpub.; see Fig. 2) has shown that under normal conditions the MEL can be much greater than 1 kDa. All these systems are two dimensional cell files and might be "hard wired" for macromolecular trafficking due to their isolation from other cells or because of specialized functions like secretion (Lazzaro and Thomson 1996). Additionally, the observation of dextran movement between phloem companion cells and sieve elements (Kempers et al. 1993; Kempers and van Bel 1997) has reinforced the idea that it is no longer correct to assume a blanket value of around 1 kDa for all plant cells. In fact the "record" size exclusion limit under normal conditions is currently at least 10 kDa and it is likely to rise much higher as further studies are made (Kempers and van Bel 1997; see also Fisher et al. 1992; Sakuth et al. 1993).

It could be claimed that the examples given above are special cases, but experiments on thin cell layers cut from the stems of *Torenia fournieri* (L. C. Cantrill, R. L. Overall and P. B. Goodwin unpub. data; see Fig. 3) have indicated that after 48 h of culturing in vitro, epidermal cells allow the intercellular movement of 10-kDa dextrans. Furthermore, as has been demonstrated numerous times, rapid movement of viral movement proteins (Fujiwara et al. 1993; Noueiry et al. 1994; Ding et al. 1995) or of viral movement proteins and coinjected dextrans (Waigmann et al. 1994) occurs between normal cells, for example leaf mesophyll. This indicates that these cells contain a mechanism for rapidly increasing the MEL to the macromolecular level. Such a mechanism may be a common feature of many cells and central to day-to-day regulation within the plant, including responding to stress. Other workers have shown that a small proportion (around 10%) of control injections of dextran probes result in cell-to-cell movement of these macromolecules (Vaquero et al. 1994; Lucas et al. 1995; Ding et al. 1996). Under induced stress conditions, the MEL has been shown to increase (Erwee and Goodwin 1984; Cleland et al. 1994). With all the information emerging on endogenous protein trafficking during shoot development (Mezitt and Lucas 1996) and, in particular, in the KNOTTED 1 system (Lucas et al. 1995), it would not be surprising to find evidence for enlargement of plasmodesmal pores and protein trafficking during wound recov-

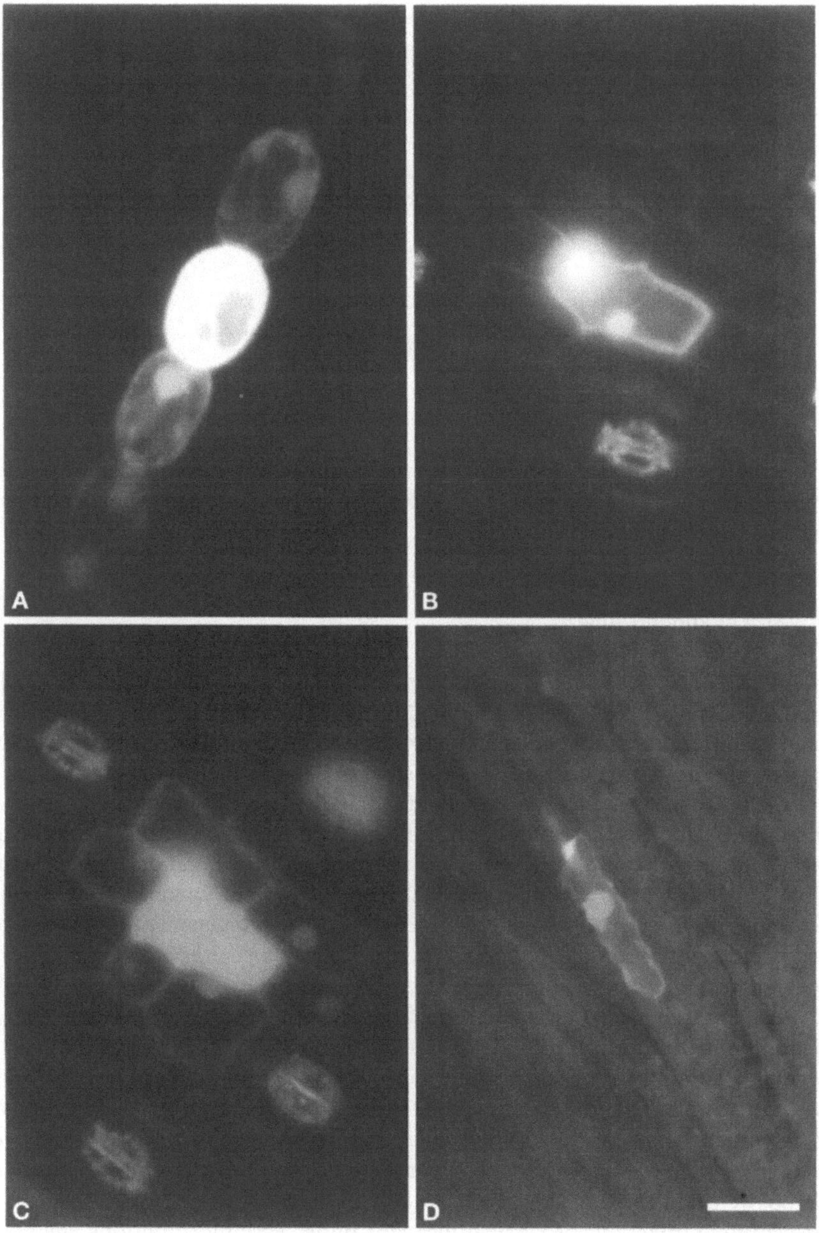

Fig. 2 A–D. Fluorescent probes of different sizes are able to move from cell to cell in different areas of *Tradescantia virginiana* and thus define communication domains that might be physiologically or developmentally significant. **A** FITC-dextran (10 kDa) injected into stamen hair cell with subsequent movement to neighbouring cells. **B** Microinjection of the same FITC-dextran into leaf epidermis without any intercellular movement. **C** CF (376 Da) injected into a leaf epidermal cells showing movement from cell to cell. **D** retention of FITC-dextran (10 kDa) microinjected into petal epidermis. *Bar* 50 μm

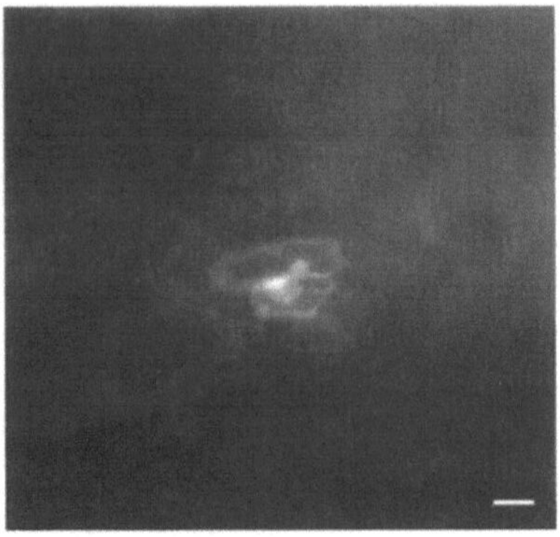

Fig. 3. FITC-labelled dextran (10 kDa) (FD 10000) microinjected into the epidermis of a cultured stem segment of *Torenia fournieri*. Cell-to-cell movement of the probe occurs between these cells after 48 h of culturing and is associated with early events in adventitious shoot regeneration. *Bar* 20 μm

ery. If experimental stresses can influence macromolecular movement between cells, we must be careful in assigning MELs in the macromolecular range to particular physiological or developmental events.

6.4 Size of Annulus Tube

Fluorescent tracers have been essential to determining the channel diameter of plasmodesmata. Using these to determine the MEL of plasmodesmata in a particular tissue and then combining the probe size with substructural dimensions taken from EM images of plasmodesmata, it has been possible to develop theoretical models for the size of the pores through plasmodesmata. Historically, there has been much controversy about the location of these pores or channels within the substructure of the plasmodesm (Gunning and Overall 1983) and there continue to be differences in the interpretation of this substructure (Ding et al. 1992b; Overall and Blackman 1996). The model of Gunning and Overall (1983), where a partly occluded cytoplasmic annulus was predicted to be the location of the pores, was quickly supported by microinjection evidence (Tucker 1982; Erwee and Goodwin 1983; Goodwin 1983; Goodwin and Lyndon 1983) and later reinforced by modelling of the tracer molecules (Terry and Robards 1987). A channel radius of around 1.5 nm was estimated and it was calculated that molecules of around 1 nm radius would move fairly rapidly from cell to cell through these channels (Terry and Robards 1987). These dimensions corresponded very well with the dimensions of the largest molecules that were known to move from cell to cell (Tucker 1982; Goodwin 1983) and also paralleled cell-to-cell communication in animals (Loewenstein 1979).

According to injections of labelled dextrans under varying conditions, the size of the pores through plasmodesmata has the potential to be much larger than these early values. For example, cells which allow movement of 10-kDa dextrans (Wolf et al. 1989; Cleland et al. 1994; Kempers and van Bel 1997; L. C. Cantrill, R. L. Overall and P. B. Goodwin unpub.) have a potential pore radius of greater than 2.4 nm (Wolf et al. 1989). After cytochalasin D and profilin treatments of tobacco mesophyll cells, 20-kDa dextrans were able to move from cell to cell (Ding et al. 1996) and similar sizes were observed by Waigmann et al. (1994) and Lucas et al. (1995), suggesting a pore radius of at least 3 nm (Wolf and Lucas 1994).

7 Concluding Remarks

What are the prospects for this technique? Although technically demanding, when proper precautions are taken, fluorochromes are able to reveal information not otherwise available on the physiology of plasmodesmata. Current research using this technique, discussed in later chapters of this book, is beginning to reveal how plant tissues and organs function. The full unravelling of this fascinating story will involve a major contribution from techniques involving the use of fluorochromes.

References

Baron-Epel O, Hernandez D, Jiang L-W, Meiners S, Schindler M (1988) Dynamic continuity of cytoplasmic and membrane compartments between plant cells. J Cell Biol 106:715–721

Beffagna N, Romani G (1988) Effects of two plasmalemma ATPase inhibitors on H^+ extrusion and intracellular pH in *Elodea densa* leaves. J Exp Bot 39:1033–1043

Bergmans A, de Boer D, van Bel A, van der Schoot C (1993) The initiation and development of *Iris* flowers: permeability changes in the apex symplasm. Flowering Newsl 16:19–26

Bryant PJ, Fraser SE (1988) Wound healing, cell communication, and DNA synthesis during imaginal disc regeneration in *Drosophila*. Dev Biol 127:197–208

Cleland RE, Fujiwara T, Lucas WJ (1994) Plasmodesmal-mediated cell-to-cell transport in wheat roots is modulated by anaerobic stress. Protoplasma 178:81–85

Cole L, Coleman J, Evans D, Hawes C (1990) Internalisation of fluorescein isothiocyanate-dextran by suspension-cultured plant cells. J Cell Sci 96:721–730

Cole L, Coleman J, Kearns A, Morgan G, Hawes C (1991) The organic anion transport inhibitor, probenecid, inhibits the transport of Lucifer Yellow at the plasma membrane and the tonoplast in suspension-cultured plant cells J Cell Sci 99:545–555

Derrick PM, Barker H, Oparka KJ (1992) Increase in plasmodesmatal permeability during cell-to-cell spread of tobacco rattle virus from individually inoculated cells. Plant Cell 4:1405–1412

Ding B, Haudenshield JS, Hull RJ, Wolf S, Lucas WJ (1992a) Secondary plasmodesmata are specific sites of localisation of the tobacco mosaic virus movement protein in transgenic tobacco plants. Plant Cell 4:925–928

Ding B, Turgeon R, Parthasarathy MV (1992b) Substructure of freeze-substituted plasmodesmata. Protoplasma 169:28–41

Ding B, Haudenshield JS, Willmitzer L, Lucas WJ (1993) Correlation between arrested secondary development and onset of accelerated leaf senescence in yeast acid invertase transgenic tobacco plants. Plant J 4:179–189

Ding B, Qiubo L, Nguyen L, Palukaitis P, Lucas WJ (1995) Cucumber mosaic virus 3a protein potentiates cell-to-cell trafficking of CMV RNA in tobacco plants. Virology 207:345–353

Ding B, Kwon M-O, Warnberg L (1996) Evidence that actin filaments are involved in controlling the permeability of plasmodesmata in tobacco mesophyll. Plant J 10:157–164

Duckett CM, Oparka KJ, Prior DAM, Dolan L, Roberts K (1994) Dye-coupling in the root epidermis of *Arabidopsis* is progressively reduced during development. Development 120:3247–3255

Erwee MG, Goodwin PB (1983) Characterisation of the *Egeria densa* Planch. leaf symplast: inhibition of the intercellular movement of fluorescent probes by group II ions. Planta 158:320–328

Erwee MG, Goodwin PB (1984) Characterisation of the *Egeria densa* symplast: response to plasmolysis, deplasmolysis and to aromatic amino acids. Protoplasma 122:162–168

Erwee MG, Goodwin PB (1985) Symplast domains in extrastelar tissues of *Egeria densa* Planch. Planta 163:9–19

Erwee MG, Goodwin PB, van Bel AJE (1985) Cell-cell communication in the leaves of *Commelina cyanea* and other plants. Plant Cell Environ 8:173–178

Fisher DB, Wu Y, Ku MSB (1992) Turnover of soluble proteins in the wheat sieve tube. Plant Physiology 100:1433–1441

Fisher DG (1988) Movement of lucifer yellow in leaves of *Coleus blumei* Benth. Plant Cell Environ 11: 639–644

Fujiwara T, Giesman-Cookmeyer D, Ding B, Lommel SA, Lucas WJ (1993) Cell-to-cell trafficking of macromolecules through plasmodesmata potentiated by the red clover necrotic mosaic virus movement protein. Plant Cell 5: 1783–1794

Gamalei YV, Fromm J, Krabel D, Eschrich W (1994) Chloroplast movement as response to wounding in *Elodea canadensis*. J Plant Physiol 144: 518–524

Goodwin PB (1983) Molecular size limit for movement in the symplast of the *Elodea* leaf. Planta 157: 124–130

Goodwin PB, Erwee MG (1985) Intercellular transport studied by microinjection methods. In: AW Robards (ed) Botanical microscopy. Oxford University Press, Oxford, pp 335–358

Goodwin PB, Erwee MG (1990) Transport across the tonoplast in relation to cell to cell communication in plants. In: Beilby MJ, Walker NA, Smith JR (eds) Membrane transport in plants and fungi. University of Sydney, Sydney, pp 303–306

Goodwin PB, Lyndon RF (1983) Synchronisation of cell division during transition to flowering in *Silene* apices not due to increased symplast permeability. Protoplasma 116: 219–222

Goodwin PB, Shepherd V, Erwee MG (1990) Compartmentation of fluorescent tracers injected into the epidermal cells of *Egeria densa* leaves. Planta 181: 129–136

Grabski S, de Feijter AW, Schindler M (1993) Endoplasmic reticulum forms a dynamic continuum for lipid diffusion between contiguous soybean root cells. Plant Cell 5: 25–28

Gunning BES, Overall RL (1983) Plasmodesmata and cell-to-cell transport in plants. BioScience 33: 260–265

Haughland RP (1989) Handbook of fluorescent probes and research chemicals, 5th edn. Molecular Probes, Eugene, Oregon

Haughland RP (1996) Handbook of fluorescent probes and research chemicals, 6th edn. Molecular Probes, Eugene, Oregon

Hawes C, Crooks K, Coleman J, Satiat-Jeunemaitre B (1995) Endocytosis in plants: fact or artefact. Plant Cell Environ 18: 1245–1252

Hepler PK, Callaham DA (1987) Free calcium increases during anaphase in stamen hair cells of *Tradescantia*. Plant Cell Biol 105: 2137–2143

Holdaway-Clarke TL, Walker NA, Overall RL (1996) Measurement of the electrical resistance of plasmodesmata and membranes of corn suspension-culture cells. Planta 199: 537–544

Jøergensen KE, Møller JV (1979) Use of flexible polymers as probes of glomerular pore size. Am J Physiol 236: F103–F111

Kempers R, van Bel AJE (1997) Symplasmic connections between sieve element and companion cell in the stem phloem of *Vicia faba* L. have a molecular exclusion limit of at least 10 kDa. Planta 201: 195–210

Kempers R, Prior DAM, van Bel AJE, Oparka KJ (1993) Plasmodesmata between sieve element and companion cell of extrafascicular stem phloem of *Curcurbita maxima* permit the passage of 3-kDa fluorescent probes. Plant J 4: 567–575

Kikuyama M, Hara Y, Shimada K, Yamamoto K, Hiramoto Y (1992) Intercellular transport of macromolecules in *Nitella*. Plant Cell Physiol 33: 413–417

Kost B, Galli A, Potrykus I, Neuhaus G (1995) High efficiency and stable transformation by optimised DNA microinjection into *Nicotiana tabacum* protoplasts. J Exp Bot 46: 1157–1167

Lazzaro MD, Thomson WW (1996) The vacuolar-tubular continuum in living trichomes of chickpea (*Cicer arietinum*) provides a rapid means of solute delivery from base to tip. Protoplasma 193: 181–190

Loewenstein WR (1979) Junctional intercellular communication and the control of growth. Biochim Biophys Acta 560: 1–65

Lucas WJ, Bouché-Pillon S, Jackson DP, Nguyen L, Baker L, Ding B, Hake S (1995) Selective trafficking of KNOTTED 1 homeodomain protein and its mRNA through plasmodesmata. Science 270: 1980–1983

Madore MA, Oross JW, Lucas WJ (1986) Symplastic transport in *Ipomea tricolor* source leaves. Plant Physiol 82: 432–442

McDonald R, Wang HL, Patrick JW, Offler T. (1995) The cellular pathway of sucrose transport in developing cotyledons of *Vicia faba* L. and *Phaseolus vulgaris* L.: a physiological assessment. Planta 196: 659–667

McLean BG, Hempel FD, Zambryski PC (1997) Plant intercellular communication via plasmodesmata. Plant Cell 9: 1043–1054

Mezitt LA, Lucas WJ (1996) Plasmodesmal cell-to-cell transport of proteins and nucleic acids. Plant Mol Biol 32: 251–273

Nicolson GL, Dulski KM, Trosko JE (1988) Loss of intercellular junctional communication correlates with metastatic potential in mammary adenocarcinoma cells. Proc Natl Acad Sci USA 85: 473–476

Noueiry AO, Lucas WJ, Gilbertson RL (1994) Two nuclear proteins of a plant DNA virus coordinate nuclear and plasmodesmal transport. Cell 76:925–932

Obermeyer G, Weisenseel MH (1995) Introduction of impermeable molecules into pollen grains by electroporation. Protoplasma 187:132–137

O'Driscoll D, Wilson G, Steer MW (1991) Lucifer Yellow and fluorescein isothiocyanate uptake by cells of *Morinda citrifolia* in suspension cultures is not confined to the endocytic pathway. J Cell Sci 100:237–241

O'Driscoll D, Read SM, Steer MW (1993) Determination of cell-wall porosity by microscopy: walls of cultured cells and pollen tubes. Acta Bot Neerl 42:237–244

Oparka KJ (1991) Uptake and compartmentation of fluorescent probes by plant cells. J Exp Bot 42: 565–579

Oparka KJ, Hawes C (1992) Vacuolar sequestration of fluorescent probes in plant cells: a review. J Microsc 166:15–27

Oparka KJ, Prior DAM (1988) Movement of Lucifer Yellow CH in potato storage tissues: a comparison of symplastic and apoplastic transport. Planta 176:533–540

Oparka KJ, Prior DAM (1992) Direct evidence for pressure-generated closure of plasmodesmata. Plant J 2:741–750

Oparka KJ, Read ND (1994) The use of fluorescent probes for studies of living plant cells. In: Harris N, Oparka KJ (eds) Plant cell biology: a practical approach. IRL Press, Oxford, pp 27–50

Oparka KJ, Murphy R, Derrick PM, Prior DAM, Smith JAC (1990) Modification of the pressure-probe technique permits controlled intracellular microinjection of fluorescent probes. J Cell Sci 98: 539–544

Oparka KJ, Murant EA, Wright KM, Prior DAM (1991) The drug probenecid inhibits the vacuolar accumulation of fluorescent anions in onion epidermal cells. J Cell Sci 99:557–563

Oparka KJ, Prior DAM, Wright KM (1995) Symplastic communication between primary and developing lateral roots of *Arabidopsis thaliana*. J Exp Bot 46:187–197

Overall RL, Blackman LM (1996) A model of the macromolecular structure of plasmodesmata. Trends Plant Sci 1:307–311

Overall RL, Gunning BES (1982) Intercellular communication in *Azolla* roots: II. Electrical coupling. Protoplasma 111:151–160

Overall RL, Wolfe J, Gunning BES (1982) Intercellular communication in *Azolla* roots: I Ultrastructure of plasmodesmata. Protoplasma 111:134–150

Palevitz BA, Hepler PK (1985) Changes in dye coupling of stomatal guard cells of *Allium* and *Commelina* demonstrated by microinjection of Lucifer Yellow. Planta 164:473–479

Plieth C, Hansen U-P (1996) Methodological aspects of pressure loading of Fura-2 into Characean cells. J Exp Bot 47:1601–1612

Poitevin E, Wahl P (1988) Study of the translational diffusion of macromolecules in beads of gel chromatography by the FRAP method. Biophys Chem 31:247–258

Saito O, Ohi A, Matsuoka H (1996) Microinjection of fluroescence dye in a plant cell and its intercellular translocation using a multi-channel microelectrode system. Biochimica et Biophysica Acta 1289:1–14

Sakuth T, Schobert C, Pescvaradi A, Eichholz A, Komor E, Orlich G (1993) Specific proteins in the sieve-tube exudate of *Ricinus communis* L seedlings: separation, characterisation and in-vivo labelling. Planta 191:207–213

Shepherd VA, Goodwin PB (1992a) Seasonal patterns of cell-to-cell communication in *Chara corallina* Klein ex Willd. I. Cell-to-cell communication in vegetative lateral branches during winter and spring. Plant Cell Environ 15:137–150

Shepherd VA, Goodwin PB (1992b) Seasonal patterns of cell-to-cell communication in *Chara corallina* Klein ex Willd. II. Cell-to-cell communication during the development of antheridia. Plant Cell Environ 15:151–162

Simpson I (1978) Labelling of small molecules with fluorescein. Anal Biochem 89:304–305

Spanswick RM (1972) Electrical coupling between cells of higher plants: a direct demonstration of intercellular communication. Planta 102:215–227

Squire PG (1981) Calculation of hydrodynamic parameters of random coil polymers from size exclusion chromatography and comparison with parameters by conventional methods. J Chromatogr 210:433–442

Steinbiss H-H, Stabel P (1983) Protoplast derived tobacco cells can survive capillary microinjection of the fluorescent dye Lucifer Yellow. Protoplasma 116:223–227

Stewart WW (1978) Functional connections between cells as revealed by dye coupling with a highly fluorescent napthalimide tracer. Cell 14:741–759

Stewart WW (1981) Lucifer dyes – highly fluorescent dyes for biological tracing. Nature 292:17–21

Terry BR, Robards AW (1987) Hydrodynamic radius alone governs the mobility of molecules through plasmodesmata. Planta 171:145–157

Tucker EB (1982) Translocation in the staminal hairs of *Setcreasea purpurea*. I. A study of cell ultrastructure and cell-to-cell passage of molecular probes. Protoplasma 113:193–201
Tucker EB (1993) Azide treatment enhances cell-to-cell diffusion in staminal hairs of *Setcreasea purpurea*. Protoplasma 174:45–49
Tucker EB, Boss WF (1996) Mastoparan-induced intracellular Ca2+ fluxes may regulate cell-to-cell communication in plants. Plant Physiol 111:459–467
Tucker EB, Spanswick RM (1985) Translocation in the staminal hairs of *Setcreasea purpurea*. II. Kinetics of intercellular transport. Protoplasma 128:167–172
Tyree MT, Tammes PM (1975) Translocation of uranin in the symplasm of staminal hairs of *Tradescantia*. Can J Bot 53:2038–2046
van Bel AJE, van Rijen HVM (1994) Microelectrode-recorded development of the symplasmic autonomy of the sieve element/companion cell complex in the stem phloem of *Lupinus luteus* L. Planta 192:165–175
van der Schoot C, Lucas WJ (1995) Microinjection and the study of tissue patterning in plant apices In: Maliga P, Klessig DF, Cashmore AR, Gruissem W, Varner JE (eds) Methods in plant molecular biology, a laboratory course manual. Cold Spring Harbor Laboratory Press, USA, pp 173–189
van der Schoot C, van Bel AJE (1989) Glass microelectrode measurements of sieve tube membrane potentials in internode discs and petiole strips of tomato (*Solanum lycopersicum* L.). Protoplasma 149:144–154
van der Schoot C, van Bel AJE (1990) Mapping membrane potential differences and dye coupling in internodal tissues of tomato (*Solanum lycopersicum* L.). Planta 182:9–21
van der Schoot C, Dietrich M, Storms M, Verbeke JA, Lucas WJ (1995) Establishment of a cell-to-cell communication pathway between separate carpels during gynoecium development. Planta 195: 450–455
Vaquero C, Turner PA Demangeat G, Sanz A, Serra MT, Roberts K, Garcia-Luque I (1994) The 3a protein from cucumber mosaic virus increases the gating capacity of plasmodesmata in transgenic tobacco plants. J Gen Virol 75:3193–3197
Waigmann E, Zambryski P (1995) Tobacco mosaic virus movement protein-mediated protein transport between trichome cells. Plant Cell 7:2069–2079
Waigmann E, Lucas WJ, Citovsky V, Zambryski P (1994) Direct functional assay for tobacco mosaic virus cell-to-cell movement protein and identification of a domain involved in increasing plasmodesmal permeability. Proc Natl Acad Sci USA 91:1433–1437
Warner AE, Bate CM (1987) Techniques for dye injection and cell labelling In: Standen NB, Gray PTA, Whitaker MJ (eds) Microelectrode techniques. The Company of Biologists, Cambridge, UK, pp 169–185
Wolf S, Lucas WJ (1994) Virus movement proteins and other molecular probes of plasmodesmal function. Plant Cell Environ 17:573–585
Wolf S, Deom CM, Beachy R, Lucas WJ (1989) Movement protein of tobacco mosaic virus modifies plasmodesmatal size exclusion limit. Science 246:377–379
Wolf S, Deom CM, Beachy R, Lucas WJ (1991) Plasmodesmatal function is probed using transgenic tobacco plants that express a virus movement protein. Plant Cell 3:593–604
Wright KM, Oparka KJ (1994) Physicochemical properties alone do not predict the movement and compartmentation of fluorescent xenobiotics. J Exp Bot 45:35–44
Wright KM, Oparka KJ (1996) The fluorescent probe HPTS as a phloem-mobile, symplastic tracer: an evaluation using confocal laser scanning microscopy. J Exp Bot 47:439–445
Yang SJ, Li MY, Zhang XY (1995) Changes in plasmodesmatal permeability of the stamen hairs of *Setcreasea purpurea* during development. Acta Phytophysiol Sin 21:355–362
Zhang D, Wadsworth P, Hepler PK (1990) Microtubule dynamics in living dividing plant cell: confocal imaging of microinjected fluorescent brain tubulin. Proc Natl Acad Sci USA 87:8820–8824

CHAPTER 6

Use of GFP-Tagged Viruses in Plasmodesmal Research

K. J. OPARKA · A. G. ROBERTS · S. SANTA CRUZ

Unit of Cell Biology, Department of Virology,
Scottish Crop Research Institute, Invergowrie, Dundee DD2 5DA, UK

1 Introduction

The insertion of the gene for the green fluorescent protein (GFP) from the jellyfish *Aequorea victoria* into the genome of a number of plant viruses is proving to be a powerful tool in the study of virus-plasmodesmata interactions. Initially, the *gfp* gene was introduced as a separate open-reading frame into the genome of potato virus X (PVX), allowing the expression of 'free' GFP in the cytosol of virus-infected cells (Baulcombe et al. 1995). This initial study of GFP, expressed from a virus vector, demonstrated that GFP could be used as a non-invasive marker of viral infections, allowing the detection of infected cells at both the whole-plant and single-cell level. However, not all viruses are amenable to this technology and some have proved problematical due to either a lack of full-length infectious clones (see Oparka et al. 1996a), or because of constraints on genome function or virus assembly due to the insertion of the additional *gfp* gene. For example, in the case of the spherical viruses cucumber mosaic virus (CMV; Canto et al. 1997), and brome mosaic virus (BMV; Rao 1997), 'free' GFP can be expressed from the complete viral genome by omitting one of the existing viral genes and replacing it with the *gfp* gene. However, this may result in loss of virus movement unless a complementation strategy is used (e.g. Canto et al., 1997). Free GFP has also been expressed from the tubule-forming virus cowpea mosaic virus (CPMV) by replacing either the coat protein gene or movement protein gene with *gfp* (Wellink et al. 1997). Rod-shaped plant RNA viruses appear to offer greater potential over spherical viruses, as insertion of an additional gene in the former does not appear to prevent particle assembly or virus movement (Santa Cruz et al. 1996). To date, GFP has been expressed successfully from the rod-shaped viruses PVX (Baulcombe et al. 1995; Oparka et al. 1995, 1996a, b; Santa Cruz et al. 1996), tobacco mosaic virus (TMV; Heinlein et al. 1995; Padgett et al. 1996; Chen et al. 1997) and tobacco etch virus (TEV; Schaad et al. 1997).

Despite the above problems, the number of studies in which GFP has been used as a marker of viral infection continues to grow as methods are found for inserting the *gfp* gene into an increasing number of plant viruses. This rapidly expanding area of research has been reviewed recently (Oparka et al. 1995; 1996a). We appreciate that by the time this chapter appears, the number of examples in which GFP has been expressed from viral vectors will have increased substantially. The purpose of this chapter is not to review further the increasing use of GFP in virus-transport studies but to give examples, from our own laboratory, in which the use of GFP-tagged viruses has begun to increase our understanding of plasmodesmal function, in particular the ways in which specific viral proteins interact with plasmodesmata. Concerning the latter topic, the reader is also referred to Chapter 17 on long-distance virus movement. Throughout this chapter, we will place emphasis on those methods for imaging GFP that have proved to be particularly useful in the study of virus-plasmodesmata interactions.

2 Methods

2.1 Virus Constructs

The work described in this chapter involves GFP-tagged PVX and TMV. For both these viruses, cDNA clones of the viral genomes were engineered to direct the expression of

GFP during infection of plants. The GFP is expressed either as a free, cytosolic, protein (Baulcombe et al. 1995) or as a fusion to a virus-coded protein such as the coat protein (CP; Santa Cruz et al. 1996) or movement protein (MP; Heinlein et al. 1995). These GFP-expressing viral vectors allow the tagged virus to be tracked non-invasively during the infection process. Where GFP is expressed as a fusion to a viral protein, additional information can be obtained on the temporal control of viral gene expression and also on the subcellular localization of the viral protein. Another application of the PVX vector described in this chapter is as an expression system for the production of fusions between GFP and proteins from heterologous viruses. This approach is particularly useful for the investigation of proteins from viruses for which infectious clones are not available or which are not suitable for genetic manipulation themselves. The use of an episomal viral vector for gene delivery has the advantage of allowing rapid and high-level foreign gene expression when compared to stable transformation techniques.

The wild-type (wt) GFP described by Chalfie et al. (1994) was found to have a cryptic splice site, recognized in plants, that prevented high-level expression of the GFP in plant cells (Haseloff and Amos 1995). Recently, the *gfp* gene has been modified to remove the cryptic splice site. However, in the case of virus vectors, transcriptional splicing does not present a problem, as the majority of plant RNA viruses do not utilize the nucleus for their replication cycle. In most of the work described here, the original GFP described by Chalfie et al. (1994) was utilized. However, it should be noted that modifications to the GFP molecule have resulted in red wavelength-shifted excitation mutations of GFP that are more suitable for imaging with an argon laser (Heim et al. 1995). Unfortunately, these new GFPs that are both brighter, and mature more rapidly, than the original (Cubitt et al. 1995), are no longer visible at whole-plant level under ultraviolet illumination. The choice of GFP for use with viral vectors will be dictated to a large extent by the virus under study. For example, the capacity of the virus to replicate in host cells will affect the amount of GFP produced per cell. Similarly, production of GFP as a fusion protein, for example to a viral MP that is highly targeted, may require maximal fluorescence from a small region of the cell.

2.2 Imaging GFP

Following infection of leaves with virus vectors expressing GFP, fluorescent symptoms become visible on the leaf surface. For wild-type PVX (Baulcombe et al. 1995) and TMV (Padgett et al. 1996; Oparka et al. 1997), this typically takes about 2–4 days. To image GFP in whole plants we have used long-wavelength illumination (365 nm) supplied by a Blak-Ray hand-held lamp (Ultraviolet Products Limited, Cambridge, UK). For photography, the plants were held on the stage of a photographic enlarger stand with a 35-mm camera mounted above. To remove the autofluorescence associated with chlorophyll, a green filter (Hoya, Japan) was used. It is worth noting that if wavelength-shifted excitation mutations of GFP are used, such infection sites will not be detected under UV light, and an appropriate light source is required for imaging GFP at the appropriate wavelength. In some cases, leaves can be removed and photographed successfully by holding them flat under a piece of glass on top of black felt. Such leaves can also be digitally reconstructed under the confocal laser scanning microscope (CLSM) using low-magnification lenses (see below). For photography, standard daylight films were used (e.g. Kodak Ektachrome 200) with exposures typically between 3 and 30 s.

2.3 Epifluorescence Microscopy

Any fluorescence microscope equipped with appropriate filter blocks is suitable for GFP imaging. For Nikon microscopes a UV-2A (dichroic mirror 400 nm, excitation filter 330–380 nm and barrier filter 420 nm), a B-2A (dichroic mirror 510 nm, excitation filter 450-490 nm and barrier filter 520 nm) and a FITC/TRITC combined filter (dichroic mirror 402–540 nm, excitation filter 489 / 562 nm and barrier filter 522.5 / 604 nm) have all been used successfully for wild-type GFP. Some excitation wavelengths (particularly blue) cause damage to plant cells and should be avoided, or exposure times minimized. Similarly, prolonged exposure to illumination should be avoided and controls should be performed routinely to assess the damage induced by extensive imaging (see Chap. 5). Many manufacturers now supply a range of filter sets designed for detection of the different wavelength mutations of GFP, for instance the HighQ collection of filters (Chroma Technology Corporation, Brattleboro, VT, USA) recently designed for both wild-type and the 65S, red-shifted GFP wavelengths.

2.4 Confocal Laser Scanning Microscopy

GFP can be imaged effectively under the CLSM. The excitation wavelength is typically 488 nm produced from an argon or krypton/argon laser. Although CLSM is generally regarded as most effective at high magnification (using high-numerical aperture lenses), we have successfully imaged GFP in whole tissues with lenses as low as x1 (Roberts et al. 1997). In such cases, the confocality is reduced although the out-of-focus flare is still minimized, allowing detailed images of the tissue to be obtained over large areas. In the case of leaves or roots systems, the entire organ can be mapped using a x1 lens. The individual images are stored on the CLSM software and later montaged digitally using Photoshop (Adobe, Mountain View, CA) software to reconstruct the entire leaf (see below).

2.5 Double Labelling

GFP, because of its inherent stability (Prasher 1995), lends itself to double-imaging techniques using fluorochromes of the appropriate wavelength. Texas Red (TR; Molecular Probes, Eugene, Oregon; absorption maximum 585 nm, emission maximum 603 nm), either as a free dye or as a dextran is suitable for dual imaging with GFP. Under the CLSM, the GFP (excitation at 488 nm) can be imaged simultaneously with TR (excitation at 568 nm) using a krypton/argon laser, without any spillover of signal between the two wavelengths. TR dextran can be used effectively when injected into GFP-expressing cells to examine the gating capacity of plasmodesmata in virus-infected cells (Oparka et al. 1997; see below). For antibody labelling, the red dye Cy 3 (Sigma) can be coupled to a secondary antibody, for example for callose labelling of plasmodesmata (see also below). Unfortunately, callose stains such as aniline blue do not have an excitation peak near 488 nm, making their detection impossible under the CLSM unless an additional UV laser is attached to the microscope. However, with epifluorescence microscopy, aniline blue can be used on GFP-expressing tissues. We have found that the normally employed decolorized aniline blue (0.1% in K_3PO_4) has a high pH that disrupts or interferes with GFP fluorescence, although good results were obtained with a dye solution at 0.01% in 0.07 M Sörensens phosphate buffer, pH 7.5). This maintained GFP fluorescence and also produced good callose staining. For staining the endoplasmic reticulum (ER) and mitochondria in GFP-expressing cells, the dye hexyl-rhodamine B (Grabski et al. 1993; Boevink et al. 1996) has proved effective.

TR dextran (3 kDa) is a useful xylem tracer and can be imaged in GFP-expressing leaf tissues, allowing the vein classes involved in virus exit from the phloem to be studied in detail (Roberts et al. 1997; see below). The free dye may also be used but with time this tends to leak from the xylem vessels causing the vein network to become poorly defined. Dextrans larger than 3 kDa are poorly transported by the transpiration stream. The dye is applied to the cut petiole in an Eppendorf tube and the leaf left to transpire under a lamp for approximately 30 min. The labelled leaf is then examined directly under the CLSM at 568 nm. Other double-labelling techniques are clearly possible although the most effective will be those that discriminate completely between GFP and the fluorochrome employed.

2.6 Sectioning GFP-Containing Plant Material

Because the optical sectioning capacity of a CLSM works only to a limited depth, it is sometimes necessary to section the material further to determine the cellular or subcellular location of the GFP. We have used a Vibroslice (Campden Instruments Ltd. UK) to section leaf material. In the case of tissues containing free GFP, the protein may leak from cut cells causing artefacts, although imaging below this cut layer is possible using the CLSM. When GFP-protein fusions are employed, leakage is less of a problem and high-resolution images can be obtained from cells relatively deep in the tissue. The tissue is first surrounded by 3% agar solution to hold it in place during sectioning. The sections are trimmed horizontally from the block and removed from the knife with a soft paintbrush and floated on water to remove the surrounding agar prior to mounting in water or buffer on a microscope slide. If necessary, the sections can be counterstained or treated with antibodies (see above) prior to microscopy.

3 Tracing Symplasmic Pathways with GFP-Tagged Viruses

Based on the above methods, the subsequent sections describe case histories in which GFP tagging has provided new information on plasmodesmata and their physiological and developmental regulation. Many small solutes move between plant cells via plasmodesmata (see reviews by Lucas et al. 1993; Fisher and Oparka 1996). However, for most solutes, an apoplasmic pathway is also possible, and several solutes are known to exchange between symplasm and apoplasm by specific membrane carriers (Fisher and Oparka 1996). A case in point is the sucrose molecule, which exchanges readily between apoplasm and symplasm depending on local source/sink relations (Patrick and Offler 1996). The use of conventional isotope-labelling techniques has helped to follow the macroscopic movement of radiolabelled solutes but does not permit sufficient resolution to identify sites of membrane transfer (Fisher and Oparka 1996). The use of fluorescent solutes has provided an alternative approach for following symplasmic transport (see Chap. 5), although in this case also caution must be exerted as physicochemical parameters will determine, to a large extent, the transport properties of different dye molecules (Wright et al. 1996).

When tracing symplasmic pathways, GFP-tagged plant viruses offer a unique opportunity in the study of intercellular communication patterns as viruses are known to exploit exclusively plasmodesmata for movement between cells (see Carrington et al. 1996), thus providing a demonstration of the capacity for macromolecular exchange between cells. GFP tagging is not only useful in studies of short- and long-distance virus movement but is beginning to identify specific plasmodesmal locations (domains) that some viruses are able to target while others are not. For studies of this type, viruses that express GFP as a free cytosolic protein (e.g. Baulcombe et al. 1995), or as a fusion to the viral coat protein (Santa Cruz et al. 1996; Roberts et al. 1997), are particularly useful for tracing the cellular pathway of viral movement, as these can be imaged at both the tissue and single-cell level. We have found, in the case of PVX, that GFP expressed as a coat protein fusion is brighter and more easily identified in cells than the free GFP (Roberts et al. 1997).

4 Local Virus Movement

Following manual inoculation of a leaf surface with either of the above viruses, there is a delay of approximately 2 days during which viral replication and GFP production occur. From the initial infection sites the virus moves in a radial pattern and GFP is detected in adjoining epidermal cells and underlying mesophyll cells (Fig. 1). Confirmation that viral infection sites arise from single-cell foci has been obtained by deleting individual, putative viral movement genes in the presence of the *gfp* gene. For example, deletion of the coat protein (*cp*) gene of PVX resulted in restriction of the virus (and GFP production) to single epidermal cells (Baulcombe et al. 1995; Oparka et al. 1996b).

Fig. 1 A–J. Imaging GFP in single cells and whole plants. **A** Infection site of tobacco mosaic virus (TMV) expressing a MP-GFP fusion. Note the bright halo of GFP-expressing cells at the leading edge of the infection site (see Oparka et al. 1997). *Bar* 500 μm. **B** Single-cell infection sites on the leaf epidermis. The leaf was inoculated with a GFP-tagged potato virus X (PVX) construct which lacked the coat protein (CP) gene. Note that the replicating virus is unable to escape from these single, inoculated cells. *Bar* 50 μm. **C, D** Colocalization of viral MP (expressed as a MP-GFP fusion; C) with callose localised with Cy 3-labelled antibody (**D**). Note that the MP is localized between the paired callose platelets. *Bar* 3 μm. **E, F** Injection of 10-kDa Texas Red (TR) dextran into the leading edge of a TMV infection site expressing a MP-GFP fusion (see **A**). **E** shows the fluorescent cells at the edge of the infection (488 nm) with the position of three injection sites indicated (*) while **F** shows the same injections of TR dextran at 568 nm. Note that the dextran moves to adjacent cells only in those injections within the leading edge of infection (see Oparka et al. 1997). *Bar* 500 μm. **G–J** Whole-plant imaging of the system-

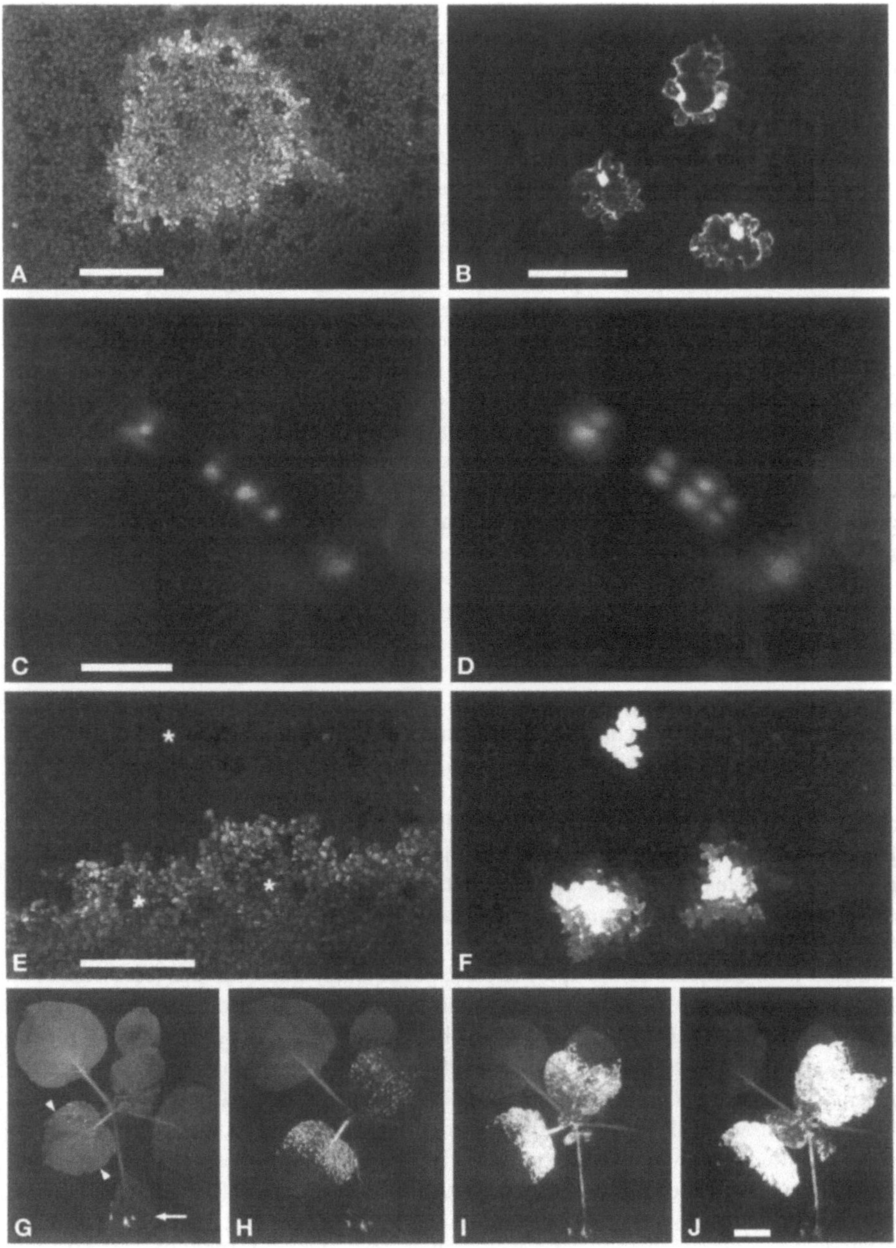

ic movement of PVX expressing GFP as a free protein. The inoculated leaf is shown in **G** [8 days post-inoculation (dpi); *arrow*] while the *darts* indicate the first fluorescent foci appearing in a sink leaf. In **H** (9 dpi), these systemic fluorescent flecks are clearly visible on the leaf lamina. In **I** (10 dpi) and **J** (12 dpi) the virus has spread from the unloading sites to invade the mesophyll by cell-to-cell movement. Note the distinct barrier to virus movement in leaves undergoing the sink/source transition (see Roberts et al. 1997). *Bar* 2.5 cm. Reproduced with permission from The American Society of Plant Physiologists

Such single cells are of interest in that they represent a useful experimental system with which to unravel the viral genes that interact with plasmodesmata to permit movement of viral RNA between cells. Without the GFP tag, such cells are impossible to locate on the leaf surface and the viral phenotype cannot be studied. By scanning the leaf surface under a x4 lens, the location of such cells can be pinpointed (Fig. 1B). Recently, we microinjected fluorescent TR dextran (10 kDa) into such single-cell foci of PVX lacking the *cp* gene and showed that the plasmodesmata of these single cells were gated to allow the passage of dextran to neighbouring cells, but not beyond (Santa Cruz et al. unpub. data). In marked contrast, microinjection of dextran into single-cell foci of PVX lacking the triple gene block (TGB), a group of genes known to be essential for virus movement (Beck et al. 1991; Morozov et al. 1997), showed that the plasmodesmata were not gated, and the dextran remained confined to the injected cell. Thus, different viral genes appear to be essential for movement, but through quite different mechanisms. In the case of the TGB of PVX, at least one or more of these genes appears to play a role in plasmodesmal gating (see also Angell et al. 1996). Interestingly, while the *cp* gene of PVX is essential for movement, and is targeted to plasmodesmata in wild-type infections (Oparka et al. 1996b), it does not perform the plasmodesmal gating function of this virus and may be involved instead in trafficking the viral RNA to the plasmodesmal pore (Oparka et al. 1996b).

5 Tagging Viral MPs with GFP

In addition to the above approach, in which viral genes were deleted in the presence of GFP, the role of viral proteins in targeting and gating plasmodesmata can be explored through the use of GFP-fusion proteins. Recently, it has been shown that fusion of the TMV MP to GFP, under control of the MP promoter, resulted in fluorescent halo-shaped infection sites that expanded radially on the leaf lamina (Heinlein et al. 1995; Padgett et al. 1996; Oparka et al. 1997). Within these infection sites the MP-GFP fusion was under a cycle of synthesis and degradation, the fluorescent halo appearing brightest where MP was being expressed maximally (Padgett et al. 1996). Within the infection site, the MP-GFP fusion was seen to associate with punctate regions on the wall and also with elements of the cytoskeleton, particularly microtubules (Heinlein et al. 1995; Padgett et al. 1996). A similar association with the cytoskeleton has been seen in isolated protoplasts (McLean et al. 1995). This has led to the suggestion that a normal feature of the TMV MP may be to hitch a ride on the cytoskeleton in order to reach the plasmodesmata (see Gilbertson and Lucas 1996; Carrington et al. 1996). An important future use of MP-GFP fusions will be to identify other host factors that may be involved in the targeting and trafficking of viral MPs to plasmodesmata. Double-labelling of the GFP-labelled plasmodesmata with antibody to callose (Cy-3-labelled secondary antibody) demonstrated that the MP was localized in the centre of the plasmodesmal pore, between the paired callose collars (Oparka et al. 1997; Fig. 1C, D), confirming previous immunoelectron microscope studies that the MP of TMV localizes to the middle of plasmodesmata in cells infected with the wild-type virus (Tomenius et al. 1987).

The strong GFP signal from cells whose plasmodesmata were targeted with MP-GFP allowed injection of these cells with TR dextran to determine if the targeting and gating functions of the TMV MP were correlated spatially in an expanding infection site. TR dextran was microinjected into epidermal cells at the leading edge of expanding TMV infection sites that were expressing a MP-GFP fusion (Oparka et al. 1997). It was found that although all plasmodesmata in the infection site were targeted with MP-GFP, only those cells at the leading edge of the infection (i.e. within the fluorescent halo; see Fig. 1E, F) were gated to allow the passage of dextran. Cells towards the darker centre of the infection site showed no increase in their plasmodesmal size exclusion limit (SEL). These results suggest that in wt infection sites plasmodesmal gating is under temporal control and restricted to the outermost cells in the infection site, i.e. where MP is being maximally expressed. The clearly defined edge of the viral infection site allowed microincisions to be made through cells in front of the leading edge. These incisions halted the progress of the expanding infection indicating that replicating virus was not present in non-fluorescent cells beyond the fluorescent leading edge of the infection. Although there is a delay between the translation of GFP and the appearance of the mature protein (see Cubitt et al. 1995; Heim et al. 1995), the above results demonstrate that the presence of GFP is an accurate indicator of the presence of virus within infection sites. The development of a range of GFPs with different excitation and emission wavelengths, and also GFP molecules that are faster to mature (Heim et al. 1995), will increase the utility of this protein in real-time tracer studies.

6 Tracing Virus Movement into the Phloem

After initial movement in the epidermis and leaf mesophyll, virus encroaches the minor leaf veins by cell-to-cell movement. Within the minor veins, the virus (if it possesses the appropriate genes) gains access to the minor-vein phloem and enters the systemic transport pathway (see Chap. 17). Movement of virus within minor veins has been difficult to study as the individual cells within the minor vein are small and not easily identified in fresh tissue (Carrington et al. 1996). Although bundle-sheath cells have been microinjected to examine their gating status (Ding et al. 1992), other vascular elements within the minor vein have not. Thus, the minor-vein phloem has remained a black box regarding virus movement. However, immunolocalisation approaches have suggested that important barriers to virus movement occur at the plasmodesmal interface between bundle-sheath and phloem parenchyma (Goodrick et al. 1991; Ding et al. 1992; Wintermantel et al. 1997), and possibly around the sieve element-companion cell (SE-CC) complex (Ding et al. 1996).

MP-GFP fusions appear to hold promise in studying cell-to-cell movement within minor veins. Recently, Blackman et al. (1998) studied the location of the 3a MP of CMV using both immunolocalization of the wt MP, and also a MP-GFP fusion expressed from a PVX vector. In both cases, the 3a-MP was targeted strongly to plasmodesmata of epidermal cells and mesophyll cells. Within minor veins, the 3a-GFP was heavily localized within sieve elements and formed elaborate networks in the SE parietal layer. In addition, the 3a-GFP was concentrated around the specialized plasmodesmata that connect the SE and CC, making them highly fluorescent under the CLSM. This localization was also seen with the wt virus infection using immunogold labelling, confirming that the MP-GFP fusion accurately reflected the behaviour of the wt MP. The above results, using a MP-GFP fusion, are the first to demonstrate that a viral MP is capable of trafficking into minor-vein SEs, and implicates a role for this protein in potentiating the long-distance spread of CMV. This appears to be unlike the situation reported for the TMV MP which, in transgenic plants, was immunolocalized to plasmodesmata in mesophyll cells, and also between bundle-sheath and phloem-parenchyma elements, but not beyond (Ding et al. 1992).

In the future, GFP tagging of a range of viral MPs will be important in determining whether these proteins play a role in viral entry into the phloem, an area of research that has to date received only superficial attention (Carrington et al. 1996).

7 Tracing the Phloem-Unloading Pathways of GFP-Tagged Viruses In Leaves

Once a virus, or an infectious component of the virus, has entered the sieve elements, it is translocated over long distances in the phloem. The final destination of the virus appears to be determined predominantly by source/sink relations and the symplasmic pathways available for cell-to-cell virus movement in sink tissues (Gilbertson and Lucas, 1996; Carrington et al. 1996; Roberts et al. 1997). To date, the study of long-distance virus movement has been hindered by a lack of suitable methods for detecting virus egress from the phloem at an early stage in the systemic infection process. The detection of visible viral symptoms (e.g. chlorosis, necrosis) indicates the presence of viral infection, but these may develop some considerable time after the initial unloading of virus from the phloem. Use of β-glucuronidase as a marker protein, expressed from vi-

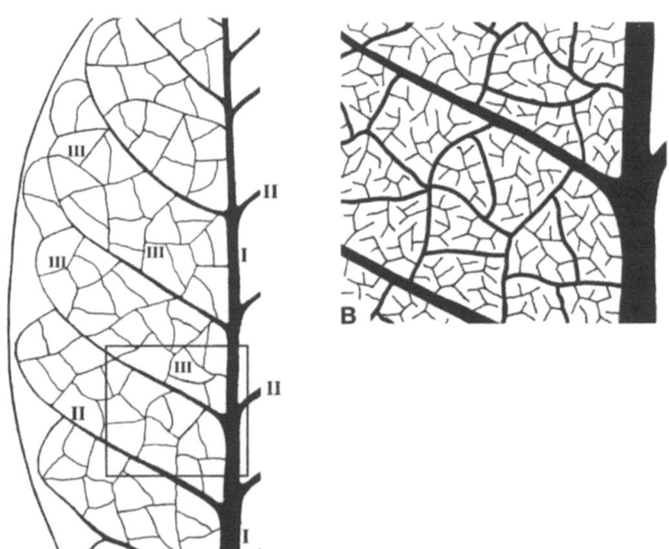

vals to class II veins. The class III veinal network, a branched veinal system that forms discrete islands on the lamina, lies between the class II veins. **B** Detail of the boxed region in **A** showing the position of the minor veins (classes IV and V) within the islands of the class III veinal network (see Roberts et al. 1997). Reproduced with permission of The American Society of Plant Physiologists ral genomes, has overcome some of these problems (Dolja et al. 1992), but this method is destructive and prone to artefacts due to the diffusion of reaction products.

The use of GFP-tagged viruses has overcome many of these problems and allowed the progress of virus infection to be monitored non-destructively. Roberts et al. (1997) recently compared the pattern of unloading of PVX expressing the GFP gene (both as a free protein and as a CP fusion) with phloem-translocated carboxyfluorescein (CF). This dye, when loaded into the phloem in the diacetate form, is cleaved by esterase activity to release the CF moiety which is both membrane- impermeant and phloem-mobile (Grignon et al. 1989). The vein architecture of the *Nicotiana benthamiana* leaf is shown in Fig. 2. In young sink leaves, CF was imported by the phloem and unloaded predominantly from the class III veinal network (Fig. 3A). The minor veins (classes IV and V) played no role in the unloading process and remained unlabelled with CF, even after prolonged unloading from the class III veins. However, the xylem of the minor veins was shown to be functional by introducing TR dextran (3 kDa) into the transpiration stream of the same leaves (see methods above). The region of the sink/source transition is shown in greater detail in Fig. 3B and shows the unloading of CF from class III veins near the transition front. These findings, using a fluorescent solute, con-

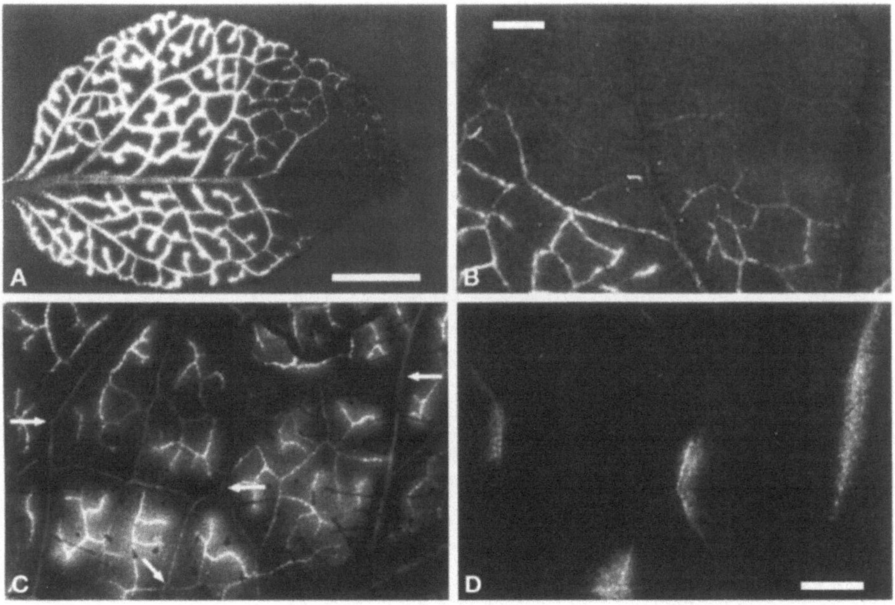

Fig. 3 A–D. Confocal imaging of whole leaves and roots. **A** Whole leaf of *Nicotiana benthamiana* which had imported carboxyfluorescein (CF) via the phloem. The dye is unloading from the class III veinal network. The image was reconstructed from several montages of overlapping regions of the leaf surface (see Roberts et al. 1997). *Bar* 0.5 cm. Reproduced with permission of The American Society of Plant Physiologists. **B** Region of the sink/source transition. *Bar* 1.25 mm. **C, D** Double-labelling of an importing leaf which had been allowed to transpire TR dextran (**C**) and which had also been unloading GFP-tagged PVX (**D**). Note that the minor leaf veins are functional in xylem transport but that only the class III veins (*arrowed* in **C**) play a role in the phloem unloading of virus (see Roberts et al. 1997). *Bar* 500 μm

firm previous studies using autoradiographic procedures that the phloem unloading in *Nicotiana* species occurs from major vein classes, and that the minor veins are not functional in phloem transport during the import phase (Turgeon 1989). When PVX containing the *gfp* gene was inoculated onto source leaves, the first traces of virus appeared in sink leaves approximately 6 days later (Fig. 1G–J). Examination of whole leaves under a UV lamp showed that the virus exited the phloem in discrete flecks (Fig. 1H). By labelling the xylem with TR dextran, it was shown that the virus exit sites were predominantly above class III veins (Fig. 3C, D), and, to a lesser extent, from class I and II veins. However, as in the case of solute unloading, virus did not exit from minor veins.

These findings have important implications for the plasmodesmal pathways utilized by solutes and viruses in sink leaves. From the class III exit sites, virus was seen to spread into the mesophyll and eventually into the areas containing minor veins. Examination of infected sink leaves in the electron microscope showed that the minor veins contained virus particles. Thus, the pathway of solute and virus movement into minor veins, regardless of whether the leaf is a source or a sink, is from the mesophyll.

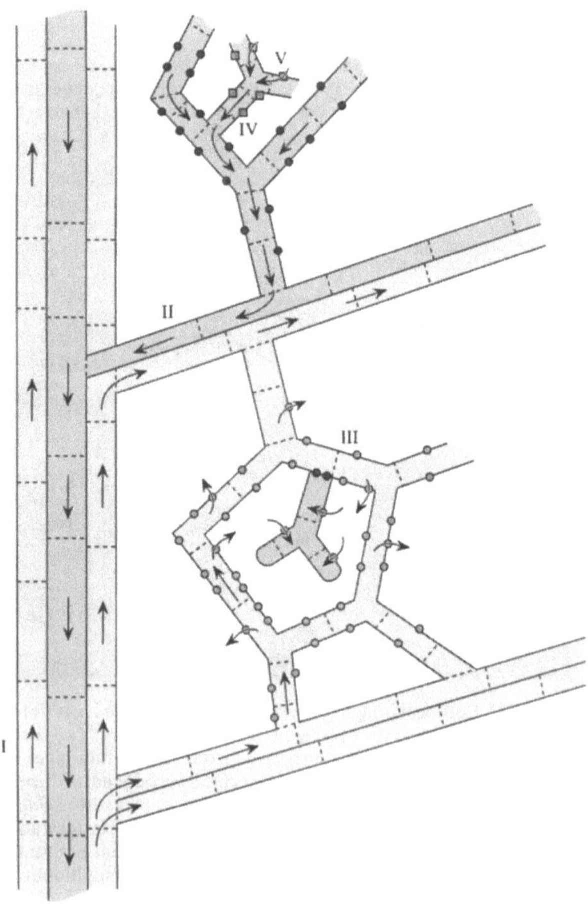

Fig. 4. Schematic model of the sink/source transition in a developing *N. benthamiana* leaf (*upward arrows* importing phloem; *downward arrows* exporting phloem). Symplasmic unloading of solutes and virus occurs from the class III vein network (*open circles*). In the sink portion of the leaf, the minor veins are immature (*dark shading*) and receive assimilate (and virus) directly from the mesophyll by cell-to-cell movement. In the source portion of the leaf, the minor veins are now mature. Switching on of sucrose transporters (*squares*) in the minor veins signals the onset of apoplasmic sucrose loading in class IV and V veins. In the source region, the class III veins function in export rather than import, which is achieved by downregulation of the plasmodesmata (*closed circles*) that connected them to the mesophyll during the import phase. Bidirectional phloem transport (in different sieve tubes) occurs in the class I vein, but it may also occur in class II and III veins as the transition front passes basipetally across different vein classes (see Roberts et al. 1997). Reproduced with permission from The American Society of Plant Physiologists

The exclusive exit of solutes and viruses from major veins raises some interesting questions concerning the regulation of plasmodesmata that connect the phloem of this veinal system with the surrounding mesophyll. If these plasmodesmata remained functional during the source (loading) stage then they would allow the leakage of loaded solutes back out into the mesophyll. Tobacco is an apoplasmic loader (Turgeon 1984) and, in other apoplasmic loading species such as *Plantago* and *Arabidopsis*, the bulk of loading occurs through the minor veins which have matured during the sink-source transition (Stadler et al. 1995; Truernit and Sauer 1995). Roberts et al. (1997) suggested that the class III vein plasmodesmata become down-regulated, or non-functional, during the leaf export phase allowing the class III phloem to function as an isolated conduit for assimilates passing into it from the minor veins. This model is shown schematically in Fig. 4. The extent to which this regulation of plasmodesmal function is developmental or physiological remains to be determined. Roberts et al. (1997) suggested that the increased turgor pressure in the class III vein phloem as a result of active loading in the nearby minor veins may have been sufficient to generate a turgor pressure differential large enough to close off the plasmodesmata between the SEs and surrounding cells (see also Oparka and Prior 1992). However, previous studies of unloading in tobacco species suggest that plasmodesmata at this interface may disappear or become greatly reduced in number (Ding et al. 1988). By whatever means, the plasmodesmata surrounding the class III phloem must switch rapidly between a functional and non-functional role. The sink-source transition may prove to be a useful model system for exploring the developmental regulation of plasmodesmata during localized changes in solute distribution.

8 Concluding Remarks

The use of GFP-tagged viruses has proved to be a beneficial tool in the study of symplasmic pathways in plants. Once in the phloem it appears that source-sink relations dictate the final destination of virus. Studies to date have indicated that the plasmodesmata utilized for solute unloading from the phloem may also be accessed by viruses when they invade sink tissues. Thus, it appears that GFP-tagged viruses will be important tools in unravelling the symplasmic pathways leading to and from the phloem (see also Chaps. 16 and 17). In the future, it will be interesting to compare the symplasmic movement of GFP-tagged viruses in species that display an apoplasmic versus symplasmic vein loading configuration and also to determine the pathways and mechanisms by which viruses enter the seeds of those plants (see Wang and Maule 1994) where a symplasmic discontinuity is thought to occur between maternal and filial generations (Patrick and Offler 1996). Unlike solutes, viruses are unable to leave the symplasm once they have been unloaded and will therefore provide an invaluable marker for the capacity for intercellular exchange between cells.

References

Aesbacher RA, Schielfelbein JW, Benfey PN (1994) The genetic and molecular basis of root development. Annu Rev Plant Physiol Plant Mol Biol 45:25–45

Angell SM, Davies C, Baulcombe DC (1996) Cell-to-cell movement of potato virus X is associated with a change in the size-exclusion limit of plasmodesmata in trichome cells of *Nicotiana clevelandii*. Virology 216:197–201

Baulcombe DC, Chapman SN, Santa Cruz S (1995) Jellyfish green fluorescent protein as a reporter for virus infections. Plant J 7:1045–1053

Beck DL, Guilford PJ, Voot DM, Andersen MJ, Forster RL (1991) Triple gene block proteins of white clover mosaic potexvirus are required for transport. Virology 183:695–702

Blackman L, Boevink P, Santa Cruz S, Palukaitis P, Oparka KJ (1998) The movement protein of cucumber mosaic virus traffics into sieve elements in minor veins of *Nicotiana clevelandii*. Plant Cell 10:525–537

Boevink P, Santa Cruz S, Hawes C, Harris N, Oparka KJ (1996) Virus-mediated delivery of the green fluorescent protein to the endoplasmic reticulum of plant cells. Plant J 10:935–941

Brown DJF (1997) Transmission of viruses by nematodes. In: Santos MSN de A, Abrantes IM de O, Brown DJF, Lemos RM (eds) An introduction to virus-vector nematodes and their associated viruses. Instituto do Ambiente e Vida, Coimbra, Portugal, pp 355–380

Canto T, Prior DAM, Hellwald K-H, Oparka KJ, Palukaitis P (1997) Characterisation of cucumber mosaic virus IV. Movement protein and coat protein are both essential for cell-to-cell movement of CMV. Virology 237:237–248

Carrington JC, Kasschau KD, Mahajan SK, Schaad MC (1996) Cell-to-cell and long-distance transport of viruses in plants. Plant Cell 8:1669–1681

Chalfie M, Tu Y, Euskirchen G, Ward WW, Prasher DC (1994) Green fluorescent protein as a marker for gene expression. Science 263:802–805

Chen J, Watanabe Y, Sako N, Ohshima K, Okada Y (1997) Complete nucleotide sequence of infectious in vitro transcripts from a full-length cDNA clone of a rakkyo strain of tobacco mosaic virus. Arch Virol 141:885–900

Cubitt AB, Heim R, Adams SR, Boyd AE, Gross LA, Tsien RY (1995) Understanding, improving and using green fluorescent proteins. Trends Biochem Sci 20:448–455

Ding B, Parthasarathy MV, Niklas K, Turgeon R (1988) A morphometric analysis of the phloem-unloading pathway in developing tobacco leaves. Planta 176:307–318

Ding B, Haudenshield JS, Hull RJ, Wolf S, Beachy RN, Lucas WJ (1992) Secondary plasmodesmata are specific sites of localization of the tobacco mosaic virus movement protein in transgenic tobacco plants. Plant Cell 4:915–928

Ding X-S, Shintaku MH, Crater SA, Nelson RS (1996) Invasion of minor veins of tobacco leaves inoculated with tobacco mosaic virus mutants defective in phloem dependent movement. Proc Natl Acad Sci USA 93:11155–11160

Dolja, VV, McBride HJ, Carrington JC (1992) Tagging of plant potyvirus replication and movement by insertion of β-glucuronidase into the viral polyprotein. Proc Natl Acad Sci USA 89:10208–10212

Duckett CM, Oparka KJ, Prior DAM, Dolan L, Roberts K (1994) Dye coupling in the root epidermis of *Arabidopsis* is progressively reduced during development. Development 120:3247–3255

Fisher DB, Oparka KJ (1996) Post-phloem transport: principles and problems. J Exp Bot 47:1141–1154

Gilbertson RL, Lucas WJ (1996) How do viruses traffic on the vascular highway? Trends Plant Sci 1: 260–268

Goodrick BJ, Kuhn CW, Hussey RS (1991) Restricted systemic movement of cowpea chlorotic mottle virus in soybean with nonnecrotic resistance. Phytopathology 81:1426–1431

Grabski S, de Feijter AW, Schindler M (1993) Endoplasmic reticulum forms a dynamic continuum for lipid diffusion between contiguous soybean root cells. Plant Cell 5:25–38

Grignon N, Touraine B, Durand M (1989) 6(5)-carboxyfluorescein as a tracer of phloem sap translocation. Am J Bot 76:871–877

Haseloff J, Amos B (1995) GFP in plants. Trends Genet 11:328–329

Heim R, Cubitt AB, Tsien RY (1995) Improved green fluorescence. Nature 373:663–664

Heinlein M, Epel BL, Padgett HS, Beachy RN (1995) Interaction of tobamovirus movement proteins with the plant cytoskeleton. Science 270:1983–1985

Lucas WJ, Ding B, van der Schoot C (1993) Plasmodesmata and the supracellular nature of plants. New Phytol 125:435–476

Matthews REF (1991) Plant Virology, 3rd edn. Academic Press, Harcourt Brace Jovanovich, New York

McLean BG, Zupan J, Zambryski PC (1995) Tobacco mosaic virus movement protein associates with the cytoskeleton in tobacco cells. Plant Cell 7:2101–2114

Morozov SY, Fedorkin ON, Jüttner G, Schiemann J, Baulcombe DC, Atabekov JG (1997) Complementation of a potato virus X mutant mediated by bombardment of plant tissues with cloned viral movement protein genes. J Gen Virol 78:2077–2083

Oparka KJ, Prior DAM (1992) Direct evidence for pressure-generated closure of plasmodesmata. Plant J 2:741–750

Oparka KJ, Duckett CM, Prior DAM, Fisher DB (1994) Real-time imaging of phloem unloading in the root tip of Arabidopsis. Plant J 6:759–766

Oparka KJ, Roberts AG, Prior DAM, Chapman S, Baulcombe D, Santa Cruz S (1995) Imaging the green fluorescent protein in plants – viruses carry the torch. Protoplasma 189:133–141

Oparka KJ, Boevink P, Santa Cruz S (1996a) Studying the movement of plant viruses using green fluorescent protein. Trends Plant Sci 1:412–418

Oparka KJ, Roberts AG, Roberts IM, Prior DAM, Santa Cruz S (1996b) Viral coat protein is targeted to, but does not gate, plasmodesmata during cell-to-cell movement of potato virus X. Plant J 10: 805–813

Oparka KJ, Prior DAM, Santa Cruz S, Padgett HS, Beachy RN (1997) Gating of epidermal plasmodesmata is restricted to the leading edge of expanding infection sites of tobacco mosaic virus. Plant J 12:781–789

Padgett HS, Epel BL, Kahn TW, Heinlein M, Watanabe Y, Beachy RN (1996) Distribution of tobamovirus movement protein in infected cells, implications for cell-to-cell spread of infection. Plant J 10 :1079–1088

Patrick JW, Offler CE (1996) Post-sieve element transport of photoassimilates in sink regions. J Exp Bot 47:1165–1177

Prasher DC (1995) Using GFP to see the light. Trends Genet 11:320–323

Rao ALN (1997) Molecular studies on bromovirus capsid protein. III. Analysis of cell-to-cell movement competence of coat protein defective variants of cowpea chlorotic mottle virus. Virology 232 :385–395

Roberts AG, Santa Cruz S, Roberts IM, Prior DAM, Turgeon R, Oparka KJ (1997) Phloem unloading in sink leaves of *Nicotiana benthamiana*: comparison of a fluorescent solute with a fluorescent virus. Plant Cell 9:1381–1397

Samuel G (1934) The movement of tobacco mosaic virus within the plant. Ann Appl Biol 21:90–111

Santa Cruz S, Chapman S, Roberts AG, Roberts IM, Prior DAM, Oparka KJ (1996) Assembly and movement of a plant virus carrying a green fluorescent protein overcoat. Proc Natl Acad Sci USA 93:6286–6290

Schaad MC, Jensen PE, Carrington JC (1997) Formation of plant RNA virus replication complexes on membranes: role of an endoplasmic reticulum-targeted viral protein. EMBO J 16:4049–4059

Schulz A (1994) Phloem transport and differential unloading in pea seedlings after source and sink manipulations. Planta 192:239–248

Stadler R, Brandner J, Schulz A, Gahrtz M, Sauer N (1995) Phloem loading by the PmSUC2 sucrose carrier from *Plantago major* occurs into companion cells. Plant Cell 7:1545–1554

Tomenius K, Clapham D, Meshi T (1987) Localization by immunogold cytochemistry of the virus-coded 30K protein in plasmodesmata of leaves infected with tobacco mosaic virus. Virology 160: 363–371

Truernit E, Sauer N (1995) The promoter of the *Arabidopsis thaliana* SUC2 sucrose-H^+ symporter gene directs expression of β-glucuronidase to the phloem: evidence for phloem loading and unloading by SUC2. Planta 196:564–570

Turgeon R (1984) Termination of nutrient import and development of vein loading capacity in albino tobacco leaves. Plant Physiol 76:45–48

Turgeon R (1989) The sink-source transition in leaves. Annu Rev Plant Physiol Plant Mol Biol 40: 119–138

Wang D, Maule AJ (1994) A model for seed transmission of a plant virus: genetic and structural analyses of pea embryo invasion by pea seed-borne mosaic virus. Plant Cell 6:777–787

Wellink JV, Bertens P, van Lent J, Goldbach R, van Kammen A (1997) Studies on the movement of cowpea mosaic virus using the jellyfish green fluorescent protein. Abstracts molecular mechanisms in the replicative cycle of viruses in plants. EMBO workshop. Las Navas, del Marques, Spain, p 52

Wintermantel WM, Banerjee N, Oliver JC, Paolillo DJ, Zaitlin M (1997) Cucumber mosaic virus is restricted from entering minor veins in transgenic tobacco exhibiting replicase-mediated resistance. Virology 231:248–257

Wright KM, Oparka KJ (1997) Metabolic inhibitors induce symplastic movement of solutes from the transport phloem of *Arabidopsis* roots. J Exp Bot 48:1807–1814

Wright KM, Horobin R, Oparka KJ (1996) Phloem mobility of fluorescent xenobiotics in *Arabidopsis* in relation to their physicochemical properties. J Exp Bot 47:1779–1787

Chapter 7

Evolution of Plasmodesmata

M. E. Cook[1] · L. E. Graham[2]

[1] Department of Biological Sciences, Illinois State University, Campus Box 4120, Normal, Illinois 61790-4120, USA
[2] Department of Botany, University of Wisconsin, Birge Hall, 430 Lincoln Drive, Madison, Wisconsin 53706, USA

1 Introduction

In the past 20 years, improved methods of preservation, microscopy, and data analysis have lead to increased understanding of the ultrastructure of the plasmodesmata of flowering plants. Features such as proteinaceous particles, spokes, and cell wall specializations have been described. In addition, we have begun to understand the molecular components of plasmodesmata. Dynamic activity of plasmodesmata is thought to be associated with the molecules myosin, actin, and possibly centrin. Because of these advances (reviewed elsewhere in this Volume), we can now more fully appreciate the complexity of higher plant intercellular connections. Moreover, this wealth of new information provides numerous characters that can potentially be used to trace the evolution of complex plasmodesmata. In this chapter, we will review briefly the types of intercellular connections found in diverse multicellular organisms, including the green algae most closely related to plants, and the seedless plants. We will then consider whether comparisons between the intercellular connections of these diverse taxa and those of higher plants may lead to an understanding of the evolution of complex plant plasmodesmata.

2 Diversity of Intercellular Connections

Plant plasmodesmata are plasma membrane-lined channels that provide cytoplasmic continuity between the cells of an integrated organism (Niklas 1989; Kaplan and Hagemann 1991; Lucas and Wolf 1993). These intercellular connections are thought to allow simple diffusion and complex molecular trafficking (Lucas et al. 1993, 1995; Epel 1994; Lucas 1995; Mezitt and Lucas 1996; Ding 1997; Ghoshroy et al. 1997). Similar structures have been reported in green and brown algae and in some fungi. Different intercellular connections have been reported in other multicellular organisms, including animals, other fungi, red algae, and filamentous cyanobacteria (Fig. 1). Intercellular cytoplasmic continuity is apparently not maintained in red algae and cyanobacteria, however, so in these taxa it is uncertain whether the connecting structures serve as channels for intercellular transport and communication. In our brief review of the diversity of intercellular connections, we will concentrate on studies that have been completed since Marchant's (1976) review of this topic. We will focus especially on lower plants and charophycean algal relatives of plants, since intercellular connections in these organisms are most likely to be relevant to the evolution of higher plant plasmodesmata.

2.1 Cyanobacteria

Filamentous cyanobacteria have intercellular connections that have been termed microplasmodesmata due to their small size (Fig. 2). These are more numerous in species that form heterocysts, specialized cells that are the site of nitrogen fixation, than in non-heterocyst-forming species (Giddings and Staehelin 1981). As a few of the vegetative cells in a filament differentiate into heterocysts, their cell ends constrict, reducing the number of microplasmodesmata from 200 or 300 to about 50 in heterocyst end

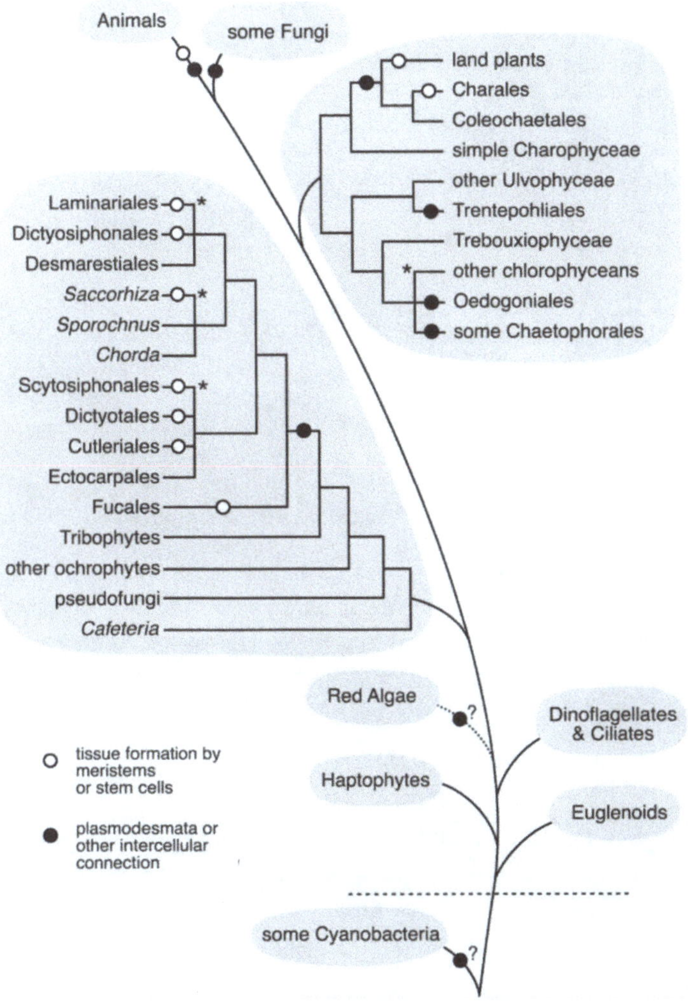

Fig. 1. Diagram showing occurrence of plasmodesmata and other intercellular connections in diverse taxa. This phylogeny represents a synthesis of molecular sequence data (primarily 18S rDNA) generated for the nucleocytoplasmic component (not plastids) of major eukaryotic lineages (Mishler et al. 1994; Cavalier-Smith and Chao 1996), upon which is mapped the occurrence of intercellular connections, as well as presence of tissues formed by meristems or stem cells. Intercellular connections are absent from early divergent eukaryotes, and are polyphyletic among higher eukaryotes. Phylogenetic relationships of red algae to other eukaryotic groups are still controversial, and it is unclear if red algal pit plugs and cyanobacterial microplasmodesmata are truly intercellular connections. Tissues formed by meristems have arisen numerous times, and always in association with intercellular connections. *The large shaded region on the left* represents a clade known as the heterokonts (the Heterokonta of Cavalier-Smith and Chao 1996), for which single-celled flagellates such as *Cafeteria* are basal. Plasmodesmata in this group occur only in the brown algae, and are absent from other autotrophic heterokonts (termed ochrophytes by Cavalier-Smith and Chao 1996). *The large shaded region on the right* represents major clades of green algae (chlorophytes). Green flagellates (known as prasinophytes) that are putatively related to early divergent ancestral forms are not shown. Unresolved regions of this phylogeny are indicated with an asterisk. (Original diagram by L. W. Wilcox and L. E. Graham)

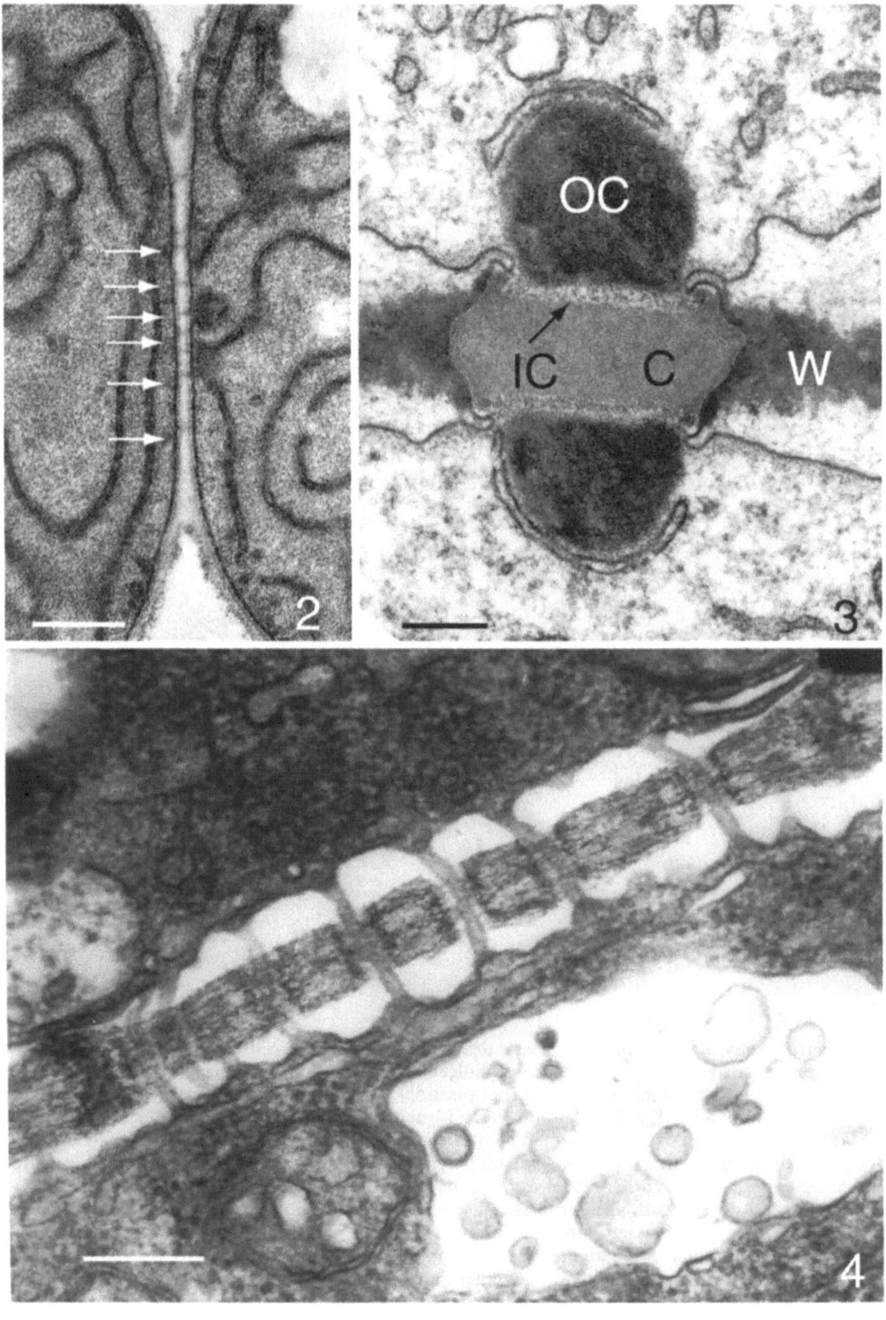
2
OC
IC
C
W
3
4

Fig. 2. Microplasmodesmata (*arrows*) traverse the wall between two vegetative cells of the filamentous cyanobacterium *Anabaena cylindrica*. *Bar* 250 nm. (Giddings and Staehelin 1978). Reprinted by permission of Gustav Fischer Verlag

Fig. 3. Pit plug of the red alga *Bossiella orbigniana* (Corallinales). The plasma membrane traverses the pit connection between the plug core (*C*) and the cell wall (*W*) and is invaginated near the plug cap. The thin inner cap layer (*IC*) is electron lucent, while the dome-shaped outer cap layer (*OC*) is electron opaque. A typical cap membrane would pass between the two cap layers and connect the plasma membrane invaginations, but cap membranes are not present in this taxon or in other members of the Corallinales. *Bar* 250 nm. (Pueschel 1987). Reprinted by permission of Journal of Phycology

Fig. 4. Plasmodesmata between meristematic cells of the brown alga *Cutleria hancockii* are not traversed by endoplasmic reticulum. *Bar* 200 nm. (Unpublished micrograph courtesy J.W. La Claire II)

walls of *Anabaena cylindrica* (Giddings and Staehelin 1978). According to estimates from thin sections and freeze-fracture electron microscopy, these intercellular connections are 10-20 nm in outer diameter (Marchant 1976; Giddings and Staehelin 1978). At the upper size limit it is possible that microplasmodesmata could have a plasma membrane lining (Lucas et al. 1993), but because the minimum diameter of lipid bilayer vesicles is greater than 20 nm, these channels may be too small to accommodate a plasma membrane (Giddings and Staehelin 1978). Nevertheless, freeze-fracture studies reveal pits in plasma membranes at their interface with microplasmodesmata, demonstrating that these minute channels do somehow connect the plasma membranes of neighbouring cells (Giddings and Staehelin 1978, 1981). Evidence that microplasmodesmata may facilitate intercellular communication and transport of molecules such as carbon and nitrogen has been reviewed previously (Marchant 1976; Giddings and Staehelin 1978, 1981). How microplasmodesmata might accomplish these tasks remains a mystery. It has been suggested that if a plasma membrane does not traverse the wall between cyanobacterial cells, the microplasmodesmal channel may be composed of proteinaceous particles, as is the case with connexin proteins in the gap junctions of animals (Giddings and Staehelin 1978, Lucas et al. 1993).

2.2 Red Algae

Cell division in the red algae takes place via furrowing, during which a passage termed a pit connection is left open in the cell wall of most species. Plasma membrane lines the pit connection, but the cytoplasm of the two cells is continuous only briefly, until the opening is filled with a structure called a pit plug (see Pueschel 1990 for review). Pit-plug ultrastructure is variable, and has been shown to correlate with phylogeny (Pueschel and Cole 1982; Pueschel 1987). Common to all pit plugs is a core formed from tubular membranes and granular protein (Pueschel 1980, 1990). Most pit plugs have a cap membrane that covers the plug and that has been shown by thin sections and freeze-fracture studies to be continuous with the plasma membrane, thereby rendering the pit plug extracellular (Pueschel 1977, 1990). Cap membranes are absent in a few orders of red algae, including the Bangiales and the Corallinales (Fig. 3); (Pueschel 1987; see Pueschel 1990, Table I for complete list).

The plug core may be covered by a flat or dome shaped cap that is composed of two layers in some orders, with the cap membrane (if present) found between these layers (Pueschel 1990). In plugs with two cap layers, the layer next to the cytoplasm contains polysaccharides and is thought to be formed by vesicles from the endomembrane system, while the inner layer is poorly understood (Pueschel 1980, 1990). Pit plugs may be removed under certain developmental circumstances, but there is no clear evidence that pit connections function in intercellular transport when plugs are in place, and the structural variability of pit plugs makes a universal explanation for their function difficult (Pueschel 1990). It would seem more likely for intercellular transport to take place in pit connections without cap membranes than in those that have them, but the presence of a plug still makes the possibility of transport problematic. It has been suggested that pit plugs with no plug caps and no cap membranes, typical of the primitive order Bangiales, may represent the ancestral condition (Pueschel 1987); (see also Chap. 12).

2.3 Fungi and Animals

Cell division in most fungi, like that in the red algae, occurs via furrowing that is incomplete, leaving an open, plasma membrane-lined septal pore (Pueschel 1977). Fungal pore structure is variable, and was shown to be a useful taxonomic character (Moore 1978; Suh et al. 1993). Unlike red algal pit connections, fungal pores are only rarely occluded by a plug (Pueschel 1977). Any resemblance between plugged fungal pores and pit plugs of red algae is considered to be superficial, and not indicative of homologous structures in these taxa (Moore and Marchant 1972; Pueschel 1977). Plasmodesmata are present during earliest development of walls formed by furrowing and (in a few taxa) in walls formed by centripetal fusion of vesicles (Marchant 1976). Some fungal plasmodesmata are traversed by ER, and therefore resemble those of higher plants (Marchant 1976). It has been suggested that these plasmodesmata may be the result of viral transfer of plant genes to fungi that infect plants (Lucas et al. 1993); (a more detailed description of the septal pore of basidiomycetes is given in Chap. 8).

Animal cells are interconnected at gap junctions, regions of the plasma membrane that incorporate numerous connexons, which are doughnut-shaped structures composed of six transmembrane protein subunits (Alberts et al. 1989). When connexons from the plasma membrane of one cell are aligned with those from the plasma membrane of a neighbouring cell, pores that are functionally 1.5 nm in diameter are formed, allowing small molecules to pass from one cell to another (Alberts et al. 1989). The proteins that compose the subunits of connexons are known as connexins. A family of at least thirteen connexin proteins of varying molecular weights is associated with intercellular connections in animals (Goodenough et al. 1996; Nicholson and Bruzzone 1997).

2.4 Brown Algae

Plasmodesmata are a common feature of the brown algae (Phaeophyceae) (Marchant 1976; La Claire 1981). Because plasmodesmata are so widespread among the brown algae, and because this group is thought to be monophyletic (including the lineages from Fucales to Laminariales), it is likely that plasmodesmata arose once, early in the diversification of brown algae (Fig. 1). Brown algal plasmodesmata are present even in the earliest stages of cytokinesis, both in taxa that form walls by furrowing (La Claire 1981), and in those in that form cell plates by vesicle fusion (Rawlence 1973). Entrapment of ER is reportedly not involved in brown algal cytokinesis (La Claire 1981). Most brown algal plasmodesmata appear to be simple plasma membrane-lined channels (Fig. 4); (Bisalputra 1966; Sideman and Scheirer 1977; La Claire 1981; Schmitz and Kühn 1982). Nevertheless, Marchant (1976) reports that desmotubules are visible in some published micrographs of *Laminaria groenlandica*, and ER also occasionally traverses plasmodesmata of *Alaria marginata* (Schmitz and Srivastava 1975). Similarly, we see desmotubules in cross-sectional views of several plasmodesmata in a micrograph of *Laminaria saccharina* published by Sideman and Scheirer (1977, Fig. 9). Thus there is some evidence that, at least in the Laminariales, brown algae have the ability to form plasmodesmata that are traversed by ER. In addition, we note in cross sectional views of several plasmodesmata in a micrograph of *L. saccharina* published by Schmitz and Kühn (1982), that there are small spoke-like structures radiating out from the plasma membrane, between the plasmodesma and the cell wall, a feature found in plasmodesmata of *Chara* and plants (Badelt et al. 1994; Franceschi et al. 1994; Cook et al. 1997).

2.5 Green Algae and Plants

Unlike fungi and brown algae, in which plasmodesmata may occur in cell walls formed by furrowing, the formation of plasmodesmata in the green algae is always associated with cytokinesis involving a cell plate, which is formed by means of a phycoplast or a phragmoplast (Stewart et al. 1973). Among the non-charophycean green algae, plasmodesmata are present in the Trentepohliales, the Oedogoniales, and some Chaetophorales (Fraser and Gunning 1969; Floyd et al. 1971; Stewart et al. 1973; Chapman and Good 1978; Chapman and Henk 1986). Differences among plasmodesmata in various chlorophycean lineages suggest multiple origins of chlorophycean plasmodesmata. Because branching patterns within the Chlorophyceae are not well resolved (Fig. 1, asterisk), it is unclear how many times plasmodesmata arose within chlorophyceans. *Trentepohlia* is unusual in that its plasmodesmata are found only in the centre of the wall and not at the edges (Fig. 5); (Stewart et al. 1973; Graham 1993). *Bulbochaete* (Oedogoniales) has been shown to have plasmodesmata that have no internal ER, but that do possess particles associated with the plasma membrane lining the plasmodesma (Fraser and Gunning 1969). Plasmodesmata with internal ER have been observed in the chaetophoralean taxa *Uronema* (previously known as *Ulothrix*) (Fig. 6); (Floyd et al. 1971) and *Aphanochaete* (Stewart et al. 1973). Only in *Aphanochaete* has the plasmodesmal ER been shown to be continuous with cellular ER.

Molecular, biochemical, and morphological evidence indicates that the green algae most closely related to the ancestry of plants are the charophycean green algae (Graham et al. 1991; Graham 1993; Mishler et al. 1994; McCourt 1995). Multicellular members of the class Charophyceae include the filamentous Zygnematales (unbranched filaments such as *Zygnema*, *Mougeotia*, and *Spirogyra*) and the more complex charo-

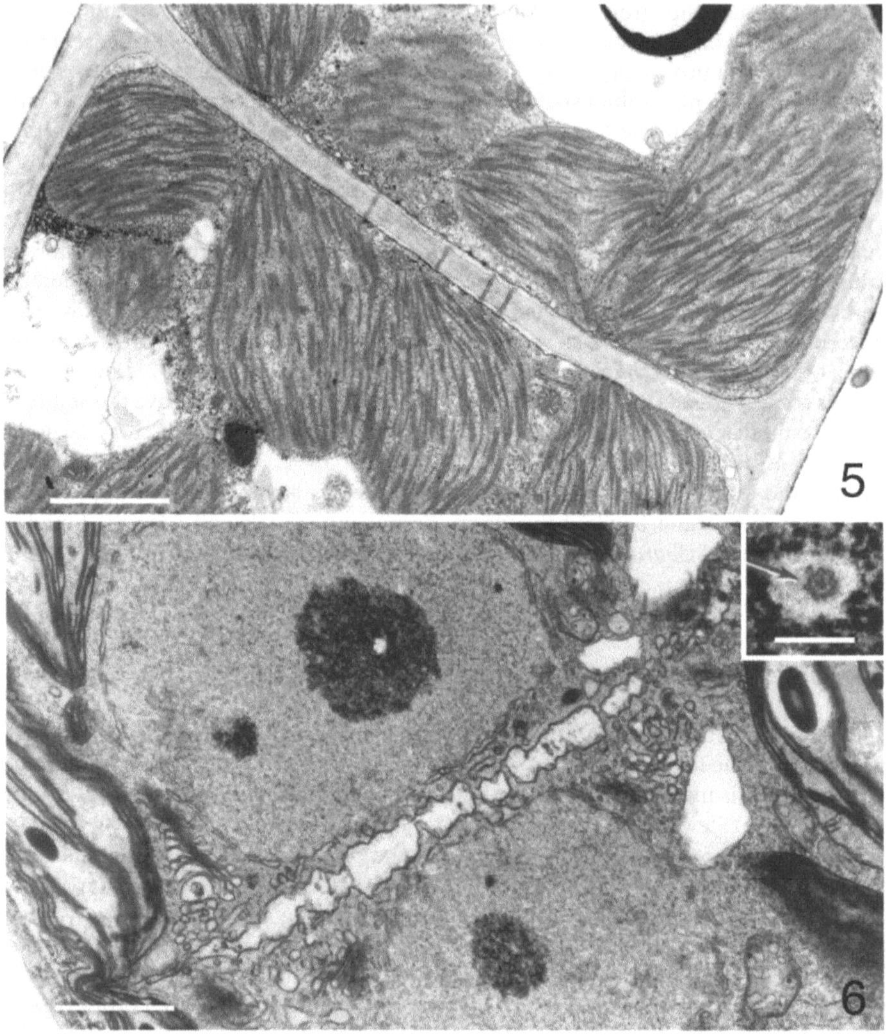

Fig. 5. Plasmodesmata are formed in the centre of the wall and not at the edges in the chlorophycean green alga *Trentepohlia*. *Bar* = 3 μm. (Graham 1993). Reprinted by permission of John Wiley & Sons, Inc.

Fig. 6. In *Uronema* (Chlorophyceae), cytokinesis involves a phycoplast and cell plate. *Bar* 2 μm. *Inset* Plasmodesmata have internal ER (arrow). *Bar* 100 nm. (Floyd et al. 1971). Reprinted by permission of Journal of Phycology

phytes *Coleochaete* and the Charales (including *Chara* and *Nitella*); (Graham 1993). In the Zygnematales, cell division occurs by centripetal furrowing, in conjunction with a small central phragmoplast and cell plate (Fowke and Pickett-Heaps 1969; Pickett-Heaps and Wetherbee 1987; McIntosh et al. 1995; Sawitzky and Grolig 1995; Nishino et al. 1996). While cytokinesis has been observed in the Zygnematales many times, no plasmodesmata have ever been reported, even in the part of the cell wall formed in association with a phragmoplast.

Plasmodesmata are formed during cytokinesis involving a phragmoplast in *Coleochaete* (Marchant and Pickett-Heaps 1973) and *Chara* (Pickett-Heaps 1967a, b; Cook et al. 1997); (Fig. 7), which are thought to be the closest algal relatives of plants (Graham 1993). Plasmodesmata in *Coleochaete* have not been shown to possess desmotubules or other internal structure (Marchant and Pickett-Heaps 1973; Marchant 1976; Graham 1993). Chemical fixation of cells in this genus is difficult, however, so further study is required. Though many charalean plasmodesmata appear to be simple plasma membrane-lined channels (Fischer and MacAlister 1975; Franceschi et al. 1994), a number of studies have revealed internal structure in charalean plasmodesmata (Spanswick and Costerton 1967; Pickett-Heaps 1967b, 1968; Kwiatkowska and Maszewski 1986; Cook et al. 1997). Most recently, plasmodesmata of *Chara zeylanica* have been demonstrated to have features that are similar to those of plant plasmodesmata, including a central structure that is connected to the plasma membrane by spoke-like structures, and smaller spoke-like structures that radiate outward from the plasma membrane (Cook et al. 1997); (Figs. 8-10). For reasons that are not understood, the internal structure of charalean plasmodesmata appears to be observed more readily in cross-sectional than in longitudinal view. (More about plasmodesmata in the green algae may be found in Chap. 12.)

The most basal extant land plants are the bryophytes, a paraphyletic grade consisting of three lineages: liverworts, hornworts, and mosses; mosses are thought to be most closely related to tracheophytes (Mishler and Churchill 1984, 1985; Waters et al. 1992; Mishler et al. 1994; Sztein et al. 1995; Lewis et al. 1997). Few studies of plasmodesmal ultrastructure have been conducted on these taxa (Schnepf and Sawidis 1991; Ligrone and Duckett 1994), but at least one relatively basal representative of each of the three clades (*Monoclea*, *Notothylas*, and *Sphagnum*, respectively) has been shown to have plasmodesmata with desmotubules (Cook et al. 1997). Other features of bryophyte plasmodesmata that are in common with those of plasmodesmata in higher plants include spokes connecting the desmotubule to the plasma membrane in the hornwort *Notothylas* and in the moss *Sphagnum*, and spokes that radiate out from the plasma membrane and connect with ring-like wall specializations in *Sphagnum* (Cook et al. 1997); (Figs. 11, 12). Similar ring-like wall specializations have been reported in ferns and in barley (Badelt et al. 1994).

3 Evolution of Plasmodesmata

In order to understand the evolutionary history of complex higher plant plasmodesmata, we need to examine plasmodesmal structure in seedless plants and charophycean algal relatives of plants, since these taxa represent the lineage from which complex plasmodesmata arose. Variations in the intercellular connections of these taxa may provide clues about when the characteristics of higher plant plasmodesmata first appeared. It has been hypothesized that plasmodesmata without desmotubules arose first, and that subsequently the ability to entrap ER and eventually to control incorporation of ER in primary plasmodesmata evolved (Lucas et al. 1993). When these events

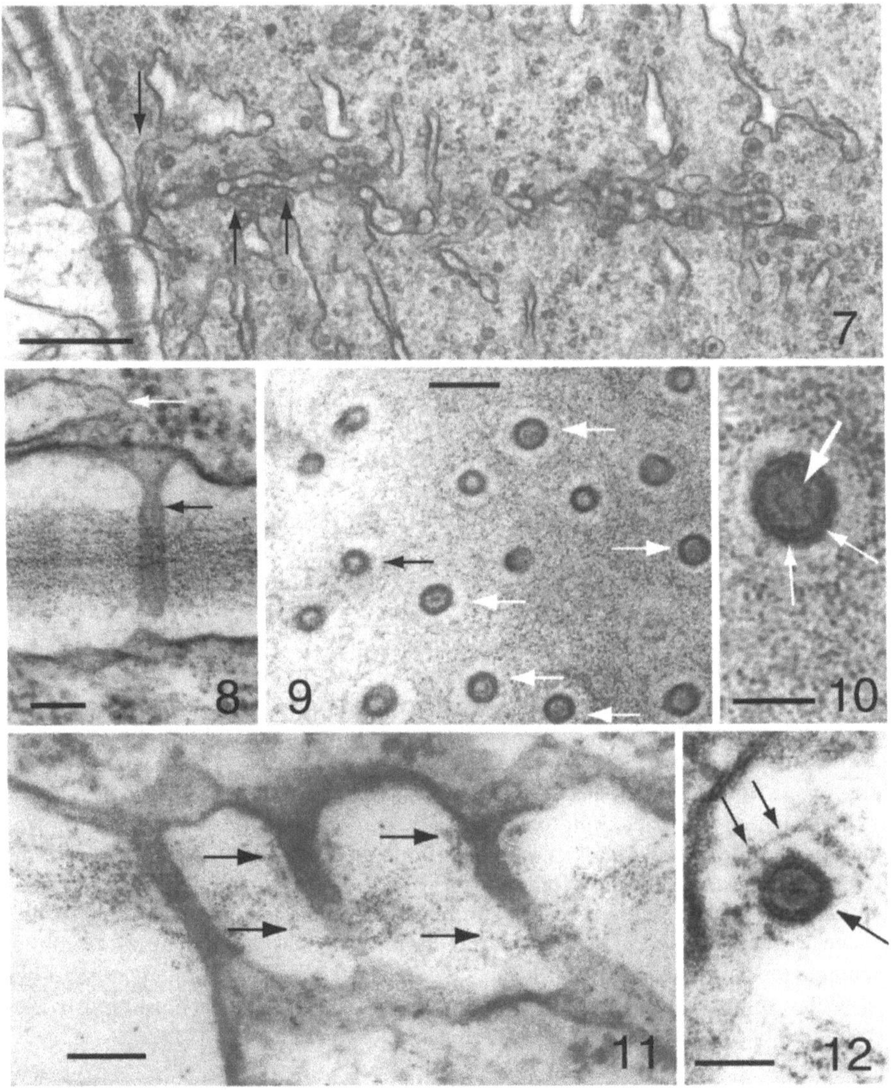

may have occurred is uncertain. Relatively few studies have been conducted on the plasmodesmata of seedless plants and charophycean algae, but available evidence indicates that plasmodesmal features such as a central structure connected to the plasma membrane by spoke-like structures arose early in the lineage leading to higher plants (Cook et al. 1997). Though it is possible that complex plasmodesmata evolved independently in advanced charophyceans and land plants, the simplest explanation for their occurrence in land plants is inheritance from ancestral advanced charophyceans. The hypothesis that plasmodesmata of charalean algae are homologous to those

Figs. 7–12. Plasmodesmata in the land plant lineage. **Fig. 7.** Cytokinesis in *Chara zeylanica* involves a phragmoplast. ER is trapped in the forming cell plate (*uppermost arrow*), and plasmodesmata are present even at this early stage of plate development (*two lower arrows*). *Bar* 500 nm. **Fig. 8.** Longitudinal view of plasmodesma in *Chara zeylanica. White arrow* indicates associated ER. *Black arrow* indicates desmotubule-like structure. *Bar* 100 nm. **Fig. 9.** Cross sectional views of plasmodesmata in *Chara zeylanica. Black arrow* indicates plasmodesmata with small structures radiating out from the plasma membrane. *White arrows* indicate plasmodesmata with internal structure. *Bar* 100 nm. **Fig. 10.** Internal structure of plasmodesmata in *Chara zeylanica* includes a central structure (*large arrow*) that appears to be connected to the plasma membrane via spoke-like structures (*small arrows*). *Bar* 50 nm. **Fig. 11.** Ring-like wall specializations (*arrows*) surround plasmodesmata in *Sphagnum fimbriatum*. *Bar* 100 nm. **Fig. 12.** Internal structure of plasmodesmata in *Sphagnum fimbriatum* includes a central structure that appears to be connected to the plasma membrane via spoke-like structures (*rightmost arrow*) and ring-like wall specializations that are attached to the plasma membrane via different spoke-like structures radiating from the plasma membrane (*two arrows on left*) *Bar* 50 nm. Figs. 7–12. (Cook et al. 1997). Reprinted by permission of American Journal of Botany

of higher plants could be tested by comparisons at the molecular level if proteins specific to plasmodesmata are eventually identified. Attempts to determine if the molecules actin, myosin, and possibly centrin are associated with the plasmodesmata of charophycean algae and lower plants, in the same way they are thought to be associated with higher plant plasmodesmata, may elucidate when dynamic aspects of plasmodesmal function may have occurred first.

The advanced charophycean green alga *Coleochaete orbicularis* may prove to be of special interest for deciphering the evolutionary history of plasmodesmata. Tubulin immunolocalization studies have demonstrated that cytokinesis in this taxon is significantly different in radial and circumferential cell divisions, even though a phragmoplast that appears to be typical of those of higher plants is associated with both types of cell division (Brown et al. 1994). Wall deposition is centrifugal in radial divisions, as is typical of plant cytokinesis, while wall deposition is centripetal in circumferential divisions and not aligned with the midline of the phragmoplast (Brown et al. 1994). Circumferential divisions in *C. orbicularis* may represent cytokinesis that is intermediate between typical plant cytokinesis and that of filamentous zygnematalean charophytes such as *Spirogyra* (Brown et al. 1994). This interpretation suggests the possibility that plasmodesmata formed during cytokinesis in circumferential divisions of *C. orbicularis* might be more primitive than those formed in radial divisions.

Since the simpler charophycean algae (Zygnematales) do not have plasmodesmata, advanced charophytes (Charales and *Coleochaete*) are the only extant organisms closely related to the ancestry of plants that may shed light on plant plasmodesmal evolution. Molecular phylogenetic evidence indicates that the class Charophyceae is related to other green algal classes (Ulvophyceae, Trebouxiophyceae, and Chlorophyceae) rather remotely, and only via single-celled flagellates. Thus, plasmodesmata almost certainly arose polyphyletically within green algae. It may be possible, however, to gain insight into the evolution of plant plasmodesmata by studying the parallel evolution of intercellular connections in more distant organisms. As Raven (1997) has pointed out, while plasmodesmata have most likely arisen many times in different lineages, genes encoding the molecules that compose plasmodesmata may have been present in ancestral unicellular organisms, where they may have encoded molecules with different functions.

Structural features such as small spokes radiating out from the plasma membrane toward the wall in some brown algal plasmodesmata; appressed ER in the plasmodesmata of some fungi, some brown algae, and some non-charophycean green algae; and particles associated with the plasma membrane lining the plasmodesma in the green alga *Bulbochaete* (Oedogoniales), are similar to those of higher plant plasmodesmata. Improved fixation may reveal additional convergence. Given the similarities between plasmodesmata in the brown algal order Laminariales and the charophycean green alga *Chara zeylanica*, it would be particularly interesting to know if there are spoke-like structures connecting the central structure with the plasma membrane in those plasmodesmata of laminarialean algae that appear to have a central structure. Studies at the molecular and gene levels are likely to be most fruitful for exploring possible convergent evolution in plasmodesmata and other intercellular connections of diverse taxa. With these methods it may be possible to test the hypotheses (see above) that fungi acquired plant plasmodesmata via viral transfer and that cyanobacterial microplasmodesmata are composed of proteins similar to the connexin proteins in animal gap junctions. These methods may also allow us to determine whether pit connections in the red algae are distantly related to intercellular connections in other taxa.

3.1 Correlation Between Plasmodesmata and Meristems

Study of convergent evolution among intercellular connections in diverse taxa may tell us not only about the evolution of plasmodesmata, but possibly also about other features that are related to the occurrence of plasmodesmata. For example, we have noticed that the ability to form plasmodesmata is correlated with the ability to form tissue by meristems (Fig. 1); (see Chap. 13). In plants, development is thought to be regulated at the cell, tissue, and organ levels via plasmodesmata in the apical meristem (Lucas et al. 1993; Duckett et al. 1994; Lucas 1995). Hence plasmodesmata are essential for development of the highly differentiated plant body. Given the frequent occurrence of meristematic tissue formation in a number of diverse taxa that possess plasmodesmata, it appears that plasmodesmata may be necessary for complex tissue formation in other taxa as well.

The early appearance of intercellular connections in the metazoan clade may have predisposed animals toward early evolutionary origin of tissues generated by specialized stem cells. Tissues produced by meristems occur in several lineages of brown algae (Sze 1993). Because branching patterns within some brown algal clades are as yet poorly resolved (Tan and Druehl, 1996); (Fig. 1, asterisk), it is unclear how many times tissues formed by meristems have arisen within phaeophytes. Presence of intercellular connections appears to be a necessary but not a sufficient condition for such tissue formation since organisms with plasmodesmata do not always possess meristems. Plasmodesmata occur in some fungi, but they are not known to be widespread, and tissues derived by meristems or stem cells are not known in this group. Claims of tissue formation in certain red algae (Bold and Wynne, 1985) require confirmation by detailed developmental analysis at the TEM level, and histogenetic meristems have not been described in this group. Some chlorophyceans (such as *Fritschiella*) may generate cell aggregates resembling tissues, but these do not seem to be associated with specialized meristems, as far as is known (Bold and Wynne, 1985).

The animal, brown algal, and green algal/land plant lineages are related only at the level of single-celled flagellates. Thus, their intercellular connections, tissues, and meristems/stem cells are of polyphyletic origin. It is possible, however, that unknown, morphologically indistinct cellular attributes that were preadaptive for the origin of intercellular connections are monophyletic, and occurred first in ancestral flagellates that diverged just prior to origin of the heterokont clade. Patterns of evolutionary occurrence of tissues and meristems/stem cells suggest that intercellular connections may be a prerequisite for the development of these tissues (necessary but not sufficient).

In the charophycean algal/land plant lineage, plasmodesmata may have preadapted the earliest land plants for the appearance of histogenetic meristems. While tissue-producing apical meristems are not present in the charophycean algae, some species of *Coleochaete* form tissue that has been said to be parenchymatous (Graham 1982), and the nodes of Charalean algae have been described as tissue resembling a meristem (Pickett-Heaps 1975). There are regions of the charalean thallus that consist of aggregations of filaments. These include internodal cells ensheathed by corticating cells (Fig. 13) and oogonia ensheathed by tube cells (Fig. 14). In both cases, the ensheathing cells grow from nearby nodes and are not sister cells with each other or with the cells they ensheathe. There are no primary (or secondary) plasmodesmata connecting en-

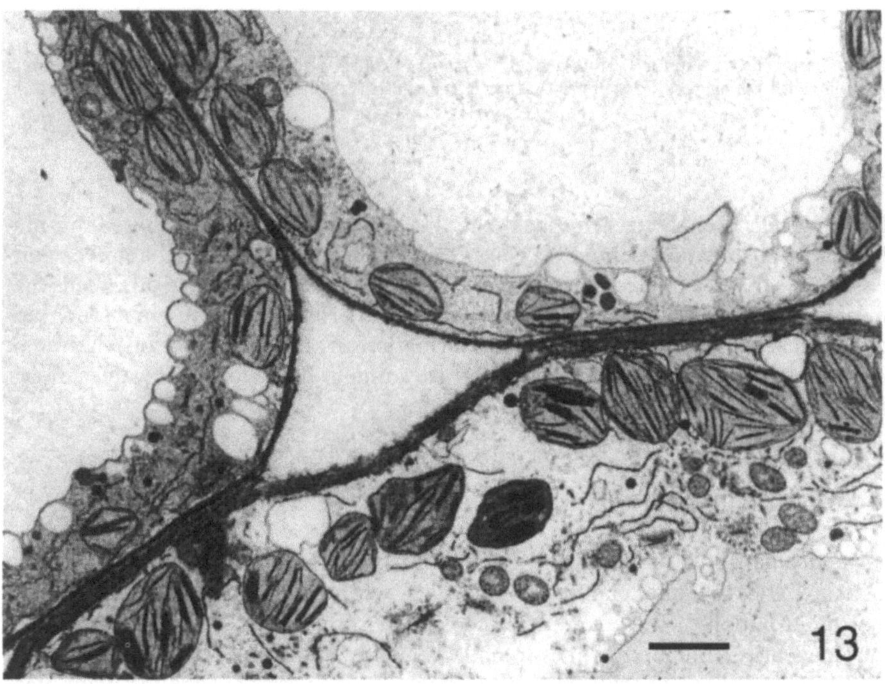

Fig. 13. Corticating cells (*top* and *left*) abut internodal cell (*bottom*) in *Chara zeylanica*. No plasmodesmata connect corticating cells with the internodal cell or with each other. *Bar* 2 μm. (Previously unpublished micrograph, M.E. Cook)

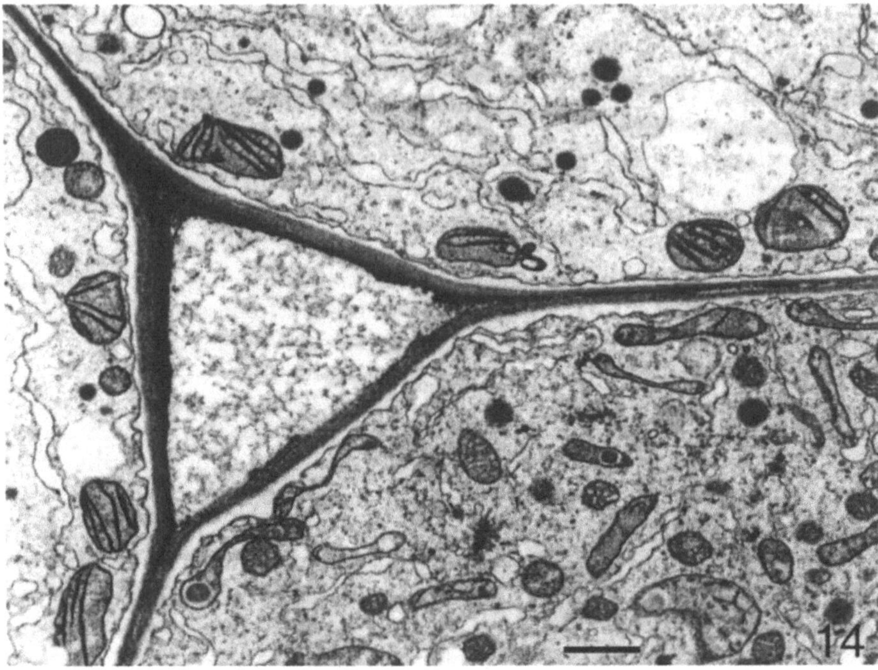

Fig 14. Tube cells (*top* and *left*) abut egg cell (*bottom*) in *Chara braunii*. No plasmodesmata connect tube cells with the egg cell or with each other. *Bar* 1 μm. (Previously unpublished micrograph, M.E. Cook)

sheathing cells directly to each other or to the cells they ensheathe in these (non-meristematic) filamentous complexes (Figs. 13, 14). The charalean node, however, consists of cells that are interconnected by plasmodesmata and give rise to other cells that continue dividing (Fritsch 1935; Pickett-Heaps 1975). Hence nodes resemble (non-apical) meristems. Study of simple tissue formation in the charophycean relatives of plants may illuminate evolutionary patterns leading to tissue formed by apical meristems in plants.

4 Concluding Remarks

Plasmodesmata of lower plants and charophycean algal relatives of plants are likely to be homologous to higher plant plasmodesmata, and hence may provide direct insight into the evolution of these complex intercellular connections. Intercellular connections have arisen numerous times in multicellular taxa that are related only through unicellular ancestors. Cytoplasmic connections in some ways resembling plant plasmodesmata have arisen in brown algae, chlorophycean green algae, and even in some fungi. Gap junctions in animals also provide cytoplasmic continuity between neighbouring cells. Microplasmodesmata in cyanobacteria and some pit connections in red algae may represent intercellular connections, though it is uncertain how cytoplasmic continuity might be maintained by these structures. If intercellular connections in diverse taxa are related at the genetic level via unicellular common ancestors, then molecular study of convergent evolution of these diverse intercellular connections may provide indirect insight into the evolution of complex higher plant plasmodesmata.

Acknowledgements. The authors thank L. W. Wilcox, T. H. Giddings, J. W. La Claire II, C. M. Pueschel, and G. L. Floyd for providing illustrations.

References

Alberts B, Bray D, Lewis J, Raff M, Roberts K, Watson JD (1989) Molecular Biology of the Cell, 2nd edn. Garland, New York
Badelt K, White, RG, Overall RL, Vesk M (1994) Ultrastructural specializations of the cell wall sleeve around plasmodesmata. Am J Bot 81:1422–1427
Bisalputra T (1966) Electron microscopic study of the protoplasmic continuity in certain brown algae. Can J Bot 44:89–93
Bold HC, Wynne MJ (1985) Introduction to the Algae 2nd Edn. Prentice-Hall, Englewood New Jersey
Brown RC, Lemmon BE, Graham LE (1994) Morphogenetic plastid migration and microtubule arrays in mitosis and cytokinesis in the green alga *Coleochaete orbicularis*. Am J Bot 81:127–133
Cavalier-Smith T, Chao EE (1996) 18S rRNA sequence of *Heterosigma carterae* (Raphidophyceae), and the phylogeny of heterokont algae (Ochrophyta). Phycologia 35:500–510
Chapman RL, Good BH (1978) Ultrastructure of plasmodesmata and cross walls in *Cephaleuros*, *Phycopeltis* and *Trentepohlia* (Chroolepidaceae; Chlorophyta). Br Phycol J 13:241–246
Chapman RL, Henk MC (1986) Phragmoplasts in cytokinesis of *Cephaleuros parasiticus* (Chlorophyta) vegetative cells. J Phycol 22:83–88
Cook ME, Graham LE, Botha CEJ, Lavin CA (1997) Comparative ultrastructure of plasmodesmata of *Chara* and selected bryophytes: toward an elucidation of the evolutionary origin of plant plasmodesmata. Am J Bot 84:1169–1178
Ding B (1997) Cell-to-cell transport of macromolecules through plasmodesmata: a novel signalling pathway in plants. Trends Cell Biol 7:5–9
Duckett CM, Oparka KJ, Prior DAM, Dolan L, Roberts K (1994) Dye-coupling in the root epidermis of *Arabidopsis* is progressively reduced during development. Development 120:3247–3255
Epel BL (1994) Plasmodesmata: composition, structure and trafficking. Plant Mol Biol 26:1343–1356
Fischer RA, MacAlister TJ (1975) A quantitative investigation of symplasmic transport in *Chara corallina*. III. An evaluation of chemical and freeze-substituting techniques in determining the in vivo condition of the plasmodesmata. Can J Bot 53:1988–1993
Floyd GL, Stewart KD, Mattox KR (1971) Cytokinesis and plasmodesmata in *Ulothrix*. J Phycol 7: 306–309
Fowke LC, Pickett-Heaps JD (1969) Cell division in *Spirogyra*. II Cytokinesis. J Phycol 5:273–281
Franceschi VR, Ding B, Lucas WJ (1994) Mechanism of plasmodesmata formation in characean algae in relation to evolution of intercellular communication in higher plants. Planta 192:347–358
Fraser TW, Gunning BES (1969) The ultrastructure of plasmodesmata in the filamentous green alga, *Bulbochaete hiloensis* (Nordst.) Tiffany. Planta 88:244–254
Fritsch FE (1935) The Structure and Reproduction of the Algae. vol I. Cambridge University Press, Cambridge
Ghoshroy S, Lartey R, Sheng J, Citovsky V (1997) Transport of proteins and nucleic acids through plasmodesmata. Annu Rev Plant Physiol Plant Mol Biol 48:27–50
Giddings TH, Staehelin LA (1978) Plasma membrane architecture of *Anabaena cylindrica*: occurrence of microplasmodesmata and changes associated with heterocyst development and the cell cycle. Cytobiologie 16:235–249
Giddings TH, Staehelin LA (1981) Observation of microplasmodesmata in both heterocyst-forming and non-heterocyst-forming filamentous cyanobacteria by freeze-fracture electron microscopy. Arch. Microbiol 129:295–298
Goodenough DA, Goliger JA, Paul DL (1996) Connexins, connexons, and intercellular communication. Annu Rev Biochem 65:475–502
Graham LE (1982) The occurrence, evolution, and phylogenetic significance of parenchyma in *Coleochaete* Bréb. (Chlorophyta). Am J Bot 69:447–454

Graham LE (1993) Origin of land plants. John Wiley, New York
Graham LE, Delwiche CF, Mishler BD (1991) Phylogenetic connections between the "green algae" and the "bryophytes". Adv in Bryol 4:213–244
Kaplan DR, Hagemann W (1991) The relationship of cell and organism in vascular plants. BioScience 41:693–703
Kwiatkowska M, Maszewski J (1986) Changes in the occurrence and ultrastructure of plasmodesmata in antheridia of *Chara vulgaris* L. during different stages of spermatogenesis. Protoplasma 132: 179–188
La Claire JW (1981) Occurrence of plasmodesmata during infurrowing in a brown alga. Biol Cell 40: 139–142
Lewis LA, Mishler BD, Vilgalys R (1997) Phylogenetic relationships of the liverworts (Hepaticae), a basal embryophyte lineage, inferred from nucleotide sequence data of the chloroplast gene *rbc*L. Mol Phylog Evol 7:377–393
Ligrone R, Duckett JG (1994) Thallus differentiation in the marchantialian liverwort *Asterella wilmsii* (Steph.) with particular reference to longitudinal arrays of endoplasmic microtubules in the inner cells. Ann Bot 73:577–586
Lucas WJ (1995) Plasmodesmata: intercellular channels for macromolecular transport in plants. Curr Opin Cell Biol 7:673–680
Lucas WJ, Wolf S (1993) Plasmodesmata: the intercellular organelles of green plants. Trends Cell Biol 3:308–315
Lucas WJ, Ding B, Van der Schoot C (1993) Plasmodesmata and the supracellular nature of plants. New Phytol 125:435–476
Lucas WJ, Bouché-Pillon S, Jackson DP, Nguyen L, Baker L, Ding B, Hake S (1995) Selective trafficking of KNOTTED1 homeodomain protein and its mRNA through plasmodesmata. Science 270: 1980–1983
Marchant HJ (1976) Plasmodesmata in algae and fungi. In: Gunning BES, Robards AW (eds) Intercellular communication in plants: studies on plasmodesmata. Springer, Berlin Heidelberg New York
Marchant HJ, Pickett-Heaps JD (1973) Mitosis and cytokinesis in *Coleochaete scutata*. J Phycol 9: 461–471
McCourt RM (1995) Green algal phylogeny. Trends Ecol Evol 10:159–163
McIntosh K, Pickett-Heaps JD, Gunning BES (1995) Cytokinesis in *Spirogyra*: integration of cleavage and cell-plate formation. Int J Plant Sci 156:1–8
Mezitt LA, Lucas WJ (1996) Plasmodesmal cell-to-cell transport of proteins and nucleic acids. Plant Mol Biol 32:251–273
Mishler BD, Churchill SP (1984) A cladistic approach to the phylogeny of the "bryophytes." Brittonia 36:406–434
Mishler BD, Churchill SP (1985) Transition to a land flora: phylogenetic relationships of the green algae and bryophytes. Cladistics 1:305–328
Mishler BD, Lewis LA, Buchheim MA, Renzaglia KS, Garbary DJ, Delwiche CF, Zechman FW, Kantz TS, Chapman RL (1994) Phylogenetic relationships of the "green algae" and "bryophytes". Ann MO Bot Gard 81:451–483
Moore RT (1978) Taxonomic significance of septal ultrastructure with particular reference to the jelly fungi. Mycologia 70:1007–1024
Moore RT, Marchant R (1972) Ultrastructural characterization of the basidiomycete septum of *Polyporus biennis*. Can J Bot 50:2463–2469
Nicholson SM, Bruzzone R (1997) Gap junctions: getting the message through. Curr Biol 7:R340–R344
Niklas KJ (1989) The cellular mechanics of plants. Am Sci 77:344–349
Nishino T, Asakawa S, Ogawa S (1996) Fine filaments observed within the cytoplasm surrounding the leading edge of the septum in telophase cells of *Spirogyra* (Zygnematales, Chlorophyta). Phycol Res 44:163–166
Pickett-Heaps JD (1967a) Ultrastructure and differentiation in *Chara* sp. I. Vegetative cells. Aust J Biol Sci 20:539–551
Pickett-Heaps JD (1967b) Ultrastructure and differentiation in *Chara* sp. II. Mitosis. Aust J Biol Sci 20:883–894
Pickett-Heaps JD (1968) Ultrastructure and differentiation in *Chara* (*fibrosa*). IV. Spermatogenesis. Aust J Biol Sci 21:655–690
Pickett-Heaps JD (1975) Green Algae. Sinauer, Sunderland Massachusetts
Pickett-Heaps JD, Wetherbee R (1987) Spindle function in the green alga *Mougeotia*: absence of anaphase A correlates with postmitotic nuclear migration. Cell Motil Cytoskeleton 7:68–77
Pueschel CM (1977) A freeze-etch study of the ultrastructure of red algal pit plugs. Protoplasma 91: 15–30
Pueschel CM (1980) A reappraisal of the cytochemical properties of rhodophycean pit plugs. Phycologia 19:210–217

Pueschel CM (1987) Absence of cap membranes as a characteristic of pit plugs of some red algal orders. J Phycol 23:150–156
Pueschel CM (1990) Cell structure. In: Cole KM, Sheath RG (eds) Biology of the Red Algae. Cambridge University Press, NY pp 7–41
Pueschel CM, Cole KM (1982) Rhodophycean pit plugs: an ultrastructural survey with taxonomic implications. Am J Bot 69:703–720
Raven JA (1997) Minireview: Multiple origins of plasmodesmata. Eur J Phycol 32:95–101
Rawlence DJ (1973) Some aspects of the ultrastructure of *Ascophyllum nodosum* (L.) Le Jolis (Phaeophyceae, Fucales) including observations on cell plate formation. Phycologia 12:17–28
Sawitzky H, Grolig F (1995) Phragmoplast of the green alga *Spirogyra* is functionally distinct from the higher plant phragmoplast. J Cell Biol 130:1359–1371
Schmitz K, Kühn R (1982) Fine structure, distribution and frequency of plasmodesmata and pits in the cortex of *Laminaria hyperborea* and *L. saccharina*. Planta 154:385–392
Schmitz K, Srivastava M (1975) On the fine structure of sieve tubes and the physiology of assimilate transport in *Alaria marginata*. Can J Bot 53:861–876
Schnepf E, Sawidis T (1991) Filament disruption in *Funaria* protonemata: occlusion of plasmodesmata. Bot Acta 104:98–102
Sideman EJ, Scheirer DC (1977) Some fine structural observations on developing and mature sieve elements in the brown alga *Laminaria saccharina*. Am J Bot 64:649–657
Spanswick RM, Costerton JWF (1967) Plasmodesmata in *Nitella translucens*: structure and electrical resistance. J Cell Sci 2:451–464
Stewart KD, Mattox KR, Floyd GL (1973) Mitosis, cytokinesis, the distribution of plasmodesmata, and other cytological characteristics in the Ulotrichales, Ulvales, and Chaetophorales: phylogenetic and taxonomic considerations. J Phycol 9:128–141
Suh S-O, Hirata A, Sugiyama J, Komagata K (1993) Septal ultrastructure of basidiomycetous yeasts and their taxonomic implications with observations on the ultrastructure of *Erythrobasidium hasegawianum* and *Sympodiomycopsis paphiopedili*. Mycologia 85:30–37
Sze P (1993) A biology of the algae. WC Brown, Dubuque, Iowa
Sztein AE, Cohen JD, Slovin JP, Cooke TJ (1995) Auxin metabolism in representative land plants. Am J Bot 82:1514–1521
Tan IH, Druehl LD (1996) A ribosomal DNA phylogeny supports the close evolutionary relationships among the Sporochnales, Desmarestiales, and Laminariales (Phaeophyceae). J Phycol 32:112–118
Waters DA., Buchheim MA, Dewey RA, Chapman RL (1992) Preliminary inferences of the phylogeny of bryophytes from nuclear-encoded ribosomal RNA sequences. Am J Bot 79:459–466

CHAPTER 8

The Perforate Septal Pore Cap of Basidiomycetes

W. H. MÜLLER* · B. M. HUMBEL · A. C. VAN AELST · T. P. VAN DER KRIFT · T. BOEKHOUT

* Utrecht University, Molecular Cell Biology, EMSA-Centraal Bureau voor Schimmelcultures, Padualaan 8, 3584 CH Utrecht, The Netherlands

1 Introduction

The interactions between cells of lower fungi may occur through plasmodesmata-like structures, which have been visualized electron microscopically in fungal septa. Developing zygospores of the zygomycetous species *Rhizopus sexualis* and *Gilbertella persicaria* contain transcellular strands in septa which separate young gametangia from the suspensors. These simple, unbranched strands occur at intervals in the gametangial wall, and resemble plasmodesmata (Hawker et al. 1966). The structures were found to be connected with the endoplasmic reticulum. Plasmodesmata-like structures have also been demonstrated in septa of hyphae of the ascomycetous yeasts *Geotrichum candidum* (Kirk and Sinclair 1966) and in *Endomycopsis fibuliger* (Takada et al. 1965). Septa of *Galactomyces geotrichum* exhibit unbranched plasmodesmata-like structures, extending through the cell wall. The width of these channels is 40–60 nm, with a desmotubule of 10 nm diameter (Marchant 1976) which connects endoplasmic reticulum on either side of the septum. Plasmodesmata-like structures occur in few fungi, while in the majority of fungi with septate hyphae the protoplasm of neighbouring cells is connected through the septal pore channel.

Cells of filamentous fungi contain all the major cytoplasmic organelles except chloroplasts, as fungi are non-photosynthetic organisms. The fungal cells are separated by septa with a central pore. The septa and associate cell structures play an important role in maintaining the cellular integrity, and allow communication between neighbouring cells (Markham 1995). Also dead or lytic cells in the hypha can be isolated by plugging the septal pore channel. This plugging mechanism seems to involve a coordinated action of organelles and cytoplasmic inclusions with the septa, but the mechanism is largely unknown. Moreover, the way in which plugging of a septal pore channel occurs is different between the ascomycetes and the basidiomycetes.

In the ascomycetes, the ultrastructure of the septum is relatively simple. The septum tapers or is of uniform thickness towards the septal pore channel. Cytoplasmic globular, electron-dense membranous particles, known as Woronin bodies, or hexagonal

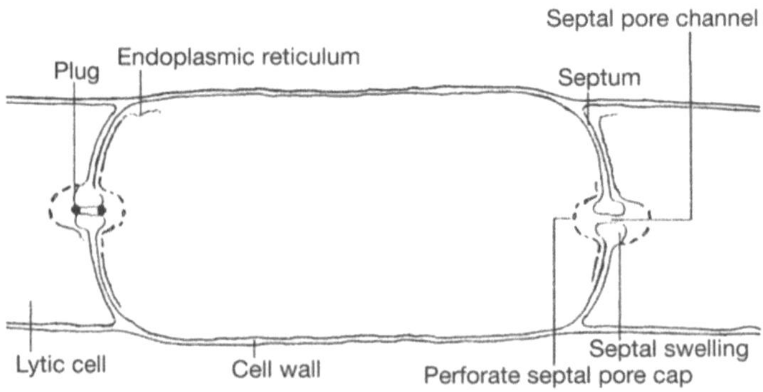

Fig. 1. Schematic presentation of a part of a hypha in a higher basidiomycete.
The septum with septal swellings is known as a dolipore septum. A perforate septal pore cap covers the opening of the septal pore channel, which may be occluded by a plug. The endoplasmic reticulum near the dolipore septum is connected with the perforate septal pore cap

crystalline structures are closely associated with the septal pore channel and may act as septal pore plugs (Kimbrough 1994; Markham 1994).

In the basidiomycetes, the septum can be more complex (Moore 1985; Markham 1994). The septum usually contains a minute central pore. In higher basidiomycetes (e.g. mushrooms), the septum around this pore is swollen, thus giving the appearance of a barrel-shaped structure known as dolipore (see Fig. 1) and characteristic for members of the higher basidiomycetes (Girbardt 1958; Moore and McAlear 1962; Bracker and Butler 1963). Dolipore septa occur in vegetative hyphae, either with or without clamp connections, and at the basal part of basidia (Bracker 1967). In many basidiomycetes, the dolipore is covered with a septal pore cap, known as parenthesome, which is in direct connection with the endoplasmic reticulum. However, some basidiomycetes with dolipore septa lack a septal pore cap, e.g. *Cystofilobasidia* and *Itersonilia* species.

2 Basidiomycete Septal Pore Cap

The septal pore cap of basidiomycetes is a distinct morphological structure of taxonomic and phylogenetic importance (McLaughlin et al. 1995). It may be perforate, cupulate, ampullate, tubular, vesiculate, non-perforate, or pauci-perforate (Moore 1985). In previous electron microscopic studies, chemical fixation and embedding in resin were used to distinguish between the various septal pore cap types. Later, more reliable electron microscopic techniques such as cryofixation and freeze substitution were used (e.g. Hoch and Howard 1981; Orlovich and Ashford 1994).

Though the fine structure of the septal pore cap of many basidiomycetes has been described, its function is still a matter of debate. There is a range of potential functions: (1) occlusion (Thielke 1972), comparable to the proposed function of Woronin bodies in *Penicillium chrysogenum* (Collinge and Markham 1987); (2) sieving of large organelles like nuclei (Wilsenach and Kessel 1965); (3) protection of the septum against damage during the process of protoplasmic streaming, with the septal pore channel opening varying from 100 nm–500 nm to ease the passage of organelles (Bracker and Butler 1964); (4) funnelling to direct small cellular organelles or components towards the septal pore channel opening (Orlovich and Ashford 1994). The different functions have been suggested because of the wide morphological variation of the septal pore cap observed in the basidiomycetes.

Detailed morphological studies are needed to understand the functions of the septal pore cap. In comparative morphological studies, the function of the septal pore cap can, for example be analyzed by using mutants for isolation and characterization of the septal pore cap matrix; immunogold labelling of the septal pore cap matrix; confocal laser scanning microscopy; and by different electron microscopy methods - each with its strengths and weaknesses.

3 Electron Microscopy of the Septal Pore Cap

Ultrastructural studies of the perforate septal pore cap as it occurs among the higher basidiomycetes revealed a considerable morphological variation. The regularly perforate septal pore cap showed holes varying between e.g. 50 nm diameter in *Nidularia confluens*, 330 nm diameter in *Ceratobasidium cornigerum* (Patton and Marchant 1978), and up to about 800 nm diameter in *Rhizoctonia solani* (Müller et al. 1995a).

Contradictory reports have been published on the connection of endoplasmic reticulum with the septal pore cap. While in e.g. *Polyporus rugulosus* the endoplasmic reticulum was connected to the perforate septal pore cap (Wilsenach and Kessel 1965), this was not observed in *Pisolithus arhizus* (Orlovich and Ashford 1994). However, in zinc iodide-osmium tetroxide stained dikaryotic hyphae of *Schizophyllum commune* the endoplasmic reticulum was found to be clearly connected with the basal part of the septal pore cap (Müller et al. 1995b). Also in species with non-perforate septal pore caps, e.g. *Epulorhiza anaticula* and *Sebacina vermifera*, the endoplasmic reticulum was connected with the septal pore cap as visualized by scanning electron microscopy (Müller et al. 1998).

Both the endoplasmic reticulum and the septal pore cap are morphologically different between the main lines of basidiomycete divergence (McLaughlin et al. 1995). For example, the endoplasmic reticulum near the dolipore septum in *Schizophyllum commune* is tubular or plate-like, while that of *Trichosporon sporotrichoides* resembles a jigsaw puzzle forming tubular structures near the septal pore channel and resulting in a relatively simple septal pore cap (Müller et al. 1995b). In *Sphaerobolus stellatus*, degenerate endoplasmic reticulum remained near septa or no endoplasmic reticulum was observed during ageing. Here, the septal pore cap was found lying freely within the cytoplasm without any connection to the endoplasmic reticulum (Dijkstra 1982).

3.1 Electron Microscopy Preparation Methods

To complement and extend the information on the septal pore cap resulting from classical thin sectioning and transmission electron microscopy (e.g. Jersild et al. 1967; Patton and Marchant 1978), we adopted several other electron microscopy methods to visualize the perforate septal pore cap of *Schizophyllum commune*. Firstly, freeze fracturing, maceration, freeze substitution, and further processing for scanning electron microscopy (Müller et al. 1994). Secondly, preferential staining of the septal pore cap in a mixture of zinc iodide-osmium tetroxide (ZIO) according to Hawes (1991), followed by freeze substitution and thick sectioning. Thirdly, cryofixation either by plunge freezing or high-pressure freezing, followed by freeze substitution (Howard and O'Donnell 1987; Humbel and Schwarz 1989) to preserve the in vivo situation as much as possible.

3.2 Septal Pore Cap of *Schizophyllum commune*

The spatial position of the septal pore cap inside the fungal cell is visualized by scanning electron microscopy. Hyphae that were cracked close to a septum showed either the dolipore septum or the septal pore cap. The perforate septal pore cap has a width of about 600 nm, and contains regularly distributed holes that vary in size (Fig. 2). The largest, apical located hole measures about 110 × 125 nm and the smallest, basal one is about 55 × 65 nm. The septal pore cap makes passage of a mitochondrion through the holes unlikely, as the diameter of the mitochondrion is about 300 nm.

Detailed features of the septal pore cap can be seen after ZIO staining, followed by thick sectioning and transmission electron microscopy (Fig. 3). This figure verifies the notion that the cap is connected to the endoplasmic reticulum. The ZIO mixture stains not only the nuclear envelope, the mitochondria, and the endoplasmic reticulum, but also the membranes of the septal pore cap. In contrast, the matrix of the septal pore cap is not stained, but the lumen of the endoplasmic reticulum is heavily stained. The septal pore cap is found to be globular to somewhat elongate and is about 570 nm high and between 455 and 570 nm wide, in agreement with the scanning electron micro-

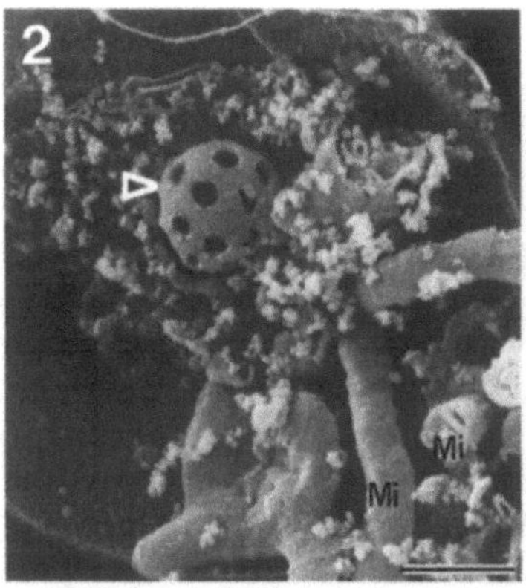

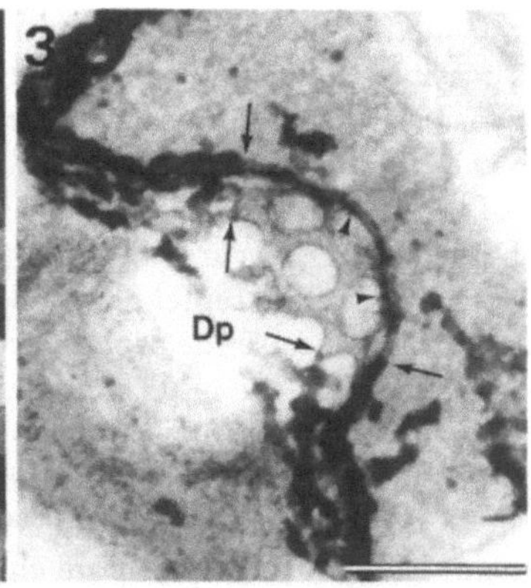

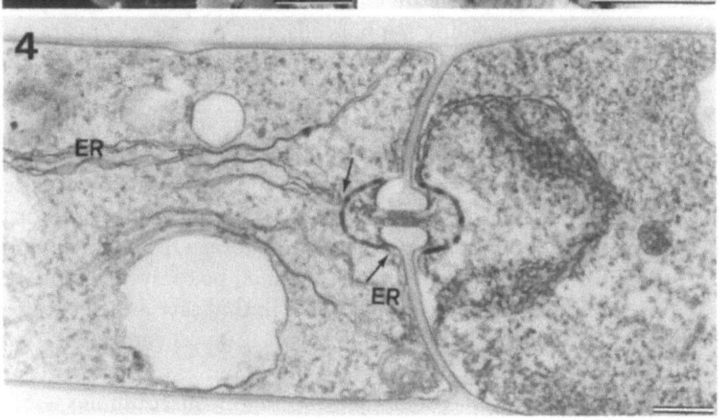

Fig. 2. Scanning electron micrograph of a fractured and macerated monokaryotic hypha of *Schizophyllum commune*. The perforate septal pore cap (*arrowhead*) covers the dolipore. Some mitochondria (*Mi*) are present in the vicinity of the septal pore cap. *Bar* 500 nm

Fig. 3. Preferentially stained and 500-nm thick-sectioned monokaryotic hypha of *Schizophyllum commune* showing densely stained endoplasmic reticulum (ER) at both sides of a less stained perforate septal pore cap (SPC). *Arrows* indicate the connection of the ER with the SPC. *Arrowheads* indicate that the inner membrane of the SPC is more densely stained than the outer membrane; *Dp* dolipore septum, which is unstained. *Bar* 500 nm

Fig. 4. Cryofixed, by plunge freezing, and freeze substituted monokaryotic hypha of *Schizophyllum commune* showing the connection of the endoplasmic reticulum (*ER*) with the top and the basal part of the septal pore cap (*arrows*) in a young cell. *Bar* 500 nm scopic measurements. The holes of the septal pore cap are round to somewhat angular in shape and vary in size; near the apex of the septal pore cap they measure about 110 × 130 nm; the basal ones are about 100 × 110 nm. The variation of holes in the septal pore cap is also shown, although not specifically described, in thick sections of a dikaryotic strain of *Schizophyllum commune* (Patton and Marchant 1978). Staining of the septal pore cap with ZIO was also reported for *Amanita rubescens* by Hawes (1981). The difference in deposition of the ZIO stain in the septal pore cap and the endoplasmic reticulum may suggest a difference in structure and function (Gilloteaux and Naud 1979).

Hyphae subjected to a combination of cryofixation and freeze substitution (Figs. 4–7) are presumed to show the in vivo ultrastructure of the perforate septal pore cap as much as possible. Connections between the septal pore cap and the endoplasmic reticulum in material fixated this way confirm the results obtained with ZIO-staining (cf. Fig. 3). The endoplasmic reticulum is found to be connected with the top and the basal part of the septal pore cap in young hyphal cells (Figs. 4 and 5), while in older cells only basal connections are observed (Figs. 3, 6 and 7). At higher magnification, the layered structure of the septal pore cap becomes evident (Figs. 5–7). An outer and an inner membrane enclose the septal pore cap matrix. Cytoplasmic, electron-dense material is situated near the inner membrane of the septal pore cap. Figure 7 shows that this inner side of the septal pore cap is connected through fibrillar structures, with the pore occlusions occurring near the opening of the septal pore channel, as observed earlier by Orlovich and Ashford (1994) in *Pisolithus arhizus*.

When the opening of the septal pore channel is occluded, the septal pore channel becomes narrower (Fig. 6). In young hyphal cells, this plugging appears to be reversible, while in mature hyphal cells it may be more permanent (Bracker 1992). Plugs block the transport of cellular constituents from one cell to another in response to stress. Interestingly, plugging also occurs at plasmodesmata and even at the disc plates of sieve tubes in higher plants (A. van Bel, pers. comm.). In filamentous fungi, the plug might originate from a coagulation of cytoplasmic components or organelles (Aylmore et al. 1984). Alternatively, our data suggest that the septal pore cap or the endoplasmic reticulum, or both, might also be the site of storage from which plugging material could be deposited in the septal pore channel, possibly through the fibrillar structures. The septal pore cap is best interpreted in that case as a separate compartment of the endoplasmic reticulum specialized in the occlusion of the septal pore channel.

After staining for sugars, tubular structures – possibly microtubules – are visible in the septal pore channel (Fig. 7). This figure shows that two tubules pass through the septal pore cap, and one of them enters the septal pore channel. These tubules might help to transport so-called motor proteins like kinesin, dynein or dynamin (Vallee and Shpetner 1990), which bind different cellular constituents. The tubules may move against the protoplasmic streaming. Transport guided or mediated by way of the tubules can be against protoplasmic streaming, as is shown by the transport of the vacuole system from one cell to a neighbouring cell in young hyphae of *Pisolithus tinctorius* (Shepherd et al. 1993). Alternatively, the endoplasmic reticulum can be associated with cytoskeletal elements, as seems likely for cells of higher plants.

ATPases may play a crucial role to generate energy required for the process of intercellular transport as ATPase activity has been reported to occur in the septal pore

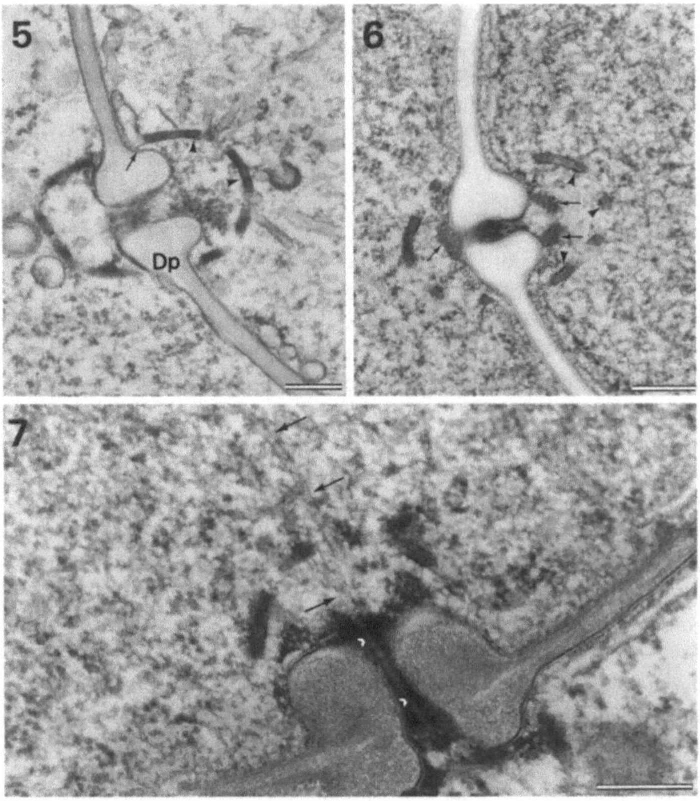

Fig. 5. High magnification of the septal pore cap (SPC) of a young dikaryotic hyphal cell of *Schizophyllum commune* after cryofixation by plunge freezing and freeze substitution. The *arrow* indicates the connection of the endoplasmic reticulum with the plasma membrane. *Arrowheads* indicate the presence of electron-dense material against the inner side of the SPC. Note that the septal pore channel is unplugged; *Dp* dolipore septum. *Bar* 250 nm

Fig. 6. High-pressure frozen and freeze-substituted monokaryotic hypha of *Schizophyllum commune* showing septal pore-occluding material on either side of the dolipore septum (*arrows*). *Arrowheads* show electron-dense material against the inner side of the septal pore cap. Note that the septal pore channel is narrower than that in Fig. 5. *Bar* 250 nm

Fig. 7. High-pressure frozen, freeze-substituted, and cytochemically sugar-stained monokaryotic hypha of *Schizophyllum commune* showing tubules (*arrows*) through the septal pore cap, and one in the septal pore channel (*arrowheads*). Note the fibrillar structures between the inner side of the right part of the perforate septal pore cap and the pore-occluding material. *Bar* 250 nm

channel of *Agaricus bisporus* (Schramm 1970). We hypothesize that ATPase is likely to be active in the septal pore channel of *S. commune* as well. Further, calcium ions which could originate from the lumen of the endoplasmic reticulum activate the ATPase in the septal pore channel of *Schizophyllum commune* hyphal cells. The septal pore cap and the endoplasmic reticulum near the septum most likely control the calcium ion homeostasis and the cellular transport between neighbouring cells in the area of the septal pore channel. Particular endoplasmic reticulum areas may be specialized for calcium storage as well as for other unknown functions, as has been reported for animal cells by Sitia and Meldolesi (1992).

4 Concluding Remarks

The function of the septal pore cap in hyphal cells remains largely unknown. The connection of the septal pore cap to the endoplasmic reticulum, its position close to the opening of the septal pore channel, and its absence in young growing apical hyphal cells, indicate the following consecutive functions of the septal pore cap in *Schizophyllum commune*: (1) a role in the transport of macromolecules between neighbouring cells, (2) a sieve function, and (3) a barrier function during cell ageing and cell lysis, inhibiting intercellular communication by plug forming. The electron-dense fibrillar structures may play an important role to transport plugging material between the septal pore cap and the opening of the septal pore channel. We hypothesize that the combined occurrence of the septal pore cap and the endoplasmic reticulum controls the calcium ion homeostasis and the intercellular transport in the septal pore channel. By further use of cytochemical, physiological, and cell-biological studies, we may elucidate the secrets of the septal pore cap in filamentous fungi.

Acknowledgements. The authors thank Prof. Dr. W. Gams and Dr. D. van der Mei (Centraalbureau voor Schimmelcultures) for critical reading of the manuscript, Prof. Dr. A. J. Verkleij (Utrecht University) for fruitful discussions, Miss L. Verspui (Utrecht University) for the preparation of the micrographs, and Dr. R. Kempers (Utrecht University) for practical working tips. This work is based on a cooperative project between the Centraalbureau voor Schimmelcultures and the Utrecht University.

References

Aylmore RC, Wakley GE, Todd NK (1984) Septal sealing in the basidiomycete *Coriolus versicolor*. J Gen Microbiol 130:2975–2982

Bracker CE (1967) Ultrastructure of fungi. Annu Rev Phytopathol 5:343–374

Bracker CE (1992) Intercellular cytoplasmic connections in fungi. In: 2nd Int Workshop on Basic and Applied Research in Plasmodesmatal Biology. Oosterbeek, The Netherlands, pp 89–93

Bracker CE, Butler EE (1963) The ultrastructure and development of septa in hyphae of *Rhizoctonia solani*. Mycologia 55:35–58

Bracker CE, Butler EE (1964) Function of the septal pore apparatus in *Rhizoctonia solani* during protoplasmic streaming. J Cell Biol 21:152–157

Collinge AJ, Markham P (1987) Response of severed *Penicillium chrysogenum* hyphae following rapid Woronin body plugging of septal pores. FEMS Microbiol Lett 40:165–168

Dijkstra MJ (1982) A cytological examination of *Sphaerobolus stellatus* fruiting bodies. Mycologia 74: 44–53

Gilloteaux J, Naud J (1979) The zinc iodide-osmium tetroxide staining of Maillet Ca^{2+}-affinity subcellular sites in a tonic smooth muscle. Histochemistry 63:227–243

Girbardt M (1958) Über die Substruktur von *Polystictus versicolor* L. Arch Mikrobiol 28:255–269

Hawes CR (1981) Applications of high voltage electron microscopy to botanical ultrastructure. Micron 12:227–257

Hawes CR (1991) Stereo-electron microscopy. In: Hall JL, Hawes CR (Eds) Electron microscopy of plant cells. Academic Press, London, pp 67–84

Hawker LE, Gooday MA, Bracker CE (1966) Plasmodesmata in fungal cell walls. Nature 212:635
Hoch HC, Howard RJ (1981) Conventional chemical fixations induce artifactual swelling of dolipore septa. Exp Mycol 5:167–172
Howard RJ, O'Donnell KL (1987) Methodology review: freeze-substitution of fungi for cytological analysis. Exp Mycol 11:250–269
Humbel BM, Schwarz H (1989) Freeze-substitution for immunochemistry. In: Verkleij AJ, Leunissen JLM (eds) Immuno-gold labelling in cell biology. CRC Press, Boca Raton, Florida, pp 115–134
Jersild R, Mishkin S, Niederpruem DJ (1967) Origin and ultrastructure of complex septa in *Schizophyllum commune* development. Arch Mikrobiol 57:20–32
Kimbrough JW (1994) Septal ultrastructure and ascomycete systematics. In: Hawksworth DL (ed) Ascomycete systematics. Plenum Press, New York, pp 127–141
Kirk BT, Sinclair JB (1966) Plasmodesmata between hyphal cells of *Geotrichum candidum*. Science 153:1646
Marchant HJ (1976) Plasmodesmata in algae and fungi. In: Gunning BES, Robards AW (eds) Intercellular communication in plants: studies on plasmodesmata. Springer, Berlin Heidelberg, pp 59–80
Markham P (1994) Occlusions of septal pores in filamentous fungi. Mycol Res 98:1089–1106
Markham P (1995) Organelles of filamentous fungi. In: Gow NAR, Gadd GM (eds) The growing fungus. Chapman and Hall, London, pp 75–98
McLaughlin, DJ, Frieders EM, Lü H (1995) A microscopist's view of heterobasidiomycete phylogeny. Stud Mycol 38:91–109
Moore RT (1985) The challenge of the dolipore/parenthesome septum. In: Moore D, Casselton LA, Wood DA, Frankland JC (eds) Developmental biology of higher fungi. Cambridge University Press, Cambridge, UK, pp 175–212
Moore RT, McAlear JM (1962) Fine structure of mycota. 7. Observations on septa of ascomycetes and basidiomycetes. Am J Bot 49:86–94
Müller WH, van Aelst AC, van der Krift TP, Boekhout T (1994) Scanning electron microscopy of the septal pore cap of the basidiomycete *Schizophyllum commune*. Can J Microbiol 40:879–883
Müller WH, van Aelst AC, van der Krift TP, Stalpers JA, Boekhout T (1995a) Septal pore caps as revealed by high-resolution scanning electron microscopy. CBS Newsl Autumn, pp 4–5
Müller WH, van Aelst AC, van der Krift TP, Boekhout T (1995b) Novel approaches to visualize the septal pore cap. Stud Mycol 38:111–117
Müller WH, Stalpers JA, van Aelst AC, van der Krift TP, Boekhout T (1998)
Field emission gun-scanning electron microscopy of septal pore caps of selected species in the *Rhizoctonia* s.l. complex. Mycologia 90:170-179
Orlovich DA, Ashford AE (1994) Structure and development of the dolipore septum in *Pisolithus tinctorius*. Protoplasma 178:66–80
Patton AM, Marchant R (1978) A mathematical analysis of dolipore/parenthesome structure in basidiomycetes. J Gen Microbiol 109:335–349
Schramm K-H (1970) Enzymmuster bei Pilzen. PH D Thesis, Freie Universität, Berlin
Shepherd VA, Orlovich DA, Ashford AE (1993) A dynamic continuum of pleomorphic tubules and vacuoles in growing hyphae of a fungus. J Cell Sci 104:495–507
Sitia R, Meldolesi J (1992) The endoplasmic reticulum: a dynamic patchwork of specialized subregions. Mol Biol Cell 3:1067–1072
Takada H, Yagi T, Hiraoka J-I (1965) Elektronenoptische Untersuchungen an *Endomycopsis fibuliger* auf festen Nährböden. Protoplasma 59:494–505
Thielke Ch (1972) Die Dolipore der Basidiomyceten. Arch Mikrobiol 82:31–37
Vallee RB, Shpetner HS (1990) Motor proteins of cytoplasmic microtubules. Annu Rev Biochem 59: 909–932
Wilsenach R, Kessel M (1965) On the function and structure of the septal pore of *Polyporus rugulosus*. J Gen Microbiol 40:397–400

CHAPTER 9

Substructure of Plasmodesmata

R. L. OVERALL

School of Biological Sciences, Macleay Bldg A12, The University of Sydney, New South Wales, 2006, Australia
Fax: +61 2 9351 4771
Email: roverall@mail.usyd.edu.au

1 Introduction

Plasmodesmata now appear to be highly dynamic structures which are able to select and transport even large molecules from cell to cell (for review, see Zambryski 1995). It has become crucial to understand the substructural architecture of plasmodesmata and the molecular interactions that control the intercellular passage of molecules and viruses. Our images of plasmodesmal structure have been necessarily static, leading to the generation of similarly static models of their architecture. The present challenge is to incorporate the emerging data on the molecular composition and operation of plasmodesmata into a dynamic model of their substructure.

2 Light Microscopy of Plasmodesmata

The first observations of plasmodesmata at the level of the light microscope are attributed to Tangl (1879). The ensuing early observations which involved swelling of the wall and, in some cases, staining were reviewed in the first book on plasmodesmata (Robards 1976; Carr 1976). The structures visible in the light microscope were no doubt pit fields or wall modifications associated with plasmodesmata. In recent years, various fluorescent probes have been used to label pit fields and, in combination with confocal laser scanning microscopy, to identify the location of single plasmodesmata. In plasmolyzed onion epidermal cells, staining of the endoplasmic reticulum (ER) with the fluorochrome $DIOC_6$ colocalizes with aniline blue-induced fluorescence of callose in the cell walls, presumably at pit fields (Oparka et al. 1994). Rhodamine phalloidin labelling results in fluorescent spots on the walls, showing actin in pit fields of epidermal peels or individual plasmodesmata in tobacco suspension culture cells (White et al. 1994). Immunofluorescence of callose (Fig. 1), cytoskeletal elements

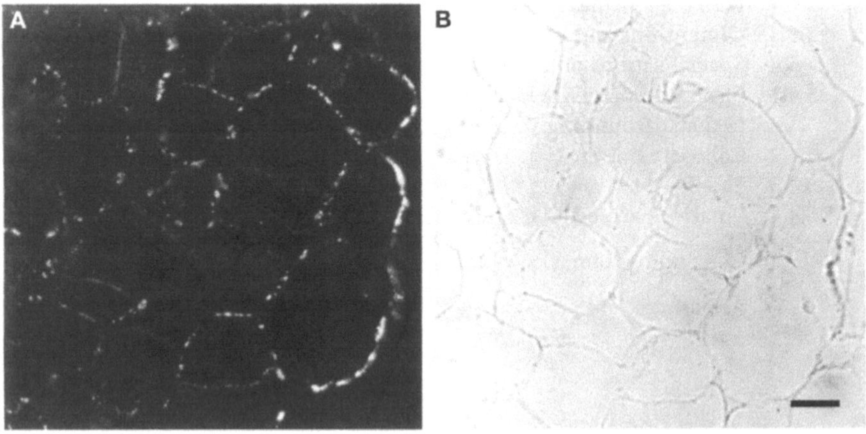

Fig. 1A, B. Immunofluorescence localization of callose in mature cells of cauliflower florets. **A** Epifluorescence image of a section labelled with an antibody to callose showing punctate label of the cell walls consistent with the distribution of plasmodesmata. **B** Corresponding interference contrast image. Bar 5 µm. (Unpublished image kindly provided by L. M. Blackman.)

(Blackman and Overall 1998) and pectin (Casero and Knox 1995) have shown punctate labelling consistent with the distribution of plasmodesmata. Recently, the viral "movement protein" which facilitates the movement of viruses through plasmodesmata has been fused to the green fluorescent protein (GFP) and expressed as an expression vector or from the whole virus (for review, see Oparka et al. 1996). The GFP-movement protein fusion localizes spectacularly to plasmodesmata and has even been seen sandwiched between paired callose platelets at the ends of single plasmodesmata (Oparka et al. 1997). While these various fluorescent tags have been useful in identifying molecular components of plasmodesmata and giving an indication of their distribution at the light microscope level, ultimately it is necessary to use the electron microscope to confirm that the labelled structures are indeed plasmodesmata.

3 Electron Microscopy of Plasmodesmata

The small size of plasmodesmata has meant that their substructure has been visible only with the electron microscope. They were first observed in the electron microscope 40 years ago (Buvat 1957), and the main features of images obtained since then have been remarkably consistent. The features visible in Fig 2A–E were clearly present in the images of Robards (1976). Even different tissue preparation techniques e.g. freeze substitution (Fig. 2E; Ding et al. 1992) yielded similar images. In a transverse section in the centre of a plasmodesma, in general, there is a central black dot of 3 nm in diameter surrounded by an electron-lucent ring 2.2 nm wide and further electron-dense material surrounded by a mottled layer of electron-lucent and electron-dense material. This mottled layer may abut directly onto the alternating outer rings of electron density, electron lucence, and electron density of the plasma membrane (Fig. 2B). Alternatively, there may be an electron-lucent region between the mottled layer and the plasma membrane which is traversed by electron-dense spokes (Fig. 2C). Often, when tannic acid has been added to the fixative, additional electron-dense material may be observed outside the plasma membrane (Fig. 3).

However, surprisingly, the interpretation of the substructure from these images is still under debate. One difficulty is that the width of the plasmodesma channel is comparable to the thickness of a thin section for electron microscopy, so that the components of the structure in longitudinal view are always superimposed within the section. In transverse view, up to one third of the plasmodesma may be superimposed in one section. In addition, our understanding of the substructure of plasmodesmata almost certainly has been stilted by artefacts induced during tissue preparation and difficulties in interpretation of staining patterns. Our interpretation of images from electron microscopy must necessarily be cautious and new technologies to avoid such artefacts must continually be sought.

4 Formation of Primary Plasmodesmata

Primary plasmodesmata are formed at cytokinesis as endoplasmic reticulum (ER) is trapped within the fusing Golgi-derived vesicles of cell wall material forming the cell plate (Hepler 1982; Staehelin and Hepler 1996). Nothing is known of how plasmodes-

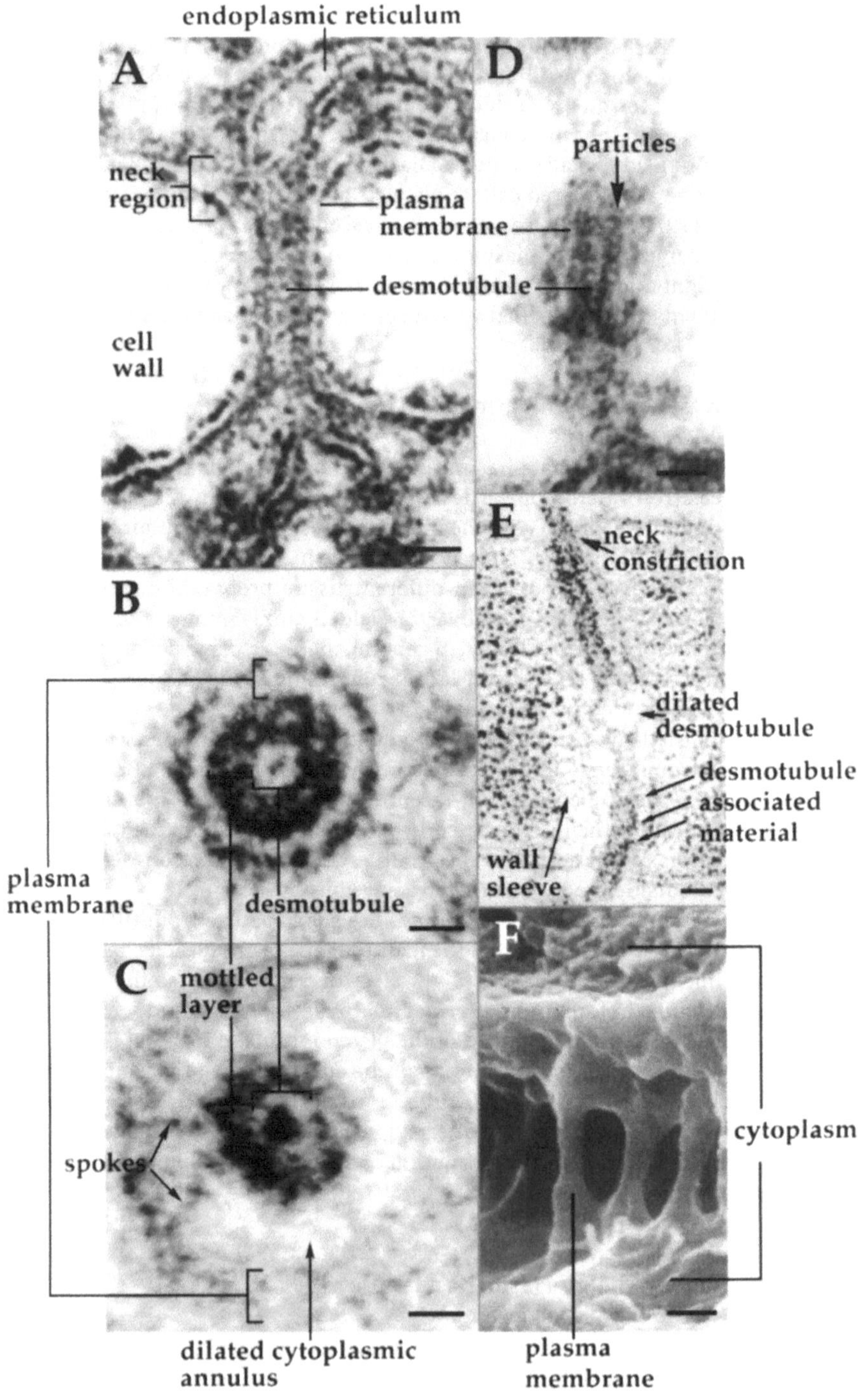
endoplasmic reticulum
A
D
particles
neck region
plasma membrane
desmotubule
cell wall
E
neck constriction
B
dilated desmotubule
desmotubule associated material
wall sleeve
plasma membrane
desmotubule
C
mottled layer
F
spokes
cytoplasm
dilated cytoplasmic annulus
plasma membrane

Fig2A–F. Plasmodesmata as seen in the electron microscope. In both longitudinal (A, D–F) and transverse (B, C) images, the plasma membrane delimits the plasmodesma. The central axial structure – the desmotubule (A–E) – is continuous with the endoplasmic reticulum (ER) of neighbouring cells; it consists of tightly furled ER membrane, although occasionally it dilates to form a lumen (E). A mottled layer outside the desmotubule (A–E) may include negatively stained electron-lucent particles (B–D). Sometimes there is a dilated cytoplasmic annulus outside the mottled layer (C, E), but this is often restricted at the neck region as shown in E; spokes traverse the cytoplasmic annulus (C). A–D are sections from chemically fixed material with tannic acid included in the fixative, E is freeze-substituted material and F is an image produced by high-resolution scanning electron microscopy. All images are from *Azolla* roots, except for C which is from *Egeria densa* leaves treated with the callose synthesis inhibitor DDG, and E from a *Berberis* floral nectary. Bars 25 nm (A, D, E), 10 nm (B, C), and 75 nm (F). (A, B, and D Overall et al. 1982; C J. Radford in Overall and Blackman 1996); F unpublished image kindly provided by P. Vesk and M. Vesk

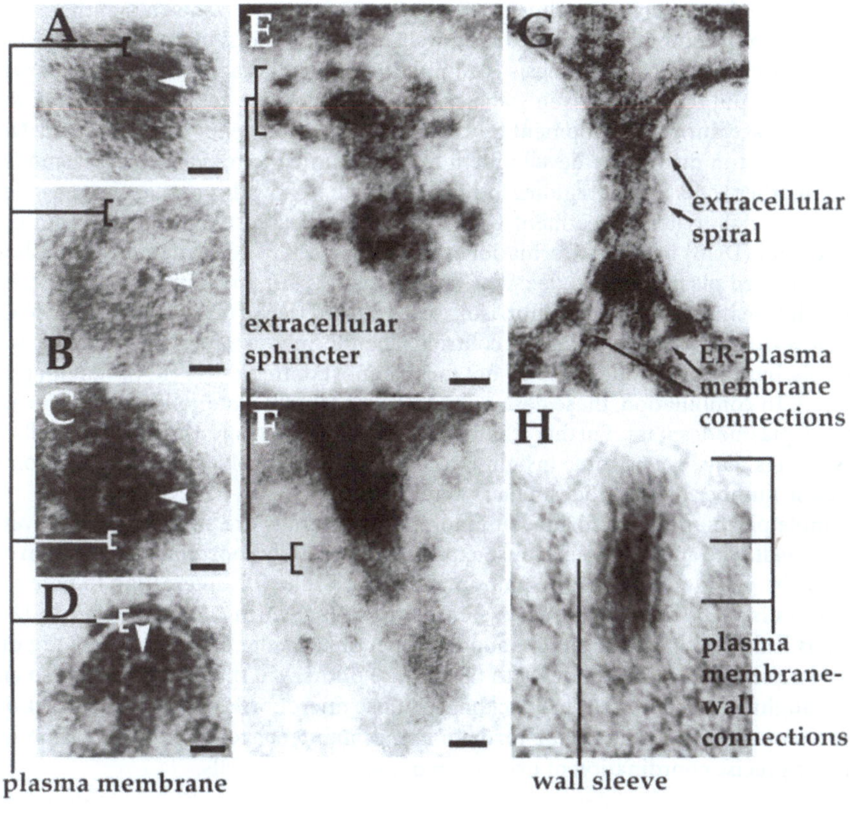

Fig 3A–H. Features of the neck region and wall sleeve of plasmodesmata seen in the electron microscope. At the neck region, the plasma membrane and the tightly furled ER membrane (*arrowheads*) of the desmotubule funnel out into the cytoplasm and ER lumen, respectively (A–D). Electron-dense particles encircle the necks of some plasmodesmata forming putative extracellular sphincters (E–G), and electron-dense material may also spiral down the length of some plasmodesmata (G). At the neck region, the plasma membrane may be connected to the ER (G) and also to the surrounding cell wall (H) with electron-dense spokes. All images are of chemically fixed material with tannic acid included in the fixative. A-D and G are from *Azolla* roots, H from isolated walls of *Azolla* roots, E from *Nephrolepsis exaltata* rhizomes and F from *Spirodela oligorhiza* roots. Bars 10 nm (A–D), 20 nm (E) and 25 nm (E–H). (A–D and G, H Overall et al. 1982; E, F Badelt et al. 1994)

mal components, including specialized surrounding wall structures, are assembled in forming plasmodesmata. Hepler (1982) found that ferric chloride used as a stain for the lumen of the ER was excluded from the ER near the point of cell plate fusion with parent walls, suggesting that the ER forms a tightly furled cylinder as it passes through the plasmodesma. It is not known if the tightly furled cylinder of ER is formed because it becomes squeezed by the fusing vesicles, or if it is formed by a more active process. Proteins of the dynamin family form helical structures around tubular templates and are involved in pinching off membranes (Gu and Verma 1996). It may be that the dynamin-like molecule, phragmoplastin, which has been identified in the forming cell plate (Gu and Verma 1996) plays a role in the production of this tightly curled ER membrane tube.

In some systems there appears to be precise control over the frequency of plasmodesmata put down during cytokinesis. For example, it has been shown that there is no formation of plasmodesmata in existing walls in *Azolla* roots, but rather all plasmodesmata are inserted at cytokinesis (Gunning 1978). Remarkably, the number of plasmodesmata put down in a given wall accurately predicts the cell wall expansion that will take place during development for the correct final frequency to be obtained. New cell walls put in during the development exactly mimic the frequency of plasmodesmata in other walls in that region. Nothing is known of how this process is controlled, but recently short-term treatment with the putative cellulose biosynthesis inhibitor dichlobenil (DCB) has led to formation of thickened cell plates with increased callose and increased plasmodesmal density (Vaughn et al. 1996). In contrast, another putative cellulose biosynthesis inhibitor, isoxaben, leads to the formation of thinner plates with less than 20% of the callose of controls. The ER lines up along the cell plate as in controls, but the plasmodesmata do not form at most of these sites (K. C. Vaughn pers. comm.). In combination, these data point to a role for cell plate callose in the formation of plasmodesmata. Further characterization of the responses of plant cells to these drugs may give us some insight into the processes involved in the deposition of a precise number of plasmodesmata at cytokinesis.

Simple plasmodesmata formed at cytokinesis generally are straight tubes traversing the wall. The adjoining cells must expand in a precisely coordinated fashion in order for the plasmodesmata to remain intact and straight. It is interesting to note that in the walls of BY-2 tobacco suspension culture cells which are habituated to DCB there is little cellulose, but large amounts of pectin, and the plasmodesmata are displaced or oriented more obliquely to the surface of the wall than usual (N. Durso and K. C. Vaughn pers. comm.). It may be that cellulose microfibrils or specific interactions with them are important for the formation and maintenance of straight plasmodesmata or precise coordination of expansion of neighbouring cells.

5 Structure of Primary Plasmodesmata

5.1 Plasma Membrane

The plasma membranes of neighbouring cells, and theoretically the whole plant, are continuous through the plasmodesmata (Fig. 2A, E, F). Its trilaminar structure can be seen clearly around the outside of the plasmodesma cut in transverse section (Fig. 2B).

At the neck region of the plasmodesma, the plasma membrane funnels out and may appear diffuse in transverse section (Fig. 2C). However, there must be a molecular sieve to prevent diffusion of membrane components between adjacent cells, so that individual cell identity markers can be maintained. Proteins commonly found in the plasma membrane are excluded from the regions around the plasmodesmata (for review, see Robards and Lucas 1990; Fleurat-Lessard et al. 1995). Grabski et al. (1993) demonstrated that plasma membrane lipids are not able to diffuse between neighbouring cells. The membrane-associated particles observed at the necks of plasmodesmata using freeze-fracture electron microscopy (Willison 1976; Thomson and Platt-Aloia 1985) may act as a sieve for molecular diffusion in the plasma membrane through the plasmodesmata. Thomson and Platt-Aloia (1985) have also shown that the external face of the plasma membrane of plasmodesmata in *Tamarix* salt glands is devoid of intramembranous particles, and they suggest that this half of the membrane might be in a gel, or non-fluid state. In contrast, they have shown that the protoplasmic face of the plasma membrane within the plasmodesma is enriched in particles in agreement with the suggestions of Ding et al. (1992) and Botha et al. (1993) on the basis of computer-enhanced images of plasmodesmata. However, it should be noted that while these workers have identified the inner electron-dense layer of the plasma membrane as being wider than the external electron-dense layer, it has not been noted by numerous other workers and, indeed, Schulz (1995) specifically notes that this is not the case in plasmodesmata of *Pisum sativum* roots.

5.2 Desmotubule

5.2.1 Endoplasmic Reticulum or Proteinaceous Cytoskeletal Element?

There is a central axial structure, termed the desmotubule, inside the plasmodesma that is generally believed to be connected to the ER in the adjoining cells. There is continuing controversy over the details of its structure. Most models now interpret the desmotubule as being tightly furled ER, as was first suggested by Lopez-Saez et al. (1966), but others have interpreted the structure as more like a cytoskeletal element.

Initially, it was suggested that the plasmodesmata form as remnants of the spindle fibres, and the similarity of the mottled layer in plasmodesmata to a microtubule led a number of workers in the 1960s to suggest the desmotubule was actually a cytoplasmic microtubule (for review, see Jones 1976). This idea was dismissed rather quickly because plasmodesmata with desmotubules are seen in some algal walls which are formed with the microtubules parallel to the forming cell plate (Marchant 1976) and the desmotubules do not always maintain tubular form (O'Brien and Thimann 1967). It is interesting that tubulin has been found in protein extracts of walls containing plasmodesmata and not in extracts of walls without plasmodesmata, but the tubulin could not be convincingly localized to plasmodesmata using immunogold cytochemistry (Blackman and Overall 1998). Robards (1971) modified the cytoplasmic microtubule idea such that the desmotubule was a hollow tube continuous with the lumen of the ER and bounded by a proteinaceous wall identified as the mottled layer seen in electron micrographs. The inner electron-lucent ring was interpreted as a lumen continuous with the lumen of the ER and the central electron-dense dot was thought to be an artefact. Waigmann et al. (1997) have recently observed plasmodesmata in the trichomes of *Nicotiana clevelandii* in which the central electron-dense dot is not visible in the desmotubule. They also suggested that in that system the desmotubule may be a proteinaceous cylinder with a lumen open for transport, possibly directly connected to the cytoplasm rather than the ER lumen.

The model of Robards (1971) was put into doubt, and that of Lopez-Saez et al. was given strong support by images of tannic acid/ferric chloride-stained plasmodesmata in which the electron-lucent layer of the ER membrane appeared continuous with the electron-lucent region in the central axial structure of the plasmodesmata (Fig. 2A; Overall et al. 1982). The dimensions of the electron-dense dot in the centre of the plasmodesmata are comparable to the dimensions of appressed inner leaflets (or phospholipid head groups) of ER membrane and the central electron-lucent ring has the same dimension as the electron-lucent layer in the ER membrane (or combined hydrocarbon tails of the lipid bilayer; Overall et al. 1982). On the basis of computer-enhanced images, Ding et al. (1992) and Botha et al. (1993) proposed that this central electron-dense dot represented proteinaceous particles, but they did not reconcile this suggestion with considerations of the staining pattern expected for the phospholipid head groups of the inner leaflet of the ER membrane bilayer, nor of what the staining pattern would be when the desmotubule is expanded. They also identified fine electron-dense strands which crossed the electron-lucent layer of the desmotubule. At the necks of plasmodesmata, the electron-lucent ring of the ER increases its diameter, as does the central black dot as it opens into the lumen of the ER (Fig. 3A-D). Occasionally, the electron-lucent layer at the neck region of the desmotubule can be observed surrounded by an electron-dense layer of the dimensions expected for the outer phospholipid head groups of the ER (Fig. 3B). The absence of the mottled layer or electron-dense particles embedded in the ER membrane, as proposed by Ding et al. (1992), suggests that they are not integral to the desmotubule structure. Contrary to earlier ideas about lipid packing, it has been shown that lipids similar to those in ER would be able to pack into a structure with a tight curvature like that required in the desmotubule (Overall et al. 1982).

Tilney et al. (1991) concluded that the central axial structure in the plasmodesmata of gametophytes of *Onoclea* was composed mainly of proteins, as digestion with a protease removed the structure but it remained intact following treatment with detergents. They proposed that the desmotubule develops as a tightly furled ER membrane and that intramembranous proteins are transported in the plane of the ER membrane into the desmotubule region and become cross-linked, forming a stable cytoskeletal-like element. At the very least they concluded that a lipid-based desmotubule must be stabilized by associated proteins. In direct contrast to these observations, however, Turner et al. (1994) showed that extraction with membrane-solubilizing detergents removed the plasma membrane and the desmotubule from the plasmodesmata of wall fragments from maize root tips. While heavy proteolysis removed the entire plasmodesma, leaving only a hole in the wall, limited proteolysis extracted material from the neck regions of the plasmodesmata but there were no discernible effects in the cores of the plasmodesmata. It is difficult to reconcile these conflicting observations, but the prediction by Tilney et al. (1991) on the basis of their model that diffusion of membrane lipids through the plasmodesmata would take place in the plasma membrane and not the desmotubule was not borne out by results of Grabski et al. (1993). They found that lipids can diffuse between cells in the ER but not in the plasma membrane, pointing to a membranous desmotubule continuous with the ER. A further observation of Tilney et al. (1991), namely that the desmotubule breaks its connection readily from the ER during plasmolysis, has also been contradicted subsequently. The fluorescent probe $DiOC_6$ has been used to demonstrate that the ER is continuous between adjacent cells at pit fields and remains tightly associated with plasmodesmata during at least the initial stages of plasmolysis (Oparka et al. 1994).

5.2.2 Transport Through the Desmotubule?

While the general interpretation of the desmotubule is a tightly furled cylinder of ER that does not function as a pathway for intercellular transport (Overall et al. 1982), it must be remembered that the images of closed desmotubules have been obtained following various tissue preparation techniques which could well stimulate the closure of such a channel or, alternatively, such a channel may be opened only momentarily during specific transport. Dilated portions of desmotubules have often been observed (Fig. 2E; Robinson-Beers and Evert 1991; Glockmann and Kollmann 1996; Waigmann et al. 1997), usually in the centre of plasmodesmata. The desmotubules without central electron-dense dots observed by Waigmann et al. (1997) in *Nicotiana clevelandii* trichomes may well simply be an ER tubule which is not tightly furled, but rather has a central lumen continuous with the ER. It has been thought generally that constrictions in the dimensions of the plasmodesmata at the neck regions would act as a barrier to the opening of the desmotubule, but it now appears that this neck constriction is an artefact (Radford et al. 1998). The potential for ER to act as intercellular channel for photosynthates (Gamalei et al. 1994) or water (Zhang and Tyerman 1997) has refocused interest on the possible role of ER in intercellular transport. Lazzaro and Thomson (1996) have observed a dynamic vacuolar-tubular network staining with Lucifer Yellow CH in trichomes of chickpea that is connected between adjacent cells via fluorescent strands across the intervening cell wall. They proposed that the vacuolar-tubular network in that system was continuous with the desmotubule and that intercellular transport of its contents occurred via a lumen in the desmotubule. The desmotubule may be a dynamic continuity of intracellular membranous networks such as ER or vacuolar-tubular systems which make wholesale movements through the plasmodesmata, possibly powered by an actin-myosin system.

5.3 Cytoplasmic Annulus

The cytoplasmic annulus is defined as the space between the desmotubule and the plasma membrane and it has been thought to be the major pathway for transport of molecules from cell to cell (Gunning and Overall 1983). There is continuing controversy over the details of any molecular components within the cytoplasmic annulus and the dimensions and geometry of the cytoplasmic annulus that is available for transport.

5.3.1 Interpretation of the Mottled Layer Outside the Desmotubule: a Possible Role for Actin

The mottled layer surrounding the electron-lucent layer of the desmotubule shows various patterns of electron-dense and electron-lucent regions. The interpretation of the architecture of this region has been controversial largely because of uncertainties about the mode of action of stains. Often this mottled layer abuts directly onto the plasma membrane, as in Fig. 2B, but when the cytoplasmic annulus is expanded, there is an electron-lucent region between the mottled layer and the plasma membrane, as in Fig. 2C.

The interpretation of the mottled layer as the outer wall of the desmotubule (Robards 1971; Olesen 1979) has largely been discounted due to the arguments presented above. Overall et al. (1982) suggested that the electron-lucent regions in the mottled layer of plasmodesmata from *Azolla* roots may represent negatively stained particles. In some transverse sections of plasmodesmata, one or possibly two adjacent electron-lucent regions appear clearly as circles 4.5 nm in diameter and outlined by electron-dense material (Fig. 2B, C). These electron-lucent circles appear to be spheres, as three such electron-lucent circles were visible in one particularly fortuitous longitudinal oblique section of a plasmodesma (Fig. 2D). The fact that all of these spheres are not clearly visible in a transverse section suggests that they are not positioned side by side, but rather form some sort of spiral or other longitudinal arrangement. The occasional clearly demarcated sphere suggests either that any superimposed spheres within the section are in exact register or that it is not superimposed by another sphere within the plane of the section. Since only two of the total of nine particles thought to encircle the perimeter of the desmotubule (Overall et al. 1982) are clear, it appears that the remaining seven are superimposed by other particles within the section but out of register (Fig. 2B). This would suggest that within the section thickness, say 70 nm, the particles make approximately one and three quarter complete spirals around the desmotubule, that is one spiral every 40 nm. The three clear particles in Fig. 2D suggest that the oblique section of the plasmodesma was cut by chance at the angle of the spiral of particles. The spiral of material around the desmotubule of a plasmodesma from a *Berberis* floral nectary (Fig. 2E) may well be composed of electron-lucent particles as observed in *Azolla* plasmodesmata. It is interesting that striations of particles on the surface of the desmotubule at an angle of 20–30° with reference to the transverse axis of the desmotubule have been proposed by Ding et al. (1992) and Botha et al. (1993). Ding et al. (1992) obtained remarkably similar images to that in Fig. 2B following freeze substitution of leaves of *Nicotiana tabacum* with and without tannic acid (their Fig. 6a, b). However, their transverse sections did not include any clearly demarcated electron-lucent spheres as in Fig. 2B, but they were almost identical to Figs. 12-14 in Overall et al. (1982). Perhaps the section thickness in these cases is such that the details of all the individual particles are obscured by other particles that are superimposed but out of register within the section.

Most of the images presented in Ding et al. (1992) were manipulated by computer image processing in general to enhance the "particulate nature of plasmodesmata". This image manipulation led Ding et al. (1992) to conclude that the mottled layer consisted of electron-dense particles of 3 nm diameter and embedded in the lipid on the cytoplasmic face of the desmotubule, with similar electron-dense particles embedded in the inner face of the plasma membrane. The electron-lucent regions of the mottled layer were interpreted as spaces between these electron-dense particles. However, this interpretation is difficult to reconcile with images such as Fig. 2C, in which the electron-lucent areas of the mottled layer are clearly visible outlined by a thin electron-dense layer, and yet any electron-dense particles on the inner leaflet of the plasma membrane are separated from the mottled layer by an obvious cytoplasmic annulus. The mottled layers in the plasmodesmata shown in Fig. 2B, C are approximately 9 nm in diameter, comparable to the 10-nm diameter mottled layer observed by Ding et al. (1992). If the interpretation of Ding et al. were correct, then there would be only 3 nm electron-dense particles surrounding the desmotubule before the cytoplasmic annulus in plasmodesmata with an expanded cytoplasmic annulus such as in Fig. 2C. Even in the image provided by Ding et al. (1992) of a plasmodesma with an expanded cytoplasmic annulus (their Fig. 4b), however, the desmotubule is surrounded by an electron-dense or mottled layer of 10 nm!

Botha et al. (1993) have also used computer image enhancement on images of plasmodesmata at the Kranz mesophyll-bundle sheath interface in *Themeda triandra.* They presented images of these plasmodesmata where the cytoplasmic annulus was constricted as in Fig. 2B. While the false colour images they produced were intriguing, they did not provide any new insights on the structure of the mottled layer. Indeed the usefulness of computer-generated image enhancement to gain information that is not already visible in sections has to be questioned. Clearly, progress will require techniques for generating 3-D computer models of the structure of plasmodesmata based upon the known staining properties of various structures throughout the thickness of a section.

The electron-lucent particles in the mottled layer may well be actin. Using immunogold cytochemistry and rhodamine phalloidin labelling in the confocal microscope, White et al. (1994) demonstrated that actin is a component of plasmodesmata not only at their necks but also along their length. Actin filaments that have been chemically fixed in the presence of 0.2% tannic acid appear as electron-lucent particles of 5.5–7 nm diameter surrounded by an electron-dense coat making their total diameter 11–17 nm (Maciver et al. 1991). These dimensions are comparable to the dimensions quoted for the electron-lucent particles and surrounding electron-dense material in the mottled layer (Overall et al. 1992). Usually only one or possibly two gold particles were observed in any transverse section of a plasmodesma. As the antibody is able to recognize only actin that is at the surface of the section, this observation is consistent with the suggestion that the electron-lucent particles form a spiral around the desmotubule. The close association of actin, ER, and the plasma membrane has been observed previously (Lichtscheidl et al. 1990) and in retrospect it is not surprising that it is also present in the plasmodesmata.

5.3.2 Spokes Connect the Mottled Layer with the Plasma Membrane

Electron-dense spokes have been observed to connect the mottled layer to the inner leaflet of the plasma membrane where there is a clear electron-lucent cytoplasmic annulus (Fig. 2C). These spokes have been observed in a wide variety of plasmodesmata in transverse (e.g. Burgess 1971; Tilney et al 1991; Ding et al. 1992; Schulz 1995; Cook et al. 1997) and longitudinal sections (Ding et al. 1992). The spokes are clearly visible in expanded cytoplasmic annuli, although it is possible that they are also present in a shortened form or parallel to the axis of the plasmodesma when no clear electron lucent cytoplasmic annulus can be seen. These spokes may be myosin which is expected to have a close association with ER (Liebe and Quader 1994) and actin microfilaments, and in addition there is high ATPase activity at the plasmodesmata (Robards and Lucas 1990). Indeed, myosin has recently been identified in the plasmodesmata of the green alga *Chara corallina* (Blackman and Overall 1998) and higher plants (Radford and White 1998).

5.3.3 Dimensions and Regulation of the Transport Pathway Through the Cytoplasmic Annulus

Since the cytoplasmic annulus was thought to be the major pathway for intercellular transport of molecules (Gunning and Overall 1983), much attention has been focused on the proposed dimensions of the pathways available for transport. Overall et al. (1982) argued that even though plasmodesmata may have expanded cytoplasmic annuli in the centre, in general there was a tight association of the plasma membrane to the mottled layer at the neck region in a construction termed the neck constriction. Transport was proposed to be restricted to channels surrounding electron-lucent particles arranged side by side in the mottled layer. These spaces were calculated to be in the order of 3 nm^2, comparable to the then known size exclusion limit for plasmodesmata in the order of 1-kDa (Goodwin 1983; Tucker 1982). If it is accepted that these particles are arranged in a spiral, then the transport pathway would form a spiral of at least the thickness of the particles, approximately 6 nm.

In a very thorough study, Schulz (1995) demonstrated that a 1-h treatment of *Pisum sativum* roots with 350 mM mannitol, a treatment known to transiently stimulate phloem unloading in this system, dramatically increased the width of plasmodesmata compared with those in roots not treated with mannitol, or those treated with mannitol for 3 h. In all treatments, the dimensions of the desmotubule and the associated mottled layer remained constant and it was the cytoplasmic annulus that increased dramatically with the short mannitol treatment, suggesting that this pathway was used for transport during phloem unloading. The spokes connecting the mottled layer to the plasma membrane remained intact during the widening of the cytoplasmic annulus. These findings corroborated the conclusion of Bret-Harte and Silk (1994) that sucrose flow through plasmodesma pore sizes of 3 nm^2 was not sufficient to meet the carbon demands of root tips.

Subsequently, it has been demonstrated that the neck constrictions in the roots of *Allium cepa* L. are an artefact of callose deposition around the neck stimulated by tissue preparation and chemical fixation (Radford et al. 1998). Callose is a β-1, 3-glucan which is rapidly deposited between the plasma membrane and the wall in response to wounding (Sec. 5.4.2). In roots treated with 2-deoxy-D-glucose (DDG), an inhibitor of callose synthesis, prior to and during fixation, the plasmodesmata have open funnel-shaped necks, but the plasmodesmata in roots prepared without DDG had clear neck constrictions (Radford et al. 1998). The open neck configuration seen in Fig. 2C is in a plasmodesma of an *Egeria densa* leaf treated with DDG. It maybe that the 1-h mannitol treatment in the work of Schulz (1995) served to protect the plasmodesmata from the damaging effects of chemical fixation.

In any event, the width of the cytoplasmic lumen in these open plasmodesmata between the mottled layer and the plasma membrane calculated from the images and measurements in Schulz (1995) and Radford et al. (1998) is 7–14 nm. If the mottled layer represents a spiral of particles, then the pathway available for transport through the cytoplasmic lumen between the particles could be 13–20 nm wide! This pathway would clearly allow passage of molecules much larger than 1 kDa. In recent years, some cells have been shown to have much larger size exclusion limits. For example, dextrans of at least 10 kDa pass between companion cells and sieve elements (Kempers and van Bel 1997), 7-kDa dextrans pass plasmodesmata in tobacco trichomes (Waigmann and Zambryski 1995), and 40-kDa molecules pass plasmodesmata in *Nitella* (Kikuyama et al. 1992). Furthermore, virally encoded proteins assist the intercellular passage of macromolecules up to 30-kDa (for review, see Ghoshroy et al. 1997). Certainly molecules with a Stokes radius of larger than the 3 nm calculated for a 20-kDa dextran (Wolf and Lucas 1994) would be expected to pass the cytoplasmic channel.

If this larger cytoplasmic annulus is to be considered the in vivo state of transporting plasmodesmata, the question arises as to why molecules larger than 1 kDa do not routinely pass through plasmodesmata. One possibility is that the techniques used to study plasmodesmal transport such as microinjection stimulate a wound response and close the plasmodesmata. A second possibility is that there is some as yet unidentified molecular sieve at the necks of plasmodesmata that regulates the intercellular passage of molecules. Such a proposal is supported by the observation that plasmodesmata actually pass larger molecules following severe plasmolysis (Erwee and Goodwin 1984), suggesting that some delicate molecular sieving mechanism at the neck may have been damaged.

A third possibility is that the dimensions of this pathway are constantly being modified. A possible site for this modulation could be around the neck region where electron-dense material has been observed connecting the plasma membrane and the ER as it enters the plasmodesma (Fig. 3G). Intercellular communication is inhibited with increasing intracellular concentration of calcium (Erwee and Goodwin 1983) and inositol 1, 4, 5-triphosphate (Tucker 1988), and in some systems increased with increases in intracellular ATP (Tucker 1993; Cleland et al. 1994). This suggests that a contractile calcium-binding protein may be involved in the regulation of intercellular communication. Indeed, the calcium-binding phosphoprotein centrin which forms filamentous structures that contract rapidly with increases in calcium concentration and require ATP for extension (Salisbury 1995) appears to be localized to the neck regions of plasmodesmata (Blackman et al. 1999). Centrin may modulate the pathway for transport through the cytoplasmic sleeve by controlling the space between the plasma membrane and the ER at the neck region.

Alternatively, the dimensions of the pathways through the plasmodesmata may be regulated by an actin-myosin system. Treatment with the actin-disrupting agent, cytochalasin B, dramatically increased the dimensions of the cytoplasmic annulus in one of the three species studied (White et al. 1994), suggesting that actin may play a role in maintaining the precise dimensions of this transport pathway through plasmodesmata. Perturbation of the actin filaments in tobacco mesophyll cells by injections of cytochalasin D or profilin allowed the intercellular passage of a 20-kDa dextran normally excluded (Ding et al. 1996). The increase in dimensions of plasmodesmata observed following papain treatment (Tilney et al. 1991) may have been due to disruption of the actin and myosin involved in maintaining integrity of the pathway.

5.4 Specializations of the Wall Sleeve

The sleeve of wall material surrounding the plasmodesmata must be quite specialized in order to accommodate, or even modulate, changes in dimensions of plasmodesmata such as described above. There is often an electron-lucent region in the wall surrounding the plasmodesmata (Figs. 2E and 3H). Such electron-lucent wall sleeves are not visible in other longitudinal sections of plasmodesmata (Figs. 2A, D, 3E–G) because in these preparations the walls had been digested out after fixation in order to enhance the details of the plasmodesma substructure.

5.4.1 Plasmodesmata Are Firmly Anchored to the Surrounding Wall

Fine electron-dense spokes are sometimes observed to connect the plasma membrane to the surrounding wall material (Fig. 3H; Cook et al. 1997) clearly at the neck region but also down the length of the plasmodesma. These connections presumably anchor the plasmodesmata in the walls following plasmolysis (Tilney et al. 1991; Oparka et al. 1994) or wall isolation (Tilney et al. 1991; Turner et al. 1994). Of course, they may also be involved in regulating the dimensions of the plasmodesmal channel.

5.4.2 Callose

Callose, a β-1,3-glucan, is deposited between the plasma membrane and the plant wall and appears electron-lucent in electron micrographs (for review, see Stone and Clarke 1992). Callose is deposited at plasmodesmata in response to wounding or chemical fixation (Smith and McCully 1977; Galway and McCully 1987; Northcote et al. 1989; Balestrini et al. 1994; Turner et al. 1994; Radford et al. 1998), and Delmer et al. (1993) have localized the callose synthase to plasmodesmata. The deposition of callose around plasmodesmata has been thought to restrict transport through them (Olesen and Robards 1990), thereby serving as a crude sphincter mechanism which is deposited very rapidly but takes hours to disappear. One possible site of callose deposition is the raised electron-lucent collar surrounding the necks of some plasmodesmata (Olesen and Robards 1990). This collar is associated with constricted necks within the plasmodesmata (see Radford et al. 1998). Turner et al. (1994) found that antibodies to callose did not label these raised collars but rather the wall regions between them. In addition, following treatment with callose-degrading enzymes there was no labelling of callose, but the raised collars were not removed. Turner et al. (1994) concluded that the cell wall around the plasmodesmata was structurally subdivided, with an inner amorphous collar surrounded by a peripheral zone in which callose is interspersed with the fibrillar wall material. On the basis of differential extractions, they concluded that the inner amorphous zone was a carbohydrate held in position by protein links. This contradicts the findings of Radford et al. (1998) in which inhibition of callose deposition with DDG prevented the formation of raised collars.

5.4.3 Pectin

Pectins play a role in cell wall hydration, adhesion of adjoining cells, and wall plasticity (Vreeland et al. 1989). Unesterified pectins which are found in the middle lamella are located to the plasmodesmata in tomato pericarp cells (Casero and Knox 1995). Low esterified pectin surrounds the plasmodesmata in ripening apples and are not involved in calcium cross-bridging, but are probably surrounded by a cationic environment (Roy et al. 1997). These modifications are thought to protect the wall around the plasmodesmata from hydrolytic enzyme breakdown during ripening. Similar cellulase-resistant wall tubes have been observed around barley aleurone plasmodesmata (Taiz and Jones 1973). During protoplast formation, cells may remain attached at plasmodesmata (for review, see Carr 1976), suggesting that this protective wall sleeve may be a common feature of plasmodesmata. It remains to be demonstrated if and how this pectin sleeve plays a role in modifying the dimensions of the plasmodesmal pathway.

5.4.4 Putative Extracellular Sphincters

Inclusion of tannic acid in the fixative has highlighted a number of presumably proteinaceous components just outside the plasma membrane of the plasmodesmata (Figs. 2D, 3E–G). They have now been observed by a number of workers in a variety of plant tissues (Olesen 1979; Overall et al. 1982; Mollenhauer and Morré 1987; Olesen and Robards 1990; Tilney et al. 1991; Robinson-Beers and Evert 1991; Badelt et al. 1994; Cook et al. 1997) and can also be discerned, although not as clearly, in material prepared without chemical fixation or tannic acid (Badelt et al. 1994). Rings of electron-dense material encircle the necks of some plasmodesmata and occasionally they can be clearly discerned to comprise definite particles (Fig. 3E) which, in turn, appear to have a clear substructure (Fig. 3F). Sometimes the ring is closely associated with the plasma membrane (Fig. 2C). In other images, the ring is expanded out from the plasma membrane and connected to it via spokes or strings (Figs. 3E, F). This arrangement of particles is reminiscent of the open and closed rings of particles observed by Willison (1976) around the necks of plasmodesmata following freeze fracture. A second major structure observed is formed by electron-dense spirals which encircle the length of the plasmodesma (Fig. 3G), possibly anchoring at the ring of particles or on the plasma membrane around the necks of the plasmodesma (Badelt et al. 1994).

Nothing is known of the molecular composition of these structures. Olesen and Robards (1990) suggested that the electron-dense particles may be callose synthase complexes, but this is contradicted by the deposition of callose in the root cap cells of onion roots in which these electron-dense particles are absent (Radford et al. 1998). The particle rings and spiral, either individually or together, could contract to constrict the cytoplasmic annulus. Alternatively, if the spiral were a fixed length, it could be made to wrap more tightly or loosely around the plasmodesma if its ends at the

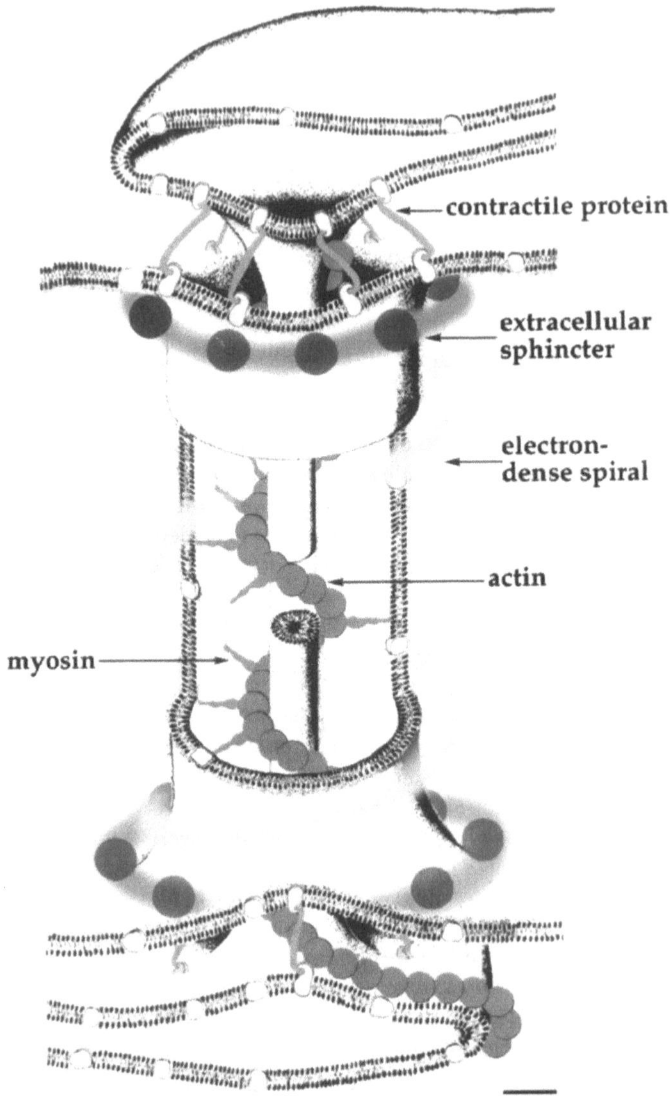

Fig 4. Model of a plasmodesma. Actin and myosin, which are helically arranged around the desmotubule, connect the plasma membrane to the desmotubule. Contractile proteins (possibly centrin) link the plasma membrane to the endoplasmic reticulum (ER) via anchoring proteins at the neck of the plasmodesma. These anchoring proteins within the membrane may have a close association with the putative extracellular sphincter. These extracellular structures may provide anchoring points to the wall for the components of the plasmodesma. The extracellular sphincters at the neck and the extracellular spiral could also be involved in modulating the outer dimensions of the plasmodesma. Transport of specific macromolecules could occur via the actin-myosin motile system or via a dilated ER. The size exclusion limit for possible transport between cells would depend on the dimension of the gap between the desmotubule and the plasma membrane as determined by contractile proteins, or by an as yet unidentified molecular sieve. Bar 10 nm. (Overall and Blackman 1996)

necks of the plasmodesma were wound in opposite directions. This could occur by a fixed attachment to the ring of particles which could spin in opposite directions driven by a molecular motor. Alternatively, the ring of particles could remain fixed and the point of attachment of the spiral could move around the rings in opposite directions at each end of the plasmodesma. There is a difficulty in considering the molecular motors operating in the apoplasm, and it may be that the ring of particles themselves are transmembrane proteins as in Fig. 3G or are connected to transmembrane proteins. The cytochalasin treatment that expanded plasmodesmata also caused the loss of these extracellular structures (White et al. 1994), suggesting that the actin-myosin complex is connected in some way via transmembrane components to the wall sleeve. A more radical proposal would be that these extracellular components actually contain actin. The high ATPase concentration in the plasmodesmata (Robards and Lucas 1990) may be involved in powering the dynamics of any proposed motor, but until the molecular details of these structures are resolved, these ideas must remain as fanciful speculation!

6 Concluding Remarks

The static models of plasmodesmata generated largely on the basis of electron microscopy (e.g. Gunning and Overall 1983) must now incorporate the emerging molecular information about plasmodesmal components. Fig. 4 is an attempt to do this, incorporating the various components of plasmodesmal substructure presented in this chapter. This model is no doubt a gross oversimplification of what could well turn out to be a complex interaction of as yet unknown molecular components. Further progress in our understanding of the substructure of plasmodesmata and how these assemblages function in the transport of viruses and macromolecules awaits further molecular characterization of these components.

References

Badelt K, White RG, Overall RL, Vesk M (1994) Ultrastructural specialisations of the cell wall sleeve around plasmodesmata. Am J Bot 81:1422–1427

Balestrini R, Romera C, Puigdomenech P, Bonfante P (1994) Location of a cell-wall hydroxyproline-rich glycoprotein, cellulose and beta-1, 3-glucans in apical and differentiated regions of maize mycorrhizal roots. Planta 195:201–209

Blackman LM, Harper JDI, Overall RL (1999) Localization of a centrin-like protein to higher plant plasmodesmata. Eur J Cell Biol 78:297–304

Blackman LM, Overall RL (1998) Immunolocalisation of the cytoskeleton to plasmodesmata of *Chara corallina*. Plant J 14:733–741

Botha CEJ, Hartley BJ, Cross RHM (1993) The ultrastructure and computer-enhanced digital image analysis of plasmodesmata at the Kranz mesophyll-bundle sheath interface of *Themeda triandra* var. imberbis (Retz) A. Camus in conventionally fixed leaf blades. Ann Bot 72:255–261

Bret-Harte MS, Silk WK (1994) Nonvascular, symplastic diffusion of sucrose cannot satisfy the carbon demands of growth in the primary root tip of *Zea mays* L. Plant Physiol 105:19–33

Burgess J (1971) Observations on structure and differentiation in plasmodesmata. Protoplasma 73: 83–95

Buvat R (1957) L'infrastructure des plasmodesmes et la continuité des cytoplasmes. C R Acad Sci Paris D 245:198–201

Carr DJ (1976) Plasmodesmata. In: Gunning BES, Robards AW (eds) Intercellular communication in plants: studies on plasmodesmata. Springer, Berlin Heidelberg New York, pp 243–289

Casero PJ, Knox JP (1995) The monoclonal antibody JIM5 indicates patterns of pectin deposition in relation to pit fields at the plasma-membrane-face of tomato pericarp cell walls. Protoplasma 188: 133–137
Cleland RE, Fujiwara T, Lucas WJ (1994) Plasmodesmal-mediated cell-to-cell transport in wheat roots is modulated by anaerobic stress. Protoplasma 178:81–85
Cook ME, Graham LE, Botha CEJ, Lavin CA (1997) Comparative ultrastructure of plasmodesmata of *Chara* and selected bryophytes: toward an elucidation of the evolutionary origin of plant plasmodesmata. Am J Bot 84:1169–1178
Delmer DP, Volokita M, Solomon M, Fritz U, Delphendahl W, Herth W (1993) A monoclonal antibody recognises a 65-kDa higher plant membrane polypeptide which undergoes cation-dependent association with callose synthase in vitro and colocalizes with sites of high callose deposition in vivo. Protoplasma 176:33–42
Ding B, Turgeon R, Parthasarathy MV (1992) Substructure of freeze-substituted plasmodesmata. Protoplasma 169:28–41
Ding B, Kwon M-O, Warnberg L (1996) Evidence that actin filaments are involved in controlling the permeability of plasmodesmata in tobacco mesophyll. Plant J 10:157–164
Erwee MG, Goodwin PB (1983) Characterisation of *Egeria densa* Planch. leaf symplast: inhibition of the intercellular movement of fluorescent probes by group II ions. Planta 158:320–328
Erwee MG, Goodwin PB (1984) Characterisation of *Egeria densa* symplast: response to plasmolysis, deplasmolysis and to aromatic amino acids. Protoplasma 122:162–168
Fleurat-Lessard P, Bouché-Pillon S, Leloup C, Lucas WJ, Serrano R, Bonnemain JL (1995) Absence of plasma membrane H^+-ATPase in plasmodesmata located in pit-fields of the young reactive pulvinus of *Mimosa pudica* L. Protoplasma 188:180–185
Galway ME, McCully ME (1987) The time course of the induction of callose in wounded pea roots. Protoplasma 139:77–91
Gamalei YV, van Bel AJE, Pakhova MV, Sjutkina AV (1994) Effects of temperature on the conformation of the endoplasmic reticulum and on starch accumulation in leaves with the symplasmic minor-vein configuration. Planta 194:443–453
Ghoshroy S, Lartey R, Sheng J, Citovsky V (1997) Transport of proteins and nucleic acids through plasmodesmata. Annu Rev Plant Physiol Plant Mol Biol 48:27–50
Glockmann C, Kollmann R (1996) Structure and development of cell connections in the phloem of *Metasequoia glyptostroboides* needles I. Ultrastructural aspects of modified primary plasmodesmata in Strasburger cells. Protoplasma 193:191–203
Goodwin PB (1983) Molecular size limit for movement in the symplast of the *Elodea* leaf. Planta 157: 124–130
Grabski S, de Feijter AW, Schindler M (1993) Endoplasmic reticulum forms a dynamic continuum for lipid diffusion between contiguous soybean root cells. Plant Cell 5:25–38
Gu X, Verma DPS (1996) Phragmoplastin, a dynamin-like protein associated with cell plate formation in plants. EMBO J 15:695–704
Gunning BES (1978) Age-related and origin-related control of the numbers of plasmodesmata in cell walls of developing *Azolla* roots. Planta 143:181–190
Gunning BES, Overall RL (1983) Plasmodesmata and cell-to-cell transport in plants. Bioscience 33: 260–265
Hepler PK (1982) Endoplasmic reticulum in the formation of the cell plate and plasmodesmata. Protoplasma 111:121–133
Jones MJK (1976) The origin and development of plasmodesmata. In: Gunning BES, Robards AW (eds) Intercellular communication in plants: studies on plasmodesmata. Springer Verlag, Berlin Heidelberg New York, pp 81–103
Kempers R, van Bel AJE (1997) Symplasmic connections between sieve element and companion cell in the stem phloem of *Vicia faba* L. have a molecular exclusion limit of at least 10-kDa. Planta 201: 195–210
Kikuyama M, Hara Y, Shimada K, Yamamoto K, Hiramoto Y (1992) Intercellular transport of macromolecules in *Nitella*. Plant Cell Physiol 33:413–417
Lazzaro MD, Thomson WW (1996) The vacuolar-tubular continuum in living trichomes of chickpea (*Cicer arietinum*) provides a rapid means of solute delivery from base to tip. Protoplasma 193: 181–190
Lichtscheidl IK, Lancelle SA, Hepler PK (1990) Actin-endoplasmic reticulum complexes in *Drosera*. Their structural relationship with the plasmalemma, nucleus, and organelles in cells prepared by high pressure freezing. Protoplasma 155:116–126
Liebe S, Quader H (1994) Myosin in onion (*Allium cepa*) bulb scale epidermal cells: involvement in dynamics of organelles and endoplasmic reticulum. Physiol Plant 90:114–124
Lopez-Saez JF, Gimenez-Martin G, Risueno MC (1966) Fine structure of the plasmodesm. Protoplasma 61:81–84

Maciver SK, Wachsstock SH, Schwarz WH, Pollard TD (1991) The actin filament severing protein actophorin promotes the formation of rigid bundles of actin filaments crosslinked with alpha-actinin. J Cell Biol 115:1621–1628

Marchant HJ (1976) Plasmodesmata in algae and fungi. In: Gunning BES, Robards AW (eds) Intercellular communication in plants: studies on plasmodesmata. Springer, Berlin Heidelberg New York, pp 59–78

Mollenhauer H, Morré J (1987) Some unusual staining properties of tannic acid in plants. Histochemistry 88:17–22

Northcote DH, Davey R, Lay J (1989) Use of antisera to localise callose, xylan and arabinogalactan in the cell-plate, primary and secondary walls of plant cells. Planta 178:353–366

O'Brien TP, Thimann KV (1967) Observations on the fine structure of the oat coleoptile. II. The parenchyma cells of the apex. Protoplasma 63:417–442

Olesen P (1979) The neck constriction in plasmodesmata. Evidence for a peripheral sphincter-like structure revealed by fixation with tannic acid. Planta 144:349–358

Olesen P, Robards AW (1990) The neck region of plasmodesmata: general architecture and some functional aspects. In: Robards AW, Lucas WJ, Pitts JD, Jongsma HJ, Spray DC (eds) Parallels in cell to cell junctions in plants and animals. Springer, Berlin Heidelberg New York, pp 145–170

Oparka KJ, Prior AM, Crawford JW (1994) Behaviour of plasma membrane, cortical ER and plasmodesmata during plasmolysis of onion epidermal cells. Plant Cell Environ 17:163–171

Oparka KJ, Boevink P, Santa Cruz S (1996) Studying the movement of plant viruses using green fluorescent protein. Trends Plant Sci 1:403–441

Oparka KJ, Prior DAM, Santa Cruz S, Padgett HS, Beachy RN (1997) Gating of epidermal plasmodesmata is restricted to the leading edge of expanding infection sites of tobacco mosaic virus (TMV). Plant J 12:781–789

Overall RL, Blackman LM (1996) A model of the macromolecular structure of plasmodesmata. Trends Plant Sci 1:307–311

Overall RL, Wolf J, Gunning BES (1982) Intercellular communication in *Azolla* roots I. Ultrastructure of plasmodesmata. Protoplasma 111:134–150

Radford JE, White RG (1998) Localization of a myosin-like protein to plasmodesmata. Plant J 14: 743–750

Radford JE, Vesk M, Overall RL (1998) Callose deposition at plasmodesmata. Protoplasma 201:30–37

Robards AW (1971) The ultrastructure of plasmodesmata. Protoplasma 72:315–323

Robards AW (1976) Plasmodesmata in higher plants. In: Gunning BES Robards AW (eds) Intercellular communication in plants: studies on plasmodesmata. Springer, Berlin Heidelberg New York, pp 15–53

Robards AW, Lucas WJ (1990) Plasmodesmata. Annu Rev Plant Physiol Mol Biol 41:369–419

Robinson-Beers K, Evert RF (1991) Fine structure of plasmodesma in mature leaves of sugarcane. Planta 184:308–318

Roy S, Watada AE, Wergin WP (1997) Characterisation of the cell wall microdomain surrounding plasmodesmata in apple fruit. Plant Physiol 144:539–547

Salisbury JL (1995) Centrin, centrosomes and mitotic spindle poles. Curr Opin Cell Bio 7:39–45

Schulz A (1995) Plasmodesmal widening accompanies the short-term increase in symplasmic phloem unloading in pea root tips under osmotic stress. Protoplasma 188:22–37

Smith MM, McCully ME (1977) Mild temperature "stress" and callose synthesis. Planta 136:65–70

Staehelin LA, Hepler PK (1996) Cytokinesis in higher plants. Cell 84:821–824

Stone BA, Clarke AE (1992) Chemistry and biology of (1–3)-b-glucans. La Trobe University Press, Melbourne, Australia

Taiz L, Jones RL (1973) Plasmodesmata and an associated cell wall component in barley aleurone tissue. Am J Bot 60:67–75

Tangl E (1879) öber offene Kommunikation zwischen den Zellen des Endosperms einiger Samen. Jb Wiss Bot 12:170–190

Thomson WW, Platt-Aloia K (1985) The ultrastructure of plasmodesmata of the salt glands of *Tamarix* as revealed by transmission and freeze-fracture electron microscopy. Protoplasma 125:13–23

Tilney LG, Cooke TJ, Connelly PS, Tilney MS (1991) The structure of plasmodesmata as revealed by plasmolysis, detergent extraction, and protease digestion. J Cell Sci 112:739–747

Tucker EB (1988) Inositol bisphosphate and inositol trisphosphate inhibit cell-to-cell passage of carboxyfluorescein in staminal hairs of *Setcresea purpurea*. Planta 174:358–363

Tucker EB (1982) Translocation in the staminal hairs of *Setcreasea purpurea* I. A study of the cell ultrastructure and cell-to-cell passage of molecular probes. Protoplasma 113:193–210

Tucker EB (1993) Azide treatment enhances cell-to-cell diffusion in staminal hairs of *Setcreasea purpurea*. Protoplasma 174:45–49

Turner A, Wells B, Roberts K (1994) Plasmodesmata of maize root tips: structure and composition. J Cell Sci 107:3351–3361

Vaughn KC, Hoffman JC, Hahn MG, Staehelin LA (1996) The herbicide dichlobenil disrupts cell plate formation: immunogold characterisation. Protoplasma 194:117–132
Vreeland V, Morse DR, Robichaux RH, Millar KL, Hau S-ST, Laetsch WM (1989) Pectate distribution and esterification in *Dubautia* leaves and soybean nodules, studies with fluorescent hybridisation probe. Planta 177:435–446
Waigmann E, Zambryski P (1995) Tobacco mosaic virus movement protein-mediated protein transport between trichome cells. Plant Cell 7:2069–2079
Waigmann E, Turner A, Peart J, Roberts K, Zambryski P (1997) Ultrastructural analysis of leaf trichome plasmodesmata reveals major differences from mesophyll plasmodesmata. Planta 203: 75–84
White RG, Badelt K, Overall RL, Vesk M (1994) Actin associated with plasmodesmata. Protoplasma 180:169–184
Willison JHM (1976) Plasmodesmata: a freeze-fracture view. Can J Bot 54:2842–2847
Wolf S, Lucas WJ (1994) Virus movement proteins and other molecular probes of plasmodesmal function. Plant Cell Environ 17:573–585
Zambryski P (1995) Plasmodesmata: plant channels for molecules on the move. Science 270: 1943–1944
Zhang WH, Tyerman SD (1997) Effect of low oxygen concentration on the electrical properties of cortical cells of wheat roots. J Plant Physiol 150:567–572

Chapter 10

Multimorphology and Nomenclature of Plasmodesmata in Higher Plants

R. Kollmann[1] · C. Glockmann

[1] Botanisches Institut der Christian-Albrechts-Universität, D-24098 Kiel, Germany

1 Introduction

In higher plants, channels for intercellular communication undergo various changes in number, structure and function in the course of cell growth and differentiation. Primary plasmodesmata formed in the cell plate during cytokinesis persist in the growing cell wall despite changes in structure and distribution depending on the pattern of cell differentiation. During cell growth and cell division, the number of primarily formed cell connections per unit wall area more and more decreases in the expanding cell walls, while it remains nearly unchanged in those walls showing less or no expansion (Juniper and Barlow 1969; Juniper 1977; Gunning 1978). Progressive dilution sooner or later results in symplasmic isolation, unless additional cell connections are established by either (1) modification of primary plasmodesmata and/or (2) a complete de novo formation of so-called secondary plasmodesmata across the existing cell wall (Fig. 1). As will be shown in the present chapter, both mechanisms have evolved in plants, resulting in the establishment of plasmodesmata with specific structures and probably special functions.

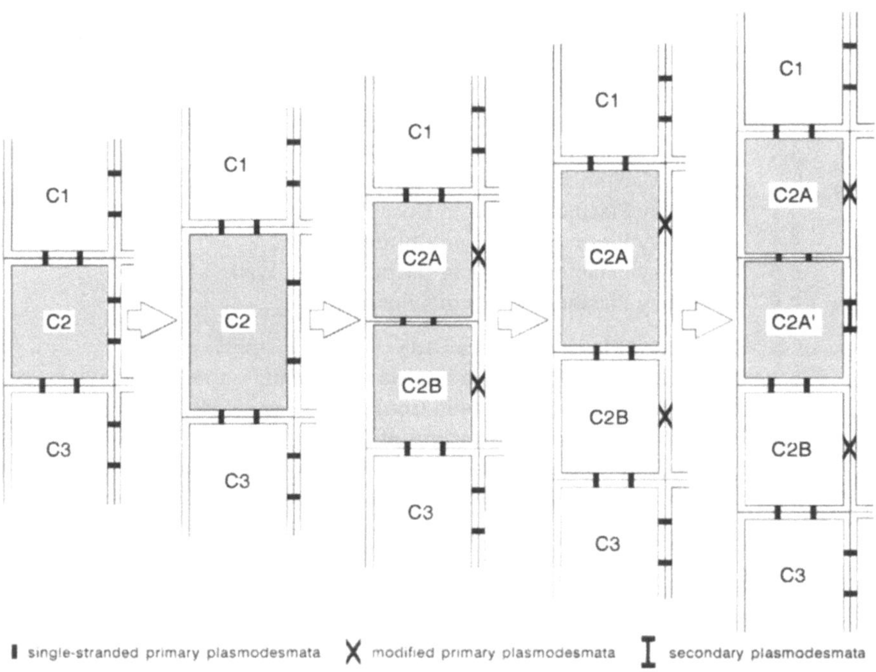

Fig. 1. Changes in intercellular connections during cell growth. In the expanding longitudinal cell wall, the number of primary plasmodesmata per unit wall area decreases during cell division. Loss of primary cell connections is compensated for by modification of single-stranded primary plasmodesmata (*C2A; C2B*) and by formation of secondary plasmodesmata (*C2A'*). In the less expanding transverse wall, structure and distribution of the primarily formed plasmodesmata remain unchanged

2 Indirect Proof for the Formation of New Cell Connections in Existing Cell Walls

Botanists were concerned with questions regarding the formation of symplasmic cell connections in existing walls as soon as the "offene Communicationen zwischen den Zellen" had been discovered by Tangl (1879) and were named plasmodesmata by Strasburger (1901); (cf. Russow 1883; Kienitz-Gerloff 1891). Strasburger (1901) was the first to point out that the plasmodesmata between epidermis and subepidermal tissue where cell divisions are lacking must have been formed de novo in existing cell walls. A comparable situation exists, according to Strasburger (1901), with respect to the radial walls of cambium descendants, the walls of laticifers, and all those non-division walls at the interface of cells and tissues of various origin. In the present-day literature, Strasburger's concept has been confirmed repeatedly in various systems such as nematode-induced giant cells (Jones 1976; Jones and Dropkin 1976), developing *Sphagnum* leaflets (Schnepf and Sych 1983), and elongating root cells (Seagull 1983); (cf. Cooke et al. 1996). As for the structure of the putatively de novo formed plasmodesmata, investigations of Barnett (1987a, b) disclosed highly branched cell connections in the pit membranes of the radial walls of developing fibre tracheids. Branched plasmodesmata have also been described at the interface between bundle-sheath and vascular parenchyma cells of C_3 plants where de novo formation in the non-division walls was expected (Ding and Lucas 1996). Furthermore, this branched type of cell connection is observed regularly in all those longitudinal cell walls having been extended during growth of a plant organ, in contrast to predominantly single-stranded plasmodesmata of transverse cell walls which show less growth by extension (Fig. 1; Kollmann et al. 1996).

Branched plasmodesmata are thus the characteristic form of cell connections in cell walls with continuing growth by extension; they differ remarkably from single unbranched plasmodesmata in a newly formed division wall. The mode of formation of branched plasmodesmata, however, cannot possibly be deduced from structural criteria only, and the question remains whether modified primary plasmodesmata or secondary plasmodesmata that are formed completely de novo are involved. In the following sections, the situation with regard to a valid terminology will be described more precisely.

3 Primary Plasmodesmata

The structure of primary plasmodesmata depends on the formation mechanism in the cell plate during cytokinesis (Hepler 1982). Tubular structures of the endoplasmic reticulum (ER) crossing the phragmoplast in the equatorial plane of a dividing cell become entrapped in the growing cell plate by fusing Golgi vesicles that deliver the cell wall material (Fig. 2A–C). During this process the ER tubules become more or less constricted and transform into a desmotubule surrounded by a cytoplasmic sleeve, the latter being demarcated from the cell wall by the plasmalemma that is originally derived from the Golgi membrane. The primarily formed cell connections are straight, single strands with a diameter of about 20–50 nm traversing the young cell wall randomly (Fig. 4A). The desmotubule is always continuous with the endomembrane system of the adjacent cells and the cytoplasmic strands may or may not develop a

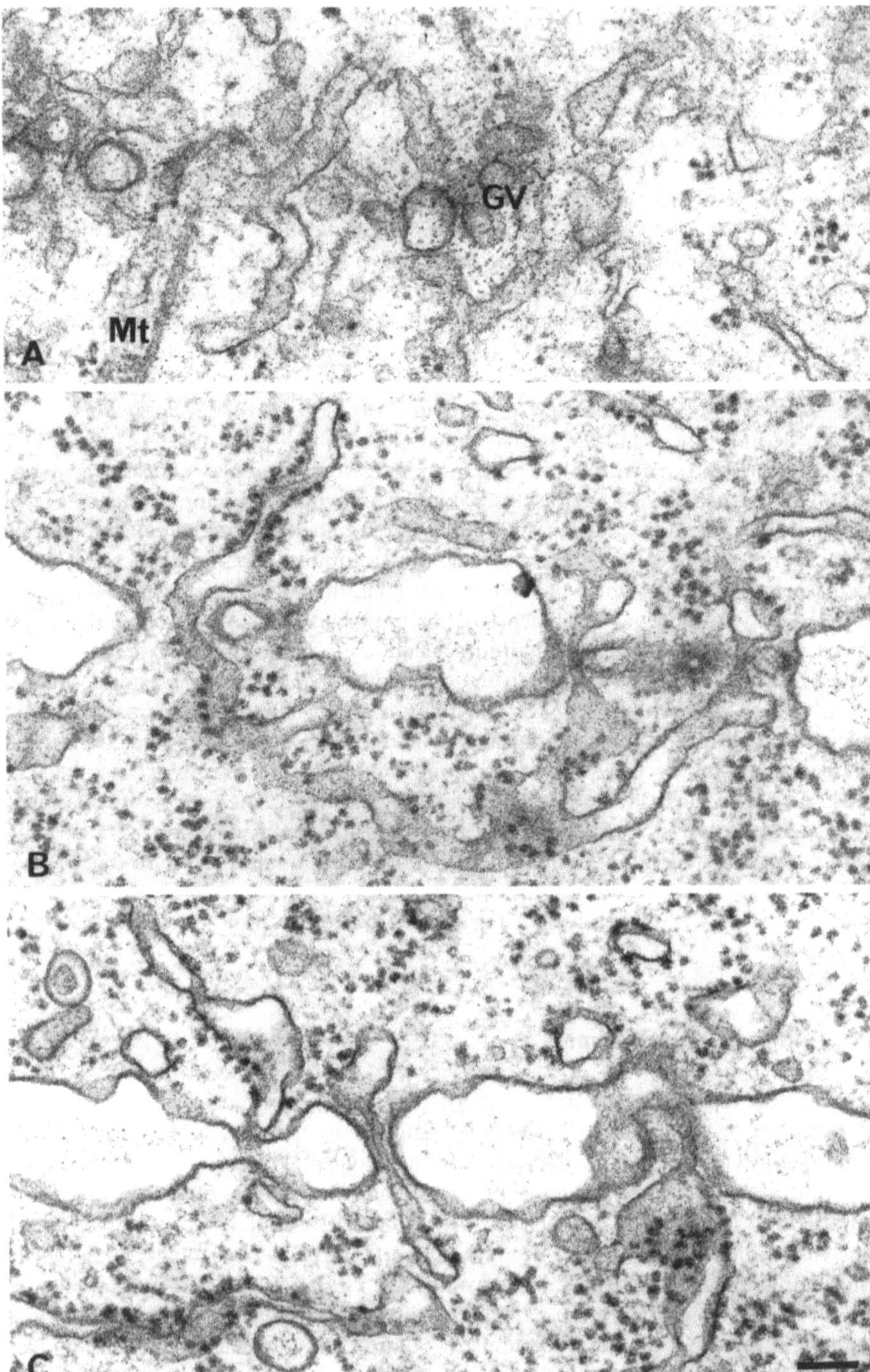
GV
Mt
A
B
C

Fig. 2A–C. Establishment of primary plasmodesmata by ER entrapment in the growing cell plate of dividing *Solanum nigrum* protoplasts (A) and Strasburger cells in developing *Metasequoia* needles (B, C). A Accumulation of Golgi vesicles (*GV*) of the phragmoplast between ER tubules and microtubules (*Mt*) crossing the equatorial plane. B Smooth domains of the ER crossing the growing cell plate indicate areas where primary plasmodesmata are formed. C Developing primary plasmodesmata with constricted ER tubules (desmotubules) in a growing cell plate. Magnifications: A–C 73 000; *bar* 0.1 μm. (A K. Ehlers)

neck constriction at their orifices (Robards and Lucas 1990). (Cell and tissue-specific variations in the structural organization of primary plasmodesmata will further be dealt with in Sect. 5.2. For further details on the substructure of plasmodesmata see Chaps. 2, 3, 9).

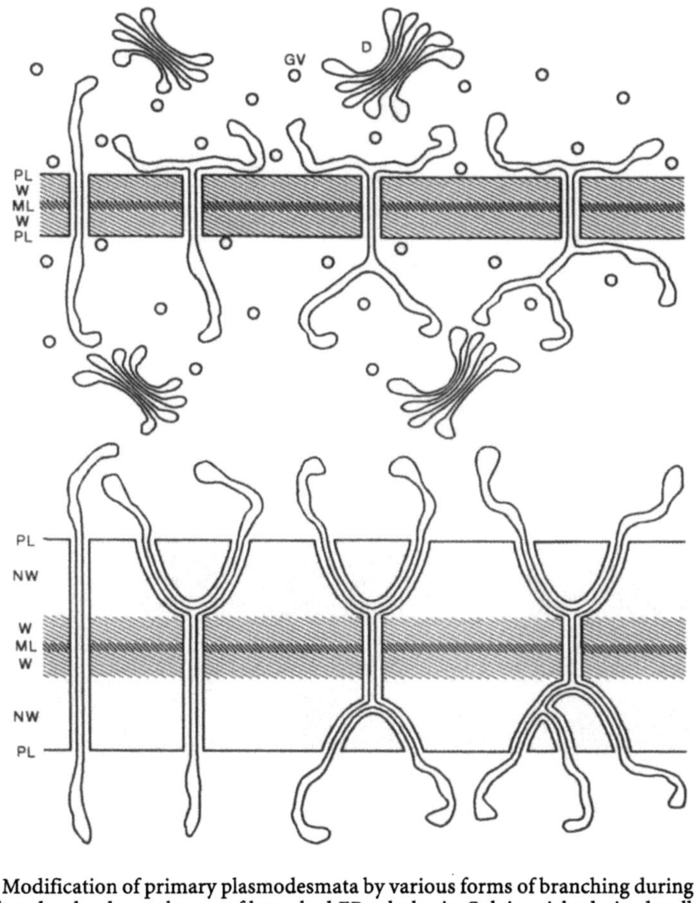

Fig. 3. Modification of primary plasmodesmata by various forms of branching during wall thickening. Branches develop by enclosure of branched ER tubules in Golgi vesicle-derived wall material. At the *left* a single-stranded but elongated plasmodesma; *PL* plasmalemma; *W* wall layers; *ML* middle lamella; *NW* new wall layers; *D* dictyosome; *GV* Golgi vesicles. (Ehlers and Kollmann 1996)

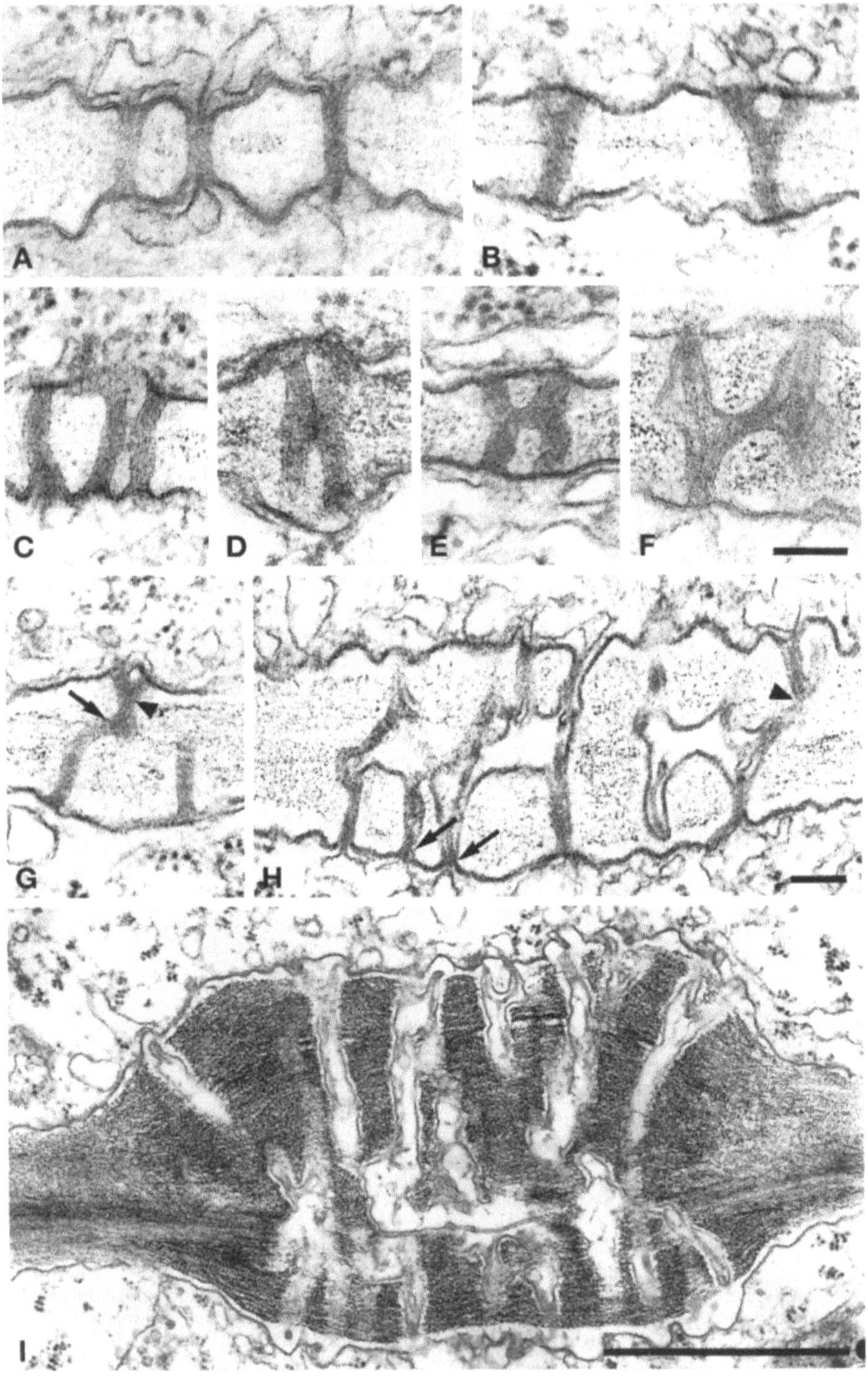
A
B
C
D
E
F
G
H
I

3.1 Modification of Primary Plasmodesmata

In the course of cell development, the unbranched primary plasmodesmata of the newly formed cell wall may undergo further structural modifications (Ehlers and Kollmann 1996). As the cell wall grows in thickness, straight or branched ER tubules continuous with the desmotubule become gradually embedded by Golgi-derived wall material, giving rise to the formation of either simple elongated or branched primary plasmodesmata, respectively (Figs. 3, 4A, B). This accounts for the occurrence of various types of branched primary plasmodesmata next to unbranched plasmodesmata in a division wall.

Another type of modification most likely results from fusion of adjacent plasmodesmata (Fig. 4C–F; Ding et al. 1992; Lucas et al. 1993; Lucas and Gilbertson 1994; Ding and Lucas 1996; Kollmann et al. 1996). Although the fusion mechanism has not been elucidated conclusively, a local wall disintegration in the mid-region of the cell wall resulting in the formation of median cavities is likely to be involved in the process. H-shaped plasmodesmata formed in this way (Fig. 4E, F) may undergo additional branching in younger wall parts (Fig. 4G, H). Based on their origin, these multiply branched cell connections undoubtedly represent modified primary plasmodesmata. This even holds for those complex plasmodesmata which are characteristic for mature Strasburger cells in conifers (Fig. 4I; Sect. 5.1.1).

4 Secondary Plasmodesmata

The principal difference between primary and secondary plasmodesmata pertains to their origin. While primary plasmodesmata are formed in the cell plate during cytokinesis from the very beginning and may eventually become modified during wall differentiation, secondary plasmodesmata are completely formed de novo in existing cell walls in places where cell connections did not exist before (Jones 1976; Lucas et al. 1993; Ehlers and Kollmann 1996).

As the origin of plasmodesmata cannot be deduced from their structure and position within the cell wall in most cases, a direct proof for the formation of secondary plasmodesmata needs special prerequisites as for the selection of the objects to be studied. This will be shown in the following sections.

Fig. 4A–I. Various forms of primary plasmodesmata in differentiating Strasburger cells of young (**A–H**) and fully developed (**I**) needles of *Metasequoia*. **A** Single-stranded plasmodesmata without neck constrictions in a newly formed cell wall. Desmotubules with ER continuity between the adjacent cells are visible. **B** Early branching in the latest deposited wall part of the upper cell. **C–F** Origin of branched plasmodesmata by fusion of two adjacent single-stranded plasmodesmata (**C, D**) forming a so-called H-shaped plasmodesma (**E, F**); the future median cavity (see **H**) is not yet dilated. **G** Primary plasmodesma branching in the mid region (*arrow*) and in younger layers of the wall (*arrowhead*). **H** Two multiply branched plasmodesmata with extended median cavities. Desmotubules, ER continuity between the adjacent cells, and neck constrictions at the orifices of some branches (*arrows*) can be discerned. The *arrowhead* points to a secondary branching. **I** Multiply branched primary plasmodesmata between mature Strasburger cells. Extending from a median cavity the plasmodesmal strands branch repeatedly in the locally thickened cell wall. Narrow and distended desmotubules as well as neck constrictions at the orifices of single branches are obvious. Magnifications: **A–F** 100 000; **G, H** 72 000; *bars* 0.1 μm; **I** 33 000; *bar* 1 μm

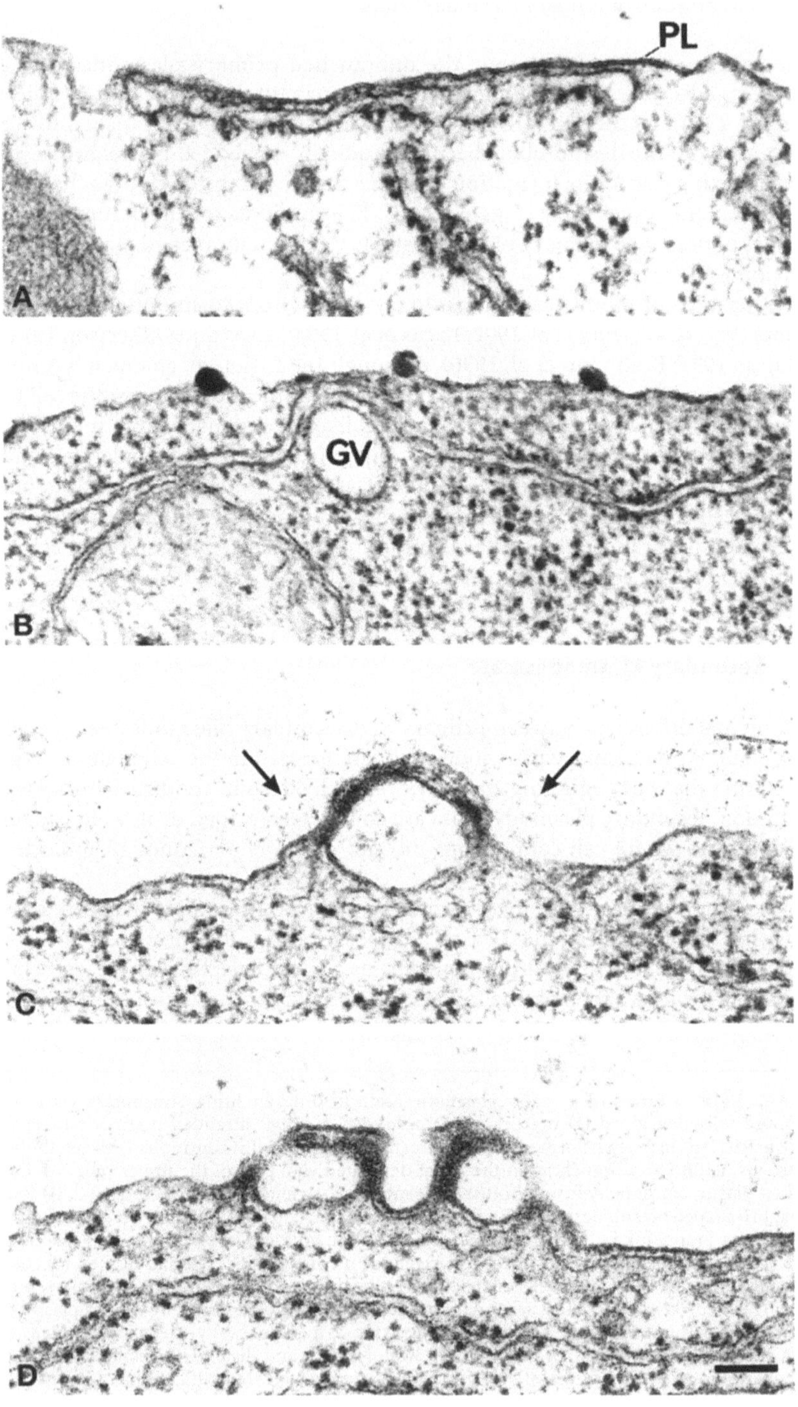
PL
A
GV
B
C
D

Fig. 5A–D. Development of outer-wall plasmodesmata during wall regeneration in cultured protoplasts of *Solanum nigrum*. **A** Protoplast surface without cell wall 1 h after isolation in culture. Close ER/plasmalemma (*PL*) contact. **B** Protoplast surface as in **A**. Golgi vesicle (*GV*) associated with the ER/plasmalemma contact indicates the area where outer-wall plasmodesmata are formed. **C** Outer-wall plasmodesma enclosed in wall material (*arrows*) of a 3-day-old protoplast in culture. The desmotubule is connected with the ER. **D** Multiply branched outer-wall plasmodesma enclosed in wall material of a 4-day-old protoplast in culture. Magnification: **A–D** 80 000; *bar* 0.1 μm. (Ehlers and Kollmann 1996)

4.1 Secondary Plasmodesmata in Cell and Tissue Cultures

Regenerating protoplasts produce half plasmodesmata in their new outer cell walls 3–7 days after isolation in culture medium (Fig. 5). The building mechanism corresponds essentially with the process described for the formation of primary plasmodesmata in division walls: tubular strands of the ER are entrapped by Golgi vesicle-derived cell wall material (Monzer 1990, 1991; Ehlers and Kollmann 1996). In dependence on the structure of the ER enclosed, different configurations of half plasmodesmata are formed in the outer cell wall. As these discontinuous plasmodesmata are exposed only to the culture medium, they have been termed outer-wall plasmodesmata (Ehlers and Kollmann 1996). When regenerating cells come into intimate contact with one another in the culture medium, the opposite outer-wall plasmodesmata fuse, thus forming continuous intercellular connections between adjacent cells. The cytoplasmic

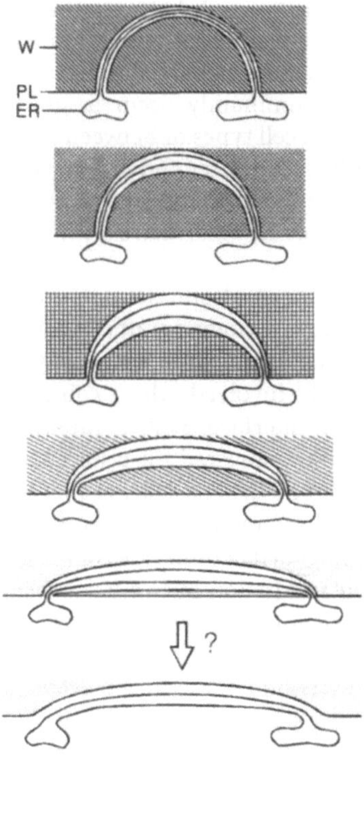

Fig. 6. Schematic representation of the degradation process of an outer-wall plasmodesma in a cultured regenerating cell. During cell growth the arch-shaped half plasmodesma becomes stretched and inflated in the expanding wall with loosened matrix and will be subsequently reintegrated in the protoplast. The whole process represents the inverse way of outer-wall plasmodesmata formation; *W* wall; *PL* plasmalemma. (Ehlers et al. 1996)

bridges emerging in this way in non-division (i.e. fusion) walls represent true secondary plasmodesmata and are usually branched with a more or less extended median cavity.

Those outer-wall plasmodesmata which do not take part in the formation of secondary plasmodesmata disappear in 7- to 9-day-old cell cultures. The selective disintegration of outer-wall plasmodesmata which is obviously regulated by ubiquitination of plasmodesmal proteins can be regarded as an inverse way of their formation (Fig. 6): due to loosening of the wall matrix during wall growth, plasmodesmata inflate within the expanding cell wall and reintegrate into the cytoplasm (Ehlers et al. 1996).

The mode of formation and disintegration of outer-wall plasmodesmata underlines the special role of the cell wall in structuring and maintaining the delicate plasmodesmal structure. The plant cell wall should be regarded as a dynamic rather than a static extracellular component with essential functions in the regulation of intercellular communication. This will be further elucidated in the following section.

4.2 Secondary Plasmodesmata in Graft Unions and the Mechanism of Their Formation in Fusion Walls

Occurrence of secondary plasmodesmata in graft unions has been demonstrated using heterografts of plants with species-specific subcellular markers (Kollmann and Glockmann 1985). The interspecific cell connections in the fusion walls between cells of the graft partners are usually branched plasmodesmata with extended median cavities. The same type of plasmodesmata occurs at the interfaces of homografts and autografts. Apart from continuous branched plasmodesmata, half plasmodesmata extending across the wall part of only one cell partner were observed. Such half plasmodesmata predominantly occur at graft interfaces of incompatible heterografts, between different cell types or between cells at different stages of differentiation (Kollmann et al. 1985). Diameter and fine structure of the plasmodesmal branches in graft unions resemble those found in cell cultures (Monzer 1991; Ehlers and Kollmann 1996) and are thus in accordance with the structure of primary plasmodesmata (cf. Sect. 4.1).

The mechanism of plasmodesmata formation in graft unions has been disclosed with the aid of in vitro heterografts (Fig. 7; Kollmann and Glockmann 1991). A prerequisite for the establishment of secondary cell connections is the dedifferentiation of both partner cells involved, which coincides with a striking modification of the cell walls at the graft interface. By local thinning and loosening of opposite wall parts, cell membranes with associated ER cisternae of juxtaposed cells come in close contact, enabling membrane fusions. During subsequent thickening of the modified wall parts by wall material secreted from Golgi vesicles, cytoplasmic strands with enclosed ER tubules are embedded at both sides of the graft interface. The whole process resembles the formation and fusion of two half plasmodesmata between undifferentiated cells such as the formation of secondary plasmodesmata in cultured cells (cf. Sect. 4.1). In conclusion, the formation of primary and secondary plasmodesmata are essentially similar with exception of the fusion events of the latter in non-division walls.

The following sections (4.3–4.6) will show that branching with a more or less extended median cavity, with some exceptions (see e.g. Sect. 4.3), is a characteristic

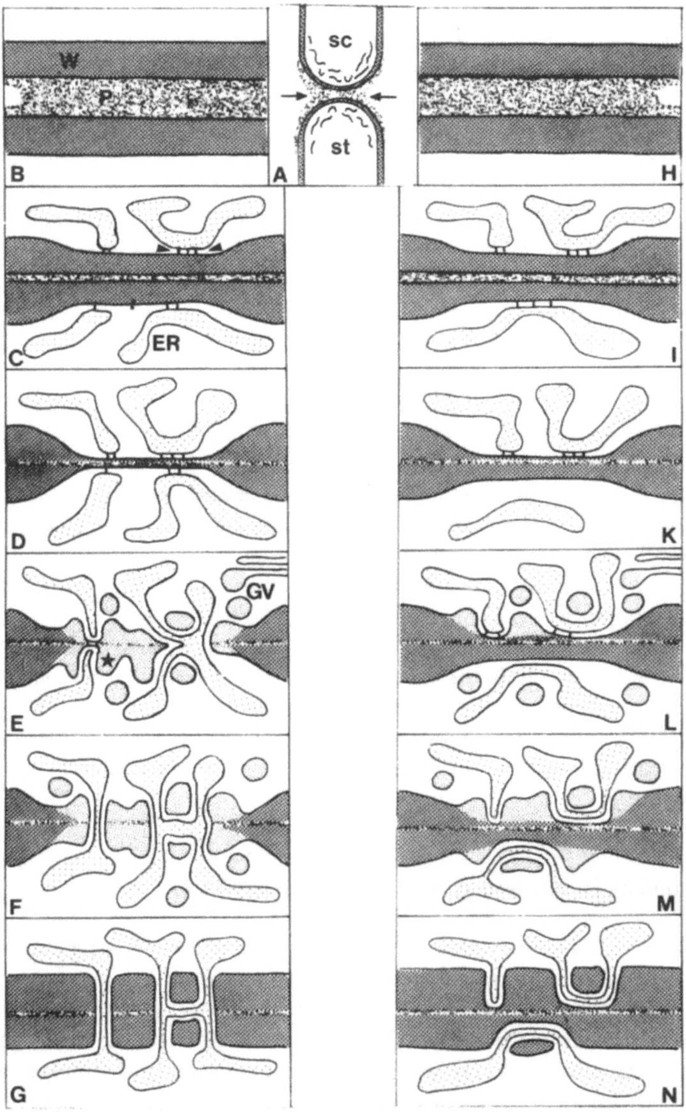

Fig. 7A–N. Formation of secondary plasmodesmata at graft interfaces. **A** Approaching callus cells of scion (*sc*) and stock (*st*). *Arrows* indicate region where secondary plasmodesmata will be formed. **B, H** Details from **A** with cell wall (*W*) and pectic layer (*P*) in between. **C–G** Formation of continuous secondary plasmodesmata by fusion of plasmalemma and ER at regions where wall parts of opposite cells have been thinned synchronously (**C–E**). Rebuilding of branched and single-stranded plasmodesmata (**E–G**) by embedding into Golgi vesicle (*GV*)-derived wall material (*asterisk*). **I–N** Formation of discontinuous (half) plasmodesmata at regions where wall parts of opposite cells have been thinned asynchronuosly. *Arrowheads* in **C** indicate unidentified 5-nm particles interconnecting ER and cell membranes. (Kollmann and Glockmann 1991)

structure of secondary plasmodesmata. This resemblance is strong evidence for a universal mechanism of secondary formation of cell connections in existing walls of any growing cell where wall thinning takes place during cell expansion.

Furthermore, the uniform substructure of cytoplasmic strands of primary and secondary plasmodesmata might be due to a uniform mode of construction: In both cases, cytoplasmic strands with enclosed ER tubules are constricted by surrounding cell wall material up to a minimal diameter. The latter might be limited by the physico-chemical characteristics of the plasmalemma and ER membranes (Overall et al. 1982), resulting in a consistent 20–50-nm plasmodesmal strand with a desmotubule of about 15 nm.

However, several questions remain to be answered concerning the regulation of plasmodesmata formation in existing cell walls. The mechanism of cell wall modification at places where secondary plasmodesmata will be established is unknown so far. Loosening of the wall matrix (hemicellulosic xyloglucans) by an auxin-induced cellulase (1,4-β-glucanase) (Hayashi et al. 1984; Fry 1989) in connection with local wall expansion and/or changes in wall porosity (Baron-Epel et al. 1988; Mc Cann et al. 1990) may be engaged. The close contact of ER and plasmalemma in those regions where wall modification and formation of plasmodesmata take place assigns a special role to the ER in wall modification, comparable to the events in the secondary formation of sieve pores in brown algae (Schmitz and Srivastava 1974) and of plasmodesmata between pollen mother cells (Cheng et al. 1987).

An important question which has not been answered so far concerns the cooperation between adjacent cells in the processes of synchronous wall thinning and concerted formation of opposite half plasmodesmata. How are these processes coordinated? A recognition system and exchange of information between cells across the cell wall has been postulated (Jeffree and Yeoman 1983); however, experimental proofs for this are still lacking.

4.3 Secondary Plasmodesmata in Host-Parasite Interfaces

In some instances, the interrelationship between parasitic flowering plants and their hosts resembles a heterograft system: cells of taxonomically different organisms come in close contact with one another. The way in which the partners grow together differs, however, with the various parasites. Concerted coalescence of the heterogeneous tissues, as in compatible grafts, is based on the cooperation of dedifferentiated callus cells. A similar degree of dedifferentiation of the cells that came in contact is the prerequisite for the formation of secondary plasmodesmata in non-division walls.

A comparable situation exists for some root parasites, for instance *Striga* or *Orobanche*. When *Striga gesneroides* parasitizes on a dicotyledoneous plant such as *Pisum sativum*, the host responds to the attack by an enormous cell proliferation in all tissues of the root around the parasitic intrusive organ. Between the intermingling undifferentiated cells of host and parasite, symplasmic cell connections are formed secondarily in the contact walls. In contrast to graft interfaces, the secondary plasmodesmata here are predominantly single-stranded. Noteworthy, in *Zea mays* – a principal host of other *Striga* species which does not show any comparable cell reactivation and dedifferentiation in the contact area with the parasite – interspecific plasmodesmata have not been observed (Dörr 1996a, b).

With the shoot parasite *Cuscuta* growing on a compatible host plant, structural interrelations between the partners differ from those of *Striga*. The parasite penetrates by means of a defined haustorium deeply into the differentiated tissue of the host in response to which the latter does not show any noticeable cell reactivity. At the apex of the parasitic intrusive organ, so-called searching hyphae grow out penetrating the host's tissues inter- and intracellularly up to the vascular bundle. At the tip of the searching hyphae, outer-wall plasmodesmata – 30–40 nm in diameter – are formed in the very thin wall of the parasite cell (Dörr 1969, 1987; Dawson et al. 1994). The outer-wall plasmodesmata of intracellular hyphae, being surrounded by the cytoplasm of the host cell, interconnect the protoplasts of host and parasite cell symplasmically. Most of these interspecific cell connections subsequently become plugged with wall material deposited by the host cell. Only a few plasmodesmata become elongated inside the newly formed host cell wall, thus being complete secondary plasmodesmata between host and parasite cells. In contrast, the outer-wall plasmodesmata, formed in the intercellularly growing hyphae parts never become connected with the differentiated host cell symplasmically. They remain half plasmodesmata, extending across the parasitic cell wall alone.

The observations with the intercellularly growing *Cuscuta* hyphae and those with *Striga* confirm the general concept as before mentioned (Sects. 4.1, 4.2) according to which complete secondary plasmodesmata can only be established between undifferentiated cells or cells showing the same stage of dedifferentiation. (With the intracellularly growing *Cuscuta* hyphae a special situation exists in so far as outer-wall plasmodesmata in the parasitic wall part interconnect the host cytoplasm directly.)

So far, *Cuscuta* is the most thoroughly investigated parasitic flowering plant with regard to secondary plasmodesmata in host-parasite interrelationships: In three *Cuscuta* species on five different host plants interspecific plasmodesmata have been observed unequivocally (Dörr 1987; Dawson et al. 1994). The success of these investigations was due to easy identification of the characteristic *Cuscuta* cells within the host tissue. With other parasitic plants, discrimination between the partner cells can only be accomplished with species-specific subcellular markers, as has been shown with heterografts (Sect. 4.2). In this manner, symplasmic cell connections so far have been proved for *Orobanche crenata* on *Vicia narbonensis* (Dörr and Kollmann 1995; Dörr 1996a, b), *Striga gesneroides* on *Pisum sativum* (Dörr 1996a, b) and *Pilostyles hamiltonii* (Rafflesiaceae) on *Daviesia preisii* (Dell et al. 1982). Observations with the dwarf mistletoes *Arceuthobium pusillum* on *Picea mariana* (Tainter 1971) and *A. occidentale* on *Pisum sabiniana* (Alosi and Calvin 1985) remain uncertain as long as host and parasite cells have not been identified unequivocally. It has to be noted that the secondary – branched or unbranched – cell connections at host-parasite interfaces have been localized in thinned areas of the contact wall in all systems examined so far (except for *Cuscuta*). This localization matches the arrangement of cell connections at graft interfaces and will most likely be associated with the special mechanism of secondary plasmodesmata formation (Sect. 4.2).

Next to higher-plant parasites, interspecific symplasmic cell connections at host-parasite interfaces have also been detected in various mycoparasitic relationships (Hoch 1977a, b, 1978; Bauer and Oberwinkler 1990). As the cytoplasmic bridges between fungal cells described differ in size (14 nm-1 μm) and substructure from plasmodesmata, any comparison with higher plant plasmodesmata is inappropriate (cf. also Chaps. 7, 8).

4.4 Secondary Plasmodesmata in Chimeras

Chimeras have often been used for investigations on secondary plasmodesmata in non-divison cell walls. Results of earlier light and electron microscopic studies , however, have not been unequivocal (cf. the observations of Buder 1911; Hume 1913; Burgess 1972; see also Jones 1976). A clear-cut decision as to the existence of cell connections between heterotypic cells in chimeras was only recently made possible by the use of plant pairs with genotype-specific subcellular markers, a method already successfully employed with heterografts and host-parasite relations. In this way, interspecific plasmodesmata have been proved to exist in heterospecific periclinal, sectorial and mericlinal chimeras of *Solanum nigrum* (+) *Solanum tuberosum* obtained by cell grafting in protoplast cocultures (Binding et al. 1987). In the chimeral leaves, secondary plasmodesmata were found in the non-division walls between epidermis and mesophyll cells of a periclinal chimera as well as between heterotypic mesophyll cells of sectorial and mericlinal chimeras, respectively. The single and branched cell connections showed the typical plasmodesmal substructure containing desmotubules connected with the ER, median cavities, and neck constrictions.

A detailed analysis of the interspecific cell connections in a chimera was performed with *Laburnocytisus adamii* (Steinberg and Kollmann 1994). In this well-known monekto-periclinal chimera, an epidermis of *Cytisus purpureus* overlays the tissue of *Laburnum anagyroides*. The interspecific plasmodesmata between the genotypically different cells of L1 and L2 of the chimeral leaves occur in pit fields. In contrast with earlier observations (Burgess 1972; see also Jones 1976), most of these secondarily formed plasmodesmata interconnect the heterotypic cells directly and are branched with median cavities. A few half plasmodesmata always exist in the subepidermal (*Laburnum*) part of the contact wall which emphasizes their secondary origin. A quantitative study on the branching pattern and on cross-sections of plasmodesmata at different planes, carried out with interspecific and intraspecific cell connections, yielded some unexpected results. Whereas the intraspecific cell connections of the parental species (*Cytisus, Laburnum*) usually show a symmetrical branching pattern, the interspecific plasmodesmata of the chimera are more asymmetrically branched with the primary and secondary branchings mainly in the subepidermal *Laburnum* wall part. Accordingly, the plasmodesmal branches at the epidermal side represent a bottle neck for the symplasmic continuity between L1 and L2 in *Laburnocytisus*, while in the parental species the symplasmic connections are well tuned in size and structure across the cell wall between epidermis and subepidermal cells. The functionality of the discrepancy between the interspecific and intraspecific cell connections is open to debate. It might reflect a way in which the partner cells control symplasmic transport. Insufficient coordination and cooperation during the establishment of cell connections between the heterotypic cells due to incompatibility should also be taken into consideration.

4.5 Secondary Plasmodesmata in Carpel-Fusion Walls

Competence of plant cells to postgenital fusion was basically shown in cell and tissue culture (Sect. 4.1) as well as in graft-union formation (Sect. 4.2). Fusion will be found either between cells of different origin (e.g. graft interfaces) or between cells with a common descent (e.g. proliferating callus cells in tissue culture). Most likely, postgenital fusion is a normal event in cell growth and development of any plant.

A well-known example for postgenital cell fusion is the ovary development of flowering plants. Coalescence of carpels takes place during the ontogeny of the gynoecium and involves redifferentiation of closely attached epidermal cells of the carpels (Baum 1948; Boeke 1971, 1973 a, b; van der Schoot et al. 1995). In the differentiated ovary, the suture, where tissues have been grown together, is usually difficult to locate. Exceptions are plants with subcellular epidermal markers such as the monekto-periclinal chimera *Laburnocytisus* (Sect. 4.4). In the thinned wall areas of the fused cells, "structures ..., that closely resemble the plasmodesmata" were detected (Boeke and van Vliet 1979), which must have been formed secondarily in the fusion wall. More convincing proof of secondary plasmodesmata at the interface of coalesced carpels comes from studies on the ovary development of *Catharanthus* (van der Schoot et al. 1995), where simple and branched plasmodesmata connect the fused epidermal cells symplasmically at an early stage of cell redifferentiation. The functioning of these secondary plasmodesmata showing the normal fine structure was proved with dye-coupling experiments.

4.6 Secondary Plasmodesmata in Tyloses

Formation of tyloses is a further example for postgenital fusion of like cells in a plant. Tyloses are proliferations of vascular parenchyma cells extending into the lumina of vessel elements. Where tyloses come in close contact, fusion of cell walls is accomplished and common pits are being formed (Esau 1948). However, the postgenital fusion sites are difficult to identify, as tyloses often divide within the vessel lumen. Only where tyloses make contact to other vascular parenchyma cells across a pit membrane can a postgenital wall fusion be discerned unequivocally. Within these interfaces, branched plasmodesmata with extended median cavities (several μm in diameter) have been detected (Czaninski 1974). Branching and extension of the median cavity are in accordance with the cell connections at graft interfaces (Sect. 4.2) thus complying with secondary formation as described for non-division walls.

5 Specialized Plasmodesmata

Primary and secondary plasmodesmata may undergo essential modifications during cell differentiation associated with their special function in intercellular trafficking. Most of these specialized cell connections are characterized by symmetrical or asymmetrical branching depending on the specific cell type involved. Only those specialized plasmodesmata will be reviewed that exhibit lasting structural modifications. (For transient substructural changes, see Chap. 14).

5.1 Specialized Plasmodesmata in Phloem Loading

In leaves of many plant species plasmodesmata, with a different degree of branching and with more or less extended median cavities, occur along the route from mesophyll to the sieve elements (Russin and Evert 1985). Branching of the cell connections takes place during cell differentiation and coincides in angiosperm leaves with the transition from sink to source conditions of a developing leaf (Ding et al. 1992; Moore et al. 1992; Gagnon and Beebe 1996; Volk et al. 1996; for further structural and functional characteristics of the specialized cell connection which most likely are involved in phloem loading, see Chap. 15).

The origin of the branched plasmodesmata in angiosperm leaves needs further elucidation. At the interface between bundle sheath and vascular tissue, true secondary plasmodesmata have to be expected as, in most cases, they interconnect cells of different origin in a non-division wall. (In the leaves of dicotyledons and some monocotyledons the bundle sheath is regarded as being part of the ground tissue, whilst the vascular elements originate from the procambium: Esau 1965). The branched plasmodesmata, however, between mesophyll cells, between mesophyll and bundle-sheath cells, between bundle-sheath cells, or between vascular cells, may either be modified primary plasmodesmata or de novo formed, truly secondary plasmodesmata. The clustered arrangement of the cytoplasmic strands (Russin and Evert 1985; Volk et al. 1996), and an obvious correlation between branching and wall thickening (Russin and Evert 1985), is to be regarded as a strong evidence in favour of modified primary plasmodesmata. To elucidate the arguments further, the highly differentiated cell connections between Strasburger cells, the Strasburger / sieve-cell connections in conifer needles and plasmodesma-pore units between companion cells and sieve elements in angiosperms will be discussed in the following sections.

5.1.1 Specialized Plasmodesmata in Conifer Phloem

The complex plasmodesmata between mature Strasburger cells present the most elaborate intercellular connections described hitherto in plants. In needles of *Pinus* (Carde 1974) and *Metasequoia* (Glockmann and Kollmann 1996) multiply branched plasmodesmata with extended cavities in various planes of conspicuous wall thickenings have been described (Fig. 8). The substructure of these branched plasmodesmata differs from that of "normal" plasmodesmata in many respects (Glockmann and Kollmann 1996). The single branches with constricted neck regions of about 42 nm and sleeve regions expanded up to 100 nm are interconnected by cavities of variable extensions. Desmotubules, continuous with the ER of the adjacent cells, often branch and inflate within the sleeve region and cavities. With this fine structure single plasmodesmal branches are very similar to conifer sieve pores at early stages of development, indicating a common way of origin of both types of cell connections.

The complex cell connections obviously originate from single unbranched plasmodesmata which are randomly distributed in young cell walls of immature Strasburger cells. A reconstruction (Fig. 4) of the events suggests a drastic transformation of the primary plasmodesmata (Sect. 3.1). A progressive enclosure of branched ER structures during wall thickening results in the highly branched cell connections. Multiple branching might be regarded as a most efficient mechanism to compensate for the reduction in density of single primary plasmodesmata during wall expansion and cell division in differentiating phloem tissue and for a higher demand of solute passage in the conifer needle. Whether these structural changes in intercellular connections are related to the sink-to-source transition, as suggested in angiosperm leaves (Volk et al. 1996), is likely, but has not been demonstrated so far.

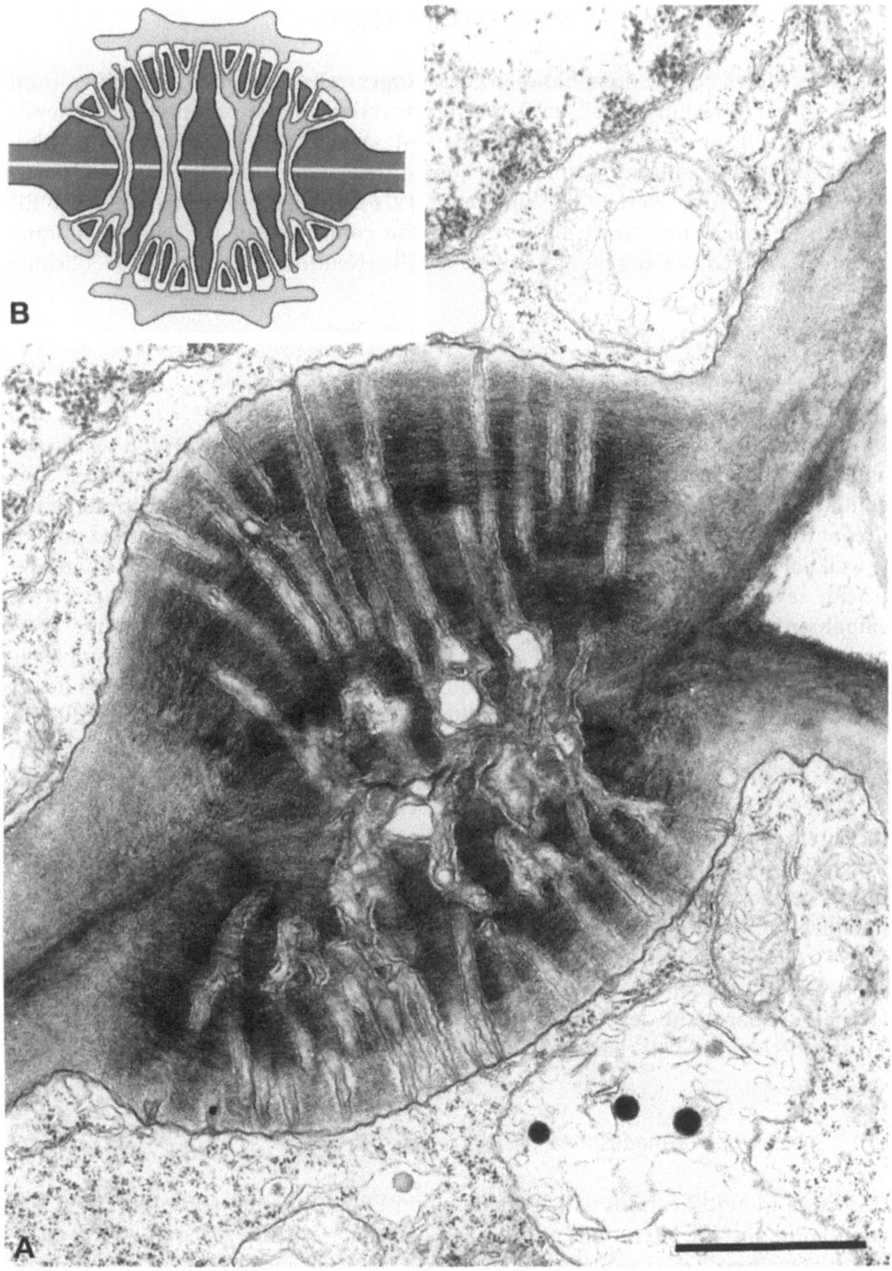

Fig. 8A, B. Multiply branched modified primary plasmodesmata in conspicuous wall thickenings between mature Strasburger cells of Metasequoia. The numerous plasmodesmal branches are interconnected by extended cavities in various planes of the wall thickenings. Each branch shows a distinct neck constriction at its orifice to the cytoplasm and a desmotubule which is inflated within the cavities. **B** Diagrammatic reconstruction based on serial sections of a plasmodesmal complex showing the position of four formerly single-stranded primary plasmodesmata in the region of the middle lamella. Magnification of **A** 25 000; *bar* 1 μm. (Glockmann and Kollmann 1996)

5.1.2 Plasmodesma-Pore Connections of Conifer Phloem

The intercellular connections between Strasburger cells and sieve cells differ from those between Strasburger cells in that cell connections in the wall part of the developing sieve cell transform to the more enlarged pores, each with several ER tubules. Pores and branched plasmodesmata are joined by extended median cavities (Carde 1974; Neuberger and Evert 1975; Sauter et al. 1976). Tubules and cisternae of the ER traverse the cell connections and establish the continuity of the endomembrane systems between Strasburger cells and sieve cells (Neuberger and Evert 1975; Schulz 1990, 1992).

5.1.3 Plasmodesma-Pore Connections of Angiosperm Phloem

The symplasmic connections between sieve-tube members and companion cells in the angiosperm phloem are basically comparable to the sieve cell / Strasburger cell connections in conifers. The main difference pertains to the number and structure of the pores at the sieve-tube side of the cell connections. There is only one enlarged pore in the wall part of the sieve tube interconnected via a more or less extended median cavity with several branched plasmodesmata in dome-shaped wall thickenings at the companion cell side (Wooding and Northcote 1965; Behnke 1975; Deshpande 1975; Esau and Thorsch 1985; Russin and Evert 1985; Volk et al. 1996). The whole complex is referred to as pore / plasmodesma unit (PPU) (van Bel and Kempers 1997). Tubules of the ER travers the PPU continuously (Kollmann 1973; Behnke 1975; Russin and Evert 1985; Ding et al. 1993; Volk et al. 1996). Within the pores P-protein filaments occasionally occur (Kollmann 1973). In contrast to the plasmodesmata, the pore channel is entirely lined by a callose cylinder of varying thickness, thus narrowing the pore sometimes up to the diameter of plasmodesmata. (For further species-specific details of the PPU structure, see Chap. 15.)

PPUs develop from single unbranched plasmodesmata in the division wall between immature companion cells and sieve tube members (Esau and Thorsch 1985; Volk et al. 1996). As with Strasburger cells, the branching process is connected with local wall depositions (Deshpande 1975; Esau and Thorsch 1985). The mechanism of pore formation is similar to the development of true sieve pores (Deshpande 1975). In conclusion, the PPUs exhibit all characteristics of modified primary plasmodesmata.

5.2 Specialized Plasmodesmata in Grass Leaves

In addition to modifications which are based on ER entrapment and plasmodesmata fusion, some specific alterations of the plasmodesmal structure in leaves of C_3 and C_4 grasses have been described (Evert et al. 1977; Hattersley and Browning 1981; Robinson-Beers and Evert 1991a; Botha 1992; Botha et al. 1993. In this section the situation in leaves of *Saccharum* will be dealt with exemplarily. For a more detailed analysis of plasmodesmal substructure and discussion on structure/function relationships of plasmodesmata in grass leaves, see Chap. 15.). In mature leaves of sugarcane, plasmodesmata show substructural cell-specific variations between different cell combinations (Robinson-Beers and Evert 1991a). Plasmodesmata between vascular parenchyma cells represent the "normal" type as they have neck constrictions at both orifices and an entirely constricted desmotubule. Those plasmodesmata interconnecting mesophyll cells show electron dense sphincters at both orifices with desmotubule constrictions in these regions only. Plasmodesmata between (Kranz) mesophyll cells and bundle-sheath cells as well as between bundle-sheath cells possess sphincters, with desmotubule constrictions in addition to an extra constriction of the plasmodesmal sleeve and the desmotubule in the median part of the cell wall where a suberin lamella is deposited. The functional significance of these particular modifications of plasmodesmata which occur during sink-to-source transition in the developing leaves (Robinson-Beers and Evert 1991b) is still open to question. Different control functions in symplasmic transport between mesophyll and vascular tissue across the bundle sheath are obvious (cf. Schulz 1998).

6 Concluding Remarks

Growth and development of a multicellular plant organism require a permanent formation of new cell walls and cell-specific modifications of existing walls. Dynamics of the extracellular matrix bring about alterations of intercellular communication during cell differentiation which may reflect the ability of the cell to regulate exchange of matter and information, varying with the demands of the developmental stage.

Alterations of intercellular communication in the course of cell growth, cell division and cell differentiation comprise the establishment of primary plasmodesmata in the cell plate, their subsequent modification and / or a completely new formation of secondary plasmodesmata in the existing and growing cell wall.

While the mechanism of formation of primary plasmodesmata in division walls has been well known for a long time, the situation for secondary plasmodesmata was less clear. A secondary modification of primary plasmodesmata has not been considered at all. Meanwhile the formation of secondary plasmodesmata has been cleared up exemplarily by the use of graft unions and in cell cultures. As for the modification of primary plasmodesmata, a model was presented recently on the basis of studies on cultured cells (Ehlers and Kollmann 1996). The three ways by which cell connections are formed – (1) establishment and (2) modification of primary plasmodesmata as well as (3) fully de novo formation of plasmodesmata (secondary plasmodesmata) – are remarkably common in the essential steps. This provides strong evidence for the concept that symplasmic connections in plant cells are established following a general principle.

Alternatives for the mechanism of secondary plasmodesmata formation and of plasmodesmal branching (Jones 1976; Juniper 1977; Lucas and Gilbertson 1994; Ding and Lucas 1996) cannot entirely be ruled out, although experimental proof is lacking so far. The most favoured model to date (Jones 1976), according to which secondary plasmodesmata develop by penetrating existing walls from one or both cell sides by a locally restricted enzymatic wall-degrading process, leaves some questions open. (1) The process requires a sophisticated recognition system between abutting cells across a thick wall. (2) The occurrence of discontinuous plasmodesmata that never cross the middle lamella is difficult to reconcile with this concept. (3) It is also difficult to conceive why cell connections of quite different origin – either passively entrapped into a growing cell plate or actively penetrating through an existing wall – show the same fine structure and nearly the same diameter. Constriction of cytoplasmic strands up to a minimal diameter is most easily explained by the entrapment mechanism.

It should be pointed out that the mechanism of plasmodesmal branching advanced here is in accordance with a former hypothesis regarding the formation of branched plasmodesmata in a growing cell wall (Krull 1960). According to Krull's model, the median cavity of a branched plasmodesma represents that part of the primary plasmodesma which has been formed in the cell plate and became dilated during wall expansion, while the plasmodesmal branches are cytoplasmic strands being established during wall thickening. It is therefore tempting to speculate that the mechanism of wall modification and of ER entrapment, leading to the branching of plasmodesmata, generally holds for the establishment of new cell connections in any growing plant cell wall.

The results of studies on primary plasmodesmata modification and on secondary plasmodesmata formation are summarized in a diagram (Fig. 9) showing the different ways by which branched plasmodesmata may be formed. In all cases, modification of existing cell walls – localized thinning and rebuilding of wall parts – is the prerequisite for the formation of median cavities and emerging plasmodesmal branches. While the mode of cavity formation differs with primary and secondary plasmodesmata – dilation of primary plasmodesmata during wall expansion with the first, and fusion of opposite half plasmodesmata with the latter – the branching process follows one common mechanism. In their final stage of development, branched plasmodesmata derived from primary or secondary cell connections show an identical structure and cannot be discriminated any further.

The terminology of the different types of plasmodesmata described so far in the literature is not always consistent. A majority of researchers designate all branched cell connections as being secondary plasmodesmata as opposed to the non-branched simple primary plasmodesmata (e.g. Ding et al. 1992; Lucas et al. 1993; Lucas and Gilbertson 1994; Ding and Lucas 1996). This terminology, however, is inconsequent, as stated by the same authors (cf. Ding et al. 1992). Certainly, single-stranded plasmodesmata represent primary plasmodesmata in most cases; they have, however, also been proved to occur, though infrequently, in non-division walls containing secondary plasmodesmata only (graft union, host-parasite contact, chimera, carpel fusion). Moreover, branched plasmodesmata, as was shown, not only are formed de novo as secondary plasmodesmata but also occur as a result of modification of primary plasmodesmata. Finally, when a division wall stretches during expansion of cultured cells primary and secondary cell connections emerge side by side within one and the same wall (Ehlers and Kollmann 1996).

To avoid problems with the terminology, it is recommended to use the term secondary plasmodesmata only for those cell connections which have been formed completely de novo in non-division walls. This would include plasmodesmata between cells which evidently have been fused postcytokinetically (e.g. in graft unions, chimeras, host-parasite interactions or, in general, with cell, tissue, and organ fusions). In all the other situations, where the origin of the cell connections is uncertain, the term *branched plasmodesmata* versus *unbranched plasmodesmata* should be used because these terms are always correct for both, the primary and secondary plasmodesmata.

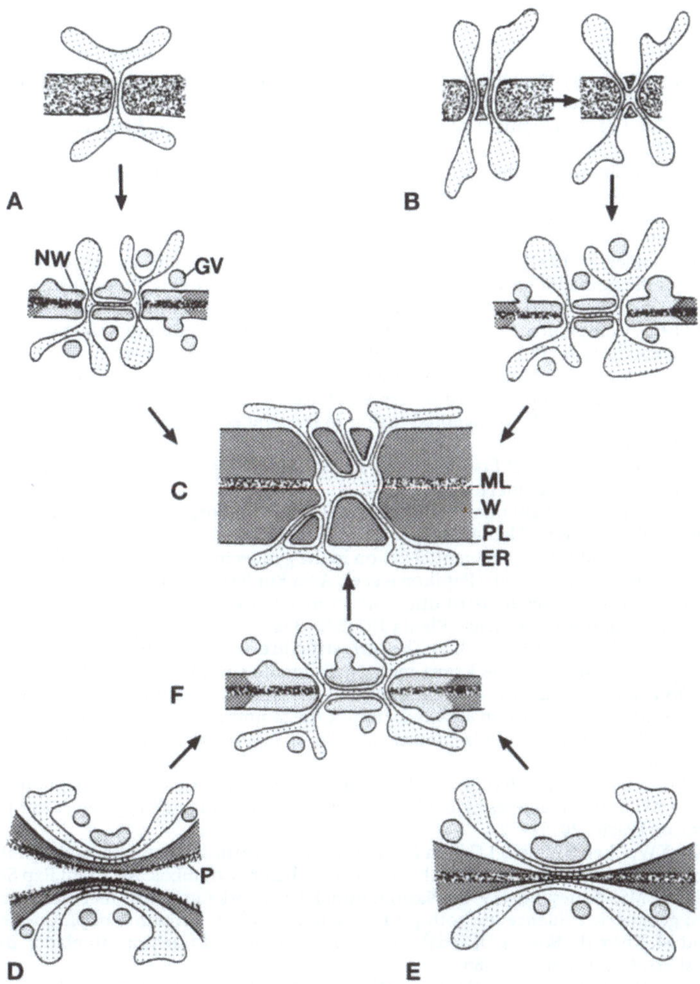

Fig. 9A–E. Diagram illustrating four different ways by which branched plasmodesmata are established in growing plant cell walls. **A, B** Modification of primary plasmodesmata: During cell growth median cavities are formed in the expanding wall by dilation of single-stranded (**A**) or fused primary plasmodesmata (**B**) in the region of the middle lamella and primary wall parts. In the course of wall thickening branching of plasmodesmata takes place by enclosure of branched ER tubules (**C**). **D, E** Secondary plasmodesmata are established by fusion of opposite half plasmodesmata either in fusion walls (**D**) or in locally thinned areas of existing walls during cell expansion (**E**), each resulting in the formation of median cavities (**F**). Further branching follows the same mechanism of ER entrapment as described for the modification of primary plasmodesmata. In their final stage of development (**C**), branched plasmodesmata derived from primary or secondary cell connections show an identical structure and cannot further be discriminated; *GV*, Golgi vesicles; *NW* newly deposited wall layers; *ML* middle lamella and primary wall parts; *W* wall layers; *PL* plasmalemma; *P* pectic layer

References

Alosi MC, Calvin CL (1985) The ultrastructure of dwarf mistletoe (*Arceuthobium* spp.) sinker cells in the region of the host secondary vasculature. Can J Bot 63:889–898

Barnett JR (1987a) The development of fibre-tracheid pit membranes in *Pyrus communis* L. IAWA Bull 8:134–142

Barnett JR (1987b) Changes in the distribution of plasmodesmata in developing fibre-tracheid pit membranes of *Sorbus aucuparia* L. Ann Bot 59:269–279

Baron-Epel O, Gharyal PK, Schindler M (1988) Pectins as mediators of wall porosity in soybean cells. Planta 175:389–395

Bauer R, Oberwinkler F (1990) Direct cytoplasm-cytoplasm connection: an unusual host-parasite interaction of the tremelloid mycoparasite *Tetragoniomyces uliginosus*. Protoplasma 154:157–160

Baum H (1948) Über die postgenitale Verwachsung in Karpellen. Österr Bot Z 95:86–94

Behnke H-D (1975) Companion cells and transfer cells. In: Aronoff S, Dainty J, Gorham PR, Srivastava LM, Swanson CA (eds) Phloem transport. Plenum, New York, pp 153–175

Binding H, Witt D, Monzer J, Mordhorst G, Kollmann R (1987) Plant cell graft chimeras obtained by co-culture of isolated protoplasts. Protoplasma 141:64–73

Boeke JH (1971) Location of the postgenital fusion in the gynoecium of *Capsella bursa-pastoris* (L.) Acta Bot Neerl 20:570–576

Boeke JH (1973a) The postgenital fusion in the gynoecium of *Trifolium repens* L.: light and electron microscopical aspects. Acta Bot Neerl 22:503–509

Boeke JH (1973b) The use of light microscopy versus electron microscopy for the location of postgenital fusions in plants. Proc K Ned Akad Wet Ser C 76:528–535

Boeke JH, van Vliet GJCM (1979) Postgenital fusion in the gynoecium of the periclinal chimera *Laburnocytisus adamii* (Poit.) Schneid. (Papilionaceae). Acta Bot Neerl 28:159–167

Botha CEJ (1992) Plasmodesmatal distribution, structure and frequency in relation to assimilation in C_3 and C_4 grasses in southern Africa. Planta 187:348–358

Botha CEJ, Hartley BJ, Cross RHM (1993) The ultrastructure and computer-enhanced digital image analysis of plasmodesmata at the Kranz mesophyll-bundle sheath interface of *Themeda triandra* var. *imberbis* (Retz) A. Camus in conventionally-fixed leaf blades. Ann Bot 72:255–261

Buder J (1911) Studien an *Laburnum adami*. Z Indukt Abstammungs Vererbungsl 5:209–284

Burgess J (1972) The occurrence of plasmodesmata-like structures in a non-division wall. Protoplasma 74:449–458

Carde J-P (1974) Le tissu de transfert (= cellules de Strasburger) dans les aiguilles du pin maritime (*Pinus pinaster* AIT.). II. Caractères cytochimiques et infrastructuraux de la paroi et des plasmodesmes. J Microsc (Paris) 20:51–72

Cheng KC, Nie XW, Chen SW, Jian LC, Sun LH, Sun DL (1987) Studies on the secondary formation of plasmodesmata between the pollen mother cells of lily before cytomixis. Acta Biol Exp Sin 20:1–11

Cooke TJ, Tilney MS, Tilney LG (1996) Plasmodesmatal networks in apical meristems and mature structures: geometric evidence for both primary and secondary formation of plasmodesmata. In: Smallwood M, Knox JP, Bowles DJ (eds) Membranes: specialized functions in plants. BIOS Scientific Publishers, Oxford, pp 471–488

Czaninski Y (1974) Cytologie Végétale. Formation des thylles dans le xylème de *Daucus carota* L. Etude ultrastructurale. C R Acad Sci Paris Ser D 278:253–256

Dawson JH, Musselman LJ, Wolswinkel P, Dörr I (1994) Biology and control of *Cuscuta*. Rev Weed Sci 6:265–317

Dell B, Kuo J, Burbidge AH (1982) Anatomy of *Pilostyles hamiltonii* C.A. Gardner (Rafflesiaceae) in stems of *Daviesia*. Aust J Bot 30:1–9

Deshpande BP (1975) Differentiation of the sieve plate of *Cucurbita*. A further view. Ann Bot 39: 1015–1022

Ding B, Lucas WJ (1996) Secondary plasmodesmata: biogenesis, special functions and evolution. In: Smallwood M, Knox JP, Bowles DJ (eds) Membranes: specialized functions in plants. BIOS Scientific Publishers, Oxford, pp 489–506

Ding B, Haudenshield JS, Hull RJ, Wolf S, Beachy RN, Lucas WJ (1992) Secondary plasmodesmata are specific sites of localization of the tobacco mosaic virus movement protein in transgenic tobacco plants. Plant Cell 4:915–928

Ding B, Haudenshield JS, Willmitzer L, Lucas WJ (1993) Correlation between arrested secondary plasmodesmal development and onset of accelerated leaf senescence in yeast acid invertase transgenic tobacco plants. Plant J 4:179–189

Dörr I (1969) Feinstruktur intrazellulär wachsender *Cuscuta*-Hyphen. Protoplasma 67:123–137

Dörr I (1987) The haustorium of *Cuscuta* – new structural results. In: Weber HC, Forstreuter W (eds) Parasitic flowering plants. Proc 4th Int Symp on parasitic flowering plants, Marburg, Germany, 1987, pp 163–170

Dörr I (1996a) Interspecific plasmodesmata between higher parasites and their host plants. In: Third Int Worksh on basic and applied research in plasmodesmatal biology, Zichron-Yakov, Israel, March 10–16, 1996, pp 120–122
Dörr I (1996b) New results on interspecific bridges between parasites and their hosts. In: Moreno T, Cubero JI, Berner D, Joel D, Musselman LJ, Parker C (eds) Advances in parasitic plant research. Proc 6th Int Parasitic Weed Symp, Cordoba, Spain, April 16–18, 1996, pp 195–201
Dörr I, Kollmann R (1995) Symplasmic sieve element continuity between *Orobanche* and its host. Bot Acta 108:47–55
Ehlers K, Kollmann R (1996) Formation of branched plasmodesmata in regenerating *Solanum nigrum*-protoplasts. Planta 199:126–138
Ehlers K, Schulz M, Kollmann R (1996) Subcellular localization of ubiquitin in plant protoplasts and the function of ubiquitin in selective degradation of outer-wall plasmodesmata in regenerating protoplasts. Planta 199:139–151
Esau K (1948) Anatomic effects of the viruses of Pierce's disease and phony peach. Hilgardia 18: 423–482
Esau K (1965) Plant anatomy. John Wiley, New York
Esau K, Thorsch J (1985) Sieve plate pores and plasmodesmata, the communication channels of the symplast: ultrastructural aspects and developmental relations. Am J Bot 72:1641–1653
Evert RF, Eschrich W, Heyser W (1977) Distribution and structure of the plasmodesmata in mesophyll and bundle-sheath cells of *Zea mays* L. Planta 136:77–89
Fry SC (1989) Cellulases, hemicelluloses and auxin-stimulated growth: a possible relationship. Physiol Plant 75:532–536
Gagnon M-J, Beebe DU (1996) Minor vein differentiation and the development of specialized plasmodesmata between companion cells and contiguous cells in expanding leaves of *Moricandia arvensis* (L.) DC. (Brassicaceae). Int J Plant Sci 157:685–697
Glockmann C, Kollmann R (1996) Structure and development of cell connections in phloem cells of *Metasequoia glyptostroboides* needles. I. Ultrastructural aspects of modified primary plasmodesmata in Strasburger cells. Protoplasma 193:191–203
Gunning BES (1978) Age-related and origin-related control of the numbers of plasmodesmata in the walls of developing *Azolla* roots. Planta 143:181–190
Hattersley PW, Browning AJ (1981) Occurrence of the suberised lamella in leaves of grasses of different photosynthetic types. I. In parenchymatous bundle-sheaths and PCR ("Kranz") sheaths. Protoplasma 109:371–401
Hayashi T, Wong Y-S, MacLachlan G (1984) Pea xyloglucan and cellulose. II. Hydrolysis by pea endo-1, 4-β-glucanases. Plant Physiol 75:605–610
Hepler PK (1982) Endoplasmic reticulum in the formation of the cell plate and plasmodesmata. Protoplasma 111:121–133
Hoch HC (1977a) Mycoparasitic relationships. III. Parasitism of *Physalospora obtusa* by *Calcarisporium parasiticum*. Can J Bot 55:198–207
Hoch HC (1977b) Mycoparasitic relationships: *Gonatobotrys simplex* parasitic on *Alternaria tenuis*. Phytopathology 67:309–314
Hoch HC (1978) Mycoparasitic relationships. IV. *Stephanoma phaeospora* parasitic on a species of *Fusarium*. Mycologia 70:370–379
Hume M (1913) On the presence of connecting threads in graft hybrids. New Phytol 12:216–220
Jeffree CE, Yeoman MM (1983) Development of intercellular connections between opposing cells in a graft union. New Phytol 93:491–509
Jones MGK (1976) The origin and development of plasmodesmata. In: Gunning BES, Robards AW (eds) Intercellular communication in plants: studies on plasmodesmata. Springer, Berlin Heidelberg New York, pp 81–105
Jones MGK, Dropkin VH (1976) Scanning electron microscopy of nematode-induced giant transfer cells. Cytobios 15:149–161
Juniper BE (1977) Some speculations on the possible roles of the plasmodesmata in the control of differentiation. Theor Biol 66:583–592
Juniper BE, Barlow PW (1969) The distribution of plasmodesmata in the root tip of maize. Planta 89: 352–360
Kienitz-Gerloff F (1891) Die Protoplasmaverbindungen zwischen benachbarten Gewebselementen in der Pflanze. Bot Z 49:32–46
Kollmann R (1973) Cytologie des Phloems. In: Hirsch GC, Ruska H, Sitte P (eds) Grundlagen der Cytologie. VEB Gustav Fischer, Jena, pp 479–504
Kollmann R, Glockmann C (1985) Studies on graft unions. I. Plasmodesmata between cells of plants belonging to different unrelated taxa. Protoplasma 124:224–235
Kollmann R, Glockmann C (1991) Studies on graft unions. III. On the mechanism of secondary formation of plasmodesmata at the graft interface. Protoplasma 165:71–85
Kollmann R, Yang S, Glockmann C (1985) Studies on graft unions. II. Continuous and half plasmodes-

mata in different regions of the graft interface. Protoplasma 126:19–29

Kollmann R, Ehlers K, Glockmann C, Schulz M (1996) Plasmodesmata in cell growth and differentiation. In: Third Int Worksh on basic and applied research in plasmodesmatal biology, Zichron-Yakov, Israel, March 10–16, 1996, pp 68–73

Krull R (1960) Untersuchungen über den Bau und die Entwicklung der Plasmodesmen im Rindenparenchym von *Viscum album*. Planta 55:598–629

Lucas WJ, Gilbertson RL (1994) Plasmodesmata in relation to viral movement within leaf tissues. Annu Rev Phytopathol 32:387–411

Lucas WJ, Ding B, van der Schoot C (1993) Plasmodesmata and the supracellular nature of plants. New Phytol 125:435–476

McCann MC, Wells B, Roberts K (1990) Direct visualization of cross-links in the primary plant cell wall. J Cell Sci 96:323–334

Monzer J (1990) Secondary formation of plasmodesmata in cultured cells. In: Robards AW, Jongsma H, Lucas WJ, Pitts J, Spray D (eds) Parallels in cell-to-cell junctions in plants and animals. NATO ASI Ser H 46. Springer, Berlin Heidelberg New York, pp185–197

Monzer J (1991) Ultrastructure of secondary plasmodesmata formation in regenerating *Solanum nigrum*-protoplast cultures. Protoplasma 165:86–95

Moore PJ, Fenczik CA, Deom CM, Beachy RN (1992) Developmental changes in plasmodesmata in transgenic tobacco expressing the movement protein of tobacco mosaic virus. Protoplasma 170:115–127

Neuberger DS, Evert RF (1975) Structure and development of sieve areas in the hypocotyl of *Pinus resinosa*. Protoplasma 84:109–125

Overall RL, Wolfe J, Gunning BES (1982) Intercellular communication in *Azolla* roots: I. Ultrastructure of plasmodesmata. Protoplasma 111:134–150

Robards AW, Lucas WJ (1990) Plasmodesmata. Annu Rev Plant Physiol Plant Mol Biol 41:369–419

Robinson-Beers K, Evert RF (1991a) Fine structure of plasmodesmata in mature leaves of sugarcane. Planta 184:307–318

Robinson-Beers K, Evert RF (1991b) Structure and development of plasmodesmata at the mesophyll/bundle sheath cell interface of sugarcane leaves. In: Bonnemain JL, Delrot S, Lucas WJ, Dainty J (eds) Recent advances in phloem transport and assimilate compartmentation. Ouest Editions, Nantes, pp 116–122

Russin WA, Evert RF (1985) Studies on the leaf of *Populus deltoides* (Salicaceae): ultrastructure, plasmodesmatal frequency, and solute concentrations. Am J Bot 72:1232–1247

Russow E (1883) Über die Perforation der Zellwand und den Zusammenhang der Protoplasmakörper benachbarter Zellen. Sitzungsber Dorp Naturf-Ges 6:15

Sauter JJ, Dörr I, Kollmann R (1976) The ultrastructure of Strasburger cells (= albuminous cells) in the secondary phloem of *Pinus nigra* var. *austriaca* (Hoess) Badoux. Protoplasma 88:31–49

Schmitz K, Srivastava LM (1974) Fine structure and development of sieve tubes in *Laminaria groenlandica* Rosenv. Cytobiologie 10:66–87

Schnepf E, Sych A (1983) Distribution of plasmodesmata in developing *Sphagnum* leaflets. Protoplasma 116:51–56

Schulz A (1990) Conifers. In: Behnke H-D, Sjolund RD (eds) Sieve elements – comparative structure, induction and development. Springer, Berlin Heidelberg New York, pp 63–88

Schulz A (1992) Living sieve cells of conifers as visualized by confocal, laser-scanning fluorescence microscopy. Protoplasma 166:153–164

Schulz A (1998) Phloem. Structure related to function. Prog Bot 59:429–475

Seagull RW (1983) Differences in the frequency and disposition of plasmodesmata resulting from root cell elongation. Planta 159:497–504

Steinberg G, Kollmann R (1994) A quantitative analysis of the interspecific plasmodesmata in the non-division walls of the plant chimera *Laburnocytisus adamii* (Poit.) Schneid. Planta 192:75–83

Strasburger E (1901) Über Plasmaverbindungen pflanzlicher Zellen. Jb Wiss Bot 36:493–610

Tainter FH (1971) The ultrastructure of *Arceuthobium pusillum*. Can J Bot 49:1615–1622

Tangl E (1879) Über offene Communicationen zwischen den Zellen des Endosperms einiger Samen. Jb Wiss Bot 12:170–190

van Bel AJE, Kempers R (1997) The pore/plasmodesm unit; key element in the interplay between sieve element and companion cell. Prog Bot 58:278–291

van der Schoot C, Dietrich MA, Storms M, Verbeke JA, Lucas WJ (1995) Establishment of a cell-to-cell communication pathway between separate carpels during gynoecium development. Planta 195:450–455

Volk GM, Turgeon R, Beebe DU (1996) Secondary plasmodesmata formation in the minor-vein phloem of *Cucumis melo* L. and *Cucurbita pepo* L. Planta 199:425–432

Wooding FBP, Northcote DH (1965) The fine structure and development of the companion cell of the phloem of *Acer pseudoplatanus*. J Cell Biol 24:117–128

CHAPTER 11

Physiological Control of Plasmodesmal Gating

A. SCHULZ

Department of Plant Biology,
The Royal Veterinary and Agricultural University,
DK-1871 Frederiksberg C, Copenhagen, Denmark

1 Introduction

From the start, a freshly divided plant cell is linked to its neighbours by plasmodesmata, offering a symplasmic pathway for the exchange of solutes. During differentiation as well as at maturity, however, the properties of adjacent cells, including the active set of enzymes and the metabolite levels, may be different. Dissimilarity and autonomy of individual cells can develop only if plasmodesmal passage is controlled. Indeed, a transient uncoupling of cells was shown for neighbouring cells that renewed mitotic activity asynchronously (Ehlers and Kollmann 1996).

Control of plasmodesmal transport also includes the dilation of plasmodesmata. Dilation may be non-specific, admitting the passage of larger-than-normal molecules, or specific, then potentiating the selective passage of certain molecules. For the individual cell, control of transport through plasmodesmata is equally as important as that across the plasma membrane, both regulating the quality and quantity of substances approaching and leaving a cell in concert.

Plasmodesmal gating implies that the supracellular nature of the plant (see Lucas et al. 1993) is realized as a flickering pattern rather than as a stable condition within the plant body. The intimacy of intercellular contact between neighbouring cells changes in space and time throughout plant development. The extremes of symplasmic cell contact realized in angiosperms are the fertilized egg cell (zygote) and mature leaf stomata on the one hand, both totally isolated from abutting cells, and the mature, syncytium-like sieve tube system intimately interlinked by sieve pores on the other.

In the present chapter, gating is defined as a transient change of plasmodesmal conductivity. Definitive changes in the plasmodesmal conductivity which lead to more permanent symplasmic domains within a plant and allow for preferential exchange pathways are treated elsewhere (Chaps. 13, 14).

2 Nature of Plasmodesmal Gating

2.1 Terminology

"Gate" has an ambiguous meaning. It is the opening in a wall, hedge or similar border, but also the barrier(s) closing this opening at the same time. To gate a person means that she or he is not allowed to leave (or enter) an area delimited by such a border. Since this implies the existence of a gate keeper who controls the persons passing the gate, the term gating is commonly used in the context of intercellular transport for the specific interaction between a molecule and a plasmodesma which leads to an active dilation of the plasmodesma. Thus, molecules or viral nucleoproteins which are larger than the size exclusion limit (SEL) can pass. However, passage of smaller solutes is also controlled by the plasmodesmal width. This fact is neglected in the common usage of gating.

In order to exemplify all gating functions of a plasmodesma, the image of a lock in a river may be more adequate than of a gate in a wall. By active changes, the lock permits the selective passage of ships, but it also determines the quantity of water going through. Openings within the lock barrier are regulated in such a way that the water levels above and below the lock are appropriate. Overall water movement across the lock barrier is passive, i.e. follows the inherent gradient of the river bed. This compares well with plasmodesmal gating, since the width of plasmodesmata determines the quantity of solutes entering or leaving a cell. The transport of permeant solutes (below the SEL) through the plasmodesmal channel is passive in the sense that it is not driven by a change of the plasmodesma but by diffusion or bulk flow between coupled cells.

As in the case of a river lock, also other substances are swept through the plasmodesmal channel together with a specific molecule that is gated selectively. Injection of a movement protein (MP) into mesophyll cells effected the increase of the SEL so that large fluorescing dextrans could pass non-specifically (Wolf et al. 1989).

In the context of this chapter, the different cases of plasmodesmal gating will be distinguished by adding non-selective and selective, respectively. Non-selective gating means a change in the plasmodesmal passage area, such that the amount of permeant substances is regulated, and selective gating means the active adaptation of the plasmodesma to the molecule to enable its passage. Increase of the SEL under selective gating provides the decisive evidence that it is the plasmodesmal channel that is changed and not (or not only) the conformation of the permeant molecule.

2.2 Plasmodesmal Gateways

Based on the present knowledge of plasmodesmal structure (Chaps. 2, 9, 10) two principal, tortuous pathways can be envisaged in intercellular transport: (1) the cytosolic pathway following the cytoplasmic sleeve between plasma membrane and desmotubule, and (2) the intratubular pathway via the desmotubular core which is contiguous with the intracisternal lumen of the endoplasmic reticulum (ER), i.e. has the potential to link the ER of both cells (Fig. 1). According to Schnepf's compartmentation theorem, the ER and intratubular lumen are non-plasmic compartments (cf. Sitte 1998). In order to emphasize that the cytoplasmic and non-plasmic pathway are under cellular control, they are termed cytoplasmic and intratubular gateway in the present chapter. If both are functional, plasmodesmata offer the possibility of unidirectional or bidirectional transport between neighbouring cells.

2.2.1 Cytosolic Continuum Through the Cytoplasmic Sleeve

The cytoplasmic gateway is delimited by the inner leaflet of the plasma membrane and by the outer leaflet of the desmotubule. Though continuous, the plasma membrane does not show lateral diffusion of compounds from cell to cell, but it retains its respective cellular affiliation (Grabski et al. 1993). Plasmodesma models derived from electron microscopy (EM) show gaps in the range of 3 nm between the proteins within the cytoplasmic gateway (see Chaps. 2, 3 and 9). Narrowing of the cytoplasmic gateway either extends over the entire length of the short and straight plasmodesmata in meristematic root or immature leaf tissue (e.g., Overall et al. 1982; Ding et al. 1992; Glockmann and Kollmann 1996, their Fig. 6) or occurs only at the orifices of mature unbranched (Fig. 1; Olesen and Robards 1990) and branched plasmodesmata as neck regions (Chap. 10).

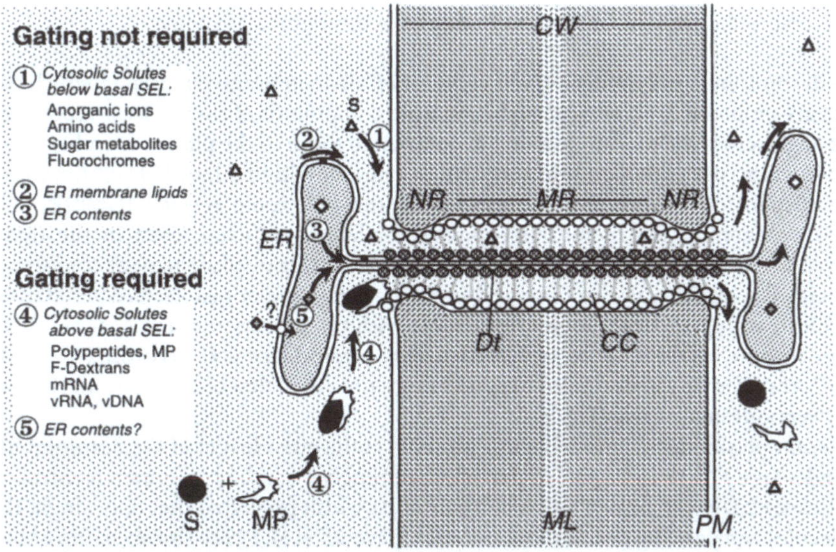

Fig. 1. Schematic drawing of the structure of a plasmodesma (after Ding et al. 1992) indicating the plasmodesmal gateways (cytoplasmic sleeve and ER-desmotubule-ER lumen, respectively) involved in solute transfer. Active change of the plasmodesma (gating) is not required for the permeant substances (*1–3*), but for substances above the SEL (*4*). Active uptake of solutes into the ER (*5*) is not yet confirmed. *CC* Central cavity; *CW* cell wall; *Dt* desmotubule; *ML* middle lamella; *MP* movement protein; *MR* middle region; *NR* neck region, *PM* plasma membrane; *s* solutes below SEL; *S* solutes above SEL

In the interface between (Kranz) mesophyll and bundle-sheath cells of grasses, the cytoplasmic gateway of plasmodesmata is furthermore constricted in the middle by a suberin lamella (median constriction according to Botha et al. 1993; see also Evert et al. 1985). These plasmodesmata display an even higher degree in complexity by additional proteinaceous structures between plasma membrane and desmotubule at their orifices, called internal sphincters. In sugarcane, the sphincters (not depicted in Fig. 1) characterize the plasmodesmata that link all mesophyll cells but the vascular parenchyma (Robinson-Beers and Evert 1991).

The narrow sections of the cytoplasmic gateway constitute a bottleneck and, thus, the rate-controlling step in symplasmic transport. Conspicuously, these bottle necks have a constant diameter of 25 to 40 nm in most plants (Overall et al. 1982; Botha et al. 1993; Schulz 1995), while the unconstricted regions show more variation in size. According to Overall et al. (1982), the plasmodesmal channel cannot be constricted further due to the molecular architecture of its membranes and proteins. The 3-nm gaps determined ultrastructurally for the distance between the inner-plasmodesmal proteins are in the same range as the basal SEL shown by water-soluble fluorescent markers for several cell types, i.e. a value between 400 and 1000 Da (see Chap. 5 and Erwee and Goodwin 1985), as is exemplified in young cortex tissue of a root for fluorescein (332 Da) using the fluorescence recovery after photobleaching (FRAP) technique (Fig. 2).

2.2.2 Non-Plasmic (ER) Continuum Through the Desmotubular Lumen

The desmotubule is a tightly curved cylinder of ER-membrane material in which proteins are embedded. Fluorescent-labelled lipid compounds of the ER were reported to exchange across the plasmodesmata in contrast to those of the plasma membrane. After bleaching the endomembrane system of one cell within a labelled cell file from a root suspension culture, the fluorescence in that cell could recover (Grabski et al. 1993). Apparently, the membrane lipids show lateral diffusion even in the tightly curved ER cylinder constituting the desmotubule. Such a displacement of compounds associated with ER was also observed in the cortex tissue of pea roots. After staining the ER in young cortical tissue of pea roots with the endomembrane-specific $DiOC_6$ fluorochrome, dye exchange occurred across plasmodesmata. According to FRAP, $DiOC_6$ transfer via the desmotubule takes more time than that of cytosolic markers via the plasmodesmal sleeve (Fig. 2).

Recent plasmodesma models show proteins in the outer leaflet of the desmotubular cylinder (see Fig. 1) and a row of particles seemingly occluding its lumen (central rod particles; cf. Ding et al. 1992). Calculations of its width predicted that only single H_2O molecules can pass the desmotubular lumen (Overall et al. 1982). However, the EM appearance of the desmotubule in standard situations might be the locked state due to the fixation process. In recent papers, the space between central particles was calculated to be in the range of 2.5–3 nm (Botha et al. 1993), and open lumina were also envisioned (Waigmann et al. 1997). It is certainly conceivable that the intratubular gateway is under tight control. In grass leaves, the sphincters located at the neck region were discussed as being involved in such a control (Evert et al. 1977; Robinson-Beers and Evert 1991).

Functional evidence for a solute exchange via the desmotubular lumen was given for chickpea trichomes by Lazzaro and Thomson only recently (1996). Using confocal laser scanning microscopy (CLSM), they demonstrated that Lucifer Yellow (LYCH), taken up in the basal cell of a trichome, moves inside a "vacuolar-tubular system" into the terminal cell. Continuity from cell to cell was provided by strands connecting the stained tubules/vacuoles with the plasmodesmata. Apoplasmic dye diffusion was slower, thus confirming a symplasmic LYCH transport that was accelerated by intracellular streaming and fusion of the stained tubules (Lazzaro and Thomson 1996). In view of the versatility of the endomembrane system and its division in different subdomains (Staehelin 1997), the special vacuoles and tubules documented in chickpea trichomes certainly belong to the ER system.

3 Methods to Study Plasmodesmal Gating

The mechanisms of plasmodesmal gating are largely unresolved. Since plasmodesmata respond immediately to any microenvironmental changes, most preparation methods induce artefacts which may obscure any indication of plasmodesmal gating. Therefore, technical improvements continue to be a challenge for structural and functional studies (see Chaps. 3, 5 and 6).

3.1 Structural Investigations

The crucial point for structural investigations into plasmodesmal gating is the time needed for any fixation method to become effective. Neither physical nor chemical methods are able to fix neighbouring cells immediately and at the same time (cf. Johnson 1978). The sequential progress of fixation, however, may lead to closure or other responses of plasmodesmata (wound gating). Even with rapid high pressure freezing methods, any pretreament (preparatory tissue cuts and/or freeze protection steps) adds to artefact formation, while slow, chemical fixations can induce callose formation at plasmodesmata (Radford et al. 1998) and may arrest the conformation of proteins involved in gating in an unrealistic, transitory position. During penetration, membrane-perforating and -stabilizing effects of chemical fixatives compete with one another. Eventually, adsorption/precipitation of contrasting substances (e.g., tannic acid, osmium acid, $KMnO_4$, and/or uranyl acetate) after embedding may also change appearance of plasmodesmata in EM (for further discussion see Chap. 3; Ding et al. 1992; Botha et al. 1993; Schulz 1995).

3.2 Functional Investigations

It was an important breakthrough for studies on intercellular transport when Schumacher introduced fluorescein as an inert tracer for short-distant transport in trichomes and for phloem translocation (Schumacher 1933, 1936). It appears that he detected plasmodesmal gating, since he reported that the fluorescein front in one trichome cell did not immediately progress into the next one but was kept there for some time before this event occurred suddenly (Schumacher 1936). It was not until 40 years later that fluorescein and derivatives experienced a renaissance as symplasmic tracer (Chap. 5) and were used for functional studies on plasmodesmal gating (see Sect. 4.3). The drawback of these tracers is that they are not natural compounds of plants, and that their application/microinjection may induce wound responses.

Radioactive markers were rarely applied in studies on plasmodesmal functioning, although they offer a wide variety of naturally transported solutes. The reason for the reluctance to use radioactive material is that (1) diffusion within the apoplasm, (2) uptake from the apoplasm, as well as (3) phloem transport would compete with the plasmodesmal short-distance transport under investigation in most experimental setups, so that no clear-cut results can be expected. These problems were overcome directly by removing parallel phloem strands (Helder 1967; Arisz 1969). Wherever phloem unloading is symplasmic, the lateral or terminal unloading rate of radioassimilates is a measure for changes in plasmodesmal transport (Schulz 1995; Fisher and Wang 1995; Offler and Patrick 1996).

Yet another way to investigate plasmodesmal functioning is by studying the electric coupling between cells (Chap. 4). Tissue- and cell-specific differences in cell-membrane potential demonstrated that the plant symplasm is divided in domains (cf. Chap. 13 and 14). For measurement of transient changes in cell coupling, however, problems arise in all those cases where tissue preparation is needed to approach the cells in questions. Wound effects may change the cell potentials. Moreover, even when

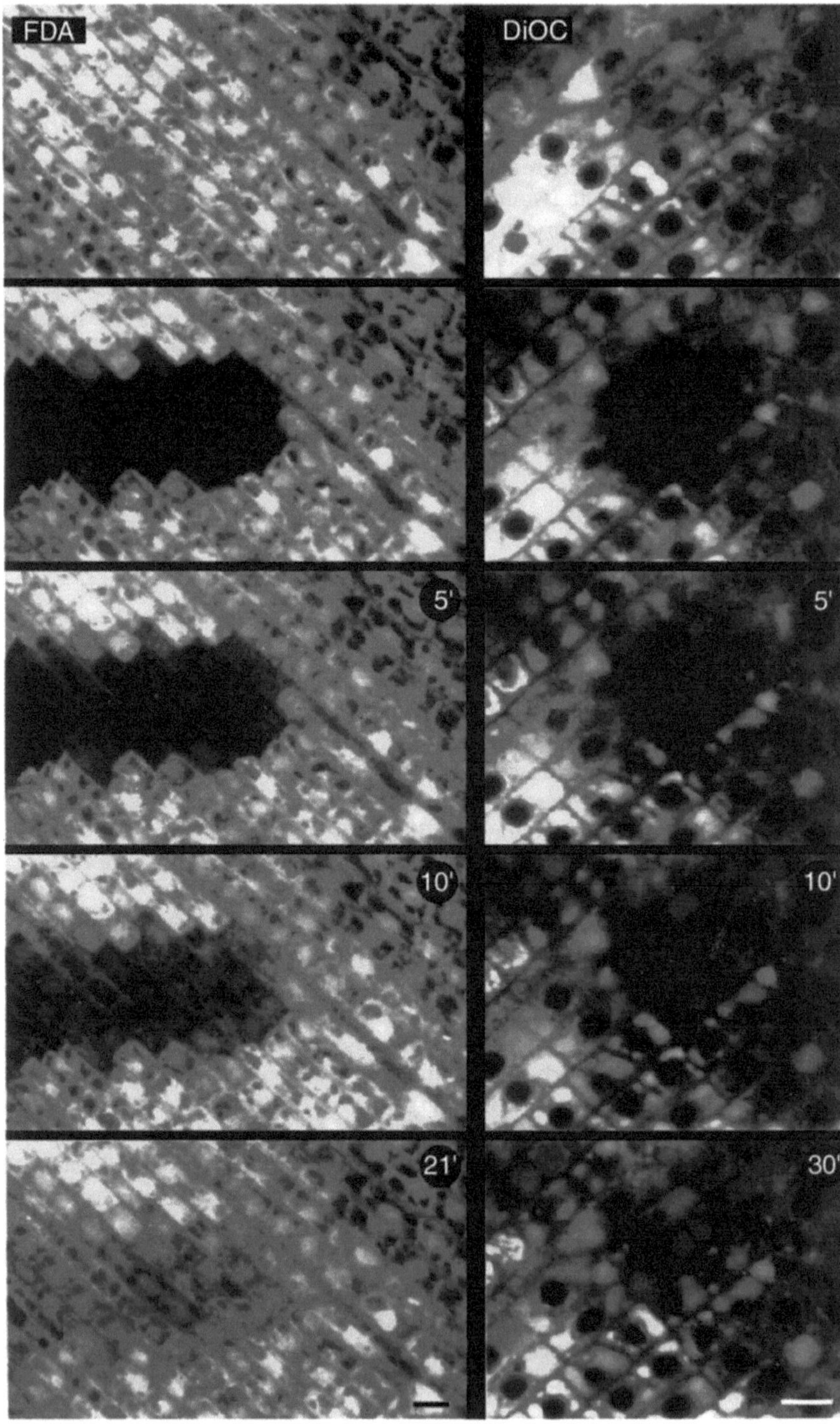
FDA
DiOC
5'
5'
10'
10'
21'
30'

Fig. 2. Fluorescence recovery after photobleaching (FRAP) experiments indicating plasmodesmal dye transport in young cortex tissue of pea roots. *Left panel* indicates fluorescence recovery of the cytosolic fluorescein, *right panel* of an ER-specific dye after bleaching of several cells, each at increasing periods after photobleaching. Root tips were incubated in fluorescein diacetate (*FDA*) for 1 min and *DiOC*$_6$ for 2 min, respectively, washed thoroughly, split in halves, and observed by confocal laser scanning microscopy (CLSM). Bar 20 μm (A. Schulz, unpubl.)

only small cell groups are treated similarly, the changes in intercellular transport of solutes and in electric coupling do not correspond (cf. Tucker 1990; Lew 1994, 1996, Holdaway-Clark et al. 1996; see also Chap. 4).

4 Control of Non-Selective Transport Through Plasmodesmata

Transport through plasmodesmata is defined as passive and non-selective, when substances below the basal SEL pass without requiring/inducing changes of the plasmodesmata, and as active and selective when substances above the SEL pass, mediated by conformational changes of the plasmodesmal architecture (see Sect. 5). As confirmed in several studies, non-selective transport is subject to gating wherever a cell needs to be isolated from intercellular transport or, conversely, depends on intercellular import. Table 1 summarizes the literature concerning plasmodesmal transport and assigns the permeant substances to the cytoplasmic and intratubular gateway, respectively. In addition, it gives the transport mechanism when known, the method by which the results were obtained, and the cases where changes in transport, i.e., non-selective gating, were observed.

4.1 The Plasmodesmal SEL Is Tissue-Specific and Depends on the Properties of the Tracer Molecule

Given their hydrodynamic radius, inorganic and organic compounds up to 1000 Da should be able to pass plasmodesmata according to a pore size of 2.5–3 nm (see Terry and Robards 1987). This SEL would allow passage of protons, most ions, amino acids, sugars, nucleotides, intermediary metabolites, and even ATP, coenzymes and peptides, but exclude proteins, enzymes, and ribonucleic acids (Goodwin 1983). Because of the need of cell autonomy and cellular isolation between dissimilar cell neighbours, an uncontrolled plasmodesmal exchange of these substances is most improbable.

Challenged by publications about cell coupling in animals, Goodwin (1976) and Tucker (1982) made use of fluorescein isothiocyanate (FITC)-labelled amino acids and peptides in order to visualize their symplasmic transport. There was no degradation of the conjugates by cytosolic enzymes (Tucker 1982; Goodwin 1983; Terry and Robards 1987). Using dyes with increasing molecular size, different size exclusion limits for plasmodesmata of *Egeria densa* and the existence of symplasmic domains in plants became evident (Erwee and Goodwin 1984, 1985; see Chap. 14). According to Terry and Robards (1987), plasmodesmata in trichomes of *Abutilon* have a somewhat larger SEL

Table 1. Modes of Intercellular Transport through Plasmodesmata

Non-selective Pd Transport				(Does not necessarily require Pd gating for transport)
Pd gateway	Permeant substance	Transport mediated by:	Indicated by:	Observed changes
Cytoplasmic sleeve	Inorganic ions Amino acids Sugar metabolites Fluorochromes	Diffusion and/or pressure flow	Intercellular transport of isotopes[a], fluorochromes[b], and FRAP[c]	Up[d]- and down[e]-regulation of SEL Widening of Pd[f]
Desmotubule	ER-membrane lipids ER- or vacuolar content	Membrane flow Diffusion	FRAP[g] LYCH transport[h]	Downregulation[i] —
Selective Pd Transport				**(Requires Pd gating for transport)**
Cytoplasmic sleeve	Polypeptides mRNA vRNA vDNA	Conformational changes of Pd proteins (? by protein-phosphorylation/dephosphorylation)	Coinjected fluorochromes[j], in situ-hybridisation[k]; DNA- or RNA-dyes[l]; MP-GFP[m] or MP-reporter protein fusion[n]	Selective passage with[o] or without[p] SEL-up-regulation
Desmotubule	? Sugar metabolites	? (+ER loading by membrane transporters)	Δψ cytosol/Pd-ER according to CLSM and EM[q]	Swelling of ER, dilation of Dt[r]

[a] Arisz and Wiersema (1966); Helder (1967); Schenk (1974); Bräutigam and Müller (1975); Goodwin (1976); Schulz (1994); Offler and Patrick (1996).
[b] Chapter 5; see also Goodwin (1976); Tucker (1982); Erwee and Goodwin (1983, 1984, 1985); Terry and Robards (1987); Tucker and Tucker (1993); Oparka et al. (1994, 1995b); Wright and Oparka (1996).
[c] Figs. 2: FDA; Baron-Epel et al. (1988).
[d] Erwee and Goodwin (1984); Tucker (1993); Cleland et al. (1994); Epel and Erlanger (1991).
[e] Erwee and Goodwin (1983, 1984); Baron-Epel et al. (1988); Tucker (1988, 1990); Tucker and Boss (1996); Oparka and Prior (1992); Ehlers and Kollmann (1996); Offler and Patrick (1996).
[f] Schulz (1995).
[g] Fig. 2: $DiOC_6$; Grabski et al. (1993).
[h] Lazzaro and Thomson (1996).
[i] Grabski et al. (1993).
[j] Wolf et al. (1989).
[k] Chapter 16; Bostwick et al. (1992); Lucas et al. (1995), Clark et al. (1997); Dannenhoffer et al. (1997), Kühn et al. (1997); Ishiwatari et al. (1998) .
[l] Fujiwara et al. (1993); Noueiry et al. (1994); Waigmann et al. (1994); Lucas et al. (1995).
[m] Chapter 6; Oparka et al. (1995a); Epel et al. (1996); Oparka et al. (1996).
[n] Waigmann and Zambryski (1995).
[o] Chapter 17; Wolf et al. (1989); Wolf and Lucas (1994); Noueiry et al. (1994).
[p] Waigmann and Zambryski (1995).
[q] Schulz (1992); Glockmann and Kollmann (1996).
[r] Schulz (1992); Gamalei et al. (1994).

than the plasmodesmata in *Egeria*. There was no specific effect of aromatic acids, either polar or hydrophobic, on the conductivity of the *Abutilon* plasmodesmata, in contrast to the situation in *Egeria* (Erwee and Goodwin 1984).

Using *Setcreasea* staminal trichomes as system, Tucker (1982; Tucker and Tucker 1993) probed a series of FITC-conjugates for intercellular transport. Movement of most markers occurred by diffusion (group I), some moved at a somewhat reduced rate (group II), and a few moved very slowly (group III). The authors argued that aromatic groups were responsible for the reduction of the transport rate. This has important consequences for the intercellular movement of phytohormones: indole acetic acid may not, but gibberellins may pass through plasmodesmata because of the respective molecular structures (Tucker and Tucker 1993). However, the generalization of amino acid transport is questionable, since several conjugates behaved differently when applied to the cut base of the stamen. Some group I ions spread slowly, and nearly immobile group III ions [FITC-Phe, -His and -$(His)_2$] permeated easily through the trichome shaft when taken up by the cut staminal end (Tucker and Tucker 1993).

The results obtained with *Nicotiana clevelandii* trichomes are confusing. In an investigation of virus spread, Waigmann and Zambryski (1995) determined a basal SEL value of about 7 kDa for FITC dextrans in control trichomes. This high value is in the range of the SEL of pore/plasmodesma contacts (SEL >10 kDa; Kempers et al. 1993). In contrast, trichomes from other species had plasmodesmal SEL values in the range of 1 kDa (Tucker 1982; Terry and Robards 1987) and those of *Nicotiana clevelandii* did not permit passage of FITC-$(Phe)_3$ (850 Da; Derrick et al. 1992). However, not the size alone but aromatic groups of the conjugate might be the determining factor for plasmodesmal mobility (cf. Tucker and Tucker 1993).

Along the postphloem pathway in developing wheat grains, size-fractionated fluorescein (F-) dextrans showed symplasmic diffusion along the grain axis up to a hydrodynamic radius of 2.0 nm (Wang and Fisher 1994b). This contrasts with 0.9 nm as the estimated hydrodynamic radius of the smallest F-conjugate immobile in *Abutilon* trichomes (FITC-Trp-Phe, 740 Da; Terry and Robards 1987). Moreover, Wang and Fisher (1994b) proposed the estimation of plasmodesmal pore size from a derived value (1.54 times the hydrodynamic radius for dextrans) and gave plasmodesmal pore diameters of 6.2 nm in wheat, double the value (3 nm) estimated for *Abutilon* trichomes (Terry and Robards 1987).

As for the properties of different tracer molecules, the dextrans generally indicate a larger SEL (in terms of Da) than amino acid conjugates of FITC. This might be related to interactions between the non-fluorescing body of the conjugates and the plasmodesmal constituents, or to the steric structure of the dextrans (Wang and Fisher 1994b). Obviously, the SEL depends on the properties of the tracer molecule and on the specific nature of the plasmodesma type involved.

4.2 Involvement of the Plasmodesmal Gateways in Short-Distance Transport Processes

Both plasmodesmal gateways have been discussed as participating in downhill and uphill transport of assimilates. Therefore, their contribution to assimilate transport will be assessed here.

4.2.1 Preference of the Cytoplasmic Gateway for Intercellular Downhill Transport

The cytoplasmic gateway of plasmodesmata appears advantageous for cytosolic solutes, since it does not involve a membrane step. Its usage can be postulated for the downhill transport of assimilates from mesophyll towards the bundle sheath of leaf minor veins and towards heterotrophic cells in general that depend upon carbohydrate import for maintenance of metabolism and respiration (trichomes, nectaries, and epidermis layers when chloroplasts are lacking). The cytoplasmic gateway is also suggested in most situations of phloem unloading. Even in those cases where the sieve element/companion cell (SE/CC) unloading itself is apoplasmic as indicated by absence or paucity of plasmodesmata, postphloem transport is mostly symplasmic, in developing fruits up to the maternal/filial interface of immature seeds (for reviews, see Patrick 1990, 1997; Patrick and Offler 1995, 1996; van Bel 1996; Schulz 1998).

4.2.2 Involvement of the Cytoplasmic Gateway in Intercellular Uphill Transport

While in most plants plasmodesmata are not involved in phloem loading, since it occurs actively across the apoplasm of phloem elements (apoplasmic loading; Delrot 1989; Stadler et al. 1995; Kühn et al. 1997), those plants translocating tri- and tetrasaccharides in the phloem apparently show an uphill transport of solutes across plasmodesmata (symplasmic loading; see van Bel 1993a). They have numerous multiply branched plasmodesmata at the interface between bundle-sheath and intermediary-type companion cells which block the backflow of the oligomers according to the "oligomer-trap hypothesis" (see Chap. 15; Turgeon 1996). Symplasmic uphill transport becomes feasible by the highly controlled domain border between bundle sheath and intermediary cells, and by the high activity of enzymes involved in raffinose and stachyose synthesis in the intermediary cell (Holthaus and Schmitz 1991; Beebe and Turgeon 1992; Turgeon 1996). Since the metabolites – starting with sucrose and ending up with stachyose – and the enzymes characteristic for symplasmic loading occur in the cytosol, sugar is supposed to move from mesophyll to sieve tubes via the cytoplasmic gateway (cf. Haritatos et al. 1996; Turgeon 1996).

4.2.3 Involvement of the Intratubular Gateway in Intercellular Transport

Usage of the ER-system – the intratubular gateway – for phloem loading was postulated by Gamalei and co-workers (1994). According to their hypothesis, assimilates approach the ER already in the photosynthetic active mesophyll cell of angiosperms and move via ER-desmotubule-ER connections towards the sieve tubes. An increase in desmotubular diameter would be an increase in transport capacity, as discussed by Gamalei et al. (1994). However, continuity of the ER pathway is not confirmed yet and the authors do not provide an answer to the question of how and where the assimilates are released from the ER into the sieve tube lumen (see Schulz 1998).

According to EM and CLSM, an intratubular gateway for assimilates might exist in gymnosperm phloem. Plasmodesmata in the collection phloem (Chap. 10) as well as the functional sieve pores in the transport phloem, are covered and traversed by ER

(Kollmann and Schumacher 1962, 1963; Schulz 1990a, 1992; Glockmann and Kollmann 1996) that was discussed as taking up sucrose actively and contributing to phloem loading and transport (Schulz 1998). It remains open to debate whether the intercellular ER in gymnosperms is subject to gating.

4.3 Observed Phenomena of Non-Selective Gating

Before putative mechanisms of non-selective gating can be discussed (Sect. 4.4), it is pertinent to review those gating phenomena which are under physiological control. These phenomena extend from a transient isolation of cells to an increase in solute exchange between cells. Evidence for a total isolation of cells is lacking due to technical limitations, i.e. closed plasmodesmata may not be watertight (movement was tested with tracers of minimally 340 Da).

4.3.1 Transient Decrease of the Size Exclusion Limit or Closure of Plasmodesmata

4.3.1.1 Incompatibility of Developmental Stages Leads to Cell Isolation

According to investigations on cell coupling in regenerating protoplast cultures, asynchronous mitotic activity in sister cells is accompanied by a transient uncoupling of the sister cells, as evidenced by exchange of LYCH (Ehlers and Kollmann 1996; see Chap. 14)

A similar transient isolation of cells was observed in young phloem after microinjection of LYCH. According to van Bel and van Rijen (1994), SE/CC mother cells become increasingly isolated from neighbouring parenchyma cells prior to their unequal division. Even a longitudinal dye transfer across the future sieve plate was prevented in immature sieve elements. Mature SE/CC complexes were isolated from the surrounding tissue, but intimately linked with one another (van Bel and van Rijen 1994). Both lateral coupling between a SE and its CC, and longitudinal coupling across the sieve plates were estimated to be subject to a SEL of more than 10 kDa in mature phloem (Kempers and van Bel 1997).

It was argued that plant cells might synchronize mitotic activity by an intercellular signal (Ehlers 1996; Ehlers and Kollmann 1996). Obviously, cells that are not prepared to enter mitosis (yet) become isolated symplasmically. This claim is supported by the isolation of immature phloem elements (van Bel and van Rijen 1994) when the sequential development of sieve tubes is considered (see Thorsch and Esau 1981a, b, c; Schulz 1986a, b, 1990b, 1993). Towards a meristem and in experimentally induced conditions (wound and graft phloem), existing sieve tubes are elongated by freshly divided SE/CC complexes. Differentiation is sequential, as is mitosis, in the axial file giving rise to a sieve tube. Contact between future sieve-tube members still consists of plasmodesmata (future sieve pores) in the prospective sieve plates. At a given moment, a mature sieve-tube member is in contact with a cell already showing specific SE characters, which cell, in turn, is in contact with a cell starting differentiation. The next cell in this file may be just entering mitosis. A transient isolation of SE/CC mother cells, as demonstrated for cambial phloem by van Bel and van Rijen (1994), is certainly required, since the untimely reception of a mitosis signal would interfere with the order of events described.

Direct and circumstantial evidence gives a time range for the transient isolation of cells of less than 1 h up to 7 h. The lower value can be derived from the growth rate of roots and the concomitant elongation of sieve tubes. It takes some 60 min from precursor division to SE maturity with open sieve pores (Schulz 1996). In protoplast cultures, uncoupling of cells was followed by coupling 7 h later (Ehlers 1996).

4.3.1.2 Stele/Cortex Coupling Is Reduced After Light Irradiation of Etiolated Maize Seedlings

Quantitative measurements of acropetal carboxyfluorescein (CF) transport showed that light pulses of different light quality inhibited mesocotyl elongation in the etiolated maize seedlings and the lateral transport of CF from the stele into the mesocotyl cortex (Epel and Erlanger 1991). The suggested involvement of phytochrome in modulating growth and development and in altering symplasmic communication remains unclear, since other factors such as the source/sink ratio have also strong influence on lateral CF unloading (Patrick and Offler 1996).

4.3.1.3 Plasmolysis Leads to Cell Isolation

Hypertonic conditions (0.7 M mannitol) led to uncoupling of cells in *Egeria densa* leaves, as visualized with CF and FITC-$(Gln)_2$. When the leaves were transferred to the standard bathing medium, dye coupling recovered within 10 min (Erwee and Goodwin 1984). These experiments suggest that the plasmodesmal gateways become mechanically disrupted or impermeable during retraction of the protoplast from the cell wall. Indeed, hypertonic treatment of pea root tips with sucrose caused disruption of a number of plasmodesmata, the only remnants of which were empty hollows in the cell wall (Schulz 1995). Plasmolysis-persistent plasmodesmata were wider in diameter than control ones and were filled with a fluffy material. Under plasmolytic conditions, both parameters affected symplasmic transport: the reduction in plasmodesmal passage area due to the fluffy material depicted in EM micrographs and the reduction in plasmodesmata number due to their breakage (cf. Schulz 1995). The rapid recovery of dye coupling (Erwee and Goodwin 1984) makes it likely that reestablished coupling occurs by persistent plasmodesmata.

Plasmolysis of onion epidermis cells involved changes of the plasma membrane, the cortical ER, and the plasmodesmal ER according to CLSM and EM (Oparka et al. 1994). It turned out that the "Hechtian strands" are $DiOC_6$-positive, i.e. the ER is not disrupted during plasmolysis. These strands led to the preexisting plasmodesmal contacts and to other attachment points at the plasma membrane. There was no staining of the cytoplasmic gateway after fluorescein diacetate treatment (Oparka et al. 1994), but this might be due to the small volume offered to any cytosolic dye in a strand.

Functional studies on intact pea seedlings with ^{14}C-labelled sucrose showed that plasmolysis caused a considerable decrease in symplasmic phloem unloading when the label was applied to the cotyledon at the same time as the plasmolyticum was added to the root tip (Schulz 1994). Concomitantly, the effective passage area was reduced as compared to controls (Schulz 1995). A 1 h pretreatment with the plasmolyticum caused the phloem import into the root tip to drop to zero (Schulz 1994).

Perfusion of the endosperm cavity of attached, developing wheat grains with sorbitol blocked postphloem transport and induced plasmolysis in nucellus and chalaza cells (Wang and Fisher 1994a). According to microanalysis of cryosectioned material, the treatment evoked a sharp drop in sucrose concentration at the vascular parenchyma/chalaza border, while there was an approximately linear decrease in sucrose concentration along the postphloem pathway in controls from the same ear, i.e., from vascular parenchyma to the endosperm cavity (Fisher and Wang 1995). Sorbitol treatment caused the vascular parenchyma to accumulate sucrose at concentrations close to the sieve-tube content (500 mM; Fisher and Wang 1995). Thus, the steep gradient of sucrose concentration across the sieve element/companion cell boundary (ca 200 mM over 1 μm) was equilibrated by sorbitol treatment.

It can be generalized that plasmodesmal blockage at one point in the symplasmic postphloem pathway causes the accumulation of transportates in the phloem parenchyma, levels out the concentration difference at the sieve tube-unloading interface, reduces the pressure gradient in the phloem system and eventually stops phloem import.

4.3.1.4 Pressure Differences Lead to Increase in Plasmodesmal Resistance and/or Cell Isolation

The effects of deliberate changes of the pressure in single trichome cells on symplasmic transport were studied by Oparka and Prior (1992). The endogenous symplasmic transport in uniseriate trichomes is from base to tip, as reflected by decreasing pressure readings. When an intermediate cell was punctured, its turgor fell to zero (Oparka and Prior 1992). LYCH applied to the tip cell was still imported into the punctured cell but did not pass beyond it. In flaccid trichomes, tracer movement to the trichome base through the punctured cell was not impeded. Oparka and Prior (1992) determined that a pressure differential of 200 kPa or more blocks intercellular transport, which is suggestive for a kind of pressure valve in plasmodesmata. Since tracer is still imported into a punctured cell (Oparka and Prior 1992), an instantaneous sealing of plasmodesmata down a sudden pressure step appears unlikely. However, downhill impedance may develop over time as indicated in a liverwort, where plasmodesmal valving occurred along or against a pressure gradient of 200 kPa (Trebacz and Fensom 1989).

The behaviour of plasmodesmata at the interface of turgor domains is of great interest for phloem transport. Depending on the sugar concentration in the sieve tubes and the apoplasmic water potential, SE and CC as a rule have a higher turgor than adjacent cells (cf. Warmbrodt 1987; Patrick 1997). It can be postulated that plasmodesmal resistance is high at all cell borders with lasting turgor differences. Where plasmodesmata are present between SE/CC complexes and surrounding cells (Kempers et al. 1998), a symplasmic leakage from sieve tubes may occur, although it should certainly be influenced by the flux rate in the axial direction (see Grimm et al. 1997). The amount of lateral leakage depends also on the frequency of, and the frictional resistance in plasmodesmata. A reduction of the lateral loss from sieve tubes along the "transport phloem" (van Bel 1993b) is indicated by the confinement of phloem tracers to the SE/CC complex along the transport phloem of roots (see Oparka et al. 1995b), by the rapidly declining lateral pressure readings from phloem to cortex of hypocotyls (Meshcheryacov et al. 1992), by the sudden drop in sucrose concentration from sieve tubes to the post phloem pathway (Fisher and Wang 1995; Pritchard 1996), and by different membrane potentials (20 mV or more) between SE/CC and phloem parenchyma cells (see van Bel 1996). Plasmodesmal resistance seems to be maintained actively. A recent investigation on transport phloem in *Arabidopsis* roots showed that metabolic inhibitors (CCCP and probenecid) induced the symplasmic leakage of CF from the metaphloem towards the adjacent cell layers (Wright and Oparka 1997). Accordingly, a short-term regulation of symplasmic unloading may be exercised actively by rapid changes of plasmodesmal conductivity (Patrick 1997).

4.3.1.5 Divalent Cations and Ammonia Lead to Cell Isolation

In their pioneering experiments, Erwee and Goodwin (1983) discovered that coinjection of divalent cations, notably Ca^{2+}, reduced the frequency of intercellular dye movement in *Egeria densa* leaves. This was an all-or-nothing effect. Once the dye had escaped from the injected cell, it moved as in controls, which suggests that Ca^{2+} is not itself plasmodesma-mobile. Microinjection of other divalent cations such as Mg^{2+} and Sr^{2+} had similar effects, while addition of Ca^{2+} to the apoplasm had no effects (Erwee and Goodwin 1983). Dye movement started to recover after 5 min and returned to normal within 30 min. Addition of inhibitors and ionophores which raise the Ca^{2+} concentration in the cytosol also reduced dye movement. The authors discussed that the Ca^{2+} effects may represent a defence mechanism. Under physical, chemical or pathogen stress, Ca^{2+} release from storage compartments such as the vacuole or the ER would isolate cells from adjacent healthy ones.

FRAP experiments on soybean microcalluses derived from root tissue also gave evidence that Ca^{2+} restricts tracer movement between cells (Baron-Epel et al. 1988). Calcium was effective only when coincubated with the ionophore A23187. Fluorescence recovery was also inhibited by a phorbol ester. Ca^{2+} and the phorbol ester could be involved in a signal transduction pathway activating a protein kinase, eventually leading to posttranslational modifications of plasmodesmal proteins (Baron-Epel et al. 1988).

In a series of experiments on *Setcreasea* trichomes, Tucker and coworkers extensively tested the influence of Ca^{2+} on intercellular transport. They coinjected CF and inositol polyphosphates of a calcium loaded chelator (CaBAPTA; Tucker 1988, 1990). Observation of a large number of fluorescein amino acid conjugates had indicated that hydrophilic anions moved rapidly through plasmodesmata while cations did not (Tucker and Tucker 1993). In keeping with these results, inositol triphosphate should move very rapidly through plasmodesmata, and Ca^{2+} should not. Both substances, however blocked plasmodesmal transport of carboxyfluorescein. CaBAPTA blocked cell-to-cell diffusion of CF in the trichomes and had a lasting effect, so that recovery of CF-transport took 2 h and more (Tucker 1990). In analogy to animal gap junctions, Tucker and Boss (1996) concluded that inositol phosphates are intercellular messengers that stimulate a release of Ca^{2+} from intracellular stores in order to regulate plasmodesmal function.

This hypothesis was supported by direct visualization of the free cytosolic Ca^{2+} concentration with the calcium green dextran and using an analogue to the wasp poison, mastoparan (Tucker and Boss 1996). Mastoparan is a cationic, amphiphilic tetradecapeptide which can activate G protein and, hence, increase the cytoplasmic concentrations of IP_3 and Ca^{2+}. Injection of mastoparan caused a transient closure of plasmodesmata that was accompanied by an increase in free calcium at the transverse wall between the injected cell and the adjacent one. The elevated Ca^{2+} level returned to normal within 90 s. Oscillations of cytosolic calcium levels were transmitted along the chain of trichome cells and accompanied by the uptake of external calcium. Inhibition of calcium uptake by La^{3+} inhibited plasmodesmal closure (Tucker and Boss 1996).

Ammonia application to the apoplasm of leaves and stems was discussed to effect an immediate closure of plasmodesmata (Fensom et al. 1990). However, the experimental setup allowed observations on the inhibition of phloem translocation only (see also Pickard and Minchin 1992b). The affected site (plasmodesmata or sieve pores) has yet to be determined before the relation of this inhibitor with plasmodesmal functioning can be assessed.

4.3.2 Transient Increase of the Plasmodesmal Size Exclusion Limit

4.3.2.1 SEL Increase During Deplasmolysis

In *Egeria* leaves, deplasmolysis led not only to a recovery of dye coupling, but even to an increase in the SEL. Substances that were immobile in control leaves became mobile after deplasmolysis (Erwee and Goodwin 1984). Simultaneously, the selectivity against aromatic groups of fluorescein conjugates disappeared and coinjection of divalent cations did not effect the closure of plasmodesmata (Erwee and Goodwin 1984). The changes in connectivity in the *Egeria* leaves lasted for more than 24 h. It may be significant in this context that viruses cannot spread in deplasmolyzed tissue for 48 h (Coutts 1978). As an extreme response to deplasmolysis, the dilation of plasmodesmata seems to affect the entire endogenous regulation of gating. Considering the functional results, EM-analysis of deplasmolyzed tissue should reveal the fate of the fluffy material observed in plasmodesmata of plasmolyzed tissue and any changes in their diameter (cf. Tilney et al. 1991; Schulz 1995).

4.3.2.2 SEL Increase After Azide and Anaerobiosis Treatment

Incubation of staminal trichomes of *Setcreasea* with azide raised the SEL, and the intercellular diffusion coefficient of fluorescein conjugates nearly doubled (Tucker 1993). The marked increase in the diffusion rate of FITC-Phe and -Try (see Sect. 4.1) and the mobility of a large octapeptide conjugate to FITC (angiotensin II; 1386 Da) suggested that the permeation pores of plasmodesmata had been enlarged under azide treatment (Tucker 1993).

Azide treatment also dilated plasmodesmata in detached roots, with an SEL increase from less than 1 to 5–7 kDa (excised roots; Cleland et al. 1994). Azide (10 mM) reduced the cellular ATP content, thus mimicking anaerobiosis. Inducing anaerobiosis by 30 min N_2 gas treatment produced similar effects. Cleland and coworkers (1994) discussed that local and/or transient anaerobiosis is a frequent event in growing roots and that the dilation of plasmodesmata may permit enhanced delivery of sugars to the affected cells where anaerobic respiration could accommodate the need of ATP.

4.3.2.3 Increase of SEL and Plasmodesmal Conductance for Intercellular Assimilate Transport

This argument focuses on the role of plasmodesmata in unloading of assimilates from the phloem. As has been pointed out earlier, the intercellular downhill transport of assimilates probably occurs via the cytoplasmic gateway wherever plasmodesmata are present between phloem and the recipient cells. Hence, dilated plasmodesmata present a reduced frictional resistance to molecules leaving the SE/CC complex.

Strong evidence for changes in the plasmodesmal passageway came from work on phloem unloading. In developing wheat grains, the plasmodesmata between the SE/CC complexes and phloem parenchyma cells coincide with a high pressure differential (>200 kPa; Fisher and Wang 1995). The plasmodesmata at this interface are expected to control symplasmic unloading (see Patrick 1990, 1997). However, they are virtually inaccessible experimentally. Therefore, most evidence for changes of plasmodesmal conductance came from plasmodesmata in the postphloem pathway (cf. Fisher and Oparka 1996).

The protophloem sieve tubes in pea root tips which lack companion cells are linked to all surrounding cells (pericycle and phloem parenchyma cells) by frequent pore/plasmodesma contacts (Schulz 1995). Phloem-imported, labelled sucrose did not appear in the root tip medium (Schulz 1994). Moreover, calculation of the sieve-element unloading rate that occurred in the apical tip segment gave a value of up to 7 μmol m^{-2} s^{-1} sucrose for the sieve-tube plasma membrane, which is much higher than any observed membrane transfer rates (Schulz 1998). Therefore, the unloading pathway was concluded to be symplasmic. The phloem import into roots increased with increasing concentrations of mannitol in the root tip medium of pea seedlings as demonstrated by the accumulation of ^{14}C originating from sucrose applied to a cotyledon (Schulz 1994). There was a linear relationship between the osmoticum in the apoplasm and the accumulation of sucrose. The root tip remained turgescent at concentrations of up to 350 mM mannitol, showing that phloem import was sufficient for osmoadaptation. A high-resolution EM analysis of the plasmodesmata between the cortex cells indicated an expansion of the neck region under 350 mM mannitol treatment (Fig. 3). According to statistical analysis, the difference in plasmodesmal width between mannitol-treated root tips and controls was highly significant. As expressed in passage area per 100 plasmodesmata, the dilation of the cytoplasmic gateway compared well with the increase in sucrose accumulation in the cortex (Fig. 4). The diameter of the desmotubule, however, remained remarkably constant (Schulz 1995).

Fig. 3A–K. Longitudinal (**A–H**) and cross sections (**I, K**) of plasmodesmata in pea root tips. **A–D** Control tissue. Plasmodesmata with neck regions (*between arrows*), straight desmotubule (*Dt*) in contact with the ER, and spokes [indicated outside the plasma membrane(PM)] between PM and Dt. Bar **H** 100 nm. **E–H** Mannitol-treated tissue. Plasmodesmata are mostly without neck constriction but with

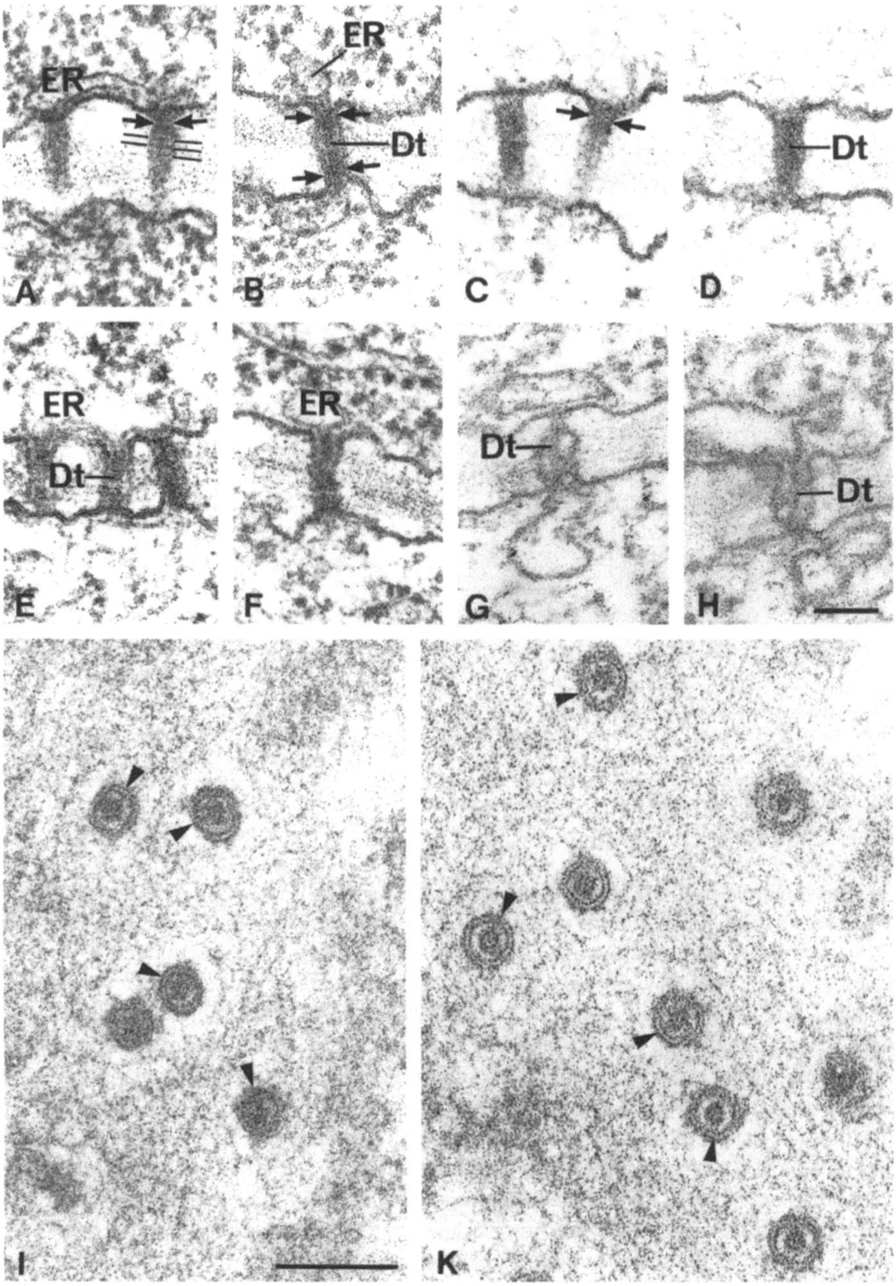

expanded orifices. Long spokes (**G, H**) connect plasma membrane and desmotubule (*Dt*). **I** Control. Cross-section of plasmodesmata with cytoplasmic sleeve obscured by spokes (*arrowheads*). Bar **K** 100 nm. **K** Mannitol-treated tissue. The cytoplasmic sleeve appears as more electron-translucent than in controls due to a greater width. Spokes (*arrowheads*) still connect plasma membrane and desmotubule. (Schulz 1995)

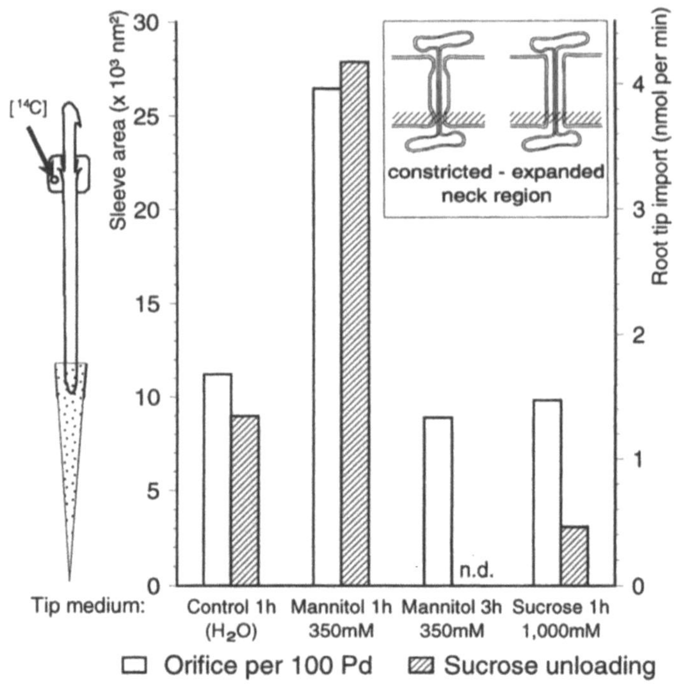

Fig. 4. Estimation of the change in orifice area per 100 plasmodesmata of cortex cells in control, mannitol-treated (350 mM) and plasmolyzed (1 M sucrose) pea roots compared to phloem unloading under the same conditions. Changes in passage area of plasmodesmal orifices are proportional to changes in unloading. Assumptions made include the relation of wall to section thickness and thus the chance to cut a plasmodesma at its orifice in cross-sections. (Schulz 1995)

For transport phloem, the lateral unloading seems to depend upon the source/sink relation in the entire phloem system. When *Phaseolus* plants were pruned to a low source/sink relation, neither did CF appeared in phloem parenchyma cells or in parenchyma further away, nor was there an accumulation of translocated ^{14}C-sucrose (Patrick and Offler 1996, Offler and Patrick 1996). Conversely, a high source/sink ratio and, accordingly, a high rate of axial assimilate transport led to CF appearing at the lateral flanks of the phloem and ^{14}C accumulation in these regions. The dependency of lateral unloading on the source/sink ratio was explained with the tightly controlled and high-resistant plasmodesmata between SE/CC complexes and phloem parenchyma (see Sect. 4.3.1.4 above; Patrick 1997). Under a high source/sink ratio, excess sucrose diffusing across the SE/CC plasma membrane (see Grimm et al. 1990) could accumulate in phloem parenchyma so that the turgor differential at this interface is reduced to a value that allows opening of plasmodesmata and release of CF from the phloem (cf. Patrick 1997).

As in pea roots, osmotic stress (here by sorbitol) induced an increase in lateral unloading from mature axial phloem of *Phaseolus* (high source/sink ratio plants) which is indicative for an osmoregulatory mechanism (Offler and Patrick 1996). In developing wheat grains, phloem unloading appeared to be less sensitive to osmotic treatment than in root tips and axial phloem (Wang and Fisher 1994a; see for a review on post-sieve-element transport in developing seeds, Patrick and Offler 1995). Modulations in plasmodesmal transport induced by osmotic stress may be tissue-specific. Cells in green leaf tissue may respond to water stress autonomously by investing part of their photosynthates in osmoregulation. Accordingly, the intercellular transport of labelled assimilates was reduced rather than increased under osmotic stress in light (Trebacz and Fensom 1989).

The application of hormones (gibberellins and cytokinins) generally led to an increase in unloading, as was shown by the accumulation of radioactively labelled sucrose as well as that of a xenobiotic amino acid (e.g., Schenk 1974; Offler and Patrick 1996; Patrick and Offler 1996). This also reflects a change in plasmodesmal conductance, provided that unloading occurs symplasmically.

4.4 Putative Gating Mechanisms and Their Control

Direct and indirect evidence for non-selective gating was summarized in Section 4.3. The size extremes documented are the blockage of plasmodesmata (for fluorescein with 332 Da and larger tracers) and a tenfold increase in SEL (for FITC dextrans of up to 10 kDa), while the visible extremes are the occlusion of the cytoplasmic gateway in plasmolyzed tissue and the loss of the neck region and dilation of the plasmodesmal diameter, respectively (Schulz 1995). More realistic evidence than by means of LM and EM may be provided by changes in the rate of intercellular transport. For symplasmic phloem unloading of sucrose in root tissue, the extremes varied between a complete arrest and an approximately threefold increase in unloading (Schulz 1994; see also Fig. 4). The increase may be underestimated, since the transport capacity of the phloem is rate-limited in unloading studies.

Closure and widening of plasmodesmata might be effected by two distinct mechanisms: (1) the mutual interaction of plasmodesmal proteins; and (2) changes within the cell wall around the plasmodesmal orifice (neck region). As a working hypothesis for (1), changes in the configuration and conformation of plasmodesmal proteins are presumed to minimize or increase the distance between plasma membrane- and desmotubular membranes. As a working hypothesis for (2), callose synthesis and breakdown at the plasmodesmal orifices are postulated to constrict and dilate the plasmodesmal channel, respectively. Are both mechanisms involved in non-selective gating and, if so, do they work synergistically or antagonistically? The discussion is confined to the cytoplasmic gateway as information on the intratubular gateway is sparse. In view of structural differences between plasmodesma types in different tissues (see Waigmann et al. 1997) and between plasmodesmata from leaves collected at different temperatures (Gamalei et al. 1994), modulations of the intratubular gateway may, however, be of significance.

4.4.1 Closure of Plasmodesmata

The mechanisms triggering the closure of plasmodesmata appear remarkably similar to those observed as wound responses in cut sieve tubes. The resemblance is not coincidental, since sieve pores can be regarded as specialized plasmodesmata (cf. Olesen and Robard 1990). The plugging of the pore lumen by filamentous P-proteins and the synthesis of callose are involved in the wound response of sieve pores (Sjölund 1997; Schulz 1998). While the aggregation of structural cytoplasmic proteins into the orifices of plasmodesmata has never been described, the presence of callose at their orifices has long be known (but see Turner et al. 1994; Radford et al. 1998). Accumulation of proteins in sieve pores is somehow comparable to the plugging of plasmodesmata by an unknown electron-dense material that was detected during tissue-domain development (Ehlers and Kollmann 1996; see also Kwiatkowska and Maszewski 1986) and in plasmolyzed tissue by EM (Tilney et al. 1991; Oparka et al. 1994; Schulz 1995). Transient interactions of plasmodesmal proteins may shut down plasmodesmata prior to permanent plugging. These interactions, however, were not substantiated by substructural changes (Ehlers and Kollmann 1996) and their visualization is questionable, since fixation procedures affect plasmodesmal structures (see Sect. 3.1).

Plasmodesmal closure was described to be triggered by an increase in cytosolic calcium (Erwee and Goodwin 1983; Baron-Epel et al. 1988; Tucker 1988, 1990; Tucker and Boss 1996). Because calcium is known to induce callose synthesis, it remains open to discussion whether calcium-dependent closure is by callose synthesis alone and/or by conformational changes of plasmodesmal proteins.

As observed in freeze-substituted root material by aniline blue fluorescence, callose synthesis starts in pit fields of cortex parenchyma at the cutting edge 5 min after wounding and continues over 1 day (Galway and McCully 1987). In glutaraldehyde-fixed material, callose may be considered as an artefact (Hughes and Gunning 1980; Radford et al. 1998). Callose synthesis is induced by Ca^{2+} influx from the apoplasm (Kauss 1985, 1996), a process in which mechanosensitive ion channels in the plasma membrane may be engaged (Pickard and Minchin 1990, 1992a, c; Garrill et al. 1996). Cation channels were detected in the plasma membrane of roots that would mediate calcium influx in response to depolarization (White 1997). Implicitly, a sudden pressure release and membrane depolarization in response to wounding might trigger callose synthesis. Callose synthase was localized at sieve pores and plasmodesmata (Delmer et al. 1993). The induction of callose synthesis seems to depend not only on Ca^{2+} but also on an endogenous activator, the vacuolar furfuryl glucoside (Ohana et al. 1992). The relative distribution of furfuryl glucoside between cytosol and vacuole may be one component of the signal transduction pathway for activation of callose synthase (Ohana et al. 1993).

These data indicate that the constriction of plasmodesmata by callose deposition is primarily a wound response of plasmodesmata. It remains questionable whether callose synthesis can shut off plasmodesmata completely. Gaps between plasmodesmal proteins were reported to be 2–3 nm wide just at the neck and middle constriction (Ding et al. 1992; Botha et al. 1993) where, for steric reasons, the minimal diameter of the plasmodesmal outline is about 30 nm (see Sect. 2.2; cf. Overall et al. 1982). A further constriction by callose appears impossible. Doubts on callose synthesis as sole mechanism of plasmodesmal closure are also cast by the finding that changes in Ca^{2+} levels do not only cause an immediate closure of plasmodesmata but also a rapid re-opening (Tucker and Boss 1996), both of which occur too rapidly to be evoked by callose synthesis and hydrolysis.

For the transduction of plasmodesmal signals, Tucker and Boss (1996) presented a model in which inositol triphosphate (IP_3) (or another yet unknown stimulus) diffuses rapidly to adjacent cells through the plasmodesmata. Here, it causes the release of Ca^{2+} from internal stores via IP_3-sensitive channels, which is followed by sequestering of internal, and uptake of external Ca^{2+} through lanthanum-sensitive Ca^{2+} channels in the plasma membrane. Initial increase of cytosolic Ca^{2+} leads to closure of plasmodesmata which reopen after a drop in free calcium (Tucker and Boss 1996). The model of a IP_3/Ca^{2+} signal model explains the oscillating effect of mastoparan and may also apply to other cases of intercellular signal transduction, e.g. in transmission of wound signals. In contrast to the findings in *Setcreasea* (Tucker and Boss 1996), investigations on cell coupling between root hair cells of *Arabidopsis* demonstrated no effect of IP_3 on electrical coupling, while injection of Ca^{2+} led to uncoupling (Lew 1994).

Accordingly, it can be speculated that a calcium increase first induces plasmodesmal proteins to close the cytoplasmic gateway. When calcium remains at a high level, with other signal substances such as furfuryl glucoside participating, callose synthesis is activated, which keeps the plasmodesmal channel constricted permanently. A lasting high cytoplasmic Ca^{2+} level was indeed reported to produce a lasting functional closure (1 h) of trichome plasmodesmata (Tucker 1990). The decisive difference between a transient and a more permanent closure of plasmodesmata would be a transient rise of free Ca^{2+} by its rapid release from, and subsequent sequestering into intracellular stores (e.g. ER cisternae), and a lasting rise brought about by a general influx of Ca^{2+} from the apoplasm, respectively.

In conclusion, changes in plasmodesmal proteins rather than callose play a pivotal role in plasmodesmal closure. These changes demand the shortening of the spokes (see Fig. 3) in order to reduce the distance between plasma membrane and desmotubule. When pulled together, the proteins embedded in the plasma membrane and the desmotubule might undergo conformational changes, thus further closing the gaps in the cytoplasmic gateway (for additional models, see McLean et al. 1997).

Immunolocalization and inhibitor experiments suggested the presence of actin, and possibly also myosin, in plasmodesmata, which would suggest the presence of a well-known motor system in plants (White et al. 1994; Ding et al. 1996; Overall and Blackman 1996). Plasmodesmal proteins showing homology to animal connexin and localized at plasmodesmata (Yahalom et al. 1991; Epel 1994) gave rise to exciting speculations about functional similarities between plasmodesmata and gap junctions. However, evidence for the presence of actin and a putative plant-connexin has failed as yet to produce a consistent model for the gating mechanism in plasmodesmata. With respect to connexin, it is difficult to understand how a protein complex that forms pores between animal cells in a perpendicular orientation to the plasma membrane would do the same job in plasmodesmata of plant cells where any pore in a similar orientation would connect apoplasmic space with the cytoplasmic sleeve.

Experiments with azide and N_2 inferred that a reduction of the ATP level leads to a dilation of plasmodesmata (Cleland et al. 1994). It is suggestive to ask whether inversely high levels of ATP induce the closure of plasmodesmata. This relationship has not been assessed yet; any indications for an increase in ATP level are lacking in most studies on plasmodesmal closure.

4.4.2 Dilation of Plasmodesmata

Functional studies have shown that plasmodesmata can dilate considerably under anaerobiosis conditions or osmotic stress (Tucker 1993; Cleland et al. 1994, Schulz 1994; Offler and Patrick 1996). Only few structural investigations demonstrated that plasmodesmata can dilate. The regions with the smallest diameters are transport bottlenecks in plasmodesmata, i.e. the neck regions and/or the middle constriction when present. In fact, measurement and statistical analysis of plasmodesmata in root apices under osmotic stress showed an expansion of the neck region up to a funnel-shaped orifice (Fig. 3E–H). Measurements of cross-sectioned plasmodesmata in controls revealed a bimodal frequency distribution of outer diameters. Neck and middle regions occurred in proportions of 40 and 60%, respectively(Schulz 1995). This proportion could be expected from the chance to cut these regions in 30-nm- thick sections. In contrast, less than 20% plasmodesmata in roots under osmotic stress fell into the neck-region class. This reflected that the neck region of more than 20% plasmodesmata was dilated (Schulz 1995). It was estimated that the passage area of plasmodesmata in treated roots was more than double that of controls (Fig. 4).

Changes in the wall microdomain at the plasmodesmal orifices and/or in the conformation of plasmodesmal proteins could be responsible for widening of the neck region, as was postulated for plasmodesmal narrowing. If the microdomain consists of callose (Radford et al. 1998; but see Turner et al. 1994; Watada and Wergin 1997), it could be loosened by callose hydrolysis. However, no callose was detectable in control roots by EM (Schulz 1995), or in similar fixed, cryosectioned and aniline blue-stained material by fluorescence microscopy (Schulz 1987). Moreover, it is unlikely that callose hydrolysis would change the plasmodesmal diameter as long as spokes, internal sphincters, or other plasmodesmal proteins maintain the distance between plasma membrane and desmotubule. In root tissue under osmotic stress, these spokes were present but were longer than in controls (cf. Fig. 3).

It appears more likely that the extension of spokes controls the dilation of plasmodesmata. Treatment with cytochalasin B led to a funnel-shaped dilation of plasmodesmata, suggesting that actin keeps plasmodesmata constricted at their orifices (White et al. 1994). Functional studies using cytochalasin D and profilin as actin-disorganizing drugs, and indicating an SEL increase up to 20 kDa confirmed these effects. This increase in SEL was inhibited by phalloidin, which stabilizes actin filaments conversely (Ding et al. 1996). Apparently, actin keeps plasmodesmata at a basal SEL value (possibly at a high ATP level; see Sect. 4.4.1). Disruption of actin as well as interactions between more specific plasmodesmal proteins, would widen the neck region, hence permitting passage of larger solutes (increase in SEL) as well as allowing a higher intercellular transport rate. Overall and Blackman (1996) discussed the participation of myosin, localized to plasmodesmata in preliminary experiments (Radford and White 1996), and of the calcium-dependent centrin in plasmodesmal gating as a more specific interaction with actin.

It can be concluded provisionally that the prime mechanism of non-selective gating is the rapid and transient interaction and conformational change of plasmodesmal proteins, probably controlled by calcium and ATP levels. Decrease of the plasmodesmal permeability may be brought about by comparatively small and local differences in the calcium concentration, for instance from ER stores close to the plasmodesmata. Conversely, wound blockage of plasmodesmata may depend upon a massive influx of calcium from the apoplasm or the vacuole accompanied by callose synthesis, keeping the plasmodesmal channel constricted. Conformational changes in the plasmodesmal proteins and callose synthesis are reversible, the former changing very rapidly, the latter much more slowly.

5 Control of Selective Transport Through Plasmodesmata

Selective transport of viruses through plasmodesmata involves translation of a virus movement protein (MP) that forms a pass-key for their spread. A MP is therefore characterized by a plasmodesma-targeting signal and the ability to dilate it. In addition, MPs seem to function as a chaperon molecule. The conformation of polypeptides or nucleic acids is changed after binding to the MP, so that they become adjusted to the plasmodesmal gateway. Selective gating mediated by MPs may or may not be accompanied by an increase in the SEL (for reviews, see Lucas 1995; Ghoshroy et al. 1997; Ding 1997; McLean et al. 1997). It was hypothesized that MPs open plasmodesmata for a short time only, which allows their own rapid transport (see also Fujiwara et al. 1993; Noueiry et al. 1994) and that of fused molecules. If not bound to an MP, transport of large, slowly diffusing reporter or tracer molecules might be prevented by the length and structural complexity of the plasmodesmata between trichome cells (McLean et al. 1997; Waigmann et al. 1997). The capability to change the SEL seems to depend on the virus and host tissue studied (Wolf et al. 1989; Waigmann et al. 1994; Waigmann and Zymbryski 1995; Angell et al. 1996).

Endogenous soluble and structural phloem proteins (Chap. 16), and a membrane transporter of the phloem (Kühn et al. 1997) were shown to pass plasmodesmata. At first sight, this selective transport of macromolecules between SE and CC appears as an exception rather than a rule in symplasmic transport, owing to the fact that sieve elements are enucleate when mature and lack a translation apparatus (for reviews see Behnke 1989; Behnke and Sjolund 1990), this extreme specialization being required for long-distance transport. The cells associated with SE take over cellular tasks including protein turnover and supply of ATP, enzymes, and substrates for scavenging mechanisms (see Schulz 1998; Chap. 16). The special role of associated cells is emphasized by the pore/plasmodesma units that provide an intimate contact between both cell types with a particularly high SEL (Kempers et al. 1993; Kempers and van Bel 1997; Kempers et al. 1998; van Bel and Kempers 1997). One could argue that it is the larger width within the plasmodesmal substructures on the CC side compared with other plasmodesmata that permits the transport of proteins. Evidently however, some cytosolic contents of the CC, e.g. substrates and enzymes involved in sugar metabolism, are retained in the CC and are delivered into the SE only when sieve tubes are wounded (Lehmann 1981a, b; Haritatos et al. 1996; Turgeon 1996). This provides evidence that the pore/plasmodesma contacts are controlled on the CC side. Therefore, protein transport in intact plants through these contacts is due to selective gating. Easy access to the sieve-tube content as well as the fact that mature sieve elements are unable to synthesize proteins independently has turned the phloem into a model system for studying inter-

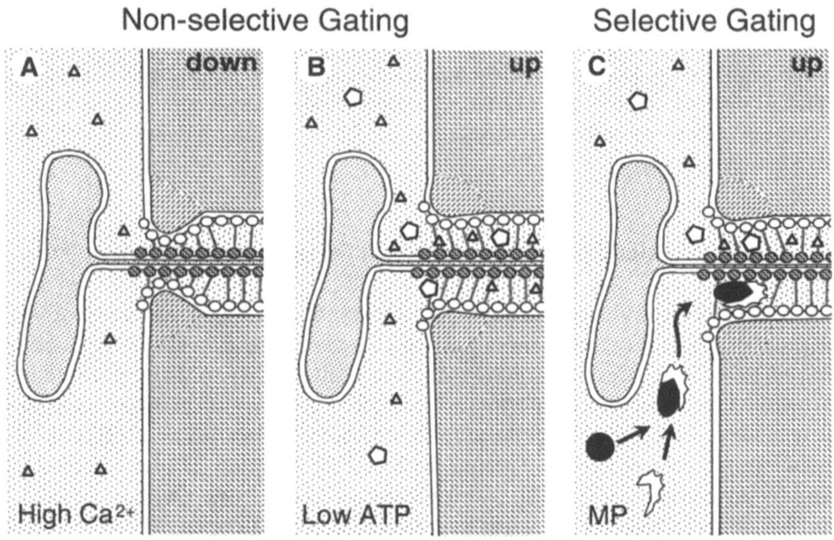

Fig. 5A–C. Schematic drawing exemplifying conditions that control up- and downregulation of plasmodesmal transport in non-selective (**A, B**) and selective (**C**) gating. **A** Interaction and conformational changes of plasmodesmal proteins rapidly close the plasmodesmal orifice at high calcium levels. Callose synthesis starts subsequently, leading to a more permanent constriction of the neck region. **B** Low ATP levels as under azide treatment or anaerobiosis trigger extension of the spokes that link plasma membrane and desmotubular proteins, and thus lead to a dilation of the cytoplasmic gateway as indicated by mobility of larger molecules through plasmodesmata. **C** Viral or endogenous movement proteins chaperone their load and are targeted to the plasmodesmal orifice. While passing the cytoplasmic gateway, they selectively dilate the space between plasma membrane and desmotubule, which may or may not be accompanied by an increase in SEL

cellular protein trafficking (Bostwick et al. 1992; Fisher et al. 1992; Sakuth et al. 1993; Ishiwatari et al. 1995, 1998; Balachandran et al. 1997; Clark et al. 1997; Dannenhoffer et al. 1997; Kühn et al. 1997; Szederkenyi et al. 1997; see Chap. 16).

6 Concluding Remarks

A summary of up- and downregulating features of plasmodesmata is shown in Fig. 5. It has emerged that non-selective gating is controlled by the cellular levels of calcium and ATP. A high calcium level will close plasmodesmata, first by interaction of the plasmodesmal proteins and then by deposition of callose. Conversely, a low level of ATP presumably induces a dilation of the cytoplasmic gateway, thus permitting an enhanced intercellular exchange rate. Explanations for the turgor-dependent pressure-valve mechanism and changes during deplasmolysis are still lacking, but can be expected from comparative functional and structural investigations.

Selective gating involves a specific interaction between an (endogenous or viral) MP that functions as chaperon for its load and targets to plasmodesmata. Selective gating might be induced by the local dephosphorylation of spokes. This could be effected by the phosphorylation of movement proteins (see e.g. Epel 1994; Sokolova et al. 1997). Dephosphorylation/phosphorylation of the MP could, in turn, be the motive force for the passage of MP through the plasmodesmal channel, together with the viral genome. Thus, the spokes would be relaxed individually and then locked again by the passing MP. Given the different ultrastructure and lengths of mesophyll and trichome plasmodesmata (Waigmann et al. 1997), this hypothesis would explain why selective movement of large molecules in trichomes is not accompanied by an increase of the SEL (see also McLean et al. 1997). The channel might already be closed again, while the large dextran is still at the plasmodesmal orifice.

A key role in a unifying mechanism of non-selective and selective dilation of plasmodesmata may be played by the dephosphorylation of plasmodesmal proteins. It is assumed that plasmodesmata are kept constricted to the basal SEL under high ATP. In non-selective gating, a reduction in cytosolic ATP would lead – possibly over several metabolic intermediates – to a dephosphorylation of plasmodesmal proteins and thus to a extension of the plasmodesmal spokes. In selective gating, each MP would directly interact with plasmodesmal proteins, thus relaxing the spokes locally. This speculative working hypothesis implies an important self-regulatory mechanism for the autonomy of plasmodesma-linked cells. In cells with a high energy status, the plasmodesmata are then constricted at the basal SEL and, hence, these cells would tend to retain all phosphorylated metabolites and to maintain metabolic independence. Conversely, plasmodesmata would dilate and the rate of assimilate exchange could increase in cells with a low energy status and/or a high ATP consumption. These cells would tend, in part, to give up autonomy for the gain in assimilates.

7 References

Angell SM, Davies C, Baulcombe DC (1996) Cell to cell movement of potato virus X is associated with a change in the size exclusion limit of plasmodesmata in trichome cells of *Nicotiana clevelandii*. Virology 216:197–201

Arisz WH (1969) Intercellular polar transport and the role of the plasmodesmata in coleoptiles and *Vallisneria* leaves. Acta Bot Neerl 18:14–38

Arisz WH, Wiersema EP (1966) Symplasmatic long distance transport in *Vallisneria* plants investigated by means of autoradiograms. Proc K Ned Akad Wet Ser C 69:223–241

Balachandran S, Xiang Y, Schobert C, Thompson GA, Lucas WJ (1997) Phloem sap proteins from *Cucurbita maxima* and *Ricinus communis* have the capacity to traffic cell to cell through plasmodesmata. Proc Natl Acad Sci 94:14150–14155

Baron-Epel O, Hernandez D, Jiang L-W, Meiners S, Schindler M (1988) Dynamic continuity of cytoplasmic and membrane compartments between plant cells. J Cell Biol 106:705–721

Beebe DU, Turgeon R (1992) Localization of galactinol, raffinose, and stachyose synthesis in *Cucurbita pepo* leaves. Planta 188:354–361

Behnke H-D (1989) Structure of the phloem. In: Baker DA, Milburn JA (eds) Transport of photoassimilates. Longman, Harlow, pp 79–137

Behnke H-D, Sjolund RD (1990) Sieve elements – comparative structure, induction and development. Springer Verlag, Berlin Heidelberg New York

Bostwick DD, Dannenhoffer JM, Skaggs MI, Lister RM, Larkins BA, Thompson GA (1992) Pumpkin phloem lectin genes are specifically expressed in companion cells. Plant Cell 4:1539–1548

Botha CEJ, Hartley BJ, Cross RHM (1993) The ultrastructure and computer-enhanced digital image analysis of plasmodesmata at the Kranz mesophyll-bundle sheath interface of *Themeda triandra* var. *imberbis* (Retz) A. Camus in conventionally-fixed leaf blades. Ann Bot 72:255–261

Bräutigam E, Müller E (1975) Transportprozesse in *Vallisneria*-Blättern und die Wirkung von Kinetin and Kolchizin. II. Symplastischer Transport von a-Aminobuttersäure in *Vallisneria*-Blättern und die Wirkung von Kinetin. Biochem Physiol Pflanzen 167:17–28

Clark AM, Jacobsen KR, Bostwick DE, Dannenhoffer JM, Skaggs MI, Thompson GA (1997) Molecular characterization of a phloem-specific gene encoding the filament protein, Phloem Protein 1 (PP1), from *Cucurbita maxima*. Plant J 12:49–61
Cleland RE, Fujiwara T, Lucas WJ (1994) Plasmodesmal-mediated cell-to-cell transport in wheat roots is modulated by anaerobic stress. Protoplasma 178:81–85
Coutts RHA (1978) Suppression of virus-induced local lesions in plasmolysed tissue. Plant Sci Lett 12 :77–85
Dannenhoffer JM, Schulz A, Bostwick DE, Skaggs, MI Thompson GA (1997) Expression of phloem lectin is developmentally linked to vascular differentiation in cucurbits. Planta 201:405–414
Delmer DP, Volokita M, Solomon M, Fritz U, Delphendahl W, Herth W (1993) A monoclonal antibody recognizes a 65 kDa higher plant membrane polypeptide which undergoes cation-dependent association with callose synthase in vitro and co-localizes with sites of high callose deposition in vivo. Protoplasma 176:33–42
Delrot S (1989) Loading of photoassimilates. In: Baker DA, Milburn JA (eds) Transport of photoassimilates.. Longman Scientific & Technical, Harlow, pp 167–205
Derrick PM, Barker H, Oparka KJ (1992) Increase in plasmodesmatal permeability during cell-to-cell spread of tobacco rattle virus from individually inoculated cells. Plant Cell 4:1405–1412
Ding B (1997) Cell-to-cell transport of macromolecules through plasmodesmata: a novel signalling pathway in plants. Trends Cell Biol 7:5–9
Ding B, Turgeon R, Parthasarathy MV (1992) Substructure of freeze-substituted plasmodesmata. Protoplasma 169:28–41
Ding B, Kwon M-O, Warnberg L (1996) Evidence that actin filaments are involved in controlling the permeability of plasmodesmata in tobacco mesophyll. Plant J 10:157–164
Ehlers K (1996) Untersuchungen zur Entstehung, Struktur, Funktion and Regulation der symplastischen Zellverbindungen bei regenerierenden *Solanum nigrum*-Protoplasten. PhD Thesis, Christian-Albrechts-Universität, Kiel
Ehlers K, Kollmann R (1996) Regulation of the symplasmic contact between physiologically different cells. III Int Worksh on Basic and Applied Research in Plasmodesmal Biology, Israel, March 10–16, 1996, pp 77–81
Epel BL (1994) Plasmodesmata: composition, structure and trafficking. Plant Mol Biol 26:1343–1356
Epel BL, Erlanger MA (1991) Light regulates symplastic communication in etiolated corn seedlings. Physiol Plant 83:149–153
Epel BL, Padgett HS, Heinlein M, Beachy RN (1996) Plant virus movement protein dynamics probed with a GFP protein fusion. Gene 173:75–79
Erwee MG, Goodwin PB (1983) Characterisation of the *Egeria densa* Planch. leaf symplast. Planta 158:320–328
Erwee MG, Goodwin PB (1984) Characterization of the *Egeria densa* Planch. leaf symplast: Response to plasmolysis, deplasmolysis and to aromatic amino acids. Protoplasma 122:162–168
Erwee MG, Goodwin PB (1985) Symplast domains in extrastelar tissues of *Egeria densa* Planch. Planta 163:9–19
Evert RF, Eschrich W, Heyser W (1977) Distribution and structure of the plasmodesmata in mesophyll and bundle-sheath cells of *Zea mays* L. Planta 136:77–89
Evert RF, Botha CEJ, Mierzwa RJ (1985) Free-space marker studies on the leaf of *Zea mays* L. Protoplasma 126:62–73
Fensom DS, Thompson RG, Caldwell CD (1990) Ammonia gas temporarily interrupts translocation of ^{11}C photosynthate in sunflower. J Exp Bot 41:11–14
Fisher DB, Oparka KJ (1996) Post-phloem transport: principles and problems. J Exp Bot 47:1141–1154
Fisher DB, Wang N (1995) Sucrose concentration gradients along the post-phloem transport pathway in the maternal tissues of developing wheat grains. Plant Physiol 109:587–592
Fisher DB, Wu Y, Ku MSB (1992) Turnover of soluble proteins in the wheat sieve tube. Plant Physiol 100:1433–1441
Fujiwara T, Giesman-Cookmeyer D, Ding B, Lommel SA, Lucas WJ (1993) Cell-to-cell trafficking of macromolecules through plasmodesmata potentiated by the red clover necrotic mosaic virus movement protein. Plant Cell 5:1783–1794
Galway ME, McCully ME (1987) The time course of the induction of callose in wounded pea roots. Protoplasma 139:77–91
Gamalei YV, van Bel AJE, Pakhomova MV, Sjutkina AV (1994) Effects of temperature on the conformation of the endoplasmic reticulum and on starch accumulation in leaves with the symplasmic minor-vein configuration. Planta 194:443–453
Garrill A, Findlay GP, Tyerman SD (1996) Mechanosensitive ion channels. In: Smallwood M, Knox JP, Bowles DJ (eds) Membranes: specialized functions in plants. Bios Scientific Publications, Oxford, pp 247–260
Ghoshroy S, Lartey R, Sheng JS, Citovsky V (1997) Transport of proteins and nucleic acids through plasmodesmata. Annu Rev Plant Physiol 48:25–48

Glockmann C, Kollmann R (1996) Structure and development of cell connections in the phloem of *Metasequoia glyptostroboides* needles. Ultrastructural aspects of modified primary plasmodesmata in Strasburger cells. Protoplasma 193:191–203
Goodwin PB (1976) Physiological and electrophysiological evidence for intercellular communication in plant symplasts. In: Gunning BES, Robards AW (eds) Intercellular communication in plants: studies on plasmodesmata. Springer, Berlin Heidelberg New York, pp 121–129
Goodwin PB (1983) Molecular size limit for movement in the symplast of the *Elodea* leaf. Planta 157: 124–130
Goodwin PB, Shepherd V, Erwee MW (1990) Compartmentation of fluorescent tracers injected into the epidermal cells of *Egeria densa* leaves. Planta 181:129–136
Grabski S, de Feijter S, Schindler M (1993) Endoplasmic reticulum forms a dynamic continuum for lipid diffusion between contiguous soybean root cells. Plant Cell 5:25–38
Grimm E, Bernhardt G, Rothe K, Jacob F (1990) Mechanism of sucrose retrieval along the phloem path - a kinetic approach. Planta 182:480–485
Grimm E, Jahnke S, Rothe K (1997) Photoassimilate translocation in the petiole of *Cyclamen* and *Primula* is independent of lateral retrieval. J Exp Bot 48:1087–1094
Haritatos E, Keller F, Turgeon R (1996) Raffinose oligosaccharide concentrations measured in individual cell and tissue types in *Cucumis melo* L leaves: implications for phloem loading. Planta 198: 614–622
Helder RJ (1967) Translocation in *Vallisneria spiralis*. In: Schumacher W (ed) Translocation in plants. Encyclopedia of Plant Physiology XIII. Springer, Berlin Heidelberg New York, pp 20–43
Holdaway-Clarke TL, Walker NA, Overall RL (1996) Measurement of the electrical resistance of plasmodesmata and membranes of corn suspension culture cells. Planta 199:537–544
Holthaus U, Schmitz K (1991) Distribution and immunolocalization of stachyose synthase in *Cucumis melo* L. Planta 185:479–486
Hughes JE, Gunning BES (1980) Glutaraldehyde-induced deposition of callose. Can J Bot 58:250–258
Ishiwatari Y, Honda C, Kawashima I, Nakamura S-I, Hirano H, Mori S, Fujiwara T, Hayashi H, Chino M (1995) Thioredoxin h is one of the major proteins in rice phloem sap. Planta 195:456–463
Ishiwatari Y, Fujiwara T, McFarland KC, Nemoto K, Hayashi H, Chino M, Lucas WJ (1998) Rice phloem thioredoxin h has the capacity to mediate its own cell-to-cell transport through plasmodesmata. Planta 205:12–22
Johnson RPC (1978) The microscopy of P-protein filaments in freeze-etched sieve pores. Planta 143: 191–205
Kauss H (1985) Callose biosynthesis as a Ca^{2+}-regulated process and possible relations to the induction of other metabolic changes. J Cell Sci suppl 2:89–103
Kauss H (1996) Callose synthesis. In: Smallwood M, Knox JP, Bowles DJ (eds) Membranes: specialized functions in plants. Bios Scientific Publications, Oxford, pp 77–92
Kempers R, van Bel AJE (1997) Symplasmic connections between sieve element and companion cell in the stem phloem of *Vicia faba* have a molecular exclusion limit of at least 10 kDa. Planta 201: 195–201
Kempers R, Prior DAM, van Bel AJE, Oparka KJ (1993) Plasmodesmata between sieve element and companion cell of extrafascicular stem phloem of *Cucurbita maxima* permit passage of 3-kDa fluorescent probes. Plant J 4:567–575
Kempers R, Ammerlaan A, van Bel AJE (1998) Symplasmic constriction and ultrastructural features of the sieve element/companion cell complex in the transport phloem of apoplasmically and symplasmically phloem-loading species. Plant Physiol 116:271–278
Kollmann R, Schumacher W (1962) Über die Feinstruktur des Phloems von *Metasequoia glyptostroboides* und seine jahreszeitlichen Veränderungen. II Vergleichende Untersuchungen der plasmatischen Verbindungsbrücken in Phloemparenchymzellen und Siebzellen. Planta 58:366–386
Kollmann R, Schumacher W (1963) Über die Feinstruktur des Phloems von *Metasequoia glyptostroboides* und seine jahreszeitlichen Veränderungen. IV Weitere Beobachtungen zum Feinbau der Plasmabrücken in den Siebzellen. Planta 60:360–389
Kühn C, Franceschi VR, Schulz A, Lemoine R, Frommer WB (1997) Macromolecular trafficking indicated by localization and turnover of sucrose transporters in enucleate sieve elements. Science 275:1298–1300
Kwiatkowska M, Maszewski J (1986) Changes in the occurrence and ultrastructure of plasmodesmata in antheridia of *Chara vulgaris* L. during different stages of spermatogenesis. Protoplasma 132: 179–188
Lazzaro MD, Thomson WW (1996) The vacuolar tubular continuum in living trichomes of chickpea (*Cicer arietinum*) provides a rapid means of solute delivery from base to tip. Protoplasma 193: 181–190
Lehmann J (1981a) Quantitative Bestimmung von Nucleotiden und Zuckerphosphaten im Phloemexsudat von *Cucurbita pepo* L. Z Pflanzenphysiol 102:415–424

Lehmann J (1981b) Versuch zur Bestimmung der Enzymverteilung zwischen Geleitzellen und Phloemparenchymzellen bei *Cucurbita pepo*. Z Pflanzenphysiol 103:323–333

Lew RR (1994) Regulation of electric coupling between *Arabidopsis* root hairs. Planta 193:67–73

Lew RR (1996) Pressure regulation of the electrical properties of growing *Arabidopsis thaliana* L. root hairs. Plant-Physiol 112:1089–1100

Lucas WJ (1995) Plasmodesmata: intercellular channels for macromolecular transport in plants. Curr Biol 7:673–680

Lucas WJ, Ding B, van der Schoot C (1993) Plasmodesmata and the supracellular nature of plants. New Phytol 125:435–476

Lucas WJ, Bouché-Pillon S, Jackson D, Nguyen L, Baker L, Ding B, Hake S (1995) Selective trafficking of KNOTTED1 homeodomain protein and its mRNA through plasmodesmata. Science 270: 1980–1983

McLean BG, Hempel FD, Zambryski PC (1997) Plant intercellular communication via plasmodesmata. Plant Cell 9:1043–1054

Meshcheryakov A, Steudle E, Komor E (1992) Gradients of turgor, osmotic pressure, and water potential in the cortex of the hypocotyl of growing *Ricinus* seedlings. Plant Physiol 98:840–852

Noueiry AO, Lucas WJ, Gilbertson RL (1994) Two proteins of a plant DNA virus coordinate nuclear and plasmodesmal transport. Cell 76:925–932

Offler CE, Patrick JW (1996) Solute transport dynamics of plasmodesmata in stems and roots. In: III Int Worksh on Basic and Applied Research in Plasmodesmal Biology, Israel, March 10-16, 1996, pp 156–161

Ohana P, Delmer DP, Volman G, Steffend J, Metthews D, Benziman M (1992) β-Furfuryl-β-glucoside: an endogenous activator of higher plant UDP-glucose: (1–3)-β-glucan synthase. Plant Physiol 98: 706–715

Ohana P, Benziman M, Delmer DP (1993) Stimulation of callose synthesis in vitro correlates with changes in intracellular distribution of the callose synthase activator β-furfuryl-β-glucoside. Plant Physiol 101:187–191

Olesen P, Robards AW (1990) The neck region in plasmodesmata: general architecture and some functional aspects. In: Robards AW, Jongsma H, Lucas WJ, Pitts J, Spray D (eds) Parallels in cell to cell junction in plants and animals. Springer, Berlin Heidelberg New York, pp 145–170

Oparka KJ, Prior DAM (1992) Direct evidence for pressure-generated closure of plasmodesmata. Plant J 2:741–750

Oparka KJ, Prior DAM, Crawford JW (1994) Behaviour of plasma membrane, cortical ER and plasmodesmata during plasmolysis of onion epidermal cells. Plant Cell Environ 17:163–171

Oparka KJ, Roberts AG, Prior DAM, Chapman S, Baulcombe D, Santa Cruz S (1995a) Imaging the green fluorescent protein in plants – viruses carry the torch. Protoplasma 189:133–141

Oparka KJ, Prior DAM, Wright KM (1995b) Symplastic communication between primary and developing lateral roots of *Arabidopsis thaliana*. J Exp Bot 46:187–197

Oparka KJ, Roberts AG, Roberts I M, Prior DAM, Santa Cruz S (1996) Viral coat protein is targeted to, but does not gate, plasmodesmata during cell-to-cell movement of potato virus X. Plant J 10: 805–813

Overall RL, Blackman LM (1996) A model of the macromolecular structure of plasmodesmata. Trends Plant Sci 1:307–311

Overall RL, Wolfe J, Gunning BES (1982) Intercellular communication in *Azolla* roots. I. Ultrastructure of plasmodesmata. Protoplasma 111:134–150

Patrick JW (1990) Sieve element unloading: cellular pathway, mechanism and control. Physiol Plant 78:298–308

Patrick JW (1997) Phloem unloading: sieve element unloading and post-phloem transport. Annu Rev Plant Physiol Plant Mol Biol 48:191–222

Patrick JW, Offler CE (1995) Post-sieve element transport of sucrose in developing seeds. Aust J Physiol 22:681–702

Patrick JW, Offler CE (1996) Post sieve element transport of photoassimilates in sink regions. J Exp Bot 47:1165–1177

Pickard WF, Minchin PEH (1990) The transient inhibition of phloem translocation in *Phaseolus vulgaris* by abrupt temperature drops, vibration, and electric shock. J Exp Bot 41:1361–1369

Pickard WF, Minchin PEH (1992a) The electroshock-induced inhibition of phloem translocation. J Exp Bot 43:409–417

Pickard WF Minchin PEH (1992b) The inhibition of phloem translocation by ammonia. J Exp Bot 43: 51–54

Pickard WF, Minchin PEH (1992c) The nature of the short-term inhibition of stem translocation produced by abrupt stimuli. Aust J Plant Physiol 19:471–480

Pritchard J (1996) Aphid stylectomy reveals an osmotic step between sieve tube and cortical cells in barley roots. J Exp Bot 47:1519–1524

Radford J, White J (1996) Preliminary localisation of myosin to plasmodesmata. In: III Int Worksh on Basic and Applied Research in Plasmodesmal Biology, Israel, March 10–16, 1996, . pp 37–38
Radford JE, Vesk M, Overall RL. (1998) Callose deposition at plasmodesmata. Protoplasma 201:30–37
Robinson-Beers K, Evert RF (1991) Fine structure of plasmodesmata in mature leaves of sugarcane. Planta 184:307–318
Sakuth T, Schobert C, Pecsvaradi A, Eichholz A, Komor E, Orlich G (1993) Specific proteins in the sieve-tube exudate of *Ricinus communis* L seedlings - separation, characterization and *in vivo* labelling. Planta 191:207–213
Schenk E (1974) Similar effects of cytokinins and carbonyl cyanide m-chlorophenylhydrazone (CCCP) on amino acid translocation in leaves of *Sagittaria graminea* Michx. Acta Bot Neerl 23:739–746
Schulz A (1986a) Wound phloem in transition to bundle phloem in primary roots of *Pisum sativum* L. I. Development of bundle-leaving wound-sieve tubes. Protoplasma 130:12–26
Schulz A (1986b) Wound phloem in transition to bundle phloem in primary roots of *Pisum sativum* L. II. The plasmatic contact between wound-sieve tubes and regular phloem. Protoplasma 130:27–40
Schulz A (1987) Sieve-element differentiation and fluoresceine translocation in wound phloem of pea roots after the complete severance of the stele. Planta 170:289–299
Schulz A (1990a) Conifers. In: Behnke H-D, Sjolund RD (eds) The sieve element - comparative structure, induction and development. Springer Berlin Heidelberg New York pp 63–88
Schulz A (1990b) Wound-sieve elements. In: Behnke H-D, Sjolund RD (eds) The sieve element - comparative structure, induction and development. Springer, Berlin Heidelberg New York, pp 199–217
Schulz A (1992) Living sieve cells of conifers as visualized by confocal, laser-scanning fluorescence microscopy. Protoplasma 166:153–164
Schulz A (1993) Sink strength - the importance of the distance between phloem and receiver cells. Plant Cell Environ 16:1031–1032
Schulz A (1994) Phloem transport and differential unloading in pea seedlings after source and sink manipulations. Planta 192:239–248
Schulz A (1995) Plasmodesmal widening accompanies the short-term increase in symplasmic phloem unloading of pea root tips under osmotic stress. Protoplasma 188:22–37
Schulz A (1996) Experimentelle Untersuchungen zur Entwicklung und Funktion der Assimilatleitbahnen in Höheren Pflanzen. Habil Thesis, Christian-Albrechts-Universität, Kiel
Schulz A (1998) The phloem. Structure related to function. Prog Bot 59:429–472
Schumacher W (1933) Untersuchungen über die Wanderung des Fluoreszeins in den Siebröhren. Jb Wiss Bot 77:685–732
Schumacher W (1936) Untersuchungen über die Wanderung des Fluoresceins in den Haaren von *Cucurbita pepo*. Jb Wiss Bot 82:507–533
Sitte P (1998) Facts and concepts in cell compartmentation. Prog Bot 59:3–45
Sjölund RD (1997) The phloem sieve element: a river runs through it. Plant Cell 9:1137–1146
Sokolova M, Prüfer D, Tacke E, Rohde W (1997) The potato leafroll virus 17k movement protein is phosphorylated by a membrane-associated protein kinase from potato with biochemical features of protein kinase c. FEBS Lett 400:201–205
Stadler R, Brandner J, Schulz A, Gahrtz M, Sauer N (1995) Phloem loading by the PmSUC2 sucrose carrier from *Plantago major* occurs into companion cells. Plant Cell 7:1545–1554
Staehelin LA (1997) The plant ER: a dynamic organelle composed of a large number of discrete functional domains. Plant J 11:1151–1165
Szederkenyi J, Komor E, Schobert C (1997) Cloning of the cDNA for glutaredoxin, an abundant sieve-tube exudate protein from *Ricinus communis* L and characterisation of the glutathione-dependent thiol-reduction system in sieve tubes. Planta 202:349–356
Terry BR, Robards AW (1987) Hydrodynamic radius alone governs the mobility of molecules through plasmodesmata. Planta 171:145–157
Thorsch J, Esau K (1981a) Changes in the endoplasmic reticulum during differentiation of a sieve element in *Gossypium hirsutum*. J Ultrastruct Res 74:183–194
Thorsch J, Esau K (1981b) Nuclear degeneration and the association of endoplasmic reticulum with the nuclear envelope and microtubules in maturing sieve elements of *Gossypium hirsutum*. J Ultrastruct Res 74:195–204
Thorsch J, Esau K (1981c) Ultrastructural studies of protophloem sieve elements in *Gossypium hirsutum*. J Ultrastruct Res 75:339–351
Tilney LG, Cooke TJ, Connelly PS, Tilney MS (1991) The structure of plasmodesmata as revealed by plasmolysis, detergent extraction, and protease digestion. J Cell Biol 112:739–747
Trebacz K, Fensom D (1989) The uptake and transport of ^{14}C in cells of *Conocephalum conicum* L. in light. J Exp Bot 40:1089–1092
Tucker EB (1982) Translocation in the staminal hairs of *Setcreasea purpurea*. I. A study of cell ultrastructure and cell-to-cell passage of molecular probes. Protoplasma 113:193–201

Tucker EB (1988) Inositol bisphosphate and inositol trisphosphate inhibit cell-to-cell passage of carboxyfluorescein in staminal hairs of *Setcreasea purpurea*. Planta 174:358–363
Tucker EB (1990) Calcium-loaded 1, 2-bis(2-aminophenoxy)ethane-N, N, N′, N′-tetra acetic acid blocks cell-to-cell diffusion of carboxyfluorescein in staminal hairs of *Setcreasea purpurea*. Planta 182:34–38
Tucker EB (1993) Azide treatment enhances cell-to-cell diffusion in staminal hairs of *Setcreasea purpurea*. Protoplasma 174:45–49
Tucker EB, Boss WF (1996) Mastoparan induced intracellular Ca^{2+} fluxes may regulate cell-to-cell communication in plants. Plant Physiol 111:459–467
Tucker EB, Tucker JE (1993) Cell-to-cell diffusion selectivity in staminal hairs of *Setcreasea purpurea*. Protoplasma 174:36–44
Turgeon R (1996) Phloem loading and plasmodesmata. Trends Plant Sci 1:418–423
Turner A, Wells B, Roberts K (1994) Plasmodesmata of maize root tips: structure and composition. J Cell Sci 107:3351–3361
van Bel AJE (1993a) Strategies of phloem loading. Annu Rev Plant Physiol Plant Mol Biol 44:253–281
van Bel AJE (1993b) The transport phloem. Specifics of its functioning. Progr Bot 54:134–150
van Bel AJE (1996) Interaction between sieve element and companion cell and the consequences for photoassimilate distribution. Two structural hardware frames with associated physiological software packages in dicotyledons? J Exp Bot 47:1129–1140
van Bel AJE, Kempers R (1997) The pore/plasmodesm unit; key element in the interplay between sieve element and companion cell. Prog Bot 58:278–291
van Bel AJE, van Rijen HVM (1994) Microelectrode-recorded development of the symplasmic autonomy of the sieve element/companion cell complex in the stem phloem of *Lupinus luteus* L. Planta 192:165–175
Waigmann E, Zambryski P (1995) Tobacco mosaic virus movement protein mediated protein transport between trichome cells. Plant Cell 7:2069–2079
Waigmann E, Lucas WJ, Citovsky V, Zambryski P (1994) Direct functional assay for tobacco mosaic virus cell-to-cell movement protein and identification of a domain involved in increasing plasmodesmal permeability. Proc Natl Acad Sci USA 91:1433–1437
Waigmann E, Turner A, Peart J, Roberts K, Zymbryski P (1997) Ultrastructural analysis of leaf trichome plasmodesmata reveals major differences from mesophyll plasmodesmata. Planta 203: 75–84
Wang N, Fisher DB (1994a) Monitoring phloem unloading and post-phloem transport by microperfusion of attached wheat grains. Plant Physiol 104:7–16
Wang N, Fisher DB (1994b) The use of fluorescent tracers to characterize the post-phloem transport in maternal tissue of developing wheat grains. Plant Physiol 104:17–27
Warmbrodt RD (1987) Solute concentrations in the phloem and apex of the root of *Zea mays*. Am J Bot 74:394–402
Watada RS, Wergin WP (1997) Characterization of the cell wall microdomain surrounding plasmodesmata in apple fruit. Plant Physiol 114:539–547
White PJ (1997) Cation channels in the plasma membrane of rye roots. J Exp Bot 48:499–514
White RG, Badelt K, Overall RL, Vesk M (1994) Actin associated with plasmodesmata. Protoplasma 180:169–184
Wolf S, Lucas WJ (1994) Virus movement proteins and other molecular probes of plasmodesmal function. Plant Cell Environ 17:573–585
Wolf S, Deom CM, Beachy RN, Lucas WJ (1989) Movement protein of tobacco mosaic virus modifies plasmodesmatal size exclusion limit. Science 246:377–379
Wright KM, Oparka KJ (1996) The fluorescent probe HPTS as a phloem mobile, symplastic tracer: an evaluation using confocal laser scanning microscopy. J Exp Bot 47:439–445
Wright KM, Oparka KJ (1997) Metabolic inhibitors induce symplastic movement of solutes from the transport phloem of *Arabidopsis* roots. J Exp Bot 48:1807–1814
Yahalom A, Warmbrodt RD, Laird DW, Traub O, Revel J-P, Willecke K, Epel BL (1991) Maize mesocotyl plasmodesmata proteins cross-react with connexin gap junction protein antibodies. Plant Cell 3:407–417

CHAPTER 12

Plasmodesmal Coupling and Cell Differentiation in Algae

M. KWIATKOWSKA

Department of Cytophysiology, University of Łodź,
Łodź, ul. Pilarskiego 14, Poland

1 Introduction

In the past few years, our knowledge about plasmodesmata and their function has grown steadily. As for plasmodesmata formation during cell division, its current image is based mainly on cell plate development in higher plants. This generalization of the plasmodesmal development and structure shuld probably be revised for algae, in which numerous types of cytoplasmic junctions, often described in fragmentary form, have been called plasmodesmata for many years. This view refers to the insight (Mezitt and Lucas 1996) that "to understand the fundamental role of plasmodesmata in plant biology, it is first necessary to describe their formation and distribution throughout the plant".

An early, detailed, systematic, and very instructive review of plasmodesmata in algae was given by Marchant (1976). The present chapter will be a brief description of plasmodesmata formation in various algae and their possible role in cell coupling and cell differentiation.

2 Plasmodesmata Formation in Algae Dividing by a Division Furrow

The occurrence of plasmodesmata during formation of a division furrow has been described for meristematic cells of the brown alga *Cutleria cylindrica* (La Claire 1981). Cytokinesis occurs after karyokinesis has been completed and becomes manifest by the emergence of a furrow in the plasma membrane growing towards the region between the two daughter nuclei. Electron-transparent vesicles are associated with the furrow, especially at the earliest observed stages of cytokinesis (Fig. 1A). Cytoplasmic channels (20–40 nm in diameter), perpendicular to the plane of cytokinesis, perforate the developing furrow. These channels are lined by furrow membranes. Cisternae of the endoplasmic reticulum (ER) are positioned parallel to the growing furrow but do not penetrate the cytoplasmic channels running through the furrow. After completion of cytokinesis, cytoplasmic strands connect the daughter cells. As cell-wall formation proceeds, cytoplasmic connections become more elongated and obtain the typical appearance of most brown algae plasmodesmata (Fig. 1B–D). La Claire (1981) supposes that the sites of these future connections have been programmed inside the plasma membranes of the furrow.

In *Fucus vesiculosus* (Phaeophyta), cytokinesis results primarily from vesicle addition to a centripetally growing furrow (Brawley et al. 1977). Fibrillar material is added to the space between the daughter cells from vesicles discharged by both cells (this plant is described in more detail in Sect. 5.5).

Cell-wall development in the filamentous green alga *Bulbochaete hiloensis* (Fraser and Gunning 1973) starts with a formation of division rings located in the parental cell during bulb formation. The lenticular appearance of each ring suggests that they develop by deposition of many vesicular packages, the contents of which are liberated by exocytosis through the plasma membrane. Outside the division ring, the parental cell wall splits. The material of the ring is stretched during cell elongation to become the first wall layer of the daughter cell. The cross-wall developing between the bulb and parental cell is penetrated by plasmodesmata. They make up the layer of electron-

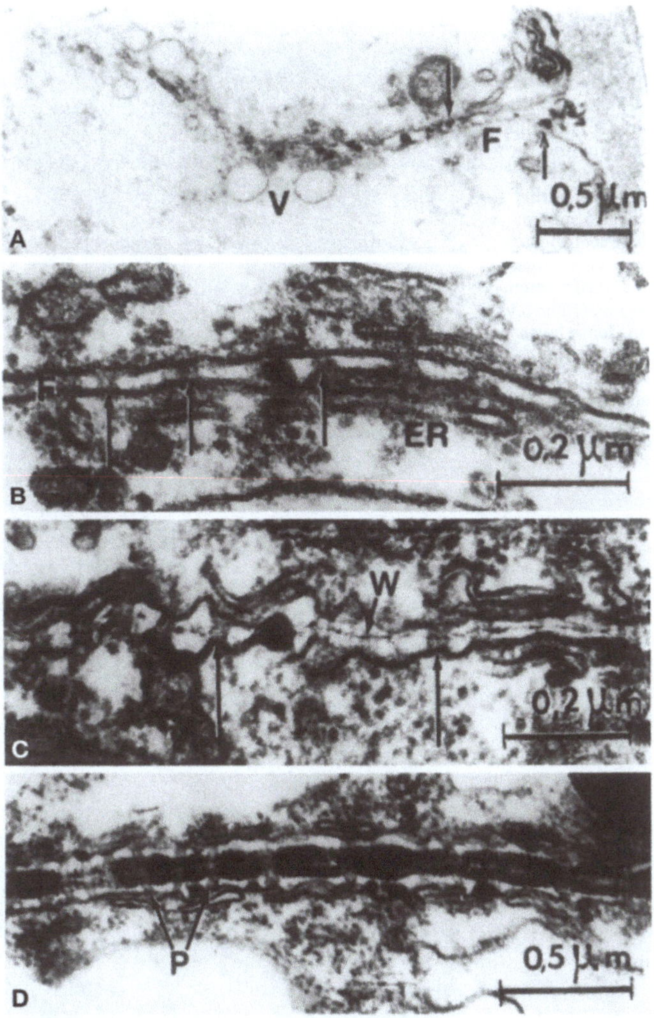

Fig. 1. A–D. EM micrographs of meristematic cells of the *Cutleria cylindrica* gametophyte. **A** Nascent furrow (*F*) with vesicles (*V*) and cytoplasmic channels perforating the furrow (*arrows*). **B** Part of a developing furrow (*F*). Cytoplasmic channels are bordered laterally by endoplasmic reticulum membranes (*ER*) continuous with the furrow membranes. **C** Early wall (*W*) deposition in the completed furrow in the spaces between the cytoplasmic channels. **D** Thicker cell wall with recognizable plasmodesmata (*P*) running through the wall. (La Claire 1981)

dense material at the region of the middle lamella in *B. hiloensis* (Fig. 2); (see also Sect. 5.4). No internal components such as a desmotubule or derivatives of the endoplasmic reticulum are visible in these plasmodesmata (Fraser and Gunning 1969), each plasmodesm consisting of a cylindrical connection between the plasma membranes of adjacent cells. The cylinder is constricted at each end by orifices which may be less than

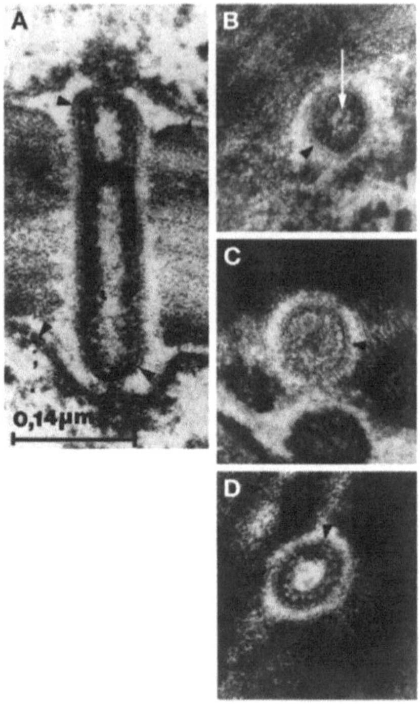

Fig. 2. A–D. EM ultrastructure of plasmodesmata in the filamentous green alga *Bulbochaete hiloensis.* **A** Median section of plasmodesmata with material deposited on the inner face of the plasma membrane (*arrowheads*). The electron-dense layer at the level of the middle lamella is also shown. **B, C, D** Transverse sections of plasmodesmata. **B** The orifice of the constricted plasma membrane (*white arrow*). **C** The level of the electron-dense layer. **D** The material on the inner face of the plasma membrane with about 12 electron-translucent centres arround the circumference. (Fraser and Gunning 1969)

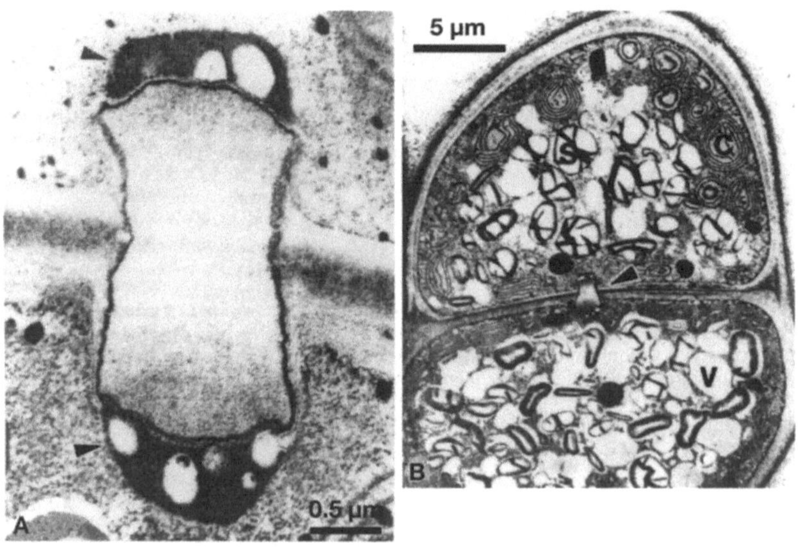

Fig. 3. A–B. Chantransia phase of *Batrachospermum breutelii.* **A** Ultrastructure of the pit plug in Chantransia phase filament with electron-dense, inflated outer cap layers (*arrowheads*). Membranes that appear between the two cap layers are continuous with the plasma membrane. **B** Line of cells with numerous chloroplasts, starch granules, small vacuoles, and pit plug (*arrowhead*) in cross-wall. (Sheath and Whittick 1995)

10 nm in diameter. Within the cylinder, the cytoplasmic face of the plasma membrane is lined with material that probably consists of helically arranged particles. The lumen here is 40–45 nm wide.

Cell division by formation of a centripetally growing furrow is also characteristic for many Rhodophyta. For example, *Batrachospermum breutelii* produces gonimoblast propagules which are divided by the simultaneous formation of division furrows (Sheath and Whittick 1995). Septa between chambers have pit plugs with two cap layers, the outer of which is dome-shaped and unevenly stained (Fig. 3).

Cleavage furrows during cell division and pit plugs also appear in *Glaucosphera vacuolata*, another Rhodophyta species (Broadwater et al. 1995). In the cytochemical studies, the pit plugs of red algae were observed to be reactive to the periodic acid Schiff procedure (PAS) specific for carbohydrates, and to the protein stain mercuric bromophenol blue. The proteolytic enzyme proteinase K appeared to remove much of the material constituting the outer cap of the pit plugs. These results, together with similarities of phosphotungstic/chromic acid staining, suggest that the outer cap layer of red algae pit plugs is composed of glycoproteins (Sheath et al. 1994; Trick and Pueschel 1990).

Fusion of previously isolated domains by new cytoplasmic connections also occurs. It has been suggested that the secondary pit formation between red alga host *Odonthalia floccosa* and red alga alloparasite *Harveyella mirabilis* may provide a symplasmic avenue for transport and communication (Wetherbee et al. 1984). It may be assumed that pit plugs appearing in other situations do not prevent intercellular exchange completely.

3 Plasmodesmata in Algae Dividing Through Phycoplast Formation

A phycoplast is a system of microtubules arranged parallel to the surface of the forming septum. A beautiful example of phycoplast formation and cytokinesis was given by Pickett-Heaps (1970) for autospore formation in the green alga *Kirchneriella lunaris* (Fig. 4) in which quadripartition of the cell occurs. The reproductive events can be divided into four successive phases. (1) The primary mitosis which forms two daughter nuclei. (2) The primary cleavage during which the primary septum shows up amongst the microtubules of a phycoplast, which elongates through them from an invagination of the plasmalemma. Cytoplasmic cleavage is generally fairly complete before the secondary nuclear division, but cytoplasmic connections are frequent. (3) The secondary mitoses when both daughter nuclei divide again synchronously. It is likely that synchronization of these divisions is conditioned by the cytoplasmic connections (see Kwiatkowska and Maszewski 1976). (4) The secondary cleavage during which the cytoplasm is partitioned around each nucleus, forming four autospores divided by two other septa. Cytokinesis is accomplished as described above, as transverse microtubules extend in between daughter nuclei and across the cell, which precedes the formation of secondary septa. Microtubules also encircle the cell before division, predicting the two planes of cytokinesis. Before being released, each autospore forms its own cell wall by secretion of cell-wall material containing vesicles.

The wall of *Haematococcus lacustris* (Girod) Rostafinski (Volvocales) is also formed centripetally during zoospore formation (Mesquita and Santos 1984). The nuclear en-

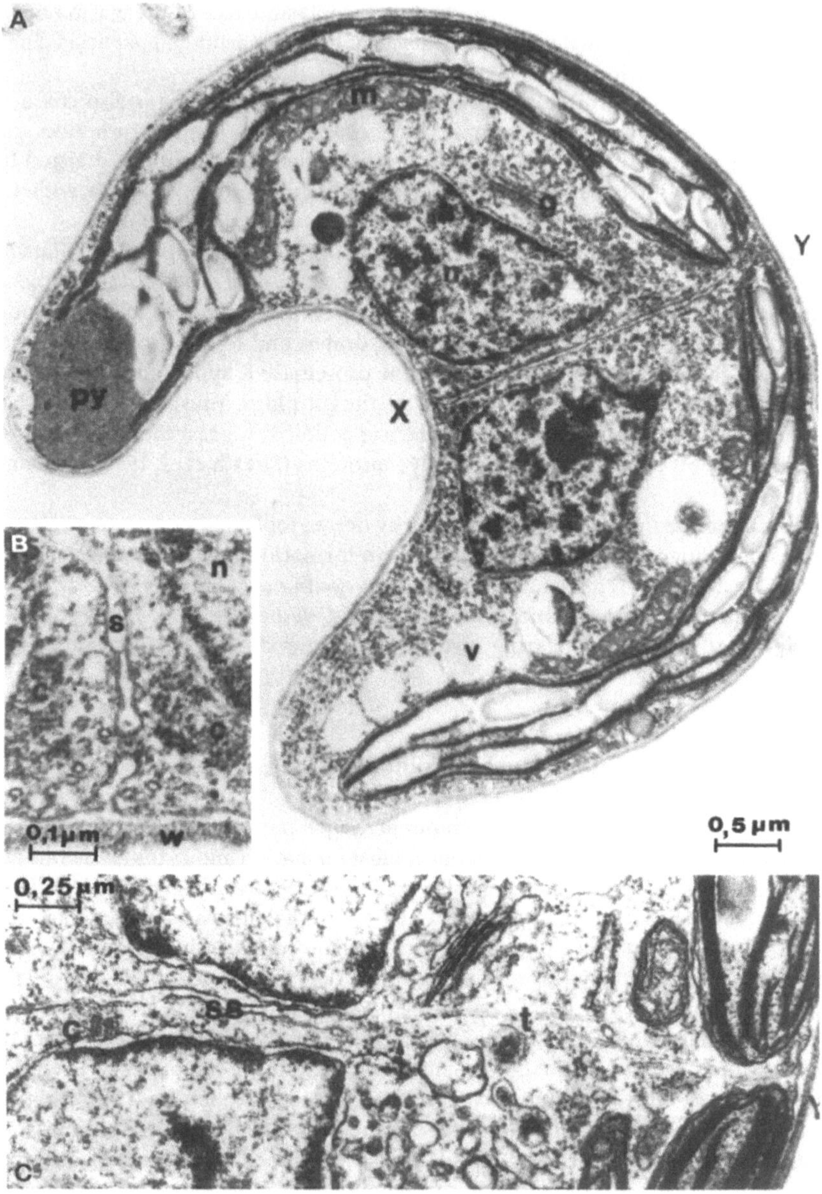

Fig. 4. A–C. EM images of mitosis and autospore formation in the green alga *Kirchneriella lunaris*. **A** A primary cleavage that has almost been completed. Vacuoles (*v*), Golgi stacks (*g*), mitochondria (*m*), a chloroplast with associated pyrenoid (*py*), and nuclei (*n*) are visible. The cross-section x-y is shown in **B**. **B** Serial section showing the septum (*s*) with a hole and situated perpendicular to the existing cell wall (*w*). **C** Secondary cleavage. Three microtubules (*t*) are visible near the chloroplast cleavage, another (*arrow*) near the septum. (Pickett-Heaps 1970)

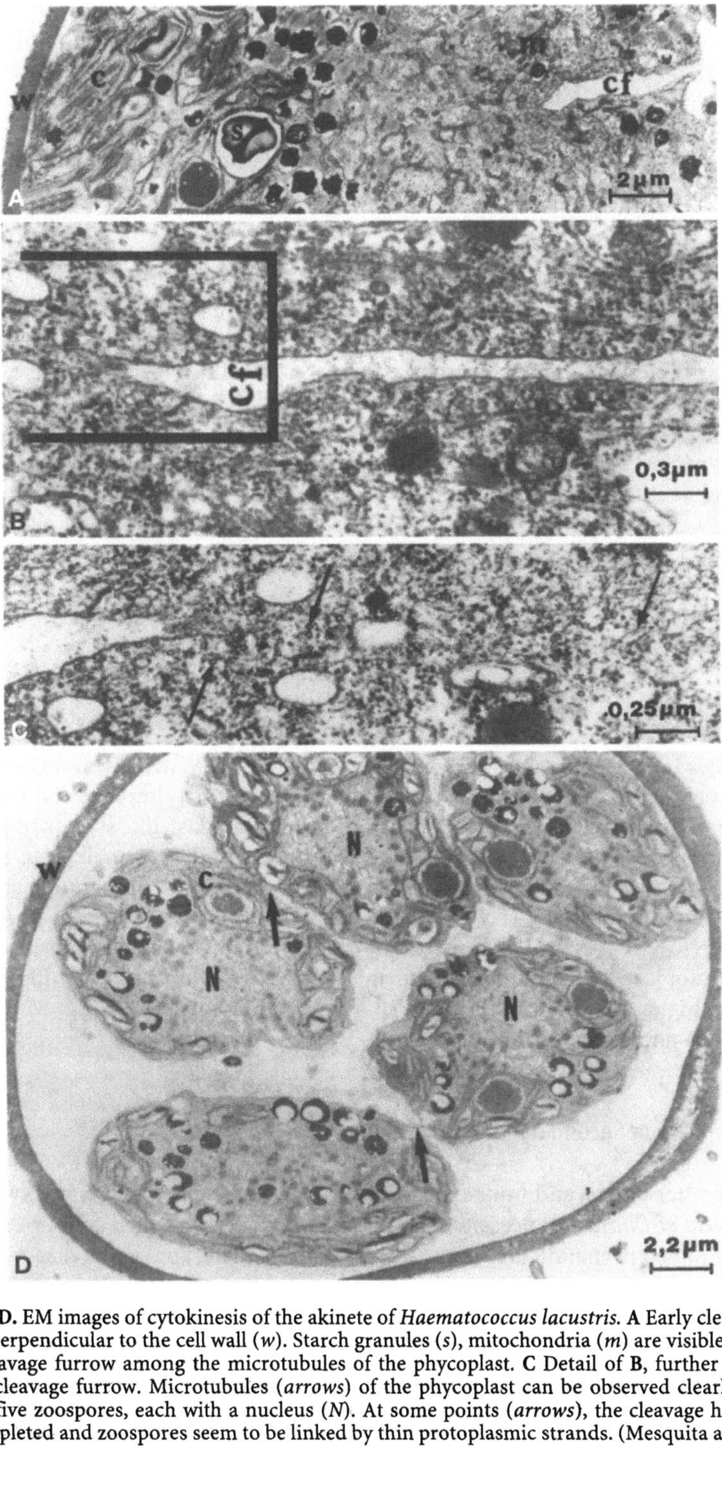

Fig. 5. A–D. EM images of cytokinesis of the akinete of *Haematococcus lacustris*. **A** Early cleavage furrow (*cf*) perpendicular to the cell wall (*w*). Starch granules (*s*), mitochondria (*m*) are visible. **B** Developing cleavage furrow among the microtubules of the phycoplast. **C** Detail of **B**, further along the growing cleavage furrow. Microtubules (*arrows*) of the phycoplast can be observed clearly. **D** Cyst showing five zoospores, each with a nucleus (*N*). At some points (*arrows*), the cleavage has not yet been completed and zoospores seem to be linked by thin protoplasmic strands. (Mesquita and Santos 1984)

velope of the telophase nucleus breaks down in the mid-region and surrounds the two daughter nuclei. Between them, a growing cleavage furrow (invagination of the plasmalemma) appears, leading to the division of the protoplast. The cleavage furrow is always associated with a microtubule system (phycoplast); (Fig. 5). Zoospores appear to be linked by thin protoplasmic strands.

During sporulation of *Neochloris minuta* and *Neochloris wimmeri*, mitosis and cytokinesis are similar to the basic types found in other chlorophycean species, with centriole replication occurring at the onset of mitosis and simultaneous centripetal cleavage along phycoplast microtubules (Kouwets 1995b). In other *Neochloris* species, however, *N. aquatica*, *N. pyrenoidosa*, *N. terrestris*, and *N. vigens*, sporulation occurs through centrifugal cleavage along phycoplast microtubules.

Cytokinesis in *Plantosphaeria gelatinosa* (Chlorophyceae) is also centrifugal. Mitosis and cytokinesis mediated by a phycoplast present in the vegetative cell may show overlap in time, so that *P. gelatinosa* is not considered to be a true coenocyte (Kouwets 1995a). Cleavage furrows apparently originate in the centre of the cell and grow out along phycoplast microtubules towards the plasma membrane in a radiating pattern. By the end of cytokinesis, the protoplast is divided in two parts, each containing a nucleus and a chloroplast with an associated pyrenoid. Cytoplasmic connections are visible between cells that have divided at separate stages.

4 Cell Division Through Phragmoplast Formation

4.1 Phragmoplast-Like Structures and the Division Furrow

Cytokinesis in the green alga *Spirogyra* (Zygnemataceae) (Sawitzky and Grolig 1995) is characterized by the centripetal growth of a septum which impinges on a centrifugally expanding telophase spindle, leading to a phragmoplast-like structure. In *Spirogyra*, cytokinesis is thus initiated by annular ingrowing of the plasmalemma, but completion occurs by cell-plate formation in association with a phragmoplast-like arrangement of microtubules (Fowke and Pickett-Heaps 1969). Cytokinesis in *Spirogyra* could therefore represent an intermediate stage in the evolutionary development of the phragmoplast, with simultaneous formation of cross-wall and plasmodesmata occurring in higher plants. As in *Zygnema* (Bakker and Lokhorst 1987), however, plasmodesmata are not visible in *Spirogyra*.

4.2 Primary or Secondary Plasmodesmata

As with higher plants, and unlike most green algae, a phragmoplast is involved in the cell division of *Chara* (Franceschi et al. 1994). This structure which contains microtubules orientated perpendicular to the developing cell plate, appears to direct cell plate formation. In higher plants, complete separation of the two daughter cells is accomplished by the controlled positioning of ER within and perpendicular to the forming cell plate (Porter and Machado 1960; Hepler 1982). As the new wall develops, this ER becomes appressed and forms the primary plasmodesmata together with the surrounding plasma membrane.

Primary plasmodesmata were found in *Chara* sp. (Pickett-Heaps 1967a, b) *Chara vulgaris* (Kwiatkowska and Maszewski 1976, 1979, 1985) and *Chara zeylanica* (Cook et al. 1996). In contrast, Franceschi et al. (1994) claim that "only when the daughter cells had separated completely were plasmodesmata formed across the division wall" in the apical zone of *Chara corallina*. Probably selective degradation takes place here, providing cell wall pores for secondary plasmodesmata that represent new cytoplasmic bridges formed across the existing cell wall partly during or after cytokinesis (Franceschi et al. 1994).

The contrasting observation may be explained as follows: primary plasmodesmata connect identical antheridial filament cells undergoing the same phases of development synchronously (Fig. 6), while some apical cells in the apex of *Chara* become polarized prior to division (Ducreux 1968). Structures accumulating in its lower part are different from those in the upper part, which results in the formation of two different cells with different future destinations. The upper cell remains an apical cell capable of further mitotic divisions. The subapical cell is divided into two different cells after a next round of polarization, with the upper one becoming a nodal cell capable of mitotic divisions and the lower one developing into an internode (Ducreux 1979). In such a situation, a temporary separation of newly formed cells by a cell wall without plasmodesmata may be necessary. After the individual features of the respective cell types are stabilized secondary symplasmic connections with limited size exclusion limits (SEL) might be established.

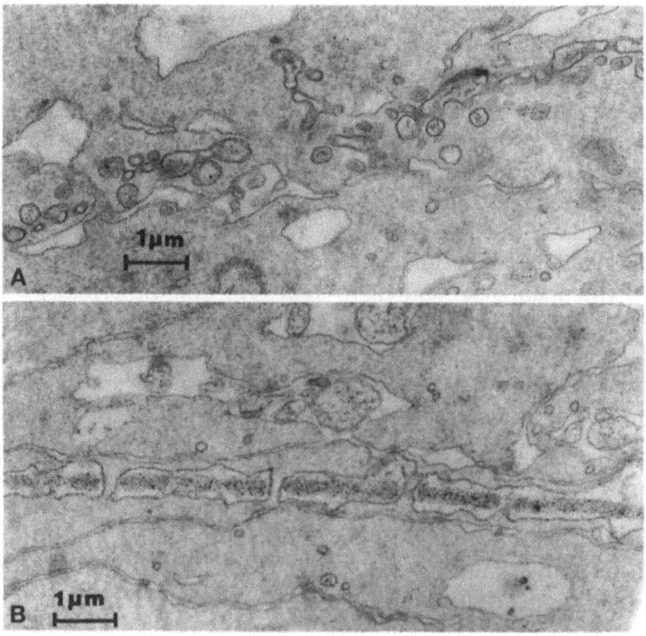

Fig. 6. A, B. The antheridial filament cells of *Chara vulgaris* in EM after staining according to Thiéry (1967). **A** Stage of cell-plate formation. **B** Cross-wall with plasmodesmata. (see also Kwiatkowska and Maszewski 1978, 1979)

5 Domain Formation

5.1 Domain Formation in Multicellular Organisms

A new class of plant genes, the supracellular control proteins (SCPs), and SCP cofactors have been hypothesized to control domain formation in multicellular organisms (Mezitt and Lucas 1996). Every SCP would contain a protein domain that allows it to interact with plasmodesmata for cell-to-cell traffic and a second domain that alters plasmodesmal SEL. According to Mezitt and Lucas (1996), SCP cofactors are involved in the delivery of SCP to the plasmodesmata and mediate general control over plasmodesmal trafficking. Expression of such a cofactor in a cell would isolate it from macromolecular signalling within its developmental domain. This cell (or cell line) might then proceed to divide, forming a supracellular domain of its own. In algae, SCPs and SCP cofactors have not yet been described, but observations on the development of multicellular algae allow the assumption that they also function in this group of plants.

5.2 Domain Formation During Spermatogenesis in Chara

The development of *Chara vulgaris* antheridia is a well-documented example of domain formation (Pickett-Heaps 1968; Kwiatkowska and Maszewski 1976, 1978, 1985, 1986; Kwiatkowska and Żylińska 1988). The antheridia of *Chara* develop from haploid nodal cells that occur at lateral branches of a thallus, referred to as pleuridia by Ducreux (1975). The lower cell following the first division of an initial cell develops into one domain, a basal cell, while the upper one transforms into another domain, the mother cell of an antheridium. The basal cell, that remains a single endopolyploidal cell (Malinowski and Maszewski 1994) adjoins the antheridium with thallus.

Precisely orientated mitotic divisions of the mother cell of the antheridium yield a sphere composed of 24 cells connected by plasmodesmata (Pickett-Heaps 1968; Kwiatkowska and Maszewski 1985; Kwiatkowska and Żylińska 1988) (Fig. 7). Eight outer shield cells envelope the antheridium, while eight median cells are transformed into secretory cells (the manubria) and eight inner cells into the primary capitular cells which give rise to derivatives of the second and third order. The development of an antheridium coincides with the disappearance of plasmodesmal connections (1) between shield cells, (2) between shield cells and the basal cell, and (3) between manubria which are torn away and become separated from each other. This leads to a break in circumferential communication in an antheridium (Maszewski and van Bel 1996) and enforces the radial orientation of symplasmic routes via the capitular cells that maintain their plasmodesmal connection with the basal cell adjoining antheridium and thallus. The intensive growth of shield cells in a tangential plane brings about the expansion of the antheridium internal space. Capitular cells bordering the internal space of an antheridium produce buds, giving rise to three to five antheridial filaments linked with each other by numerous plasmodesmata in the cell walls at their basal parts. The arrangement of such connections is determined by specific orientation of successive, cell wall separating buds initiating the filaments, which is demonstrated in Fig. 7E and F.

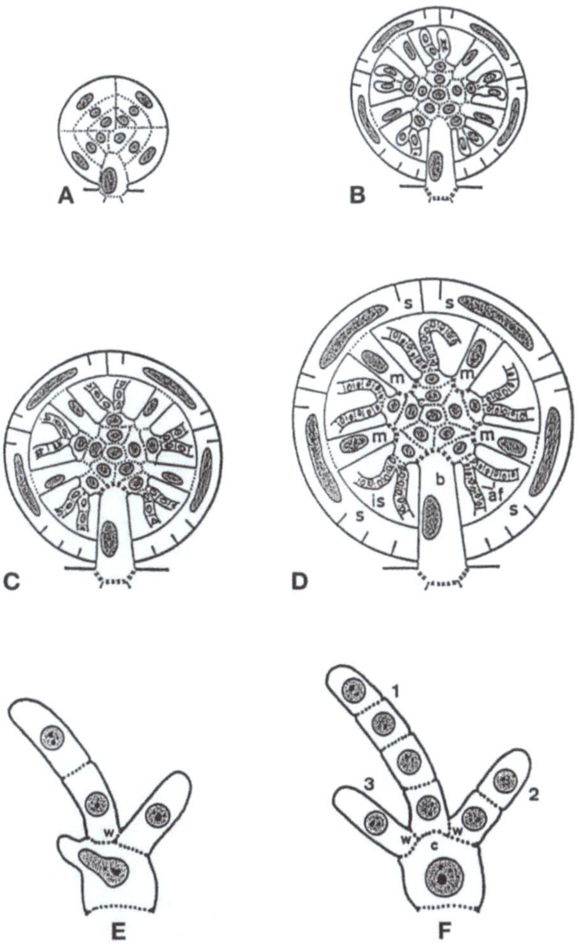

Fig. 7. A–F. Schematic illustration of changes in the arrangement of plasmodesmal connections during antheridial development. **A** 24-cell antheridial stage. **B** Early stage of initiation of antheridial filaments. **C** Antheridium during the period of proliferation of antheridial filament cells. **D** Antheridium at early stage of spermiogenesis with antheridial filaments (*af*), capitular cell (*c*), manubria (*m*), shield cells (*s*), basal cells (*b*), and internal space (*is*). Walls without plasmodesmata are indicated by *straight lines*, walls with simple plasmodesmata by *dotted lines*, and walls with complex plasmodesmata by *crosses*. **E** and **F** Schematic illustration of cell walls (*w*) in successive stages of formation of successive antheridial filaments (*1*, *2*, and *3*) originating from a budding capitular cell

Since the formation of filament cells in an antheridium occurs in a standard sequence of cell divisions, it is possible to mark off eight units connected by capitular cells. Each unit is composed of four functionally integrated domains: shield cell, manubrium, capitular cells and antheridial filament cells. The proliferation and differentiation of the haploidal antheridial filament cells are correlated with the development of endopolyploidal vegetative cells – shield cells, manubria and capitular cells (Kwiatkowska and Maszewski 1987; Kwiatkowska and Żylińska 1988; Kwiatkowska et al. 1990).

The regulatory functions of manubria are visible in the circadian rhythm of protein synthesis (Kwiatkowska et al. 1995). Manubria exhibit rapid decreases and increases in translational activity. The reaction of capitular and antheridial filament cells to day/-night changes is delayed in comparison with the reaction of manubria. It seems that manubria play the role of oscillators (starter cells) inducing the wave of changes in activity in other cells that are symplasmically connected with them.

During the proliferative stage of spermatogenesis, that is during a series of synchronous mitotic cell divisions within filaments (successive 1-, 2-, 4-, 8-, 16-, 32-, and 64-cell stages), the plasmodesmata between cells of an antheridium are plugged with an electron-dense material. The exceptions are the plasmodesmata in the walls between the

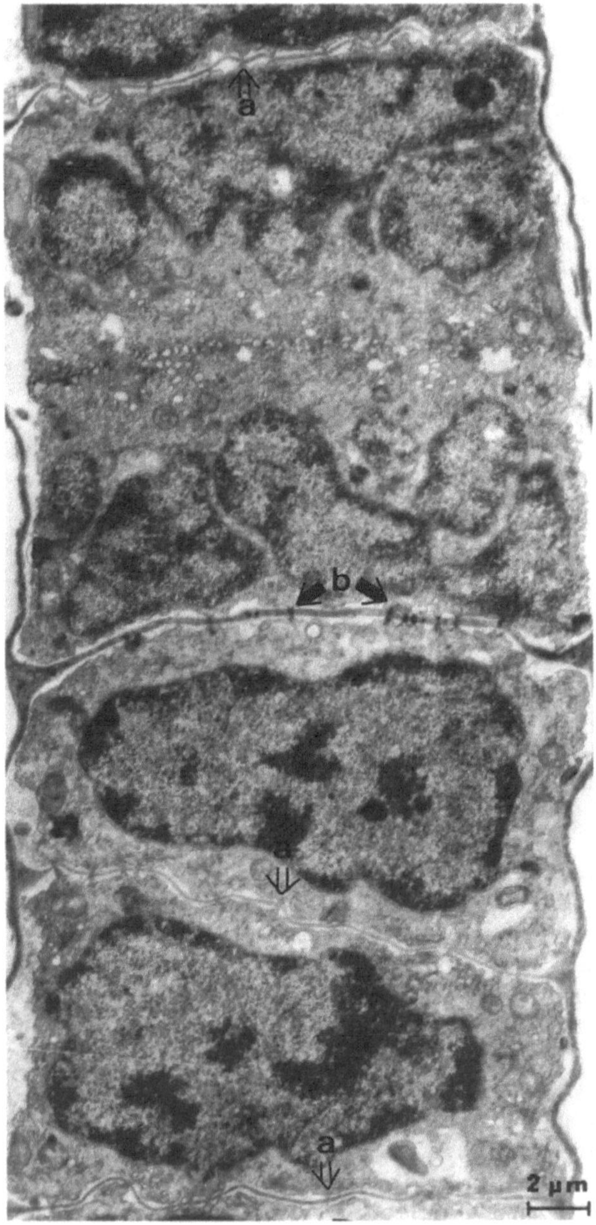

Fig. 8. Part of a longitudinal EM section through a 32-cell antheridial filament of *Chara vulgaris* with two adjacent groups of cells at different stages of development. Open plasmodesmata (*arrows marked a*) connect synchronously divided cells of the antheridial filament, while plasmodesmata plugged by an electron-dense material (*arrows marked b*) are present between asynchronously divided cells

cells within fully synchronously developing antheridial filaments (Fig. 8). Their wide and open plasmodesmal channels provide cytoplasmic continuity between neighbouring cells, and hence enable them to go through successive cell division cycles synchronously. The sporadical appearance in specific cell walls of plasmodesmata reversibly plugged with electron-dense material (Fig. 8, arrows marked b) brings about partition of the filament into two or four separate regions, indicating an independent course of division cell cycles for each of them (Kwiatkowska and Maszewski 1976, 1985, 1986). Each antheridial filament growing out of the same capitular cell also maintains its own cell division rhythm, since the plasmodesmal connections in the adjoining cell walls between basal cells of filaments are plugged with an electron-dense osmiophilic substance.

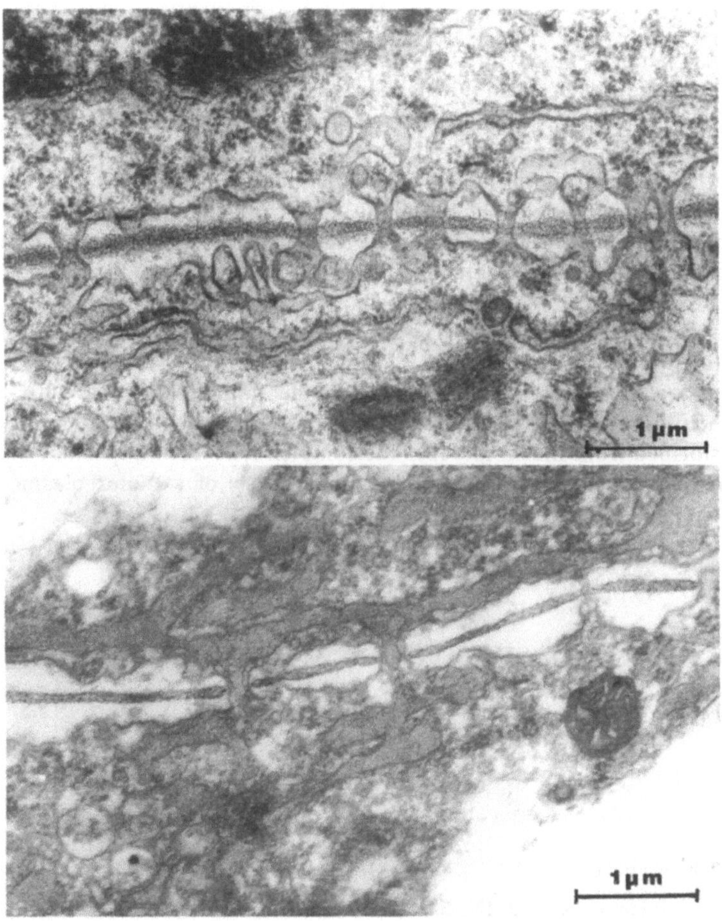

Fig. 9. A, B. Plasmodesmata of *Chara vulgaris* in EM. **A** Plasmodesmata in the walls between synchronously dividing antheridial filament cells during proliferation stage of spermatogenesis. **B** ER connections via plasmodesmata between neighbouring cells during mid-spermiogenesis

At the beginning of the second phase of spermatogenesis named spermiogenesis, i.e. at the onset of differentiation of spermatids into sperm cells, the substance which plugs plasmodesmata in cell walls located within antheridial filaments partitioning into separate regions and between basal cells of antheridial filaments was observed to disappear. During this stage in the process of development, synchronization extends from the region of a filament or its separate fragments to the whole complex of filaments within the antheridium (Kwiatkowska and Maszewski 1985, 1986). In the specific phase of spermiogenesis, probably when generative nuclear proteins (protamins) are formed, rough ER filled with a dark content connects all cells in the antheridial filament as one system of threads which pass non-appressed through wide plasmodesmata (Kwiatkowska 1996). These are the same plasmodesmata which are rarely penetrated by ER and which are formed during synchronous divisions of antheridial filaments (Fig. 9).

During the formation of intial cells of antheridial filaments, all plasmodesmata within an antheridium as well as those connecting the basal cell with the pleuridium are straight channels. This type of plasmodesmata is also characteristic for the connections between nodal cells in the thallus (Spanswick and Costerton 1967; Fischer et al. 1974). Their diameters range between about 20 and 80 nm. The narrowest plasmodesmata are found in young antheridia between closely adjacent cells. Wider plasmodesmata link the cells within the antheridial filaments. During the initial divisions of the antheridial filament cells, the plasmodesmata connecting the basal cell of an antheridium with the sub-basal cells gradually transform into more complex plasmodesmata (Fig. 10). These are also characteristic for the thallus of *Nitella translucens* (Spanswick and Costerton 1967) and are formed by the extension of the lumen of a simple plasmodesmata at the middle lamella level, and the formation of a median sinus. After the cessation of mitotic divisions in antheridial filaments of *C. vulgaris*, similar transformations are observed in plasmodesmata connecting the basal cell with capitular cells of an antheridium. Plasmodesmata connecting manubria and capitular cells are also transformed from the simple into the complex type during spermatozoid differentiation. This transformation resembles the genesis of complex plasmodesmata between particular higher plant cells (see Chap. 10)

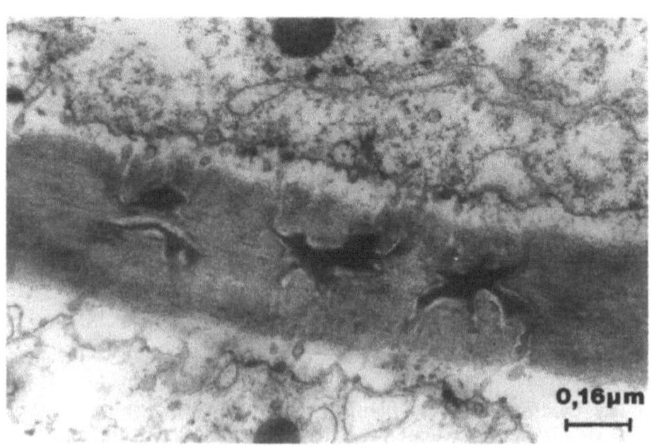

Fig. 10. EM photograph of complex plasmodesmata of *Chara vulgaris*. Cell wall between basal and sub-basal cells within an antheridium, with antheridial filaments at the 16- and 32-cell stage

5.3 Symplasmic Isolation of Chara vulgaris Antheridia

Prior to the initiation of spermiogenesis in antheridia of *C. vulgaris*, plasmodesmata between the basal cell and capitular cells, and between the basal and sub-basal cells become interrupted spontaneously (Kwiatkowska 1988). The mature antheridium of *C. corallina* is also entirely isolated from the symplasm of the thallus, as it does not allow the entry 6-carboxyfluorescein (Shepherd and Goodwin 1992). In *C. vulgaris*, this phenomenon occurs two mitotic division cycles before the formation of spermatids (Kwiatkowska 1988). The antheridium becomes a separate symplasm (supradomain) in which the extremely complicated process of spermatozoid formation takes place (Kwiatkowska 1996).

Plasmolytically induced premature symplasmic isolation of the antheridium leads to a two to four fold reduction in the length of antheridial filaments due to the omission of one to two cell cycles in the proliferative stage. Autoradiographic studies showed that induced symplasmic isolation of the antheridium results in: (1) distinctive changes in the incorporation of ^{3}H-leucine into manubria and the disappearance of the circadian rhythm of protein synthesis after 10 to 20 h (M.Kwiatkowska et al., unpubl. results); (2) the blockage of endomitotic DNA synthesis in the young manubria after 24 h and in capitular cells after 48 h; (3) inhibition of the growth and divisions of antheridial filament cells; and (4) induction of spermiogenesis in the antheridia in which the manubria attained a sufficient level of polyploidy, that is, reached the state of competence. The passage from the proliferative stage to spermiogenesis is also associated with ultrastructural changes in the manubria, suggesting that their metabolism has undergone some shift. Mitochondria show a strong condensation state, the content of secretory vesicles becomes dense, and rough ER is replaced by smooth ER (Kwiatkowska and Maszewski 1987; Kwiatkowska et al. 1990).

Other experiments have been carried out in order to understand the role of gibberellins in the process of spermatogenesis in *Chara*. Autoradiographic studies demonstrated that both natural and induced symplasmic isolation drastically decrease the entry of ^{14}C-GA_3 and ^{3}H-GA_3 into *Chara* antheridia (Kwiatkowska 1991; Kwiatkowska and Malinowski 1995). Moreover, earlier investigations have shown that only the proliferation of antheridial filament cells and not the spermiogenesis is regulated by gibberellins (Godlewski and Kwiatkowska 1980). Our most recent experiments with the use of the capillary electrophoresis system demonstrated that the young antheridia contain 5.3 times more GA_3/antheridium than antheridia at the stage of spermiogenesis (Kaźmierczak et al. 1999).

Plasmolytically induced symplasmic isolation and inhibition of GA_s synthesis by AMO1618 treatment exerts a comparable effect on the development of antheridia of *C. vulgaris*. On the basis of previous (Kwiatkowska 1995) and unpublished data, a hypothesis was put forward that premature spermiogenesis in both experiments was caused by a rapid decrease in gibberellin levels in the antheridia.

5.4 Domain Formation in Bulbochaete hiloensis (Nordst) Tiffany

Bulbochaete hiloensis exhibits a great degree of cell differentiation (Fraser and Gunning 1973). It possesses specialized basal cells formed in the germline. There are two

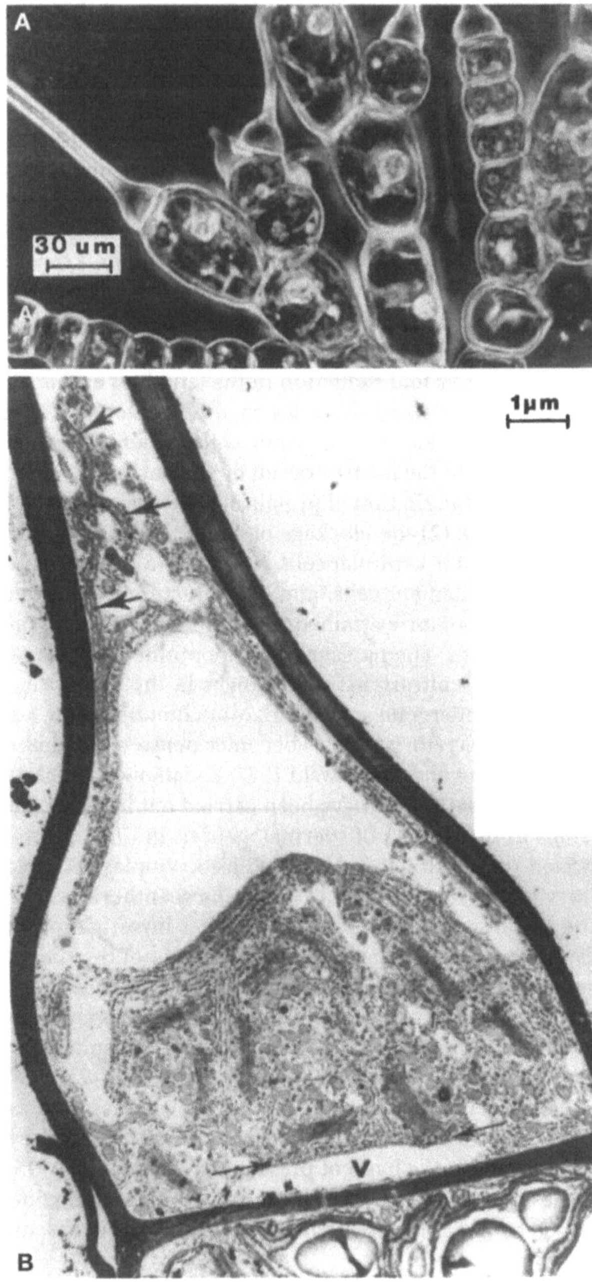

Fig. 11. A, B. *Bulbochaete hiloensis.* **A** Filaments viewed by phase contrast microscopy, with axial photosynthetic cells and parts of the chaetae showing bulbous bases and long terminal extensions. **B** Mature, active hair cell in EM. It contains a basal vacuole (*v*) positioned above the plasmodesmata and semirough ER (*small arrows*) positioned parallelly to the tonoplast on the apical face of the vacuole. The cytoplasm shows massed cisternae of ER (*large arrows*) and numerous dictyosomes. (Fraser and Gunning **A**, 1969, **B** 1973)

types of domains: axial photosynthetic cells and terminal or lateral apoplastidic hair cells (Fig. 11A). The hair cells even lack colourless remnants of plastid material and a nucleolus is absent in hair cell nuclei. Despite the absence of chloroplasts, starch reserves, and nucleolus, the ultrastructure of the cytoplasm is characteristic of intensive secretory activity. Numerous cisternae of rough endoplasmic reticulum and numerous dictyosomes are present, together with associated secretory vesicles (Fig. 11B). Substrates for the biosynthesis in these systems must enter the hair via plasmodesmata in the basal wall (Fraser and Gunning 1969; see Sect. 1.2). The hair cells produce mucilage which is excreted to the exterior through a system of pores which pierce the outer wall of the hair, making an open connection with the apoplasm. The mucus forms a thick outer coat of loosely fibrillar material.

Development of apoplastidic hair cells can to some extent be related to the nature of the cells and to the nature of the chloroplast, which is in the form of a reticulum in the parental cells. The hair cells are strictly determined in their development. They arise and differentiate, but the development of a cell lineage is precluded, since they never undergo mitosis (Fraser and Gunning 1973).

Hence, plasmodesmata linking the basal cells with the hair cells (1) allow a limited exchange of substances for the biosynthesis of mucilage, and (2) separate those domains which perform various functions and undertake different developmental routes.

5.5 Symplasmic Isolation in Multicellular *Fucus* Embryos

The first vegetative cell division in *Fucus* is an important phase of growth and differentiation, since it produces a thallus cell and rhizoid cell with very disparate functional roles (Brawley et al. 1977). Recent studies using *Fucus* embryos (Quatrano and Shaw 1997) indicate that the unfertilized egg has no cell wall, but a uniform wall is deposited over the entire egg surface within minutes after fertilization. Cell walls present 4 h after fertilization contain alginate, cellulose, and two different fucoidans (Fl and F3), each of which is a complex mixture of sulphated xylofucans. No asymmetry can be detected in the cell wall of this early embryo (Goodner and Quatrano 1993). A third fucoidan (F2) becomes sulphated within specialized Golgi-derived vesicles (F granules), starting about 8-1O h after fertilization. The F2 fucoidan is specifically stained by toluidine blue O. A vitronectin-like protein is also localized in F granules (Wagner et al. 1992).

Upon sulphation of F2, F granules are directed to a cortical target site localized at the position of subsequent polar growth. Their transport is probably microfilament-dependent as cytochalasin B disrupts the target site, prevents the localization of F granules, and prevents axis formation. Deposition of cell wall material from the F granules occurs at one pole while the zygote is still symmetrical in shape. The ensuing cell division in *Fucus* does not disrupt this asymmetry in the zygote cell wall, resulting in an embryo consisting of a rhizoid cell and a thallus cell, each with a unique type of cell wall. Only the rhizoid cell shows staining with toluidine blue O. The cell wall signal may be part of the mechanism to orientate the first asymmetric division of the embryo (Quatrano and Shaw 1997). The preservation of asymmetry may thus be related to the symplasmic interruption between the rhizoid cell and the thallus cell described by Brawley et al. (1977).

6 Concluding Remarks

Formation of cell wall is very varied in algae, and there is no strict correlation between cytokinesis type and formation of plasmodesmata. Plasmodesmata may not exist at all, they can be formed as primary plasmodesmata during cytokinesis and disappear later on, or they may form as secondary plasmodesmata after cell-wall formation. Their structure is different in different organisms.

In multicellular algae, the structure of plasmodesmata in different symplasm domains changes during ontogenesis which can be visible in EM. This proves that intercellular exchange is regulated by the plant and is controlled by supracellular morphogenetic programme similar to that in higher plants.

Acknowledgements. I would like to express my gratitude to Dr. Janusz Maszewski and Agnieszka Wojtczak MSc. for preparation of Fig. 7 and to Prof. Aart van Bel and Dr. Pim van Kesteren for voluable review remarks.

References

Bakker ME, Lokhorst GM (1987) Ultrastructure of mitosis and cytokinesis in *Zygnema* sp. (Zygnematales, Chlorophyta). Protoplasma 138:105–118
Brawley SH, Quatrano RS, Wetherbee R (1977) Fine structural studies of the gametes and embryo of *Fucus vesiculosus* L. (Phaeophyta) III. Cytokinesis and the multicellular embryo. J Cell Sci 24: 275–294
Broadwater ST, Scott JL, Goss SPA, Saunders BD (1995) Ultrastructure of vegetative organization and cell division in *Glaucosphera vacuolata* Korshikov (Porphyridiales, Rhodophyta). Phycologia 34: 351–361
Cook ME, Graham LE, Botha CEJ (1996) Toward an elucidation of the evolutionary origin of plant plasmodesmata, In: 3rd Int Worksh on basic and applied research in plasmodesmatal biology. Zichron-Gakov, Israel March 10–16, pp 26–29
Ducreux G (1968) Sur l'ultrastructure des apex des bourgeons, principal et axillaire de *Chara vulgaris*. C R Acad Sci Paris 267:163–166
Ducreux G (1975) Corrélations et morphogenèse chez le *Chara vulgaris* L. cultivé in vitro. Rev Gén Bot 82:215–357
Ducreux G (1979) Fonctionnement de l'apex et ontogenèse; essai d'analyse ultrastructurale chez *Chara vulgaris* L. Rev. Algol, NS 14:49–62
Fischer RA, Dainty J, Tyree MT (1974) A quantitative investigation of symplasmic transport in *Chara corallina*. I Ultrastructure of the nodal complex cell walls. Can J Bot 52:1209–1214
Fowke LC, Pickett-Heaps JD (1969) Cell division in *Spirogyra*. II Cytokinesis. J Phycol 5:273–281
Franceschi VR, Ding B, Lucas WJ (1994) Mechanism of plasmodesmata formation in characean algae in relation to evolution of intercellular communication in higher plants. Planta 192:347–358
Fraser TW, Gunning BES (1969) The ultrastructure of plasmodesmata in the filamentous green alga, *Bulbochaete hiloensis* (Nordst.) Tiffany. Planta 88:244–254
Fraser TW, Gunning BES (1973) Ultrastructure of the hairs of the filamentous green alga *Bulbochaete hiloensis* (Nordst.) Tiffany: an apoplastidic plant cell with a well-developed Golgi apparatus. Planta 113:1–19
Godlewski M, Kwiatkowska M (1980) Effect of gibberellic acid on the formation and development of antheridia and oogonia in *Chara vulgaris* L. Acta Soc Bot Pol 49:459–469
Goodner B, Quatrano RS (1993) *Fucus* embryogenesis: a model to study the establishment of polarity. Plant Cell 5:1471–1481
Hepler PK (1982) Endoplasmic reticulum in the formation of the cell plate and plasmodesmata. Protoplasma 111:121–133
Kaźmierczak A, Kwiatkowska M, Popłońska K (1999) GA_3 content in antheridia of *Chara vulgaris* in the proliferative stage and spermiogenesis estimated by capillary electrophoresis. Folia Histochem Cytobiol 37:49–52
Kouwets FAC (1995a) Ultrastructural aspects of the cell cycle in *Planktosphaeria gelatinosa* (Chlorophyceae), with emphasis on the flagellar apparatus of the zoospore. Arch Protistenkd 146:37–48

Kouwets FAC (1995b) Comparative ultrastructure of sporulation in six species of *Neochloris* (Chlorophyta). Phycologia 34:486–500
Kwiatkowska M (1988) Symplasmic isolation of *Chara vulgaris* antheridium and mechanisms regulating the process of spermatogenesis. Protoplasma 142:137–146
Kwiatkowska M (1991) Autoradiographic studies on the role of plasmodesmata in the transport of gibberellin. Planta 183:294–299
Kwiatkowska M (1995) Effect of symplasmic isolation and antigibberellin treatment on morphogenesis in *Chara*. Folia Histochem Cytobiol 33:133–137
Kwiatkowska M (1996) Changes in ultrastructure of cytoplasm and nucleus during spermiogenesis in *Chara vulgaris*. Folia Histochem Cytobiol 34:41–56
Kwiatkowska M, Malinowski S (1995) The influence of the disappearance of plasmodesmal connections between antheridia and thallus on ^3H-GA_3 transport. Folia Histochem Cytobiol 33:53–55
Kwiatkowska M, Maszewski J (1976) Plasmodesmata between synchronously and asynchronously developing cells of the antheridial filaments of *Chara vulgaris* L. Protoplasma 87:317–327
Kwiatkowska M, Maszewski J (1978) Ultrastructure of nuclei during different phases of the cell cycle in the antheridial filaments of *Chara vulgaris* L. Protoplasma 96:59–74
Kwiatkowska M, Maszewski J (1979) Changes in the activity of the Golgi apparatus during the cell cycle in antheridial filaments of *Chara vulgaris* L. Protoplasma 99:31–38
Kwiatkowska M, Maszewski J (1985) Changes in ultrastructure of plasmodesmata during spermatogenesis in *Chara vulgaris* L. Planta 166:46–50
Kwiatkowska M, Maszewski J (1986) Changes in the occurrence and ultrastructure of plasmodesmata in antheridia of *Chara vulgaris* L. during different stages of spermatogenesis. Protoplasma 132: 179–188
Kwiatkowska M, Maszewski J (1987) Ultrastructural and autoradiographic studies of non-generative cells in the antheridium of *Chara vulgaris*. I Manubria. Folia Histochem Cytobiol 25:215–223
Kwiatkowska M, Żylińska K (1988) Ultrastructural and autoradiographic studies of non-generative cells in the antheridium of *Chara vulgaris*. III Shield cells. Folia Histochem Cytobiol 26:41–51
Kwiatkowska M, Popłońska K, Żylińska K (1990) Biological role of endoreplication in the process of spermatogenesis in *Chara vulgaris* L. Protoplasma 155:176–187
Kwiatkowska M, Popłońska K, Gosek A, Malinowski S (1995) Circadian rhythm of protein synthesis in generative and non-generative antheridial cells of *Chara vulgaris* L. J Biol Rhythm Res 26:55–67
La Claire JW (1981) Occurrence of plasmodesmata during infurrowing in a brown alga. Biol Cell 40: 139–142
Malinowski S, Maszewski J (1994) DNA endoreplication, RNA and protein synthesis during growth and development of the antheridial basal cell in *Chara vulgaris* L. Folia Histochem Cytobiol 32: 137–142
Marchant HJ (1976) Plasmodesmata in algae and fungi. In: Gunning BES, Robards WA (eds) Intercellular communication in plants: studies on plasmodesmata. Springer Berlin Heidelberg New York, pp 59–80
Maszewski J, van Bel AJE (1996) Different patterns of intercellular transport of Lucifer Yellow in young and mature antheridia of *Chara vulgaris* L. Bot Acta 109:110–114
Mesquita JF, Santos MF (1984) Ultrastructural study of *Haematococcus lacustris* (Girod) Rostafinski (Volvocales). II Mitosis and cytokinesis. Cytologia 49:229–241
Mezitt LA, Lucas WJ (1996) Plasmodesmal cell-to-cell transport of proteins and nucleic acids. Plant Mol Biol 32:251–273
Pickett-Heaps JD (1967a) Ultrastructure and differentiation in *Chara* sp. I Vegetative cells. Aust J Biol Sci 20:539–551
Pickett-Heaps JD (1967b) Ultrastructure and differentiation in *Chara* sp. II Mitosis. Aust J Biol Sci 20:883–894
Pickett-Heaps JD (1968) Ultrastructure and differentiation in *Chara* (Fibrosa) IV Spermatogenesis. Aust J Biol Sci 21:655–690
Pickett-Heaps JD (1970) Mitosis and autospore formation in the green alga *Kirchneriella lunaris*. Protoplasma 70:325–347
Porter KR, Machado RD (1960) Studies on the endoplasmic reticulum. IV Its form and distribution during mitosis in cells of onion root tip. J Biophys Biochem Cytol 7:167–180
Quatrano RS, Shaw SL (1997) Role of the cell wall in the determination of cell polarity and the plane of cell divisions in *Fucus* embryos. Trends Plant Sci 2:15–21
Sawitzky H, Grolig F (1995) Phragmoplast of the green alga *Spirogyra* is functionally distinct from the higher plant phragmoplast. J Cell Biol 130:1359–1371
Sheath RT, Whittick A (1995) The unique gonimoblast propagules of *Batrachospermum breutelii* (Batrachospermales, Rhodophyta). Phycologia 34:33–38
Sheath RG, Whittick A, Cole KM (1994) *Rhododraparnaldia oregonica*, a new freshwater red algal genus and species intermediate between the Acrochaetiales and the Batrachospermales. Phycologia 33:1–7

Shepherd VA, Goodwin PB (1992) Seasonal patterns of cell-to-cell communication in *Chara corallina* Klein. ex Willd. II Cell-to-cell communication during the development of antheridia. Plant Cell Environ 15:151–162

Spanswick RM, Costerton JWF (1967) Plasmodesmata in *Nitella translucens*: structure and electrical resistance. J Cell Sci 2:451–464

Thiéry J (1967) Mise en évidence des polysaccharides sur coupes fines en microscopie électronique. J Microsc 6:987–1018

Trick HN, Pueschel CM (1990) Cytochemistry of pit plugs in *Bossiella californica* (Decaisne) Silva (Corallinales, Rhodophyta). Phycologia 29:403–409

Wagner VT, Brian L, Quatrano RS (1992) Role of a vitronectin-like molecule in embryo adhesion of the brown alga *Fucus*. Proc Natl Acad Sci USA 89:3644–3648

Wetherbee R, Quirk HM, Mallett JE, Ricker RW (1984) The structure and formation of host-parasite pit connections between the red alga alloparasite *Harveyella mirabilis* and its red alga host *Odonthalia floccosa*. Protoplasma 119:62–73

Chapter 13

The Symplasmic Organization of the Shoot Apical Meristem

C. van der Schoot[1] · P. Rinne[2,*]

[1] ATO-DLO, P.O. Box 17, NL-6700 AA Wageningen, The Netherlands
[2] Department of Virology, Agricultural University Wageningen, Binnenhaven 6, NL-6709 PD Wageningen, The Netherlands
* Permanent address: Department of Biology, University of Oulu, P.O. Box 3000, FIN-90401 Oulu, Finland

1 Introduction

The buildup of symplasmic organization has been a driving force in the evolution of multicellular plants. A symplasmic organization could arise when walled unicellular ancestors commenced subdividing their protoplasm by means of pore-bearing walls. By restricting the functional size of the pores, cells could develop physiological individuality without giving up cytoplasmic continuity. Further refinement of the pores, leading to the formation of true plasmodesmata, subsequently set the stage for the development of increasingly complex symplasmic organizations. Particularly the appearance of regulatory devices inside the pores, permitting a selective exchange of morphogenetic signals, might have been crucial in the evolution of higher plants (Carr 1976; Gunning and Robards 1976; Robards 1976; Lucas et al. 1993; Franceschi et al. 1994). Since all higher plants arise from the activity of morphogenetic units, the root and shoot meristems, the symplasmic organization of these meristems might play an important role in primary morphogenesis. The symplasmic organization of the shoot apical meristem (AM) channels the flow of signals dependent on the patterns in which cells divide and establish de novo contacts with neighbours. In the following, we will examine how symplasmic networking in the AM might have played a role in the emergence of higher plants, and how it functions in the coordination of morphogenesis.

2 The Shoot Apical Meristem: Structural Aspects

2.1 Evolutionary Designs

For proper development of the shoot, the AM must not only produce the cells required for growth and development, but also position them in such a way that the basic organization of the AM is maintained. The AM reconciles these tasks by confining the leaf primordia to its periphery, thereby keeping its centre unoccupied. Although this is most clearly so in higher plants, Newman (1965) pointed out that all multicellular plants of the major plant taxa have their growth centres somehow subdivided into a source region and a region of elaboration. The source is the ultimate origin of all cells, probably being identical to the initials. The region of elaboration constitutes the part of the meristem involved in the direct production of leaf buttresses.

In principle, the AM cells are spatially fixed with regard to their neighbours by an extracellular matrix (Roberts 1994), which limits their possibilities for setting up free collaborations. Since collaboration is essential for achieving a common morphogenetic goal, the pattern of cell divisions may have substantial influence on AM functioning. The division of the initial cells and their derivatives leads to specific lineage patterns. The number and position of the initial cells, and the way they restrict the plane of their divisions are dissimilar for the major taxa, i.e. ferns/lower vascular plants, gymnosperms, and angiosperms. This observation led Newman (1965) to classify their AMs as monoplex, simplex, and duplex, respectively (Fig. 1; see also discussions in Gifford and Corson 1971). The existence of taxonomic differences is informative since it indicates that the distinct AM types represent ontogenetically as well as phylogenetically robust organizations, and that each type of organization suffices for morphogenesis. The fact, however, that the major plant taxa possess shoots with distinct designs rais-

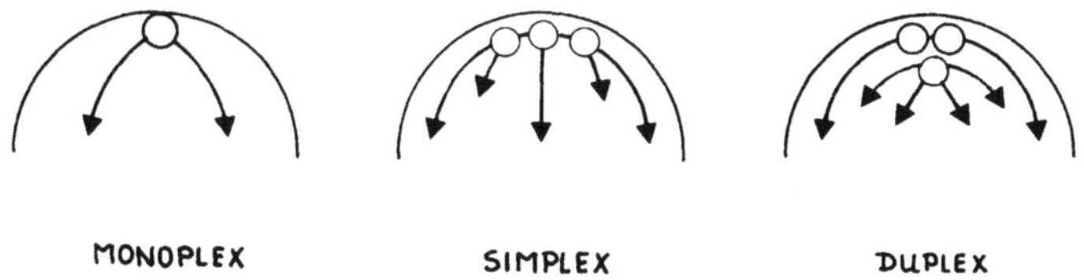

Fig. 1. Schematic representation of shoot apical meristem (AM) types: monoplex (ferns), simplex (gymnosperms) and duplex (angiosperms). In the monoplex AM, a single, usually tetrahedral, triangular initial cell gives rise to the entire shoot system. In the simplex AM, a superficial layer of initials is present. The duplex AM possesses two sets of initials which give rise to distinct growth zones. (After Newman 1965)

es an important question: to what extent does shoot design result from the type of AM organization? In the following we will therefore examine in greater detail the cell lineage patterns in two distinct AM types.

2.2 Monoplex and Duplex

The fern AM has a monoplex construction, i.e. the entire shoot system is derived from a single apical initial cell (Fig. 1). This single cell source is triangular in shape, and the specific way in which it divides by insertion of perpendicular walls to alternating sides of the triangle imposes a characteristic geometrical cell pattern on the emerging tissues (Raghavan 1989; Tilney et al. 1990; Cooke et al. 1996). Since only cytokinetically formed plasmodesmata are present in the monoplex apex, this pattern may be due to the necessity to maintain symplasmic continuity in the shoot system.

In contrast to ferns, the cells in the AM of angiosperms are produced in two structurally or geometrically distinct areas, termed tunica and corpus (Figs. 2A, 3A). This so-called duplex pattern results from the activity of two separate sets of initial cells

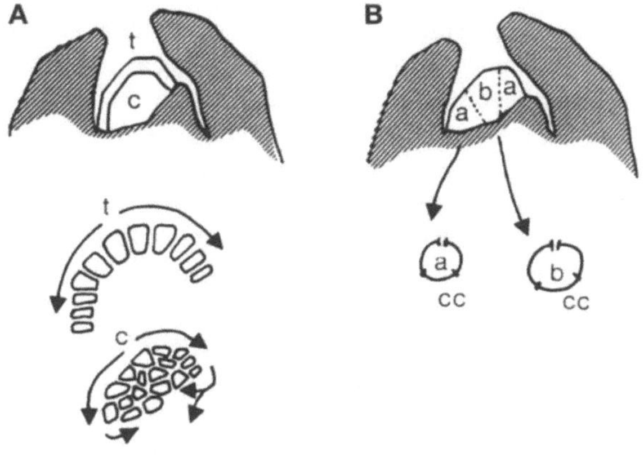

Fig. 2 A, B. Schematic representation of models defining morphogenesis at the shoot apical meristem (AM). **A** Structural organization of the AM reflects the direction of cell divisions (*arrows*), i.e. anticlinal in the tunica (*t*) and in all planes in the corpus (*c*). **B** Physiological organization is reflected in the cytohistological zonation. Metabolically active cells with high cell cycling (*cc*) are at the AM periphery (*a*) and metabolically less active cells with low cell cycling occupy the centre (*b*). (After Rinne and van der Schoot 1998)

with distinct division patterns (Figs. 1, 3A; Gifford and Corson 1971; Tilney-Bassett 1986; Steeves and Sussex 1989). The initials and their derivatives in the tunica show exclusively anticlinal divisions, thereby giving rise to a two-dimensional cell sheet that expands centrifugally. The corpus initials and their derivatives divide in all planes, resulting in the formation of a compact three-dimensional cell group (Tilney-Bassett 1986; Steeves and Sussex 1989; Sachs 1991). The expanding tunica is separated from the corpus by non-division walls which are continuously "stretched and plastered" in the horizontal plane (Roberts 1994). Within the tunica sheet and within the three-dimensional corpus symplasmic continuity of cell lineages is ensured by plasmodesmata formed in the cell plate. The periclinal walls between the tunica and the corpus do not, as a rule, possess cytokinetic plasmodesmata, and thus the numerous plasmodesmata present in this interface must have been made in a different way.

2.3 Symplasmic Unity

It may be postulated that the appearance of the duplex AM on the evolutionary stage required the exploitation of a second type of plasmodesmata, one which is formed through existing walls. The occurrence of non-cytokinetic plasmodesmata in angiosperms was proposed as early as 1901 by Strassburger (cited by Carr 1976), who observed that the radial walls of a secondary meristem, the cambium, do not arise as a result of mitosis and yet have plasmodesmata numbers similar to the walls originating from cell plates. Much later, specific cases of such "secondary plasmodesmata" (Jones 1976) in angiosperms were described for developing stem tissues (Seagull 1983; see Chap. 10), graft systems (e.g. Kollmann and Glockmann 1985), and fusing floral organs (van der Schoot et al. 1995). Their presence within the duplex AM, at the interface of tunica and corpus, has remained unnoticed until recently (van der Schoot 1994, 1996; Cooke et al. 1996; van der Schoot and Rinne 1996; Bergmans et al. 1997), although the situation is identical to that of the cambium. Our model shows, in addition, that within the layers and the corpus branches sideways are exclusively interconnected by secondary plasmodesmata (van der Schoot 1996; van der Schoot and Rinne 1996; Bergmans et al. 1997). Resumingly, the plasmodesmata at the tunica/corpus interface, and those between lineages and lineage branches, are all produced through existing walls as a normal and ongoing event during cellular proliferation at the AM. The formation of numerous plasmodesmata at every cell-cell interface of the AM, regardless of cell lineage and cell location (van der Schoot et al. 1991; van der Schoot 1996), is probably a general phenomenon in angiosperms.

The advantage of the subdivision of the AM into the two zones, tunica and corpus, was that their distinct growth dynamics allowed physical stresses to aid phyllotactic patterning by bulging and outgrowth of leaf primordia (e.g. Green 1992; Green et al. 1996). Due to the peripheral positioning of leaf primordia, not involving the central zone of initial cells and their youngest derivatives, the duplex AM gained a potential for indeterminate growth. In addition, secondary plasmodesmata permitted the presence of an entire zone of uncommitted cells (in the tunica as well as in the corpus) – instead of a single cell – in which the division frequency could become lower, thereby increasing the longevity of the AM. In contrast, ferns may be limited by the presence of a single initial cell which after a restricted number of divisions becomes senescent, resulting in a progressive decrease of plasmodesmata production and, eventually, in cessation of growth (Tilney et al. 1990). Resumingly, the presence of secondary plasmodesmata permitted the introduction of a tunica, which provided the AM of angiosperms with unique properties.

2.4 Tunica

The activity of the AM has been characterized as extended embryogenesis (Sachs 1991). From this perspective, the outer wall of the "extended embryo" bears the mark of the zygote since this wall derives directly from it, although stretched and plastered in an endless way (Roberts 1994). All other walls are created as internal walls which "septate" the original zygote (Kaplan and Hagemann 1991; Roberts 1994). One distinctive feature of these outer tunica walls is the complete absence of plasmodesmata (Bergmans et al. 1997). Although plasmodesmata frequently arise in non-division walls and even in single walls, which results in half-plasmodesmata ending blind at the middle lamella (Jones 1976; Ehlers and Kollmann 1996), the outer tunica wall forms no plasmodesmata. (Early discussions on the possible existence of epidermal ecto-plasmodesmata have been reviewed by Carr, 1976.) The lack of plasmodesmata in the outer wall of the tunica, and the presence of exclusively secondary plasmodesmata at the opposite wall, assign a unique polarity to the tunica (Bergmans et al. 1997). It may polarize the organization of the plasmamembrane, membrane-associated cytoskeletal elements, and the cortical ER/plasmodesmata complexes of the entire symplasmic continuum within the AM. This may, for example, facilitate the deposition of materials at the outer walls and the cuticle, and the import and export of macromolecules (see Sect. 3.3).

Since the embryonic wall has been shown to influence cell fate (Berger et al. 1994), it seems feasible that the outer, zygote-derived tunica wall functions somehow as a reference for cellular differentiation (Bergmans et al. 1997). The outer tunica wall may also function as a mechanical framework for the AM due to its relative thickness, whereas the fragile anticlinal walls may serve the spatial partitioning of the protoplasm of the mother cells (Fig. 3A). Although phyllotactic patterning is assisted by biophysical forces in the tunica (e.g. Green 1992; Selker et al. 1992; Green et al. 1996), the outgrowth and further development of the leaf primordia requires continuous input of morphogens. Since both, the signalling mechanisms that are utilized in spacing mechanisms such as phyllotactic patterning and those which regulate the dorsiventral development of leaves, act in a radial fashion between the AM centre and the primordia (e.g. Sussex 1952, 1989; Metford et al. 1992), they may resign in the tunica. While serving these purposes, the radially extending tunica layer could have conserved ancient signalling mechanisms, which traffic exclusively through primary, i.e. cytokinetic plasmodesmata.

2.5 Plasmodesmata: Loci and Modifications

In the duplex AM symplasmic unity was guaranteed by progressive ways to unite all cells symplasmically. However, a functional subdivision could be achieved if the two

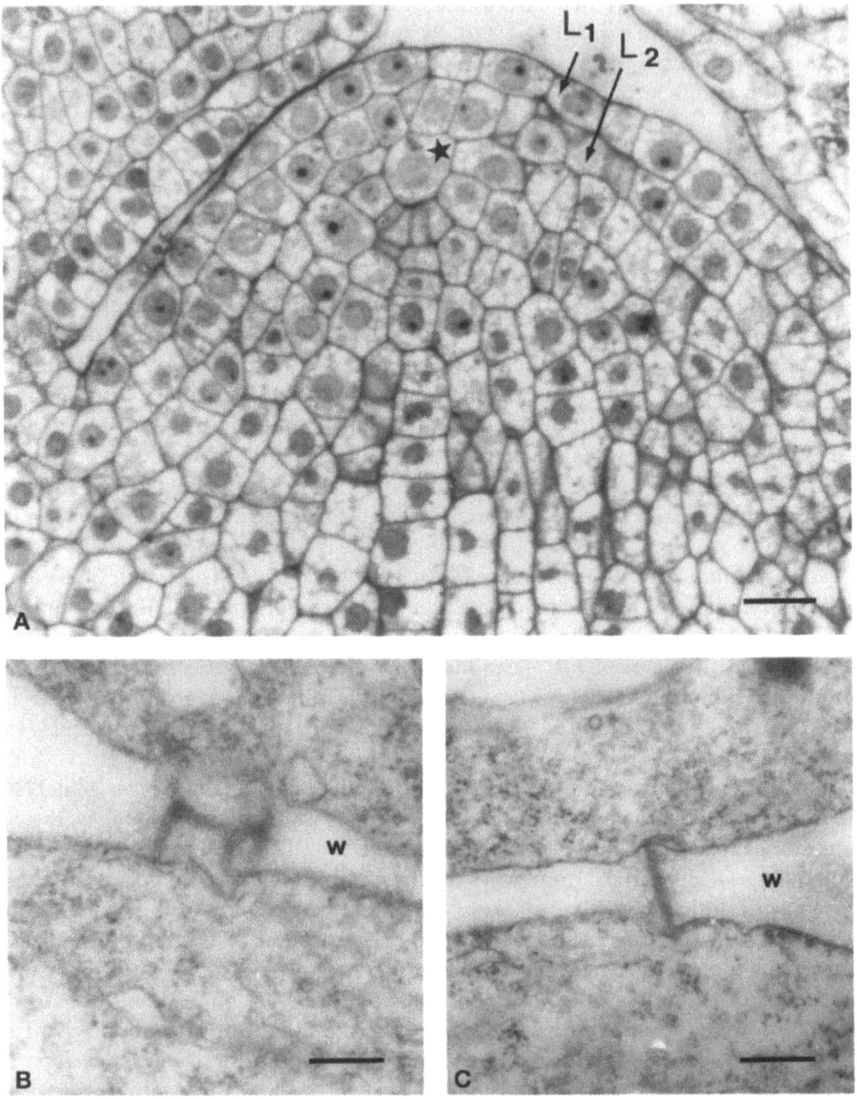

Fig. 3A–C. The inflorescence meristem (IM) of the monocot *Iris hollandica* and plasmodesmata therein. **A** A median longitudinal section through the IM shows a two-layered tunica superimposed on a corpus. Note the radiation of cell files from a corpus initial (*); L_1 outer tunica layer; L_2 second tunica layer. **B** Branched plasmodesmata, formed by the interconnection of three existing linear secondary plasmodesmata in a periclinal wall (*w*), between L_1 and L_2. **C** Linear plasmodesmata (secondary or primary) in an anticlinal wall (*w*) of the L_1. Bar equals 25 µm (**A**) and 0.25 µm (**B, C**). (Bergmans et al. 1997)

types of plasmodesmata differed in their ability to transfer morphogenetic substances or metabolites from cell-to-cell. For example, signals that are generated in the tunica and traffic through primary plasmodesmata only would be confined to the tunica as their trafficking to the subtending corpus would be prevented by the lack of primary plasmodesmata at this interface. However, even when their transfer abilities were identical, a specific distribution of plasmodesmata types within the AM could influence signal flow. This is due to the fact that primary and secondary plasmodesmata are produced by different mechanisms and in different locations, so that their frequencies might be altered independently during development. In the inflorescence meristem of *Iris* this scenario is not followed, however, and the plasmodesmal frequencies of primary and secondary plasmodesmata decrease simultaneously during the transition to floral meristem (Bergmans et al. 1997). The downregulation of plasmodesmata-forming mechanisms was restricted to the second cell layer within the AM (Fig. 4), showing that the regulation of plasmodesmal frequencies can be locus- rather than type-dependent. Such frequency changes may suffice to alter the flow of morphogenetic signals and thereby morphogenesis, since even the channelling of simple metabolites affects the morphogenetic potential of the various AM areas (Sussex 1952; Steeves and Sussex 1989; Sachs 1991).

It is commonly assumed that branching modifies plasmodesmal function, and therefore, it is important to establish whether it occurs in the AM. Observations indi-

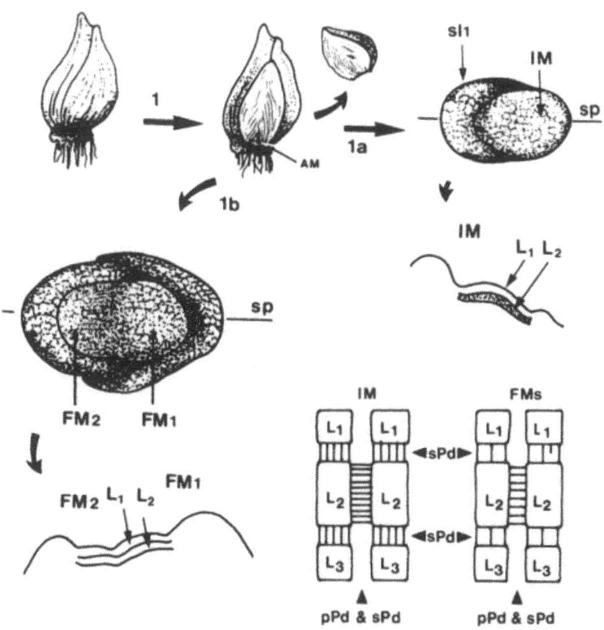

Fig. 4. An *Iris* bulb cut open longitudinally (*1*) shows the presence of the shoot apical meristem (*AM*) at the bottom of the bulb. The AM develops over time (*1a*) into an inflorescence meristem (*IM*) with a single spathe leaf (sl_1), and further (*1b*) into a double floral meristem (*FM1* and *FM2*). Schema for longitudinal sections are taken at sp (section plane). Plasmodesmata in the AM showed strong frequency shifts at specific boundaries. A modified plasmodesmogram (*bottom right*) visualizes the frequency changes occurring during the transition from IM to FM. *pPd* Primary plasmodesmata; *sPd* secondary plasmodesmata. Within the first tunica layer (L_1) and within the corpus (L_3), frequencies remained the same. In contrast, plasmodesmal frequencies among the second tunica layer (L_2), between L_2 and L_3, and between L_2 and L_1 decreased with 25, 40 and 25%, respectively. (Bergmans et al. 1997)

cate that plasmodesmata in the AM can be branched, although the majority of them is linear. In the inflorescence meristem of *Iris*, for example, virtually all plasmodesmata are linear (Fig. 3B, C; Bergmans et al. 1997). Similarly, in the floral meristem of *Helianthus annuus* L. only few branched plasmodesmata are present (Sawhney et al. 1981). In potato (*Solanum tuberosum* L.), however, the plasmodesmata between tunica and corpus tend to branch, whereas the plasmodesmata formed within layers are almost exclusively linear (C. van der Schoot, unpub. results). The lack of branched plasmodesmata in most meristems indicates that branched plasmodesmata are not required for AM functioning. The occasional branching of plasmodesmata in the AM may simply reflect the response of the aligning cells to unequal wall stretching (Fig. 5) and the necessity to maintain the symplasmic unity of the meristem. The tendency to branching between tunica and corpus and the lack of it within the tunica would then indicate that differential wall stretching is stronger between tunica and corpus than within the tunica.

It has been speculated that branched plasmodesmata have a special ability to traffic so-called informational molecules, while linear plasmodesmata do not have this ability (Ding et al. 1992, 1993). There seems to be no reason, however, why the macromolecular complex intrinsic to branched plasmodesmata could not be present in linear plasmodesmata (primary or secondary). The plasmodesmata which transport the necessary signals for the fusion of flower carpels also do not seem to require exclusive-

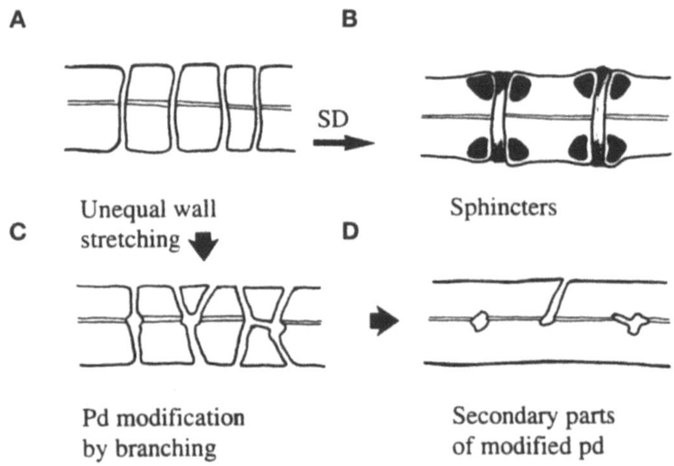

Fig. 5A–D. Schematic representation of plasmodesmata modifications in the angiosperm AM. **A** Cell-cell interfaces cemented together by a middle lamella. Plasmodesmata are linear and may represent either primary plasmodesmata formed in a cell plate or secondary plasmodesmata formed at the interface of non-doughter cells. **B** Short photoperiod (*SD*) triggers mechanisms that result in modification of plasmodesmata as observed in the birch AM (Rinne and van der Schoot 1998). The structural modifications resemble sphincters (cf. Olesen and Robards 1990) and serve the temporary symplasmic isolation of all AM cells. **C** Unequal stretching of periclinal tunica walls leads to morphological modifications of existing plasmodesmata. Plasmodesmata may form small median cavities, new branches, or interconnect two or three existing plasmodesmata. **D** Projection (from **C**) of the newly constructed parts of modified plasmodesmata. In case such additions have different functional properties, this in most cases will not lead to alterations in cell-to-cell transfer, since the new parts do not cross both cell walls entirely. Pd, plasmodesmata ly linear or branched plasmodesmata, since both morphotypes of secondary plasmodesmata are present (van der Schoot et al. 1995).

Another modification of the plasmodesmata which may occur in the AM is the formation of valve-like structures at the plasmodesmata entrances (Fig. 5; van der Schoot 1994, 1996; Rinne and van der Schoot 1998). Earlier, these structures, found in a variety of other plant parts, were depicted as flux-regulating sphincters (Olesen and Robards 1990) or as defence structures formed during a stress response (Turner et al. 1994). It has been shown with microinjection techniques that in the AM of birch they function as valves which physically close the plasmodesmata, resulting in the symplasmic isolation of all AM cells (Rinne and van der Schoot 1998). The production of valves on the plasmodesmata appears to be an integral part of the dormancy mechanism (discussed in Sect. 3.6). It remains to be seen if their occurrence is a more universal response against environmental stress.

3 Shoot Apical Meristem: Supracellular Dynamics

3.1 Physiological Barriers

Newman's classification of AM types (1965) is based on differences in the cellular construction patterns (Fig. 1). Equally important is a model describing the AM in terms of physiological patterns, known as the cytohistological zonation-model (Fig. 2B; Steeves and Sussex 1989). In most seed plants, cytohistological zones are present in the AM, which are discernible by their staining properties, organelle distributions and cell cycling times (Steeves and Sussex 1989; Sachs 1991). The zones are present as a central disc and a peripheral ring, which include several superimposed layers. A further subdivision can be found within the central zone. For example, in sunflower, the tunica layer of the central zone has a cytohistological appearance that is distinct from the underlying corpus layers within the central zone (Sawhney et al. 1981). The distinct cell groups and subgroups suggest the presence of multiple physiological barriers inside the AM (Sawhney et al. 1981; Bergmans et al. 1993). Recent evidence shows that the barriers in the tunica may result from the narrowing of the plasmodesmal channels at specific locations between the peripheral and the central area of the AM and between the tunica and the corpus (see Sect. 3.5; Rinne and van der Schoot 1998).

The physiological barriers match the division of tasks within the AM, and may separate the processes of primordium formation and self-renewal (van der Schoot 1996; van der Schoot and Rinne 1996). It would be of interest to establish whether and how this physiological subdivision relates to gene expression during morphogenesis and morphogenetic switching.

3.2 Gene Expression

During vegetative growth, gene expression patterns in the AM are relatively stable, and no major developmental shifts occur. The spatially restricted activation of homeotic genes such as *knox3* and *rs1* in maize (Jackson et al. 1994) and *nam* in petunia (Souer et al. 1996) may serve the specification of primordia and their separation from the AM centre, respectively. The expression domain of the homeobox gene *knotted 1* in maize, important for indeterminate growth, is also restricted and does not include the tunica layer and the future primordia sites as judged from in situ hybridization experiments (Jackson et al. 1994). However, the KNOTTED 1 protein is detected in the tunica, and is proposed to be imported from lower cell layers. Some other spatial differences occur in the expression of "house-hold" genes (Fleming et al. 1992, 1993; Pri-Hadash et al. 1992). Their differential expression in tunica or corpus, central or peripheral zone, reflects differences in metabolism, possibly serving the division of tasks in the AM.

The area-wise expression of various regulatory genes in the vegetative meristem may be altered during the transition to flowering (e.g. Kelly et al. 1990; Pri-Hadash et al. 1992; Menzel et al. 1996). For example, in mustard (*Sinapis alba*) the expression of the MADS box genes SAMADS A and SAMADS B is restricted to the central zone in the vegetative meristem, but after the transition it spreads out to the peripheral zone (Menzel et al. 1996). This might reflect the disappearance of the cytohistological zonation pattern and a corresponding change in the symplasmic organization of the AM. Concommitantly, the AM alters its gene expression patterns to canalize he further development of a flower by a sequential expression of regulatory genes (e.g. Coen and Meyerowitz 1991; Weigel and Clark 1996).

An important problem, which deserves attention, is how the patterns are maintained since the AM cells are collectively displaced towards the periphery, and their gene expression, cell cycling and cytological characteristics change accordingly. The persistence of these patterns implies that the cells continuously communicate their position to each other and to newcomers. Although little is known about the kind of signals that are exchanged in these interactions, recent speculations assign an important role to macromolecular transfer in development.

3.3 Macromolecular Trafficking

The transcription factor KNOTTED 1, a homeodomain protein, is suggested to move selectively through plasmodesmata in the AM of maize and facilitate also the movement of its own mRNA (Jackson et al. 1994; Lucas et al. 1995). Fluorescently tagged KNOTTED 1, microinjected into mesophyll cells of maize and tobacco, moved to all neighbouring cells, indicating that it moved through primary and secondary plasmodesmata. Movement of KNOTTED 1, therefore, seems independent of plasmodesmata type (primary, secondary), tissue status (meristematic, mature) and plant taxa (the monocotyledon maize and the dicotyledon tobacco), and demonstrates lack of selectivity. It is possible, of course, that the plasmodesmata in the mesophyll are not representative, since their major function is assimilate transfer, and that in meristematic tissues such selectivity plays a role. In the AM, the protein, but not its mRNA, is present in the tunica, whereas both are absent from the tunica derived leaf buttress (Lucas et al. 1995). Since both, tunica and leaf buttress, are connected by secondary plasmodesmata to KNOTTED 1 producing cells, selectivity may be due to physiological and physical factors that determine whether specific functions of plasmodesma are actually executed.

Also in the floral meristem macromolecular trafficking through plasmodesmata could possibly occur. This might be inferred from the observation that certain genes function non-cell-autonomously, and their downstream effects are triggered also in neighbouring cells that do not express the gene themselves (Becraft 1995). In some cases, like *deficiens* (*def*) and *globosa* (*glo*) in *Antirrhinum*, this interaction is polar in the sense that lower layers affect cells in the tunica but not vice versa (Perbal et al. 1996). In other cases, like that of *floricaula* (*flo*) in *Arabidopsis*, interactions are opposite, i.e. from the tunica towards the corpus (Furner et al. 1996). In case of FLO it is unclear whether the protein itself traffics between the tunica and the corpus (e.g. Jackson et al. 1994), or whether as yet unidentified factors mediate this non-cell-autonomous gene expression (Hake and Freeling 1986; Jackson et al. 1994; Carpenter and Coen 1995; Hantke et al. 1995; Furner et al. 1996; Perbal et al. 1996). FLO or its mediating factors may be transferred between the meristem layers via the direct symplasmic route, without passage through the cell membrane (Jackson et al. 1994; Mezitt and Lucas 1996), or diffuse through the apoplasmic space (Becraft 1995; Furner et al. 1996). The polar interaction in case of *globosa* is thought to result from the trafficking of the GLOBOSA (GLO) itself, moving via secondary plasmodesmata towards the tunica (Perbal et al. 1996). Within the tunica lateral spread of GLO was severely restricted. This was attributed to the assumptive absence of secondary plasmodesmata. However, a symplasmetric map, visualizing the distribution of plasmodesmata types in the meristem, shows that many secondary plasmodesmata are present within the tunica (Fig. 6; Bergmans et al. 1997), indicating that in principle all cells can be reached via secondary plasmodesmata. Therefore, only an set of parallel cell lineages in part of a bisemetrical meristem may have restricted movement of GLO in one direction. Macromolecular transfer through plasmodesmata in the AM thus seems feasible. It remains to be dem-

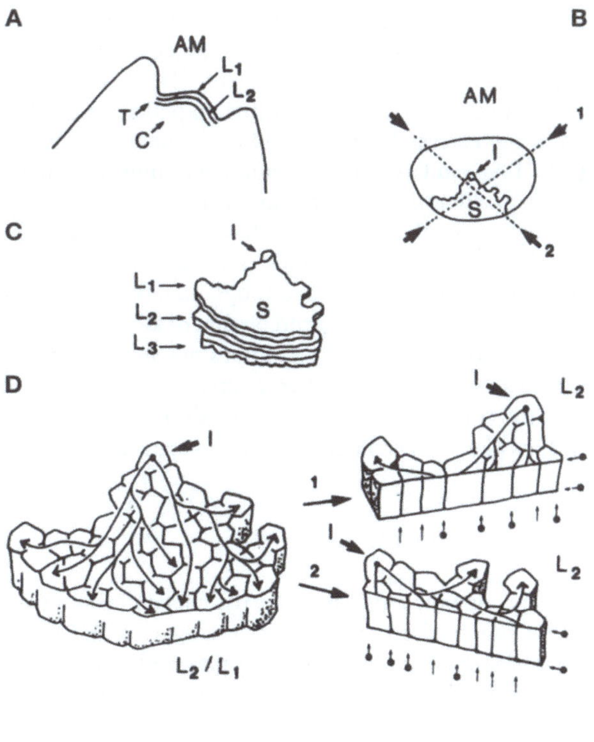

Fig. 6A–D. Construction of a symplasmetric map of the duplex shoot apical meristem (*AM*) of angiosperms. **A** Longitudinal section through the apex. *C* Corpus; *T* two-layered tunica (L_1, L_2). **B** Surface projection of the AM with initial cell (*I*), derived sector (*S*), and two longitudinal sections through the sector (*arrows 1 and 2*). **C** Reconstruction of layered sectors; L_1, L_2 tunica layers; L_3 corpus; *I* initial cell; *S* derived sector. **D** Symplasmetric map representing a sector in L_2 (in L_1 outer-wall plasmodesmata are absent). As an example, two cross-sections (*arrows 1, 2* in **B**) through layer L_2 are depicted with anticlinal and periclinal cell interfaces which possess exclusively primary (*small arrow*) or secondary (*small dotted arrow*) plasmodesmata. Note that the plasmodesmata are exclusively secondary at L_2/L_1 and between L_2 and corpus. (Bergmans et al. 1997)

onstrated, however, whether GLO, DEF and FLO themselves move from cell-to-cell instead of mediating factors and, if so, through which pathways.

Understanding how the complex interactions involved in morphogenesis lead to an integrated output requires a concept of the AM as a whole. Such concepts are available from, for example, traditional embryology which explained morphogenesis in terms of dynamic systems and morphogenetic fields.

3.4 A Morphogenetic Field

The AM resembles the morphogenetic field of animal systems (Holder 1979; Rinne and van der Schoot 1998), which is defined as "an area whose cells can cooperatively bring about a distinct set of structures" (Müller 1997). The AM, like the morphogenetic fields in animals (de Robertis et al. 1991; Müller 1997), has remarkable regenerative and regulative capacities (e.g. Steeves and Sussex 1989; Sussex 1989). For animal systems these properties have been explained in a variety of ways (see e.g. Müller 1997). According to the positional information model (PI model), gradients of morphogens specify cellular differentiation (e.g. Wolpert 1969, 1989). Gradients may be composed of immobile substances, fixed either inside the cells or at the extracellular matrix. Alternatively, gradients may be formed by morphogens which diffuse from a source to a sink region. The cells in the source-to-sink paths may consume morphogen, and thereby decrease the size of the morphogenetic field. The presence of such gradient fields prepatterns the morphogenetic field (Crick 1970; Müller 1997) and permits cells to establish their position with regard to the boundaries of the field. Dependent on their developmental history cells interpret gradients and record them as positional values. Collectively, these values could function as a tissue memory, used for regeneration and further development (Holder 1979; Wolpert 1969, 1989).

In animals and plants the diffusion paths of morphogens are assumed to be situated in the extracellular space, but gap junctions (Wolpert 1978) and plasmodesmata (Pitts 1990) suit as well the formation of morphogen gradients in a symplasmic space. In the proliferating AM cells are displaced towards the periphery and may then pass through sequential states. If each cell broadcasts signals in correspondence to its state, a process of active symplasmic conversation would continuously update the information on cellular positions in a cooperatively generated field. The behaviour of an individual cell would then depend on that of all other cells, a condition which goes beyond the PI model, however, the AM does not show a gradual change in its cellular characteristics, but instead a rather discrete zonation pattern. Like indicated earlier (Sect. 3.1), the transient closure of plasmodesmata at specific demarcation lines within the AM may keep physiologically distinct areas apart.

3.5 Symplasmic Fields

Evidence is now emerging that the coordination of morphogenesis requires the dynamic subdivision of the symplasmic space of the AM into diffusion compartments by the position-dependent closure of plasmodesmata, (van der Schoot et al. 1991; van der Schoot and Rinne 1997; Rinne and van der Schoot 1998). Symplasmic compartments were first found in mature tissues of *Elodea* and were referred to as "symplastic domains" (Erwee and Goodwin 1985). They serve the physiological stabilization of tissues by the internal redistribution of metabolites (van der Schoot and van Bel 1990; Lucas et al. 1993). Since the AM is different from mature tissues in that it is continuously renewing its cellular composition, these "domains" may be better envisaged as "fields" (van der Schoot and Rinne 1997; Rinne and van der Schoot 1998). This would also be in line with the embryological concepts (see e.g. Holder 1979). The symplasmic fields can be visualized in the actively proliferating AM by iontophoretic microinjection of Lucifer Yellow CH (LYCH), a fluorescent membrane-impermeable probe (Rinne and van der Schoot 1998). When injected into a single cell, its movement to neighbouring cells reveals the symplasmic pathways open for the diffusion of small metabolites, hormones and other regulators. In the tunica of birch and potato, the distinct symplasmic fields corresponded in position to the peripheral part and the central part of the AM (Fig. 7). The symplasmic fields appeared to be dynamic, and they showed spatiotemporal alterations which corresponded to morphogenetic events at the AM (van der Schoot and Rinne 1996, 1997; Rinne and van der Schoot 1998). These fields harbour gradients of potential morphogens small enough to move through plasmodesmata by diffusion (up to 1 kDa). The membrane potentials (E_m) recorded from the individual

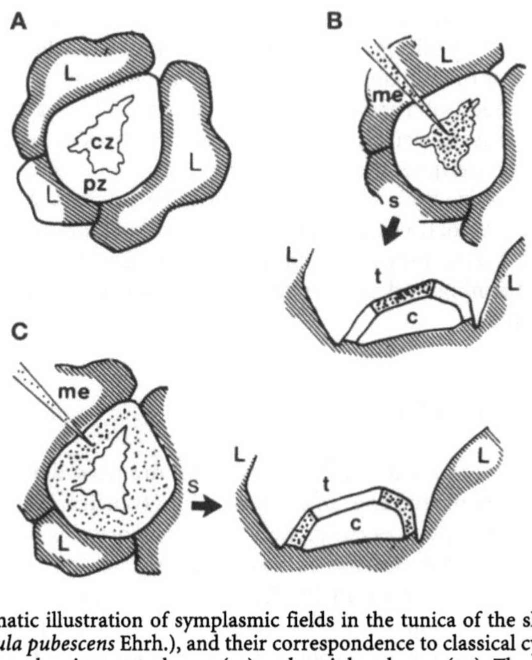

Fig. 7A–C. A schematic illustration of symplasmic fields in the tunica of the shoot apical meristem (AM) in birch (*Betula pubescens* Ehrh.), and their correspondence to classical cytohistological zones. **A** AM in surface view showing central zone (*cz*) and peripheral zone (*pz*). The precise outline of the border was revealed by microinjection experiments shown in **B** and **C**. Symplasmic fields were visualized by the cell-to-cell movement of LYCH (*dotted areas*), iontophoretically injected into a single tunica cell in the centre (**B**) or periphery (**C**) of the AM. The symplasmic field at the pz corresponded to the organogenetic ring, and the symplasmic field at the cz corresponded to the uncommitted cells at the AM centre; *s* longitudinal section through the injected AMs showing the presence of LYCH in the tunica (*t*) cells; *c* corpus; *l* leaf primordia; *me* microelectrode. (After Rinne and van der Schoot 1998)

AM cells were field-specific, demonstrating that these fields also function as metabolic working units. The plasmodesmata between the symplasmic fields and towards the primordia and the corpus were closed for LYCH. The size of the central symplasmic field was rather constant, even though the constituent cells proliferate. This means that plasmodesmata closure between the central and peripheral field is transient and position-dependent, and under the control of a yet hypothetical mechanism anchored in the initial cells. The initial cells may therefore organize the central fields of the tunica and the corpus, which collectively may represent a shoot organizer (van der Schoot and Rinne 1996; Rinne and van der Schoot 1998).

Some mutants may provide access to the mechanism which are involved in formation of symplasmic fields. The petunia mutant *no apical meristem* (*nam*) is defective in a gene, the product of which demarcates the boundaries between the primordia and the AM proper (Souer et al. 1996), and NAM may hence be involved in either the establishment of boundaries between symplasmic fields or in mediating field-field interactions. Similarly, the *Arabidopsis* mutants *Fey* and *Dip* which early in development loose the ability to maintain a central undifferentiated AM area (Medford et al. 1992), might be impaired in the mechanism which closes the plasmodesmata at the field boundaries.

The symplasmic fields harbour signal networks and may therefore be depicted as circuits. Signal coupling of these separate circuits can be achieved by the exchange of macromolecular signals after transient gating of the closed plasmodesmata, or by the exchange of lipophilic signalling molecules via the ER (Grabski et al. 1993). Cells, which are forced out of their position, will join a new subcircuit and their differentiation becomes redirected. Alternatively, cells of the central tunica field, for example, may be released towards the periphery, and thereby carry specific information into an expanding part of the peripheral field. Upon symplasmic integration into this area, they may start to organize their surroundings in order to create a primordium (van der Schoot and Rinne 1996; Rinne and van der Schoot 1998). Resumingly, the AM may harbour distinct but overlapping signal circuits, which serve the execution of the conflicting tasks of patterning and self-maintenance, and which may also function in phyllotactic patterning.

3.6 Symplasmic Field Transitions

Since the AM responds to environmental signals with dramatic developmental changes, e.g. dormancy development and flower formation, transitions in the symplasmic network of the AM are likely to occur as well. For dormancy this has been confirmed by the demonstration that the patterns of the symplasmic fields in the AM change during development of the dormant state in birch (Rinne and van der Schoot 1998) and potato (C. van der Schoot, unpub.). In birch, the symplasmic fields disappeared prior to visible cessation of growth in response to short photoperiod, suggesting a link between phytochrome action and symplasmic uncoupling of AM cells. This was due to physical blockage of all plasmodesmata in the entire AM by the formation of callose-containing sphincters (Fig. 5), thereby preventing any form of symplasmic communication. How phytochrome action relates to mechanisms that control plasmodesmata remains to be established. In the dormant potato AM, the majority of the plasmodesmata was closed by sphincters, but part was widened due to the removal of the ER (van der Schoot 1996). The latter permitted the cell-to-cell movement of LYCH-dextran molecules whose size prevents its movement through normal plasmodesmata. Such an increase in the size exclusion limit of plasmodesmata possibly creates a partly syncytial state, which renders communication ineffective due to loss of selectivity. These two examples of symplasmic non-communication during dormancy development demonstrate the sensitivity of the symplasm to environmental factors, and they support the assumption that symplasmic communication is indispensable in morphogenesis.

There is no experimental evidence so far that the symplasm alters its overall organization during the transition to flowering. Transformation of an AM into an inflorescence meristem does not alter the zonation pattern (e.g. Bernier 1962). This is expected, since an undifferentiated centre is retained as long as the inflorescence develops. Therefore, the inflorescence meristem may also possess a central and a peripheral field. When a terminal flower is formed, zonation disappears and the meristem centre is used up (Takimoto 1969). The disappearance of the zonation pattern during the transition may indicate that the symplasmic fields are reorganized and gradually used up. We suggest that a stable central symplasmic field is characteristic of an open growth process; when the AM (or inflorescence meristem) commits itself to end differentiation, becoming a flower or a thorn, the central symplasmic fields of the tunica and corpus give up their sovereignty.

4 Concluding Remarks

The various types of shoot AM – monoplex, simplex and duplex – and the specific ways in which they organize the symplasm by cell divisions, are dependent on the capacity to unite non-daughter cells by secondary plasmodesmata. Phylogenetically, the capacity to form plasmodesmata through existing walls permitted the duplex AM construction, found in all angiosperms. The emerging picture of the duplex AM is that of a dynamically organized unit, which orchestrates primary morphogenesis and simultaneously maintains itself. The implementation of these conflicting tasks is achieved by subdividing the AM by means of plasmodesmata closure into physiologically distinct symplasmic compartments, which might interact via transiently gating plasmodesmata. The challenge for the future will be to elucidate how the AM, as a symplasmically integrated whole, maintains local differences in cell physiology and gene expression, and how it alters them in response to signals from the rest of the plant or the environment during development.

Acknowledgements. We thank R.W. Goldbach (Agricultural University Wageningen) and A.D. de Boer (ATO-DLO, Wageningen) for support. The Academy of Finland is acknowledged for an 'International Cooperation Grant' to P. R.

References

Becraft PW (1995) Intercellular induction of homeotic gene expression in flower development. Trends Genet 11:253–255

Berger F, Taylor A, Brownlee C (1994) Cell fate determination by the cell wall in early *Fucus* development. Science 263:1421–1423

Bergmans A, de Boer D, van Bel A, van der Schoot C (1993) The initiation and development of *Iris* flowers: permeability changes in the apex symplasm. Flowering Newsl 16:19–26
Bergmans ACJ, de Boer AD, Derksen JWM, van der Schoot C (1997) The symplasmic coupling of L_2-cells diminishes in early floral development of *Iris*. Planta 203:245–252
Bernier G (1962) Evolution of the apical meristem of *Sinapis alba* L. (long-day plant) in long days, in short days and during the transfer from short days to long days. Caryologia 15:303–325
Carpenter R, Coen ES (1995) Transposon induced chimeras show that *floricaula*, a meristem identity gene, acts non-autonomously between cell layers. Development 121:19–26
Carr DJ (1976) Plasmodesmata in growth and development. In: Gunning BES, Robards AW (eds) Intercellular communication in plants: studies on plasmodesmata. Springer, Berlin Heidelberg New York, pp 243–289
Coen ES, Meyerowitz EM (1991) The war of the whorls: genetic interactions controlling flower development. Nature 353:31–36
Cooke TJ, Tilney MS, Tilney LG (1996) Plasmodesmatal networks in apical meristems and mature structures: geometric evidence for both primary and secondary formation of plasmodesmata. In: Smallwood M, Knox JP, Bowles DJ (eds) Membranes: specialized functions in plants. Bios Scientific Publishers, Oxford, pp 471–488
Crick FHC (1970) Diffusion in embryogenesis. Nature 225:546–548
de Robertis EM, Morita EA, Cho KWY (1991) Gradient fields and homeobox genes. Development 112:669–678
Ding B, Haudenshield JS, Hull R, Wolf S, Beachy RN, Lucas WJ (1992) Secondary plasmodesmata are specific sites of localization of the tobacco mosaic virus movement protein in transgenic tobacco plants. Plant Cell 4:915–928
Ding B, Haudenshield JS, Willmitzer L, Lucas WJ (1993) Correlation between arrested secondary plasmodesmal development and onset of accelerated leaf senescence in yeast acid invertase transgenic tobacco plants. Plant J 4:179–189
Ehlers K, Kollmann R (1996) Formation of branched plasmodesmata in regenerating *Solanum nigrum*-protoplasts. Planta 199:126–138
Erwee MG, Goodwin PB (1985) Symplast domains in extrastellar tissues of *Egeria densa* Planch. Planta 163:9–19
Fleming AJ, Mandel T, Hofmann S, Sterk P, de Vries SC, Kuhlemeier C (1992) Expression patterns of a tobacco lipid transfer protein gene within the shoot apex. Plant J 2:855–862
Fleming AJ, Mandel T, Roth I, Kuhlemeier C (1993) The patterns of gene expression in the tomato shoot apical meristem. Plant Cell 5:297–309
Franceschi VR, Ding B, Lucas WJ (1994) Mechanism of plasmodesmata formation in characean algae in relation to evolution of intercellular communication in higher plants. Planta 192:347–358
Furner IJ, Ainscough JF-X, Pumfrey JA, Petty LM (1996) Clonal analysis of the late flowering fca mutant of *Arabidopsis thaliana*: cell fate and cell autonomy. Development 122:1041–1050
Gifford EM, Corson GE (1971) The shoot apex of seed plants. Bot Rev 37:143–229
Grabski S, de Feijter AW, Schindler M (1993) Endoplasmic reticulum forms a dynamic continuum for lipid diffusion between contiguous soybean root cells. Plant Cell 5:23–35
Green PB (1992) Pattern formation in shoots: a likely role for minimum energy configurations of the tunica. Int J Plant Sci Suppl 153:59–75
Green PB, Steele CS, Rennich SC (1996) Phyllotactic patterns: a biophysical mechanism for their origin. Ann Bot 77:515–527
Gunning BES, Robards AW (1976) Plasmodesmata: current knowledge and outstanding problems. In: Gunning BES, Robards AW (eds) Intercellular communication in plants: studies on plasmodesmata. Springer, Berlin Heidelberg New York, pp 297–311
Hake S, Freeling M (1986) Analysis of genetic mosaics shows that the extra epidermal cell divisions in *Knotted* mutant maize plants are induced by adjacent mesophyll cells. Nature 320:621–623
Hantke SS, Carpenter R, Coen ES (1995) Expression of *floricaula* in single cell layers of periclinal chimeras activates downstream homeotic genes in all layers of floral meristems. Development 121: 27–35
Holder N (1979) Positional information and pattern formation in plant morphogenesis and a mechanism for the involvement of plant hormones. J Theor Biol 77:195–212
Jackson D, Veit B, Hake S (1994) Expression of maize KNOTTED *1* related homeobox genes in the shoot apical meristem predicts patterns of morphogenesis in the vegetative shoot. Development 120: 405–413
Jones MGK (1976) The origin and development of plasmodesmata. In: Gunning BES, Robards AW (eds) Intercellular communication in plants: studies on plasmodesmata. Springer, Berlin Heidelberg New York, pp 81–105
Kaplan DR, Hagemann W (1991) The relationship of cell and organism in vascular plants. Are cells the building blocks of plant form? BioScience 41:693–703

Kelly AJ, Zagotta MT, White RA, Chang C, Meeks-Wagner DR (1990) Identification of genes expressed in the tobacco shoot apex during the floral transition. Plant Cell 2:963–972

Kollmann R, Glockmann C (1985) Studies on graft unions. I. Plasmodesmata between cells of plants belonging to different unrelated taxa. Protoplasma 124:224–235

Lucas WJ, Ding B, van der Schoot C (1993) Plasmodesmata and the supracellular nature of plants. New Phytol 125:435–476

Lucas WJ, Bouché-Pillon S, Jackson DP, Nguyen L, Baker L, Ding B, Hake S (1995) Selective trafficking of KNOTTED 1 homeodomain protein and its mRNA through plasmodesmata. Science 270: 1980–1983

Medford JI, Behringer FJ, Callos JD, Feldmann KA (1992) Normal and abnormal development in the *Arabidopsis* vegetative shoot apex. Plant Cell 4:631–643

Menzel G, Apel K, Melzer S (1996) Identification of two MADS box genes that are expressed in the apical meristem of the long-day plant *Sinapis alba* in transition to flowering. Plant J 9:399–408

Mezitt LA, Lucas WJ (1996) Plasmodesmal cell-to-cell transport of proteins and nucleic acids. Plant Mol Biol 32:215–273

Müller WA (1997) Developmental biology. Springer, Berlin Heidelberg New York

Newman IV (1965) Pattern in the meristems of vascular plants. III. Pursuing the patterns in the apical meristem where no cell is a permanent cell. Linn Soc (Bot) 59:185–214

Olesen P, Robards AW (1990) The neck region of plasmodesmata: general architecture and functional aspects. In: Robards AW, Lucas WJ, Pitts JD, Jongsma HJ, Sprey DC (eds) Parallels in cell-to-cell junctions in plants and animals. Springer, Berlin Heidelberg New York, pp 145–170

Perbal M-C, Haughn G, Seadler H, Schwarz-Sommer Z (1996) Non-cell-autonomous function of the *Antirrhinum* floral homeotic proteins *DEFICIENS* and *GLOBOSA* is exerted by their polar cell-to-cell trafficking. Development 122:3433–3441

Pitts JD (1990) Junctional communication: the role of communication compartments in complex multicellular organisms. In: Robards AW, Lucas WJ, Pitts JD, Jongsma HJ, Sprey DC (eds) Parallels in cell-to-cell junctions in plants and animals. Springer, Berlin Heidelberg New York, pp 53–62

Pri-Hadash A, Hareven D, Lifschitz E (1992) A meristem-related gene from tomato encodes a dUTPase: analysis of expression in vegetative and floral meristems. Plant Cell 4:149–159

Raghavan V (1989) Developmental biology of fern gametophytes. Cambridge University Press, Cambridge

Rinne P, van der Schoot C (1998) Symplasmic fields in the tunica of the shoot apical meristem coordinate morphogenetic events. Development 125:1477–1485

Robards AW (1976) Plasmodesmata in higher plants. In: Gunning BES, Robards AW (eds) Intercellular communication in plants: studies on plasmodesmata. Springer, Berlin Heidelberg New York, pp 15–53

Roberts K (1994) The plant extracellular matrix: in a new expansive mood. Curr Opin Cell Biol 6: 688–694

Sachs T (1991) Pattern formation in plant tissues. Cambridge University Press, Cambridge

Sawhney VK, Rennie PJ, Steeves TA (1981) The ultrastructure of the central zone cells of the shoot apex of *Helianthus annuus*. Can J Bot 59:2009–2015

Seagull RW (1983) Differences in the frequency and disposition of plasmodesmata resulting from root-cell elongation. Planta 159:497–504

Selker JML, Steucek GL, Green PB (1992) Biophysical mechanisms for morphogenetic progressions at the shoot apex. Dev Biol 153:29–43

Souer E, van Houwelingen A, Kloos D, Mol J, Koes R (1996) The *no apical meristem* gene of petunia is required for pattern formation in embryos and flowers and is expressed at meristem and primordia boundaries. Cell 85:159–170

Steeves TA, Sussex IM (1989) Patterns in plant development. Cambridge University Press, Cambridge

Sussex IM (1952) Regeneration of the potato shoot apex. Nature 170:755–757

Sussex IM (1989) Developmental programming of the shoot meristem. Cell 56:225–229

Takimoto A (1969) *Pharbitis nil* Chois. In: Evans LT (ed) The induction of flowering. Some case histories. Macmillan, Melbourne, pp 90–115

Tilney LG, Cooke TJ, Connelly PS, Tilney MS (1990) The distribution of plasmodesmata and its relationship to morphogenesis in fern gametophytes. Development 110:1209–1221

Tilney-Bassett RAE (1986) Plant chimeras. Edward Arnold, London

Turner A, Wells B, Roberts K (1994) Plasmodesmata of maize root tips: structure and composition. J Cell Sci 107:3351–3361

van der Schoot C (1994) The role of symplasmic organization and cell-cell communication in dormancy. In: Lang GA (ed) 1st Int Symp on Plant Dormancy, 7–10 August 1994, Oregon State University, Corvallis, Oregon, p 30

van der Schoot C (1996) Dormancy and symplasmic networking at the shoot apical meristem. In: Lang GA (ed) Plant dormancy. CAB Int, Wallingford, pp 59–81

van der Schoot C, Rinne P (1996) Symplasmic networking and autopoiesis at the shoot apical meristem. In: Proc 10ème Réunion Groupe d'Etude de l'Arbre, 16–17 April 1996, Angers, pp 36–55
van der Schoot C, Rinne P (1997) Symplasmic networking at the shoot apical meristem. In: van Montagu M, Micol JL (eds) Worksh on plant morphogenesis, Madrid 20–22 Oct 1997. Instituto Juan March de Estudios e Investigaciones. Centro de Reuniones Internacionales Sobre Biología, no 72, pp 51–52
van der Schoot C, van Bel AJE (1990) Mapping membrane potential differences and dye-coupling in internodal tissues of tomato (*Solanum lycopersicum* L). Planta 182:9–21
van der Schoot C, Drent P, Thiel F, Clerkx A, Boekestein B (1991) Functional structure of the symplast in resting apices of *Solanum tuberosum* L. Plant Physiol Suppl 96:154
van der Schoot C, Dietrich MA, Storms M, Verbeke JA, Lucas WJ (1995) Establishment of a cell-to-cell communication pathway between separate carpels during gynoecium development. Planta 195: 450–455
Weigel D, Clark SE (1996) Sizing up the floral meristem. Plant Physiol 112:5–10
Wolpert L (1969) Positional information and the spatial pattern of cellular differentiation. J Theor Biol 25:1–47
Wolpert L (1978) Gap junctions: channels for communication in development In: Feldman J, Gilula NB, Pitts JD (eds) Intercellular junctions and synapses. Receptors and recognition, series B, vol 2. Chapman and Hall, London, pp 83–96
Wolpert L (1989) Positional information revised. Development Suppl 3–12

CHAPTER 14

The Physiological and Developmental Consequences of Plasmodesmal Connectivity

K. EHLERS · A. J. E. VAN BEL

Institut für Allgemeine Botanik und Pflanzenphysiologie,
Justus-Liebig-Universität Giessen, Senckenbergstrasse 17,
D-35390 Giessen, Germany
FAX: +49-641-99-35119
E-mail: Katrin.Ehlers@bot1.bio.uni-giessen.de

1 Introduction

As early as 1930, Münch realized the paramount significance of the symplasmic cell connections for the intercellular communication and formulated the classic concept of the continuous plant symplasm. According to Gunning (1976), plasmodesmata can be regarded as the structures which elevate a plant organism from a mere collection of individual cells to an interacting community of living protoplasts. Moreover, Lucas et al. (1993) emphasized the central role of plasmodesmal connectivity, trafficking, and regulation in physiological or developmental processes. The latter authors raised the issue that plants might be organized as supracellular, rather than multicellular organisms.

Indeed, the correct timing of the developmental processes and a coordinated physiological functioning of the mature plant in its entirety require the concerted action of the different cell types which can hardly be understood without intensive intercellular communication. At present, it is common opinion that communication between plant cells largely depends on the existence of plasmodesmata which allow electrical signalling (see Chap. 4), diffusion of small molecules (see Chap. 5), and presumably even the exchange of informational macromolecules like mRNA and proteins (see Chap. 11).

With certainty, on the other hand, a plant organism does not represent one single symplasmic network with unlimited plasmodesmal connectivity and communication. In order to guarantee the differential functioning of cell types and tissues their autonomy must be warranted to a certain degree. It is to be expected that the amount of symplasmic connectivity at the different cell interfaces is precisely controlled corresponding with the particular functional demands. Not all the cells within the plant organism are symplasmically interconnected in the same way (see Chaps. 12, 13). Plasmodesmal connectivity differs depending on the particular tissue type or on the developmental state of the plant. Given the diverse tasks of the different types of cells and tissues, it is likely that their communication and transport channels have to meet a multitude of specific requirements as well. Thus, it seems reasonable to assume that the distinct cell interfaces within the plant organism vary with respect to their plasmodesmal connectivity and the structural and functional properties of their cell connections (cf. Robinson-Beers and Evert 1991).

Over the past decade, evidence has been accumulated that the plant organism has a complex symplasmic organization. As structural and functional analyses suggest, the plant symplasm is obviously subdivided into distinct operational subunits and functions as a system of cell clusters, the so-called symplasmic domains (Erwee and Goodwin 1985). The cells within such a domain are symplasmically interconnected by plasmodesmata, show intensive communication, and cooperate as a functional unit. However, no symplasmic exchange occurs across the domain borders, as cells belonging to different domains are symplasmically isolated from each other. The distinct symplasmic domains are equipped to fulfil their respective function and therefore may differ with respect to the frequencies and the functional peculiarities of their plasmodesmata. Evidence is at hand that symplasmic domains are often established permanently, due to irreversible modifications of the cell connections (see Lucas et al. 1993; McLean et al. 1997). Particularly in the course of developmental processes, reversible modifications of plasmodesmata have also been observed which might lead to the formation of temporary symplasmic domains (e.g. Kwiatkowska and Maszewski 1985, 1986; van Bel and van Rijen 1994; Ehlers and Kollmann 1996b).

The intention of this chapter is to summarize the available information on the symplasmic organization of the plant organism with special emphasis on the physiological and developmental consequences of the plasmodesmal connectivity, the functional variability of plasmodesmata, and the occurrence of symplasmic domains.

2 The Mature Plant: Plasmodesmal Frequencies, Functional Variability of Plasmodesmata, and the Occurrence of Symplasmic Domains

In order to estimate the plasmodesmal connectivity and to calculate the functional transport capacity between the plant cells, numerous publications have dealt with structural analyses of the plasmodesmal frequencies, which normally vary between about 0.1 and 10.0 μm^{-2} cell wall interface (reviewed in Robards 1976; Robards and Lucas 1990). The rationale of counting plasmodesmata is very simple. The different frequencies of plasmodesmata between the distinct types of cells or tissues might mirror a functional capacity. The higher the number of plasmodesmata observed at a given interface, the higher the potential for symplasmic transport and intercellular communication might be. On the other hand, a small number or an absence of cell connections might indicate a bottleneck in symplasmic continuity or even symplasmic isolation.

In an attempt to visualize the available transport pathways within the plant tissues and to allow a quick interpretation of the potential symplasmic connectivity, tabulated data on plasmodesmal frequencies were transformed into two-dimensional diagrammatic presentations, the so-called plasmodesmograms (van Bel et al. 1988; Botha and van Bel 1992). In these diagrams, the striping density between individual cell types reflects the respective plasmodesmal frequencies (see Fig. 1).

However, some general problems should be considered when calculations of the symplasmic connectivity are based on the plasmodesmal frequency data (Fisher 1990; Botha 1992; Botha and van Bel 1992; see also Robards and Lucas 1990; van Bel 1992, 1993a; van Bel and Oparka 1995). One major difficulty is to measure the plasmodesmal frequencies with high accuracy. Quantifications based on the structural analyses of a few ultrathin sections are influenced by various factors like cell arrangement, cell size and cell shape, and are therefore often imprecise (Fisher 1990; Botha 1992; Botha and van Bel 1992). Moreover, plasmodesmal distribution is not homogeneous in most cases, as the cell connections are aggregated in fields, which usually results in large standard deviations in the counts of plasmodesmata (Botha 1992; Botha and van Bel 1992; van Bel 1993a). Thus, small variations in plasmodesmal frequency can hardly be detected.

Another problem is that the apparent symplasmic connectivity may differ substantially depending on the unit of frequency that was chosen for the calculation (Fisher 1990; cf. also Botha 1992; Botha and van Bel 1992). In order to facilitate the comparison of different tissues or of similar tissues of distinct plant species, plasmodesmal frequencies have often been calculated using relative units. The plasmodesmal frequencies have been computed, e.g. as the percentage of the total number of plasmodesmata at each cell interface or as the plasmodesmal density, i.e. the number of plasmodesmata μm^{-1} or μm^{-2} cell interface (van Bel et al. 1988). However, these calculations have been criticized because they do not necessarily reflect the actual symplasmic connectivity between the cells (for detailed discussion see Fisher 1990). A reliable estimation of the effective symplasmic connectivity can only be deduced from the absolute plasmodesmal frequencies, i.e. from the total number of plasmodesmata at the entire cell interface.

Plasmodesmal frequencies and functional transport capacities have been calculated for a number of highly specialized cell or tissue types, e.g. for the root endodermis (Robards and Clarkson 1976) or root hairs (Kurkova and Vakhmistrov 1984), for several glands and nectary trichomes (e.g. Gunning and Hughes 1976; Eleftheriou and Hall 1983; Sawidis et al. 1987; Robards and Stark 1988; reviewed in Eleftheriou 1990), for wood rays (Sauter and Kloth 1986), and for maturing fibres of developing cotton seeds (Ryser 1992; for reviews and other examples, see also Robards 1976; Robards and Lucas 1990). It was concluded that the number of plasmodesmata observed at the respective interfaces is adequate to cope with the symplasmic transport capacity required (reviewed in Robards and Lucas 1990; see also Gunning and Robards 1976). Plasmodesmal transport capacities of 1 to 8×10^{-19} mol plasmodesma^{-1} s^{-1} have, for example, been calculated for the cells engaged in the short-distance transport of sugars (Sauter and Kloth 1986). Van Bel (1993a) pointed out that, generally, special care should be taken when computing functional transport rates on the basis of plasmodesmal frequency data. It should be taken into account that neither the total diameter of the plasmodesmal cylinder, nor the entire cytoplasmic sleeve is available for symplasmic transport. Actual symplasmic transport presumably takes place through the eight to ten microchannels of about 2–3 nm which were left between the globular particles situated in the cytoplasmic sleeve (e.g. Overall et al. 1982; Terry and Robards 1987; Ding et al. 1992). Thus, computations which are based on the total optical diameter of plasmodesmata (20–80 nm) might overestimate the effective symplasmic transport rate up to 50-fold (van Bel 1993a).

Another fundamental point of criticism should be considered. Predictions of symplasmic transport capacities based only on plasmodesmal frequency data implicitly presume that the transport capacity of all plasmodesmata is roughly the same. Thus, the validity of the computations depends on the premise that plasmodesmata are uniform (van Bel 1992, 1993a; van Bel and Oparka 1995). In fact, however, there is a growing body of evidence that plasmodesmata are dynamic structures and that different types of cell connections vary substantially with respect to structural and/or functional properties (reviewed in Robards and Lucas 1990; Lucas et al. 1993; Epel 1994).

For example, the complex pore/plasmodesma units (PPUs) in the phloem tissue evidently have specialized structural and functional properties, although they develop from simple plasmodesmata (see Chap. 15). The molecular exclusion limit (MEL) of the PPUs was shown to lie between 10 and 40 kDa (Kempers and van Bel 1997) and is therefore considerably higher than the basal MEL or the basal size exclusion limit (SEL) of normal plasmodesmata, which are usually in the order of 0.8 to 1 kDa (reviewed in Robards and Lucas 1990; Lucas et al. 1993). The high MEL of the PPUs was supposed to be a functional adaptation serving the assumed protein transport between the companion cells and the enucleate mature sieve elements (see Chap. 16). The highly branched cell connections that develop during the sink-to-source transition in the course of leaf development in different plant species were also thought to differ from simple plasmodesmata in having special functional abilities (see Chaps. 10, 15). The specialized plasmodesmata might be needed to adjust the available symplasmic pathways to the new transport requirements occurring during the developmental reorganization of the leaf symplasm. Similarly, specialized functional properties have been suggested for the different types of plasmodesmata in the leaves of monocotyledons, particularly for the structurally complex plasmodesmata traversing the suberin lamella that often occurs in the walls of either the mestome sheath in C_3 plants or the bundle sheath in C_4 plants (see Chap. 15). Special attention has also been paid to the presence of conspicuous sphincter structures that were supposed to function in the regulation of plasmodesmal permeability (reviewed in Olesen and Robards 1990; Robards and Lucas 1990; also Robinson-Beers and Evert 1991).

Yet, plasmodesmata with specialized functional capacities do not necessarily show striking structural peculiarities. Biochemical and molecular studies have revealed plant organ- and tissue-specific differences in the composition of plasmodesmata (reviewed in Epel 1994) which might indicate functional plasmodesmal variability. Direct evidence for a diversity in plasmodesmal permeability has been presented by Erwee and Goodwin (1985), who showed that the basal SEL of the plasmodesmata varies slightly between different plant organs and tissues of *Egeria* in the range of about 0.4 and 0.7 kDa. The differences in SEL were found to correlate partly with variations in the radius of the plasmodesmal neck regions. A basal SEL well above the normal size of 0.8 kDa has been reported for the plasmodesmata of the vascular parenchyma cells in the crease regions of wheat grain (Wang and Fisher 1994). Similarly, Waigmann and Zambryski (1995) have shown a basal SEL of even 7 kDa for the plasmodesmata in tobacco trichome cells. On the other hand, evidence has been given that apparently normal plasmodesmata might be non-functional as well. The presence of plasmodesmata at a cell interface obviously does not necessarily imply actual symplasmic connectivity. Dye-coupling studies on the nectary trichomes of *Abutilon* (Terry and Robards 1987) revealed that dye-coupling occurs between all trichome cells. However, the injected fluorochrome never moves from the basal cell of the trichome into the cells of the underlying nectary, although more than 10 plasmodesmata μm^{-2} wall were observed at this particular interface (Gunning and Hughes 1976). Similar functional transport barriers, at sites where remarkable numbers of apparently normal plasmodesmata were present, showed up at five defined cell interfaces in the extrastelar tissues of *Egeria* (Erwee and Goodwin 1985) and at the interface between maturing guard cells of *Allium* and *Commelina* and the surrounding cells (see Palevitz and Hepler 1985). In addition, Epel et al. (1992) reported that movement of IAA is strongly polar within the mesocotyl of corn seedlings in spite of the approximately equal plasmodesmal frequencies at all cell interfaces in this tissue.

Besides permanent differences in plasmodesmal permeability, transient changes in plasmodesmal functioning have been observed. Gating of plasmodesmata, i.e. reversible widening or closure of the cell connections, may occur in response to endogenous signals or as a reaction to changes in the cell environment (reviewed in Robards and Lucas 1990; Lucas et al. 1993; Epel 1994; van Bel and Oparka 1995; see Chap. 11). The reversible regulation of the plasmodesmal permeability may be of such importance that gating of plasmodesmata in parenchymatous tissues might be more decisive for determining intercellular transport than the plasmodesmal frequencies (Epel et al. 1992; cf. also van Bel and Oparka 1995).

All these findings on the diversity and the dynamics of plasmodesmal functioning illustrate that estimation of the actual rate of symplasmic connectivity inferred from the number of plasmodesmata is a risky enterprise. Plasmodesmal frequency data may be helpful for detecting putative intercellular transport routes because abundance of plasmodesmata at least indicates a potential symplasmic pathway. The actual functional transport capacity, however, should be proved additionally by functional analyses like dye-coupling experiments or electrophysiological studies (see van Bel and Oparka 1995; Botha and Cross 1997). Yet, it should be noted that these methods also have limitations with respect to the assessment of the in vivo functioning (e.g. van der Schoot and van Bel 1990; Oparka and Prior 1992).

In contrast, lack of plasmodesmata at a particular cell interface can be interpreted as an unmistakable indication of a symplasmic discontinuity between the cells. Full symplasmic isolation, leading to the formation of symplasmic domains, was assumed to be indispensable to the functioning of certain specialized cell types. Latex production probably profits from the symplasmic discontinuity between laticifers and the adjacent phloem and ray parenchyma cells (De Faÿ et al. 1989). The highly specialized guard cells are a more prominent example. As has been shown in structural studies and dye-coupling experiments (Wille and Lucas 1984; Erwee et al. 1985; Palevitz and Hepler 1985; see Lucas et al. 1993), the mature guard cell pair is symplasmically fully isolated from the adjoining epidermal or subsidiary cells and presumably functions as an independent symplasmic domain. Supposedly, the symplasmic isolation is essential for the controlled ion fluxes that are required for the regulation of the osmotic potentials responsible for the reversible closure of the stomatal pore. Remarkably, symplasmic isolation has also been reported for the modified stomata of the floral nectary of *Vicia faba*, which presumably cannot regulate the pore aperture (Davis and Gunning 1992).

As for the detection of symplasmic domains, the probably best-studied tissues are those involved in phloem transport. The low plasmodesmal frequencies which were observed between the sieve element/companion cell-complexes (SE/CC complexes) and the adjacent phloem parenchyma (PP) cells in the stem transport phloem of different plant species (Hayes et al. 1985; van Bel and Kempers 1990; Kempers et al. 1998) have been interpreted as an indication of symplasmic discontinuity at this particular interface (see van Bel 1993b, 1996; van Bel and Kempers 1996). This assumption found support by functional studies such as dye-coupling experiments (van Bel and Kempers 1990; Oparka et al. 1992; van Bel and van Rijen 1994; Rhodes et al. 1996) and electrophysiological analyses, which revealed striking differences in membrane potentials between the SE/CC complex and the adjacent PP cells (van der Schoot and van Bel 1989; van Bel and Kempers 1990; van Bel and van Rijen 1994; van Bel 1996). Supposedly, adjoining SE/CC complexes, interconnected longitudinally via the sieve plates or laterally via the PPUs, form a multicellular symplasmic domain that is isolated from the surrounding PP cells (van Bel and Kempers 1990; see van Bel 1993b). To date, excised tissues have mostly been used for functional studies on the transport phloem, but this symplasmic isolation has also been observed recently to occur in intact plants (Oparka et al. 1994; Knoblauch and van Bel 1998; but see Patrick and Offler 1996). The symplasmic discontinuity might play an important role in the dual physiological functioning of the stem transport phloem. Firstly, in controlling the release of photosynthates along the phloem pathway, which is required for the growth and maintenance of the heterotrophic stem tissues including the cambium. Secondly, in facilitating the carrier-mediated retrieval of photosynthates along the phloem pathway in order to maintain the pressure gradient and to nourish the terminal sinks (see van Bel 1993b; van Bel and Oparka 1995; also van Bel 1996 for a more elaborate discussion). Moreover, it has been discussed that gating of the sporadic plasmodesmata at the interface between the SE/CC-complex domain and the PP cells might allow gradual changes in symplasmic (dis)continuity and might therefore possibly influence the balance between sugar release and retrieval (see Lucas et al. 1993; van Bel and Oparka 1995; Patrick and Offler 1996; van Bel 1996).

Special attention has also been paid to the phloem-loading zone in the leaves where assimilates are transported from the photosynthesising mesophyll into the sieve elements (SE) of the minor veins (for detailed discussion see Chap. 15). Yet, it remained a matter of debate as to whether the mode of phloem loading is universal among all plant species. Gamalei (1985, 1989) observed that the plasmodesmal frequencies at the decisive interface between the SE/CC complexes and the adjacent phloem parenchyma or bundle-sheath cells differ by a factor 10^3 in the phloem-loading zone of different plant species (<0.1 to >10 plasmodesmata μm^{-2} interface). An abundance of plasmodesmata indicates a continuous symplasmic transport pathway from the mesophyll to the SE and would therefore demand only one symplasmic domain. Paucity or absence of plasmodesmata, however, points to the existence of at least two symplasmic domains. Thus, an apoplasmic step would be required in the transport of photosynthates (see Chap. 15).

The well-examined interface between SE/CC complex and adjacent cells is only part of the whole phloem-loading pathway extending from the most distal mesophyll cells to the SEs in the minor veins. In principle, symplasmic discontinuity, which marks the borderline between different functional domains might occur at any other site on the way from the mesophyll to the SEs. Therefore, studies on the symplasmic connectivity have been extended to the entire phloem-loading zone (see the literature cited in van Bel et al. 1988; Fisher 1990; Botha 1992; Botha and van Bel 1992; cf. also van Bel 1992, 1993a for discussions about the experimental limitations). On the basis of plasmodesmal frequencies, a correlation between the symplasmic connectivity in the mesophyll tissues and the respective modes of phloem loading seems to emerge (Fig. 1). In *Populus* (Russin and Evert 1985), *Coleus* (Fisher 1986; Fig. 1A), *Cananga* (Fisher 1990), and *Moricandia* (Beebe and Evert 1992), the plasmodesmal densities between the mesophyll elements were found to be markedly higher than in *Solanum* (McCauly and Evert 1989), *Spinacia* (Warmbrodt and VanDerWoude 1990), and *Vicia* (Bourquin et al. 1990; Fig. 1B). The first group includes representatives of families which are classified as symplasmic phloem loaders, except for *Moricandia (Brassicaceae)*, while the species of the second group belong to families with an apoplasmic minor vein configuration. This suggests a limited symplasmic transport in the entire phloem-loading zone of the apoplasmic species.

A few studies have examined the occurrence of symplasmic domains using entire plant organs or even plant organisms. Erwee and Goodwin (1985) were the first to discover symplasmic domains in the extrastelar tissues of the *Egeria* plant. Van der Schoot and van Bel (1989, 1990), who carried out a systematic screening of membrane potential differences in tomato stem tissues, revealed the existence of cell clusters showing distinct membrane potentials. This finding suggested electric isolation between the adjacent cell clusters and indicated symplasmic discontinuity between different symplasmic domains, as plasmodesmata are channels for electric conduc-

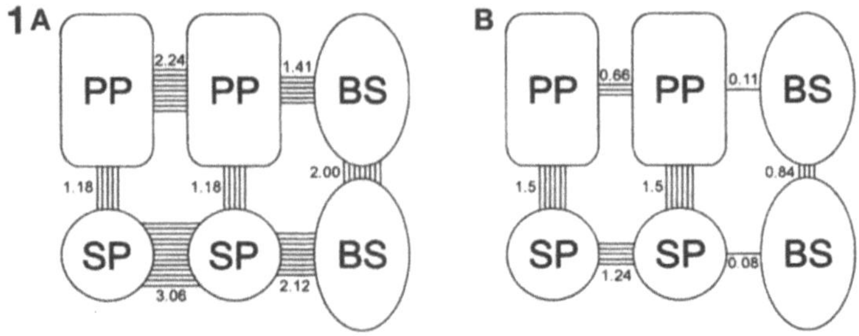

Fig. 1A, B. Plasmodesmograms of the zones around the minor veins in *Coleus* (**A** Fisher 1986) and *Vicia* (**B** Bourquin et al. 1990) which are transformations of tabulated data. The striping density between the cells reflects the respective plasmodesmal density (number of plasmodesmata μm^{-2} cell interface). *Coleus* (**A**), a species with symplasmic phloem loading, shows high plasmodesmal densities between the mesophyll elements which points to a well-equipped symplasmic transport pathway in the whole phloem-loading zone. Low plasmodesmata densities occur between the mesophyll cells of the apoplasmic phloem loader *Vicia* (**B**), which might indicate a limited symplasmic transport in the entire phloem-loading zone. *PP* Palisade parenchyma; *SP* spongy parenchyma; *BS* bundle sheath

tance (Overall and Gunning 1982). The symplasmic domains identified by differences in membrane potentials turned out to correspond with those detected by subsequent dye-coupling experiments (van der Schoot and van Bel 1990).

3 Plant Development: Changes in Plasmodesmal Frequencies and the Occurrence of Symplasmic Domains

During plant development, modifications of the symplasmic organization might require restructuring of symplasmic communication and transport pathways. This transformation probably correlates with changes in the plasmodesmal frequency due to the de novo formation of secondary plasmodesmata (Kollmann and Glockmann 1991; see Chap. 10) or to the disappearance of cell connections in the course of developmental processes (see Lucas et al. 1993).

Although different mechanisms have been discussed for the de novo formation of plasmodesmata in preexisting cell walls (cf. Jones 1976; Lucas et al. 1993; Lucas and Gilbertson 1994; and Monzer 1990, 1991; Kollmann and Glockmann 1991; Ehlers and Kollmann 1996a), there is general agreement on the fact that the formation of secondary plasmodesmata is integrated in normal plant development (for reviews see Jones 1976; Robards and Lucas 1990; Lucas et al. 1993). Based on structural and functional observations, van der Schoot et al. (1995) showed that secondary plasmodesmata are formed during the gynoecium development of *Catharanthus*, establishing a new cell-to-cell communication pathway between the fusing carpels. Schnepf and Sych (1983) and Seagull (1983) reported that plasmodesmal densities obviously do not decline as a result of cell elongation and wall expansion during the development of *Sphagnum* leaflets or in the course of root development in different species. Thus, secondary plasmodesmata formation must have been involved in the maintenance of the existing communication pathways. In contrast, Gunning (1978) found no evidence for the formation of secondary plasmodesmata in growing *Azolla* roots. He observed that each of the newly produced mesophytes forms less plasmodesmal contacts in its division wall instead, which leads to a gradual reduction of plasmodesmal frequency towards the root tip. The resulting increase in symplasmic discontinuity of the apical root cell in particular was thought to be the regulatory factor for determining the limited growth of this plant organ.

Selective loss of plasmodesmata leading to symplasmic isolation has been shown to occur during the development of reproductive cells. It turned out to be a universal rule that tissues belonging to different plant generations are not symplasmically interconnected by plasmodesmata (Carr 1976; Jones 1976; Robards 1976; Oparka and Gates 1981; Offler et al. 1989; McDonald et al. 1996 and literature cited). Consistent symplasmic discontinuity has been observed at the interface between the maternal and embryonic tissues. This suggests a compulsory apoplasmic step in the transport pathway to the growing embryo which is sometimes marked by the occurrence of transfer cells (e.g. Offler et al. 1989). Uptake of substances into the embryonic tissues occurs via the carrier-mediated apoplasmic pathway exclusively and can be precisely controlled. The strict symplasmic isolation obviously guarantees that the domains with different genetic sets are kept separated so that no mutual influence can take place. Particularly in the light of recent findings, suggesting that transport of macromolecules through plasmodesmata influences developmental processes (e.g. Lucas 1995), the symplasmic discontinuity might be imperative for the autonomous development of the plant embryo.

Symplasmic isolation might be a universal prerequisite for cell differentiation, as symplasmic discontinuity has often been found to occur simultaneously or even prior to differentiation processes. Xuhan and van Lammeren (1994), for example, observed that the plasmodesmata between the three cell layers of the developing seed coat of *Ranunculus scleratus* disappear gradually. After their symplasmic segregation, the cells of the seed coat layers go through a different development. Palevitz and Hepler (1985) showed that the functional plasmodesmata at the interfaces between the mother guard cell, or the young guard cells and the surrounding cells are selectively down-regulated as soon as the guard cells begin to swell, but well before the stomatal pore begins to open. The plasmodesmata between the stomatal precursors and the adjoining epidermal or subsidiary cells become truncated at a later developmental stage, establishing a permanent symplasmic isolation of the stoma domain (Wille and Lucas 1984; Erwee et al. 1985; Palevitz and Hepler 1985). Similarly, it has been shown that the precursors of the SE/CC complexes in the stem cambium become increasingly symplasmically isolated before the SE/CC complexes differentiate (van Bel and van Rijen 1994). The plasmodesmata of the SE/CC precursors are gradually shut off at all interfaces, beginning at the transverse walls which develop into sieve plates. After maturation, symplasmic discontinuity is obviously maintained at the interfaces with adjacent PP cells, whereas newly formed sieve pores open up between adjoining SEs. Symplasmic discontinuity, imposed by loss or reversible plugging of plasmodesmata, has also been reported to occur prior to spermatogenesis at defined stages of the development of *Chara* antheridia and is supposed to be necessary for a divergent differentiation or an asynchronous mitotic activity of adjacent cells (Kwiatkowska and Maszewski 1985, 1986; cf. also Maszewski and van Bel 1996; Chap. 12). A similar plugging of plasmodesmata was observed in structural studies on embryogenic calluses derived from immature embryos of *Molinia caerulea* (Ehlers 1998). Plugging of plasmodesmata occurs in

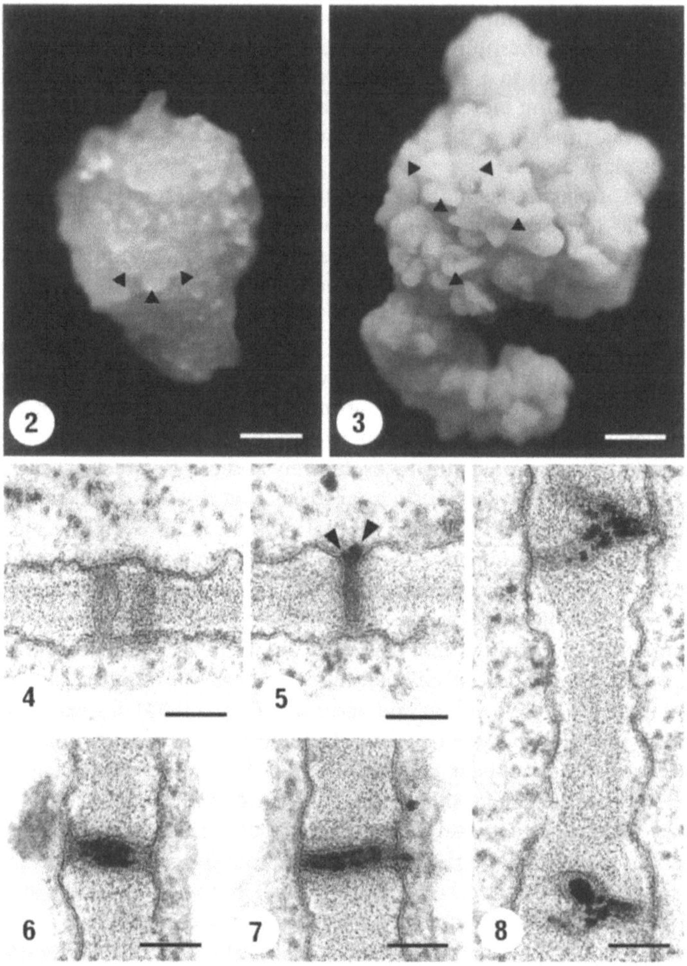

Figs. 2–9. Formation of symplasmic domains during development of embryogenic calluses of *Molinia caerulea*. The occurrence of plugged plasmodesmata indicates selective symplasmic isolation of certain cells or cell clusters

Fig. 2. Early stage in the development of an embryogenic callus. Globular, proembryogenic structures (*arrowheads*) have developed on a scutellar callus derived from immature embryos of *Molinia caerulea*. 16×. *Bar* 0.5 mm. (Kindly provided by Dr. G. Krumbiegel-Schroeren, University of Kiel, Germany)

Fig. 3. Later stage in the development of an embryogenic callus of *Molinia caerulea*. Somatic embryos (*arrowheads*) have developed from the proembryogenic structures. 16×. *Bar* 0.5 mm. (Kindly provided by Dr. G. Krumbiegel-Schroeren, University of Kiel, Germany)

Fig. 4. Inconspicuous plasmodesmata in a globular proembyogenic structure of *Molinia caerulea*. 80000×. *Bar* 0.1 µm

Figs. 5–8. Plugged plasmodesmata in a globular proembryogenic structure of *Molinia caerulea*. The plasmodesmata shown in Figs. **6, 7** were found in serial sections within the same cell wall. Note that the plugging electron-dense material is either located at the plasmodesmal neck region exclusively (**5**, *arrowheads*) or is found within almost the entire plasmodesma (**6, 7**). 80000×. *Bars* 0.1 µm

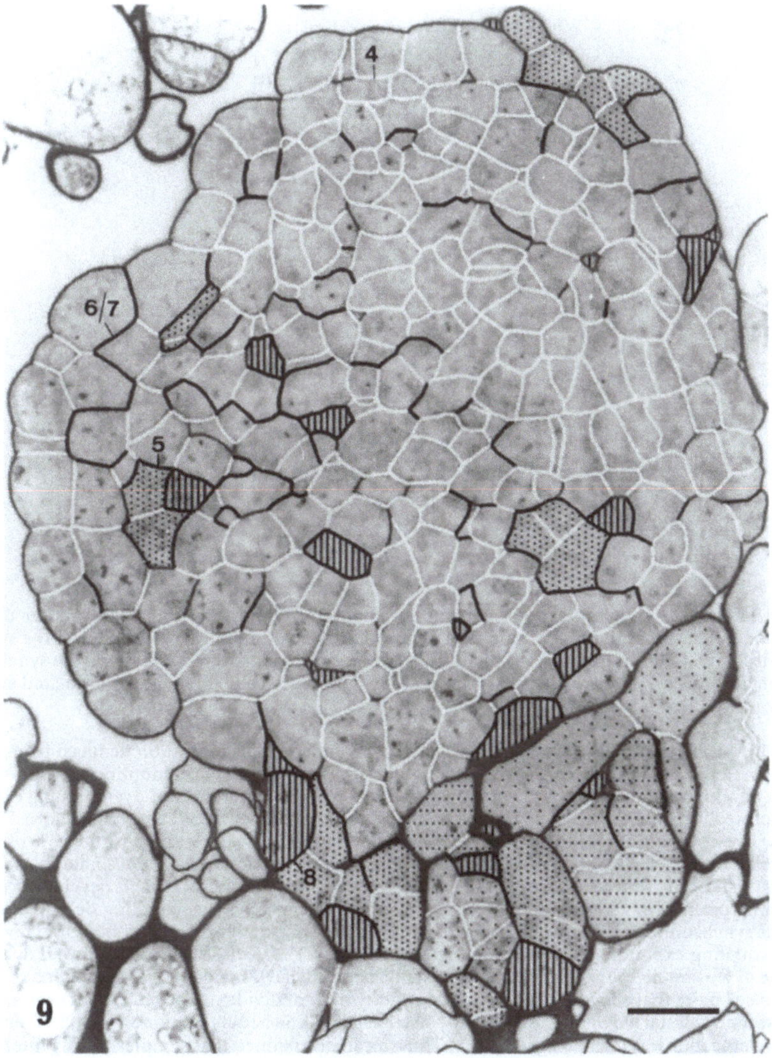

Fig. 9. EM section of a proembryogenic structure of *Molinia caerulea*, parts of which are presented in Figs. 4–8 at higher magnifications. The cells show no structural peculiarities, but conspicuous plugged plasmodesmata do occur in several cell walls, indicating a symplasmic discontinuity between particular adjacent cells. This figure visualizes the pattern of symplasmic (dis)continuity between the cells that was deduced from the analyses of 12 ultrathin sections. Cell walls in which plugged plasmodesmata were observed are presented as *black lines*. *White lines* indicate the occurrence of presumably functional plasmodesmata. No plasmodesmata were observed in the cell walls presented as *grey lines*. Obviously, certain cells or groups of cells are symplasmically isolated from the surrounding cells. The *striping pattern* marks single isolated cells, while isolated cell clusters are marked by *dotted patterns*. The formation of symplasmically isolated domains might be a prerequisite for the subsequent differentiation processes. Plasmodesmata in the walls marked with *4* to *8*, respectively, are shown enlarged in Figs. 4–8. 600×. *Bar* 20 μm

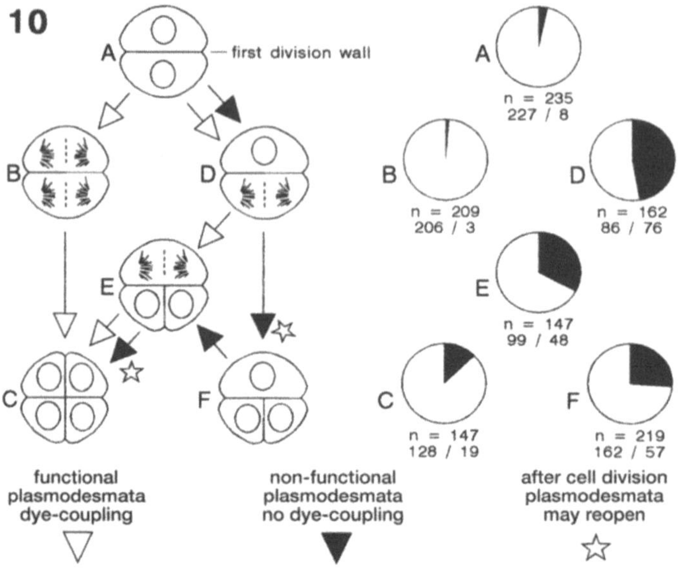

Figs. 10–14. Dye-coupling experiments on developing protoplast-derived microcalluses of *Solanum nigrum* indicate that the synchronization of mitotic activity is directly correlated with the symplasmic connectivity. Cells interconnected by functional plasmodesmata were found to divide synchronously. Asynchronous cell divisions occurred exclusively with those cell pairs that were isolated symplasmically

Fig. 10A–F. The possible developmental pathways in microcallus formation deduced from long-term observation on individual cells are shown on the *left*. Results of the dye-coupling experiments on distinct developmental stages (**A–F**) are presented on the *right*; *n* numbers of dye-coupling experiments performed. The numbers below represent the respective number of cell pairs that showed or did not show dye-coupling across the first division wall. Dye-coupling was observed with 96.6% of just divided cell pairs (**A** cf. Fig. 11), indicating the occurrence of functional plasmodesmata in their division walls. Sister cells which underwent the next mitosis synchronously (**A, B**) and developed into a four-cell microcallus (**B, C**) remained interconnected symplasmically. Dye-coupling occurred with 98.6% of the synchronously dividing cells (**B** cf. Fig. 12). If only one of the sister cells started to divide (**A, D**), dye-coupling experiments showed ambiguous results; 53% of these cell pairs showed dye-coupling, while 47% were no longer interconnected symplasmically (**D**). Long-term observations revealed that those cell pairs that showed symplasmic continuity did not divide in a strictly asynchronous way, but the second cell started to divide before the sister cell had finished its cytokinesis (**D, E**). Finally, a four-cell microcallus was formed again (**E, C**). Thus, it can be assumed that symplasmically interconnected cells show a synchronous mitotic activity in general. Strictly asynchronous mitoses and the formation of three-cell microcalluses occurred exclusively in those cell pairs that were isolated symplasmically (**D, F** cf. Fig. 13). After asynchronous mitosis, the plasmodesmata in the first division wall might reopen, as dye-coupling was observed again for 74% of the three-cell microcalluses (**F** cf. Fig. 14). However, those three-cell microcalluses that underwent a second asynchronous mitosis (**F, E, C**) remained isolated symplasmically. Symplasmic isolation therefore seems to be a prerequisite for asynchronous cell divisions

Fig. 11A–C. Dye-coupling between the sister cells of a just divided *Solanum nigrum* cell pair indicates the occurrence of functional plasmodesmata in the young division wall. Two minutes after injection of Lucifer Yellow CH into the bottom cell (marked *x* in **A**) the fluorescent dye is distributed evenly within the cytoplasm of both sister cells (**B**). After 4 h, the dye is accumulated in the vacuoles (**C**) which might be due to the activity of anion carriers in the vacuolar membrane (cf. Robinson and Hedrich 1991). 375×. *Bar* 20 µm

Fig. 11
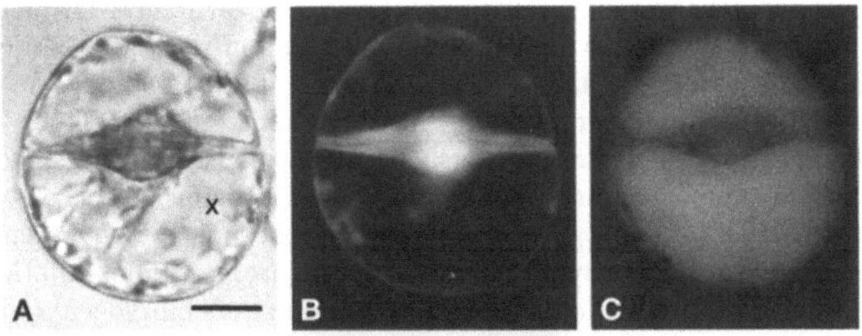

Fig. 12
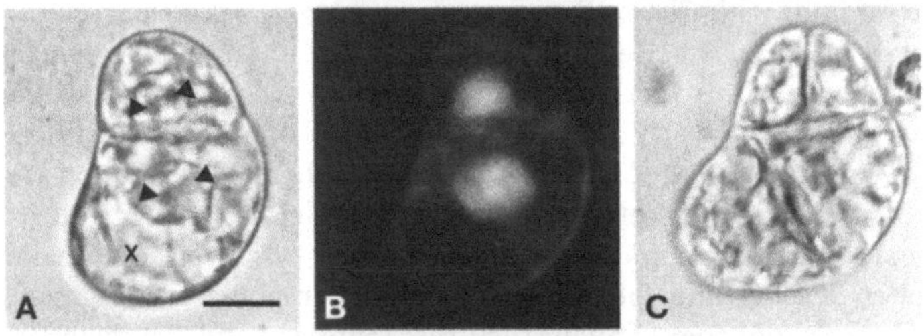

Fig. 13
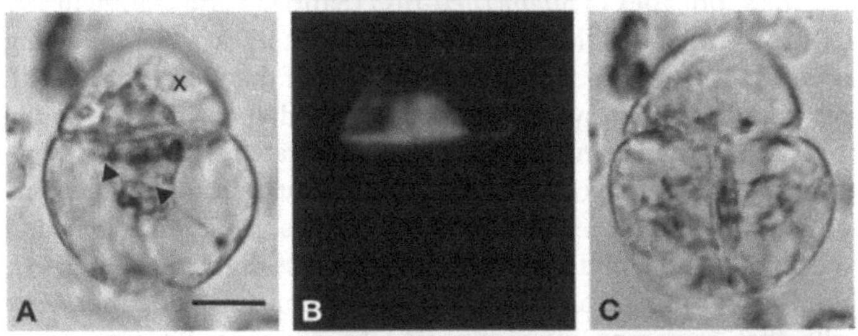

Fig. 14
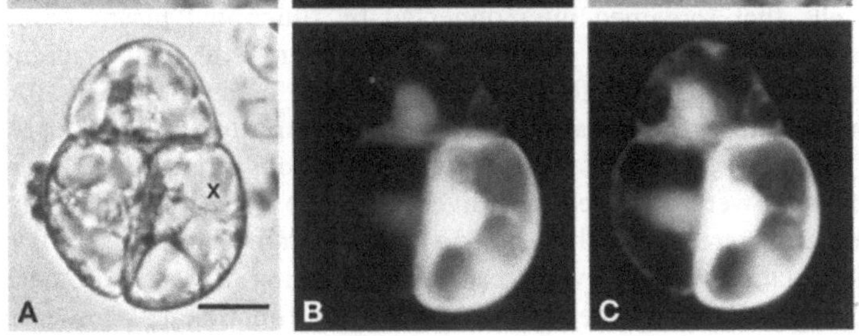

Fig. 12A–C. *Solanum nigrum* sister cells that undergo their next mitosis synchronously as indicated by two nuclei occurring in each sister cell (*arrowheads* in **A**). The cells remain interconnected symplasmically and show dye-coupling (**B**). Synchronous cell division leads to the formation of a four-cell microcallus within 4 h (**C**); *x* injected cell. 375×. *Bar* 20 μm

Fig. 13A–C. *Solanum nigrum* sister cells that undergo their next mitosis asynchronously as indicated by two nuclei occurring in the bottom cell only (*arrowheads* in **A**). These cells are symplasmically isolated from each other and do not show dye-coupling (**B**). Asynchronous mitosis leads to the formation of a three-cell microcallus within 6 h (**C**); *x* injected cell. 375×. *Bar* 20 μm

Fig. 14A–C. *Solanum nigrum* three-cell microcallus, apparently developed from a symplasmically isolated cell pair that underwent an asynchronous mitosis (**A**). The injected dye is transported across the newly formed, vertical division wall as well as across the first, horizontal division wall, indicating that the plasmodesmata in the first division wall have obviously been reopened after asynchronous division (**B, C**); *x* injected cell. 375×. *Bar* 20 μm particular cell walls of the globular proembryogenic structures formed on scutellar callus (Fig. 2) which develop rapidly into somatic embryos (Fig. 3). Selective occlusion of plasmodesmata (Figs. 4, 5–8) leads to a complicated pattern of symplasmic (dis)continuity within the proembryogenic structure, as the overall view implies (Fig. 9). Certain cells or cell clusters obviously become isolated from the surrounding cells and might therefore function as more autonomous domains. At this early stage of development, the cells do not yet show any structural peculiarities, but their symplasmic isolation might be the first symptom of the desynchronization of mitotic activity and the initiation of differentiation processes in the course of embryogenesis in agreement with the findings with *Chara* (Kwiatkowska and Maszewski 1985, 1986).

Changes in the plasmodesmal connectivity and the implicit segregation of functional domains have also been observed in other meristematic tissues and may be programmed events which play a key role in the control of growth and differentiation processes (cf. Lucas et al. 1993; McLean et al. 1997; for detailed discussion see Chap. 13). In the apical meristem of *Iris*, plasmodesmata close selectively during the transition from the vegetative to the reproductive state. Isolated symplasmic domains are formed which later develop into the respective flower components (Bergmans et al. 1993). Studies on the shoot apex of *Silene coeli-rosa* (Santiago and Goodwin 1988; cf. also Lucas et al. 1993) showed that the remarkably high mitotic activity following floral induction is associated with a gradual downregulation of the plasmodesmal SEL. Recently, a direct correlation between synchronization of cell division activity and plasmodesmal permeability in higher plants has been demonstrated by dye-coupling experiments and long-term observations carried out on individual protoplast-derived microcalluses (Ehlers and Kollmann 1996b). In these studies, functional plasmodesmata were usually observed in the division walls between just divided cell pairs (Figs. 10A, 11), but differences in symplasmic continuity may occur during further development. The sister cells that remained interconnected by functional plasmodesmata and still showed dye-coupling were always found to undergo the next cytokinesis synchronously and developed into four-cell microcalluses in any case (Figs. 10A, B, C, 12), although the mitoses did not always take place in a strictly simultaneous way (Figs. 10A, D, E, C). However, strictly asynchronous cell divisions and the formation of three-cell microcalluses were observed exclusively with those cell pairs that showed no dye-coupling and were obviously isolated symplasmically (Figs. 10A, D, F, 13). The symplasmic isolation was achieved presumably by gating of plasmodesmata, as no ultrastructural peculiarities were observed for the non-functional plasmodesmata between the asynchronously dividing sister cells so far. Closure of plasmodesmata must be reversible, because non-functional plasmodesmata in the first division walls might be reopened after asynchronous mitoses and dye-coupling might occur between all cells of the three-cell microcalluses (Figs. 10F, 14). If a second asynchronous cell division took place, the plasmodesmata remained closed (Figs. 10F, E, C). It seems to be a general rule that adjacent cells interconnected by functional plasmodesmata divide synchronously, whereas asynchronous cell divisions obviously require at least temporary symplasmic isolation and a reversible closure of plasmodesmata (see Chaps. 12, 13). It remains to be proved, of course, whether this is true in planta.

In the past years, it has often been stressed that the symplasmic transport of informational macromolecules may play a crucial role in controlling plant development and morphogenesis (e.g. Lucas 1995; Lucas et al. 1995). As the examples given above may indicate, symplasmic isolation leading to the formation of symplasmic domains is obviously also essential for controlled growth and differentiation processes. Symplasmic communication and symplasmic isolation might be regarded as the antagonists that allow the precise balance between cell cooperation and cell autonomy required for plant development and the functioning of the mature plant.

References

Beebe DU, Evert RF (1992) Photoassimilate pathway(s) and phloem loading in the leaf of *Moricandia arvensis* L. DC (Brassicaceae). Int J Plant Sci 153:61–77

Bergmans A, de Boer D, van Bel AJE, van der Schoot C (1993) The initiation and development of *Iris* flowers: permeability changes in the apex symplasm. Flowering Newsl 16:19–26

Botha CEJ (1992) Plasmodesmatal distribution, structure and frequency in relation to assimilation in C_3 and C_4 grasses in southern Africa. Planta 187:348–358

Botha CEJ, Cross RHM (1997) Plasmodesmatal frequency in relation to short-distance transport and phloem loading in leaves of barley (*Hordeum vulgare*). Phloem is not loaded directly from the symplast. Physiol Plant 99:355–362

Botha CEJ, van Bel AJE (1992) Quantification of symplastic continuity as visualised by plasmodesmograms: diagnostic value for phloem-loading pathways. Planta 187:359–366

Bourquin S, Bonnemain JL, Delrot S (1990) Inhibition of loading of ^{14}C assimilates by p-chloromercuribenzenesulfonic acid. Localization of the apoplastic pathway in *Vicia faba*. Plant Physiol 92: 97–102

Carr DJ (1976) Plasmodesmata in growth and development. In: Gunning BES, Robards AW (eds) Intercellular communication in plants: studies on plasmodesmata. Springer, Berlin Heidelberg New York, pp 243–289

Davis AR, Gunning BES (1992) The modified stomata of the floral nectary of *Vicia faba* L. 1. Development, anatomy and ultrastructure. Protoplasma 166:134–152

De Faÿ E, Sanier C, Hébant C (1989) The distribution of plasmodesmata in the phloem of *Hevea brasiliensis* in relation to laticifer loading. Protoplasma 149:155–162

Ding B, Turgeon R, Parthasarathy MV (1992) Substructure of freeze-substituted plasmodesmata. Protoplasma 169:28–41

Ehlers K (1998) The formation of symplast domains by plugging of plasmodesmata – a prerequisite for coordinated differentiation processes? Worksh on plasmodesmata and transport of plant viruses and plant macromolecules. Madrid, Spain, April 20–22, 1998, p 21

Ehlers K, Kollmann R (1996a) Formation of branched plasmodesmata in regenerating *Solanum nigrum*-protoplasts. Planta 199:126–138

Ehlers K, Kollmann R (1996b) Regulation of the symplasmic contact between physiologically different cells. 3rd Int Worksh on basic and applied research in plasmodesmatal biology. Zichron-Yakov, Israel, March 10–16, 1996, pp 77–81

Eleftheriou EP (1990) Plasmodesmatal structure and function in nectaries. In: Robards AW, Jongsma H, Lucas WJ, Pitts J, Spray D (eds) Parallels in cell-to-cell junctions in plants and animals. NATO ASI Ser, vol H 46. Springer, Berlin Heidelberg New York, pp 223–237

Eleftheriou EP, Hall JL (1983) The extrafloral nectaries of cotton. I. Fine structure of the secretory papillae. J Exp Bot 34:103–119

Epel B (1994) Plasmodesmata: composition, structure and trafficking. Plant Mol Biol 26:1343–1356

Epel BL, Warmbrodt RP, Bandurski RS (1992) Studies on the longitudinal and lateral transport of IAA in the shoots of etiolated corn seedlings. J Plant Physiol 140:310–318

Erwee MG, Goodwin PB (1985) Symplast domains in the extrastelar tissue of *Egeria densa* Planch. Planta 163:9–19

Erwee MG, Goodwin PB, van Bel AJE (1985) Cell-cell communication in the leaves of *Commelina cyanea* and other plants. Plant Cell Environ 8:173–178

Fisher DG (1986) Ultrastructure, plasmodesmatal frequency and solute concentration in green areas of variegated *Coleus blumei* Benth. leaves. Planta 169:141–152

Fisher DG (1990) Distribution of plasmodesmata in leaves. A comparison of *Cananga odorata* with other species using different measures of plasmodesmata frequency. In: Robards AW, Jongsma H, Lucas WJ, Pitts J, Spray D (eds) Parallels in cell-to-cell junctions in plants and animals. NATO ASI Ser, vol H 46. Springer, Berlin Heidelberg New York, pp 199–221

Gamalei YV (1985) Characteristics of phloem loading in woody and herbaceous plants. Fiziol Rast 32:866–876

Gamalei YV (1989) Structure and function of leaf minor veins in trees and herbs. A taxonomic review. Trees 3:96–110
Gunning BES (1976) Introduction to plasmodesmata. In: Gunning BES, Robards AW (eds) Intercellular communication in plants: studies on plasmodesmata. Springer, Berlin Heidelberg New York, pp 1–13
Gunning BES (1978) Age-related and origin-related control of the numbers of plasmodesmata in cell walls of developing *Azolla* roots. Planta 143:181–190
Gunning BES, Hughes JE (1976) Quantitative assessment of symplastic transport of pre-nectar into trichomes of *Abutilon* nectaries. Aust J Plant Physiol 3:619–637
Gunning BES, Robards AW (1976) Plasmodesmata: current knowledge and outstanding problems. In: Gunning BES, Robards AW (eds) Intercellular communication in plants: studies on plasmodesmata. Springer, Berlin Heidelberg New York, pp 203–227
Hayes PM, Offler CE, Patrick JW (1985) Cellular structures, plasma membrane surface areas and plasmodesmatal frequencies of the stem of *Phaseolus vulgaris* L. in relation to radial photosynthate transfer. Ann Bot 56:125–138
Jones MGK (1976) The origin and development of plasmodesmata. In: Gunning BES, Robards AW (eds) Intercellular communication in plants: studies on plasmodesmata. Springer, Berlin Heidelberg New York, pp 81–105
Kempers R, van Bel AJE (1997) Symplasmic connections between sieve element and companion cell in the stem phloem of *Vicia faba* have a molecular exclusion limit of at least 10 kDa. Planta 201:195–201
Kempers R, Ammerlaan A, van Bel AJE (1998) Symplasmic constriction and ultrastructural features of the sieve element/companion cell complex in the transport phloem of apoplasmically and symplasmically phloem loading species. Plant Physiol 116:271–278
Knoblauch M, van Bel AJE (1998) Sieve tubes in action. Plant Cell 10:35–50
Kollmann R, Glockmann C (1991) Studies on graft unions III. On the mechanism of secondary formation of plasmodesmata at the graft interface. Protoplasma 165:71–85
Kurkova EB, Vakhmistrov DB (1984) Distribution of plasmodesmata in the rhizodermis along the root axis in *Trianea bogotensis*. Sov Plant Physiol 31:115–119
Kwiatkowska M, Maszewski J (1985) Changes in ultrastructure of plasmodesmata during spermatogenesis in *Chara vulgaris* L. Planta 166:46–50
Kwiatkowska M, Maszewski J (1986) Changes in the occurrence and ultrastructure of plasmodesmata in antheridia of *Chara vulgaris* L. during different stages of spermatogenesis. Protoplasma 132: 179–188
Lucas WJ (1995) Plasmodesmata: intercellular channels for macromolecular transport in plants. Curr Opin Cell Biol 7:673–680
Lucas WJ, Gilbertson RL (1994) Plasmodesmata in relation to viral movement within leaf tissues. Annu Rev Phytopathol 32:387–411
Lucas WJ, Ding B, van der Schoot C (1993) Plasmodesmata and the supracellular nature of plants. New Phytol 125:435–476
Lucas WJ, Bouché-Pillon S, Jackson DP, Nguyen L, Baker L, Ding B, Hake S (1995) Selective trafficking of KNOTTED 1 homeodomain protein and its mRNA through plasmodesmata. Science 270: 1980–1983
Maszewski J, van Bel AJE (1996) Different patterns of intercellular transport of Lucifer Yellow in young and mature antheridia of *Chara vulgaris* L. Bot Acta 109:110–114
McCauley MM, Evert RF (1989) Minor veins of the potato (*Solanum tuberosum* L.) leaf: ultrastructure and plasmodesmatal frequency. Bot Gaz 150:351–368
McDonald R, Fieuw S, Patrick JW (1996) Sugar uptake by the dermal transfer cells of developing cotyledons of *Vicia faba* L. Mechanism of energy coupling. Planta 198:502–509
McLean BG, Hempel FD, Zambryski PC (1997) Plant intercellular communication via plasmodesmata. Plant Cell 9:1043–1054
Monzer J (1990) Secondary formation of plasmodesmata in cultured cells – structural and functional aspects. In: Robards AW, Jongsma H, Lucas WJ, Pitts J, Spray D (eds) Parallels in cell-to-cell junctions in plants and animals. NATO ASI Ser, vol H 46. Springer, Berlin Heidelberg New York, pp 185–197
Monzer J (1991) Ultrastructure of secondary plasmodesmata formation in regenerating *Solanum nigrum*-protoplast cultures. Protoplasma 165:86–95
Münch E (1930) Die Stoffbewegung in der Pflanze. Gustav Fischer, Jena
Offler CE, Nerlich SM, Patrick JW (1989) Pathway of photosynthate transfer in the developing seed of *Vicia faba* L. Transfer in relation to seed anatomy. J Exp Bot 40:769–780
Olesen P, Robards AW (1990) The neck region of plasmodesmata: general architecture and some functional aspects. In: Robards AW, Jongsma H, Lucas WJ, Pitts J, Spray D (eds) Parallels in cell-to-cell junctions in plants and animals. NATO ASI Ser, vol. H 46. Springer, Berlin Heidelberg New York, pp 145–170

Oparka KJ, Gates P (1981) Transport of assimilates in the developing caryopsis of rice (*Oryza sativa* L.). Planta 151:561–573
Oparka KJ, Prior DAM (1992) Direct evidence for pressure-generated closure of plasmodesmata. Plant J 2:741–750
Oparka KJ, Viola R, Wright KM, Prior DAM (1992) Sugar transport and metabolism in the potato tuber. In: Pollock CJ, Farrar JF, Gordon AJ (eds) Carbon partitioning within and between organisms. BIOS, Oxford, pp 99–114
Oparka KJ, Duckett CM, Prior DAM, Fisher DB (1994) Real-time imaging of phloem unloading in the root tip of *Arabidopsis*. Plant J 6:759–766
Overall RL, Gunning BES (1982) Intercellular communication in *Azolla* roots. II. Electrical coupling. Protoplasma 111:151–160
Overall RL, Wolfe J, Gunning BES (1982) Intercellular communication in *Azolla* roots. I. Ultrastructure of plasmodesmata. Protoplasma 111:134–150
Palevitz BA, Hepler PK (1985) Changes in dye coupling of stomatal cells of *Allium* and *Commelina* demonstrated by microinjection of Lucifer Yellow. Planta 164:473–479
Patrick JW, Offler CE (1996) Post-sieve element transport of photoassimilates in sink regions. J Exp Bot 47:1165–1177
Rhodes JD, Thain JF, Wildon DC (1996) The pathway for systemic electrical signal conduction in the wounded tomato plant. Planta 200:50–57
Robards AW (1976) Plasmodesmata in higher plants. In: Gunning BES, Robards AW (eds) Intercellular communication in plants: studies on plasmodesmata. Springer, Berlin Heidelberg New York, pp 15–57
Robards AW, Clarkson DT (1976) The role of plasmodesmata in the transport of water and nutrients across roots. In: Gunning BES, Robards AW (eds) Intercellular communication in plants: studies on plasmodesmata. Springer, Berlin Heidelberg New York, pp 181–201
Robards AW, Lucas WJ (1990) Plasmodesmata. Annu Rev Plant Physiol Plant Mol Biol 41:369–419
Robards AW, Stark M (1988) Nectar secretion in *Abutilon*: a new model. Protoplasma 142:79–91
Robinson DG, Hedrich R (1991) Vacuolar Lucifer Yellow uptake in plants: endocytosis or anion transport; a critical opinion. Bot Acta 104:257–264
Robinson-Beers K, Evert RF (1991) Fine structure of plasmodesmata in mature leaves of sugarcane. Planta 184:307–318
Russin WA, Evert RF (1985) Studies on the leaf of *Populus deltoides* (*Salicaceae*): ultrastructure, plasmodesmatal frequency, and solute concentrations. Am J Bot 72:1232–1247
Ryser U (1992) Ultrastructure of the epidermis of developing cotton (*Gossypium*) seeds: suberin, pits, plasmodesmata, and their implication for assimilate transport into cotton fibers. Am J Bot 79:14–22
Santiago JF, Goodwin PB (1988) Restricted cell/cell communication in the shoot apex of *Silene coelirosa* during the transition to flowering is associated with a high mitotic index rather than with evocation. Protoplasma 146:52–60
Sauter JJ, Kloth S (1986) Plasmodesmatal frequency and radial translocation rates in the ray cells of poplar (*Populus* × *canadensis* Moench robusta). Planta 168:377–380
Sawidis T, Eleftheriou EP, Tsekos I (1987) The floral nectaries of *Hibiscus rosa-sinensis* L. II. Plasmodesmatal frequencies. Phyton 27:155–164
Schnepf E, Sych A (1983) Distribution of plasmodesmata in developing *Sphagnum* leaflets. Protoplasma 116:51–56
Seagull RW (1983) Differences in the frequency and disposition of plasmodesmata resulting from root cell elongation. Planta 159:497–504
Terry BR, Robards AW (1987) Hydrodynamic radius alone governs the mobility of molecules through plasmodesmata. Planta 171:145–157
van Bel AJE (1992) Different phloem-loading machineries correlated with the climate. Acta Bot Neerl 41:121–141
van Bel AJE (1993a) Strategies of phloem loading. Annu Rev Plant Physiol Plant Mol Biol 44:253–281
van Bel AJE (1993b) The transport phloem. Specifics of its functioning. Prog Bot 54:134–150
van Bel AJE (1996) Interaction between sieve element and companion cell and the consequences for photoassimilate distribution. Two structural hardware frames with associated physiological software packages in dicotyledons? J Exp Bot 47:1129–1140
van Bel AJE, Kempers R (1990) Symplastic isolation of the sieve element-companion cell complex in the phloem of *Ricinus communis* and *Salix alba* stems. Planta 183:69–76
van Bel AJE, Kempers R (1996) The pore/plasmodesma unit, key element in the interplay between sieve element and companion cell. Prog Bot 58:278–291
van Bel AJE, Oparka KJ (1995) On the validity of plasmodesmograms. Bot Acta 108:174–182
van Bel AJE, van Rijen HVM (1994) Microelectrode-recorded development of the symplasmic autonomy of the sieve element/ companion cell complex in the stem phloem of *Lupinus luteus* L. Planta 192:165–175

van Bel AJE, van Kesteren WJP, Papenhuijzen C (1988) Ultrastructural indications for coexistence of symplastic and apoplastic phloem loading in *Commelina benghalensis* leaves. Differences in ontogenetic development, spatial arrangement and symplastic connections of the two sieve tubes in the minor veins. Planta 176:159–172

van der Schoot C, van Bel AJE (1989) Glass microelectrode measurements of sieve tube membrane potentials in internode discs and petiole strips of tomato (*Solanum lycopersicum* L.). Protoplasma 149:144–154

van der Schoot C, van Bel AJE (1990) Mapping membrane potential differences and dye-coupling in internodal tissues of tomato (*Solanum lycopersicum* L.). Planta 182:9–21

van der Schoot C, Dietrich MA, Storms M, Verbeke JA, Lucas WJ (1995) Establishment of a cell-to-cell communication pathway between separate carpels during gynoecium development. Planta 195: 450–455

Waigmann E, Zambryski P (1995) Tobacco mosaic virus movement protein-mediated protein transport between trichome cells. Plant Cell 7:2069–2079

Wang N, Fisher DB (1994) The use of fluorescent tracers to characterize the post-phloem pathway in maternal tissues of developing wheat grains. Plant Physiol 104:17–27

Warmbrodt R, VanDerWoude WJ (1990) Leaf of *Spinacea oleracea* (spinach): ultrastructure, and plasmodesmatal distribution and frequency, in relation to sieve-tube loading. Am J Bot 77:1361–1377

Wille AC, Lucas WJ (1984) Ultrastructural and histochemical studies on guard cells. Planta 160: 129–142

Xuhan X, van Lammeren AAM (1994) The ultrastructure of the seed coat development in *Ranunculus scleratus*. Acta Bot Neerl 43:27–37

CHAPTER 15

Plasmodesmata in the Phloem-Loading Pathway

D. U. BEEBE[1] · W. A. RUSSIN[2]

[1] Institut de recherche en biologie végétale, Université de Montréal, 4101 est, rue Sherbrooke, Montréal (Québec) H1X 2B2, Canada
[2] Department of Botany, University of Wisconsin, 430 Lincoln Drive, Madison Wisconsin 53705, USA
Tel.: (514) 872-4563, FAX: (514) 872-9406,
E-mail: beebed@irbv.umontreal.ca

1 Introduction

Twenty years ago, Brian Gunning began his analysis of the role of plasmodesmata in short-distance transport of assimilates to the phloem (Gunning 1976) by stating that the "relevant evidence is both meagre and circumstantial" for a comprehensive discussion of the subject. At that time there were very few quantitative analyses of plasmodesmata along the assimilate pathway from the mesophyll to the phloem (Geiger et al. 1973; Gunning et al. 1974; Kuo et al. 1974). There was also considerable controversy regarding the probable pathway(s) and mechanism(s) of phloem loading. Today, there is an extensive body of information on everything from plasmodesmal frequency and distribution in dicot and monocot species to detailed aspects of various phloem-loading models (Riesmeier et al. 1993, 1994; Sauer and Stolz 1994; Grusak et al. 1996; Rentsch and Frommer 1996; Kühn et al. 1997). We also know much more about the fundamentals of plasmodesmal structure (Ding et al. 1992b; Lucas et al. 1993a) and the development and possible function of secondary plasmodesmata (Monzer 1990, 1991; Kollmann and Glockmann 1991; Ding and Lucas 1996; Ehlers and Kollmann 1996; Ehlers et al. 1996; Volk et al. 1996). Analysis of alteration in carbon allocation and biomass partitioning in transgenic tobacco plants expressing native and mutated forms of the tobacco mosaic virus movement protein has led to the proposal that plasmodesmata in a leaf form a special communication network between the mesophyll and the phloem (Lucas et al. 1996).

Despite these considerable advances in our knowledge of structure/function relationships in plasmodesmal biology, we are still engaged in finding answers to basic questions that remain important to understanding the function of plasmodesmata in phloem loading. For example, a wide variety of plasmodesmal morphologies at various cell-to-cell interfaces in a given source leaf has been well documented (Evert et al. 1977; Russin and Evert 1985; Robinson-Beers and Evert 1991b; Beebe and Evert 1992; Volk et al. 1996). Yet how do these structural variations correlate with possible functional specializations in photoassimilate transport and phloem loading? Virtually all plasmodesmata share certain fundamental structural similarities, regardless of any additional complexity at a given interface. Additional modifications, either as alterations within the plasmodesma itself, e.g., presence of sphincters (Evert et al. 1977; Robinson-Beers and Evert 1991b), or in the cell wall microdomain surrounding the plasmodesma (Olesen 1979; Olesen and Robards 1990; Badelt et al. 1994; Overall and Blackman 1996), may represent a structural reflection of physiological specialization. Although this concept is intriguing, care must be exercised not to infer too much from static images of dynamic structures like plasmodesmata. Is the plasmodesmal structure shown in micrographs an accurate representation of their in vivo state? Do the methods used to preserve and prepare tissues for structural analysis induce artefacts (see Chap. 3)? We must always regard the images used for the analysis of plasmodesmal structure/function relationships, literally, with a skeptical eye.

This chapter will emphasize research from the past 20 years on the specifics of plasmodesmal structure and function along the mesophyll–phloem pathway in source leaves. We will look at how plasmodesmal structure, and its variations, are believed to function in short-distance transport of photoassimilates to leaf minor veins and in subsequent phloem loading. We focus first on the structure of plasmodesmata at the various cell-to-cell interfaces along the pathway. We then examine in more detail the role of plasmodesmata in phloem loading and what modifications in plasmodesmal structure seem associated with a particular mode of assimilate uptake. Finally, we will look briefly at a maize mutant that has provided insight into the function of plasmodesmata in phloem loading.

1.1 Models of Phloem Loading

Two models of phloem loading (Fig.1) have been proposed to explain the movement of photoassimilates from mesophyll cells (MC) to the sieve elements (SE) of minor veins (Grusak et al. 1996). Both models have in common the generally accepted concept that sugars follow a symplasmic pathway from the MC to the region of the sieve element-companion cell (SE-CC) complex (Giaquinta 1983). They differ in the specific mechanisms by which the transport sugars are delivered to the SEs. In the apoplasmic phloem-loading model (Fig. 1A), the transport sugar sucrose diffuses symplasmically along a concentration gradient from the mesophyll to the region of the SE-CC complex. The sugar is subsequently unloaded into the apoplast (cell wall continuum),

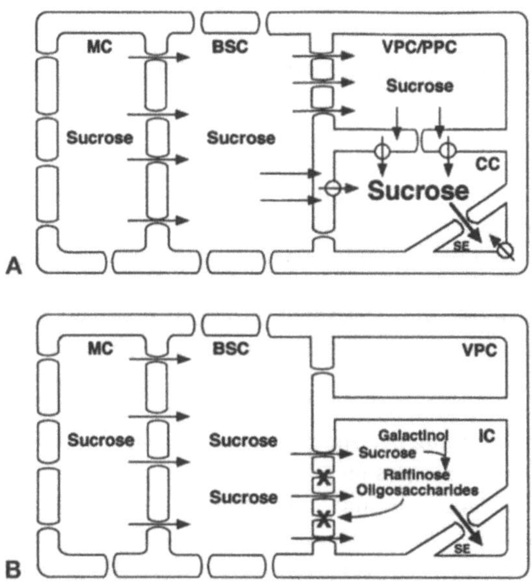

Fig. 1A, B. Schematic diagrams of phloem loading models showing the mesophyll cell (*MC*) to sieve element (*SE*) pathway. Interruptions in the cell walls indicate plasmodesmata. **A** Apoplasmic phloem-loading model showing the symplasmic movement of sucrose (*unlabeled arrows*) to the region of the sieve element-companion cell complex, followed by the unloading of the sucrose into the apoplast. Uptake of sucrose into the phloem occurs via the H^+-sucrose symport proteins [indicated in the companion cell (*CC*) wall by the *small circle pierced by the arrow*]; *BSC* bundle-sheath cell; *VPC* vascular parenchyma cell; *PPC* phloem parenchyma cell. **B** Symplasmic phloem-loading model showing the symplasmic movement of sucrose (*unlabeled arrows*) from the MC to the intermediary cell (*IC*), where the sucrose is subsequently metabolized in the synthesis of the larger raffinose oligosaccharides. The microchannels within the plasmodesmata across the bundle-sheath cell (*BSC*) interface with the IC are sufficiently reduced in diameter to prevent diffusive loss of synthesized raffinose oligosaccharides and is then loaded into the SE-CC complex in an energy-dependent fashion (Geiger 1975; Giaquinta 1983). The active uptake of sucrose is accomplished through the coupled action of an H^+-ATPase (Bouché-Pillon et al. 1994; DeWitt and Sussman 1995) and a sucrose-proton symporter protein (Bush 1992) located in the plasmalemma of the CC (Stadler et al. 1995; Stadler and Sauer 1996) or the SE (Kühn et al. 1997). Diffusive loss of loaded sucrose via plasmodesmata from the phloem is thought to be prevented either by a reduced number of plasmodesmata across the SE-CC complex interfaces with surrounding cells (Delrot 1989) or by modification of plasmodesmal structure (Beebe and Evert 1992; Gagnon and Beebe 1996b).

In the symplasmic phloem loading, or polymer trap, model (Fig. 1B; Turgeon and Gowan 1990; Turgeon 1991, 1996), sucrose synthesized in the mesophyll diffuses via plasmodesmata to the bundle-sheath cells (BSC) and then into contiguous specialized CCs called intermediary cells (IC). There, the sucrose is metabolized in the synthesis of the larger tri- and tetrasaccharides, raffinose and stachyose. These sugars are prevented from diffusing outward from the IC because plasmodesmata at the BSC/IC interface have a reduced size exclusion limit: one large enough to accommodate sucrose inflow, but small enough to prevent diffusive loss of the synthesized raffinose oligosaccharides. The downhill diffusion gradient is maintained by the depletion of sucrose during raffinose oligosaccharide synthesis.

2 Structure of Plasmodesmata Along the Phloem-Loading Pathway

2.1 General Comments

The dicotyledons represent the largest number of species whose plasmodesmata have been studied in detail. Ten sucrose-translocating species that are thought to be apoplasmic phloem loaders have been analyzed: *Amaranthus retroflexus* (Fisher and Evert 1982b), *Populus deltoides* (Russin and Evert 1985), *Beta vulgaris* (Evert and Mierzwa 1986), *Ipomea tricolor* (Madore et al. 1986), *Nicotiana tabacum* (Ding et al. 1988, 1992b, 1993), *Solanum tuberosum* (McCauley and Evert 1989), *Spinacia oleracea* (Warmbrodt and VanDerWoude 1990), *Cananga odorata* (Fisher 1990), *Sonchus oleraceus* (Fisher 1991), *Pisum sativum* (Wimmers and Turgeon 1991), and *Moricandia arvensis* (Beebe and Evert 1992). In addition, five species believed to be symplasmic phloem loaders have been examined: *Coleus* (Fisher 1986), *Cucurbita pepo* and *Cucumis melo* (Beebe et al. 1996; Volk et al. 1996; Beebe and McEvoy 1997), and *Fraxinus* and *Syringa* (Gamalei and Pakhomova 1982). One general characteristic of plasmodesmal structure along the assimilate pathway in dicot leaves is that the closer the interface is to the SE-CC complex, the more structurally complex the plasmodesmata become. This observation is not new. Gamalei and Pakhomova (1982) presented data and descriptions of plasmodesmata along the phloem-loading pathway in several dicots and concluded that this specialization may be a reflection of increasing regulation within the pathway.

With the exception of the work on *Commelina benghalensis* (van Bel and Koops 1985; van Bel et al. 1988; van Kesteren et al. 1988), the most complete studies involving plasmodesmal function and phloem loading among the monocots have been carried out in grasses (Nelson and van Bel 1998). Grasses can be separated into two general physiological groups, C_3 and C_4, and their leaves can be distinguished on the basis of distinct anatomical and ultrastructural differences, of which the structure of the bundle sheath (BS) is particularly important (Esau 1977). For example, all C_3 and some C_4 grasses have a second, inner BS layer (called the mestome sheath) composed of smaller cells that have thickened, often suberized or lignified cell walls. The grass species where all pertinent cell-to-cell interfaces have been analyzed in detail are *Zea mays* (Evert et al. 1977, 1978, 1996a; Evert and Russin 1993), *Saccharum officianarum* (Colbert and Evert 1982; Botha and Evert 1986; Robinson-Beers and Evert 1991a, b), *Themeda triandra* (Botha and Evert 1988), and *Hordeum vulgare* (Dannenhoffer and Evert 1990, 1994; Farrar et al. 1992; Evert et al. 1996; Botha and Cross 1997). Unless otherwise stated, our review of the research on monocot leaves will center on these species.

Regardless of the physiological nature of the plant, photosynthate loading and export from the leaf is accomplished, in part, by the coordinated functions of the different sized veins. Although these veins can be categorized by their size (Esau 1977; Hickey 1979), often they are classified as either major veins or minor veins according to their cellular composition, development, and specificity of function (Pristupa 1964; Evert et al. 1978, 1996a; Fisher and Evert 1982a, b; Russin and Evert 1984, 1985a; Russell and Evert 1985; McCauley and Evert 1988a, b, 1989; Fritz et al. 1989; Beebe and Evert 1989, 1992; Dannenhoffer et al. 1990; Evert and Russin 1993; Dannenhoffer and Evert 1994). Major veins are the first to be initiated and primarily function in long-distance transport of photosynthate. Minor veins are initiated later in leaf development and are primarily concerned with loading of photosynthate from the mesophyll and shunting it to the large veins for export (Turgeon 1989).

We will not attempt to produce an exhaustive review, so plasmodesmata within source leaves that have been the subject of only incidental analysis will not be addressed herein. Rather, we will summarize and discuss information provided in certain detailed analyses of the dicot and monocot species listed above. Since the focus of this chapter is the process of phloem loading we will concern ourselves mainly with plasmodesmata along the mesophyll-minor vein pathway and at cell interfaces associated with minor veins. We will begin with an examination of the relatively simple plasmodesmata at cell-to-cell interfaces within the mesophyll portion of the pathway to the minor veins.

2.2 The Mesophyll Cell–Mesophyll Cell Interface

Mesophyll plasmodesmata typically have the simplest morphology of the different forms observed within a given leaf (Fig. 2). Plasmodesmata across MC interfaces occur singly or may be aggregated in primary pit fields. The region of the wall within which the plasmodesmata are found may or may not be slightly thickened. The plasmodesmata may be unbranched or branched, and the branches may be on either or both sides of the interface, forming H-, X-, or Y-shaped groupings (see Chap. 10). A median cavity in the region of the middle lamella is often present where the plasmodesmal channel locally increases in diameter. Typically, a cytoplasmic sleeve is evident around a distinct central axial component, the desmotubule, that is in apparent continuity with the ER of contiguous MC (Fig. 2B). In certain species, e.g., the C_4 dicot *Amaranthus*, the desmotubule and the cytoplasmic sleeve stain almost equally densely near the orifices at either end of the plasmodesmata (Fisher and Evert 1982b). Neck con-

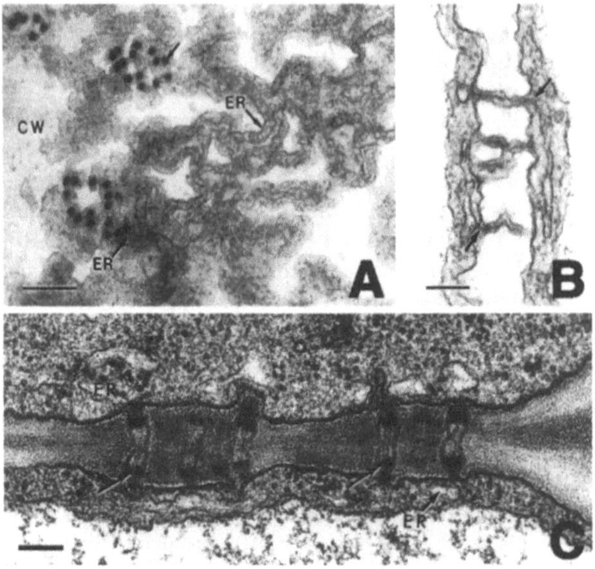

Fig. 2A–C. Plasmodesmata at mesophyll cell–mesophyll cell interfaces. **A** *Moricandia arvensis*. Glancing section along the cell wall (*CW*) passing just into the protoplast, showing plasmodesmata in transverse view and continuity of tubular endoplasmic reticulum (*ER*) network with the plasmodesmata. *Unlabeled arrow* points to desmotubule within a plasmodesma. *Bar* 0.5 μm. (Beebe and Evert 1992). **B** *Moricandia arvensis*. Longitudinal view of plasmodesmata in wall between mesophyll cells. *Unlabeled arrows* point to sites of contact between the desmotubules and endoplasmic reticulum. *Bar* 0.25 μm (Beebe and Evert 1992). **C** *Zea mays*. Longitudinal section of plasmodesmata between adjacent mesophyll cells. Plasmodesmata typically are unbranched and occur in unthickened areas of the wall. Sphincters (*unlabeled arrows*) are present at both ends of the plasmodesmata. Desmotubules are uniform in width with electron lucent lumen. *ER* Endoplasmic reticulum. *Bar* 0.14 μm. (Evert et al. 1977)

strictions, where the diameter of the plasmodesmal channel becomes reduced near the orifices of the plasmodesma as the channel opens out into the cell, may or may not be present in MC/MC plasmodesmata.

Some dicot genera, e.g., *Nicotiana* (Ding et al. 1992a, b) and *Spinacia* (Warmbrodt and VanDerWoude 1990), do have plasmodesmata with a rather complex structure across the MC interface. They occur mostly in small aggregations in slightly thickened areas of the wall, have a conspicuous median cavity, and may be extensively branched.

Plasmodesmata between MC in the C_4 grasses (Evert et al. 1977; Robinson-Beers and Evert 1991b) share unique, common ultrastructural characteristics (Fig. 2C). Typically, they are unbranched with a fairly uniform diameter for most of their length. Neck constrictions typically occur at both ends of the plasmodesma; they are uniformly present in MC/MC plasmodesmata in sugarcane leaves, but variably present in the maize leaf (Evert et al. 1977). Sphincters (electron dense, disc-shaped structures) extend from plasmalemma to desmotubule and occur at both ends of the plasmodesmal canal near the orifices (Fig. 2C). Between the sphincters the cytoplasmic sleeve is electron lucent. The desmotubule is continuous, though constricted, where it traverses the sphincters. Between sphincters, the desmotubule has a convoluted, membranous appearance and an electron lucent lumen. Plasmodesmata at this interface in maize and sugarcane leaves occupy areas of cell wall that are, at most, only slightly thicker than portions of the wall lacking plasmodesmata (Evert et al. 1977; Robinson-Beers and Evert 1991b).

2.3 The Mesophyll Cell–Bundle-Sheath Cell Interface

Plasmodesmata at the MC/BSC interface in dicots (Fig. 3) are generally similar in structure and distribution to those at the MC/MC interfaces (cf. Figs. 2B, 3A, B), al-

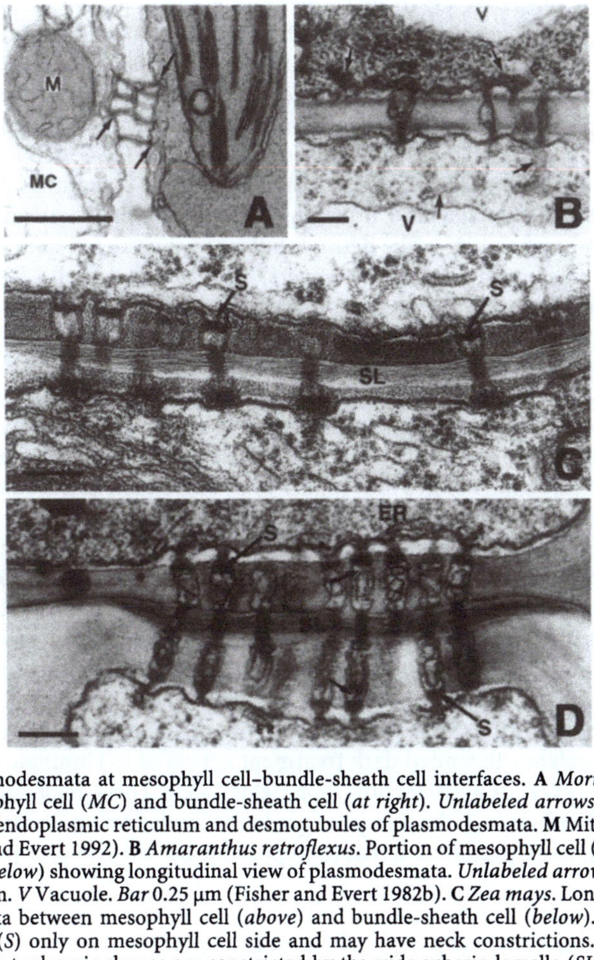

Fig. 3A–D. Plasmodesmata at mesophyll cell–bundle-sheath cell interfaces. **A** *Moricandia arvensis*. Portion of mesophyll cell (*MC*) and bundle-sheath cell (*at right*). *Unlabeled arrows* point to sites of contact between endoplasmic reticulum and desmotubules of plasmodesmata. **M** Mitochondrion. *Bar* 0.5 µm. (Beebe and Evert 1992). **B** *Amaranthus retroflexus*. Portion of mesophyll cell (*above*) and bundle-sheath cell (*below*) showing longitudinal view of plasmodesmata. *Unlabeled arrows* point to endoplasmic reticulum. *V* Vacuole. *Bar* 0.25 µm (Fisher and Evert 1982b). **C** *Zea mays*. Longitudinal section of plasmodesmata between mesophyll cell (*above*) and bundle-sheath cell (*below*). Plasmodesmata have sphincters (*S*) only on mesophyll cell side and may have neck constrictions. On the bundle-sheath-cell side cytoplasmic sleeves are constricted by the wide suberin lamella (*SL*). Desmotubules appear constricted in the region of the sphincter and suberin lamella. *Bar* 0.16 µm (Evert et al. 1977). **D** *Saccharum*. Longitudinal section of plasmodesmata between mesophyll cell (*above*) and bundle-sheath cell (*below*). Sphincters (*S*) present at both ends of plasmodesmata; plasmodesmata are distinctly narrowed where they traverse suberin lamella. Between constrictions at suberin lamella and sphincters (*unlabeled arrow*), desmotubule appears as a convoluted tubule with an electron lucent lumen. *ER* Endoplasmic reticulum. *Bar* 0.2 µm. (Robinson-Beers and Evert 1991b)

though there can be an increased amount of branching and a slight neck constriction is often present, particularly on the BSC side of the interface. In *Populus*, however, the plasmodesmata are unbranched within the BSC portion of interfaces with either palisade or paraveinal mesophyll (PVM) cells, while considerable branching is typical for the palisade and PVM sides of the interface (Russin and Evert 1985).

In many C_4 grasses plasmodesmata at the MC/BSC interface have perhaps the most complex structure and associated cell wall modifications compared to other plasmodesmata in the same species (Fig. 3C, D; Evert et al. 1977; Hattersley and Browning 1981; Robinson-Beers and Evert 1991b; Botha 1992; Botha et al. 1993). These plasmodesmata have neck constrictions on one or both sides of the interface (e.g., sugarcane, Fig. 3D; Robinson-Beers and Evert 1991b, and *Themeda*: Botha et al. 1993), sphincters at one (e.g., maize, Fig. 3C; Evert et al. 1977) or both of the orifices (e.g., sugarcane: Robinson-Beers and Evert 1991b, *Themeda*: Botha et al. 1993, *Panicum*: Botha 1992), and a convoluted, membranous-appearing desmotubule with an electron lucent lumen. Typically, there is a conspicuous suberin lamella with a multilayered appearance near the middle lamella within the cell wall on the BSC side of the interface (Fig. 3C, D). The suberin lamella is conspicuously thicker in the areas of cell wall containing plasmodesmata (Evert et al. 1977; Hattersley and Browning 1981; Robinson-Beers and Evert 1991b; Botha 1992; Botha et al. 1993). The desmotubules are constricted both where they traverse the sphincters and where they traverse the suberin lamella. The cytoplasmic sleeve varies in appearance in cross-sectional views at the different levels of the plasmodesmata (Robinson-Beers and Evert 1991b; Botha et al. 1993). At the level of the sphincters, it appears filled by the amorphous, electron-dense substance composing the sphincters. Between the sphincters and suberin lamella it appears unoccluded and free of substructural components. At the level of the suberin lamella in sugarcane (Robinson-Beers and Evert 1991b) and *Themeda* (Botha 1992), the cytoplasmic sleeve contains spherical substructures that appear to be embedded in the outer leaflet of the desmotubule. In addition, some views at the level of the suberin lamella show the presence of electron-dense, spoke-like structures that radiate outward from the spherical structures toward the plasma membrane (Robinson-Beers and Evert 1991b).

The complex structure of plasmodesmata at the BSC/MC interface tempts an inference of complex regulatory functions. Physiological studies indicate that a sharp gradient of total solute concentration (Evert et al. 1978) and of specific osmotically active metabolites (Stitt and Heldt 1985) exists across the BSC/MC interface in the maize leaf. Environmental conditions that influence the photosynthetic process also influence these solute gradients. Extended dark treatment appeared to eliminate the concentration gradients (Evert et al. 1978; Stitt and Heldt 1985). These data suggest at least a loose association between the complex gradients at the MC-BSC interface and the maintenance of these gradients across (and possibly by) the complex plasmodesmata.

2.4 The Bundle-Sheath Cell–Bundle-Sheath Cell Interface

Analysis of plasmodesmata between contiguous BSC in both dicot and monocot species is often difficult because most studies focus on the direct, radial path for assimilate movement from mesophyll to the phloem (cf. Fisher 1991). However, where these plasmodesmata have been described and illustrated, e.g., *Moricandia* (Fig 4A; Beebe

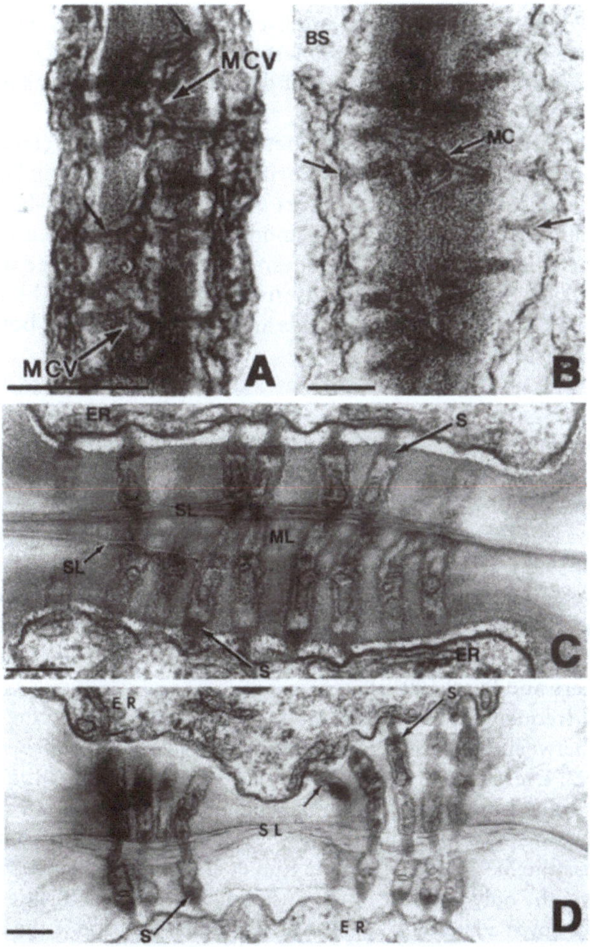

Fig. 4A–D. Plasmodesmata at bundle-sheath cell–bundle-sheath cell interfaces. **A** *Moricandia arvensis*. Transverse section through bundle-sheath cell walls, showing longitudinal views of plasmodesmata. *Unlabeled arrows* point to clearly discernible desmotubules of plasmodesmata. Note the large, extended median cavity (*MCV*). *Bar* 0.25 µm (Beebe and Evert 1992) **B** *Populus deltoides*. Longitudinal views of plasmodesmata in slightly thickened portion of cell wall between bundle-sheath cells *(BS)*. *Unlabeled arrows* point to sites of contact between endoplasmic reticulum and desmotubules of plasmodesmata. *MC* Median cavity. *Bar* 0.2 µm (Russin and Evert 1985). **C** *Saccharum*. Longitudinal section of plasmodesmata traversing the radial wall between adjacent bundle-sheath cells. Sphincters (*S*) occur at both ends of plasmodesmata. Desmotubules exhibit open lumens between the sphincters and the suberin lamella (*SL*). Plasmodesmata are narrowed and desmotubules are constricted at the level of the suberin lamella. *ER* Endoplasmic reticulum; *ML* middle lamella. *Bar* 0.2 µm. **D** *Saccharum*. Longitudinal section of plasmodesmata between bundle-sheath (*above*) and mestome-sheath (*below*) cells. Sphincters *(S)* occur at both ends of the plasmodesmata. Desmotubules are constricted both at the level of the sphincters and the level of the suberin lamella (*SL*), which occurs on the mestome-sheath-cell side of the interface. Between the sphincters and the suberin lamella the desmotubules are open. *Unlabelled arrow* points to constricted desmotubule passing through a sphincter. *ER* Endoplasmic reticulum. *Bar* 0.2 µm. (Robinson-Beers and Evert 1991b)

and Evert 1992), *Populus* (Fig. 4B; Russin and Evert 1985), and *Coleus* (Fisher 1986), the plasmodesmata are similar in structure to plasmodesmata at the MC/BSC interface in the same leaf, although the extent and size of the median cavity may be significantly greater.

In maize and sugarcane, plasmodesmata between adjacent BSCs (Fig. 4C) occur in aggregates, have neck constrictions at both orifices, and have suberin lamellae in the cell wall on both sides of the interface (Evert et al. 1977; Robinson-Beers and Evert 1991b). Suberin lamellae at the sites of plasmodesmal aggregations are significantly thicker than in the adjacent wall areas. The desmotubule is constricted in the plane of the suberin lamella (Evert et al. 1977; Robinson-Beers and Evert 1991b). In sugarcane, the plasmodesmata occur in primary pit fields and have sphincters at both ends (Robinson-Beers and Evert 1991b).

2.4.1 The Parenchymatous Bundle Sheath–Mestome Sheath Interface in Grasses

In many grass leaves, photoassimilates must cross the additional interface between the outer, parenchymatous BS and an inner mestome sheath (MS). In the C_3 grass *Bromus unioloides* (Botha 1992), and in sugarcane, one of the C_4 grasses that has an MS, plasmodesmata between two adjoining MS cells have a structure similar to that of plasmodesmata between adjacent BSCs in C_4 grass leaves (Robinson-Beers and Evert 1991b). In sugarcane, however, suberin lamellae exist only on the MS side of the interface (Fig. 4D; Robinson-Beers and Evert 1991a).

Plasmodesmal frequencies at the PBS/MS interface are similar to the frequency of plasmodesmata between BSC/BSC (Botha and Evert 1988; Robinson-Beers and Evert 1991a; Botha 1992; Evert et al. 1996b). The most detailed study of the potential for movement of assimilates through the symplasm across a MS was carried out in wheat (Kuo et al. 1974). Using plasmodesmal frequency data, the authors calculated potential sugar fluxes across the MS into the phloem. Kuo et al. (1974) concluded that the symplasmic pathway is the only possible one for assimilate traffic across the MS in wheat, and that diffusion down a lateral gradient of sugar concentration is the driving force. Such metabolite gradients have been demonstrated in the wheat leaf (Lee et al. 1980), supporting the potential symplasmic pathway across the MS into the vascular tissue.

2.5 The Bundle-Sheath Cell–Vascular Parenchyma Cell Interface

Beebe and Evert (1992) categorized minor vein parenchyma cells in the *Moricandia* leaf as either vascular parenchyma (those parenchyma cells associated spatially with tracheary elements or with both tracheary and phloem elements) or phloem parenchyma (those parenchyma cells associated with only phloem elements). Warmbrodt and VanDerWoude (1990) used similar definitions in their study of the spinach leaf. However, as these elements in the minor veins of other species cannot be identified individually as such, we will simplify our presentation and only use the term vascular parenchyma cell (VPC).

Plasmodesmata between VPC and between VPC and contiguous cells (Fig. 5), other than CCs, are generally similar in appearance to plasmodesmata at MC/MC and

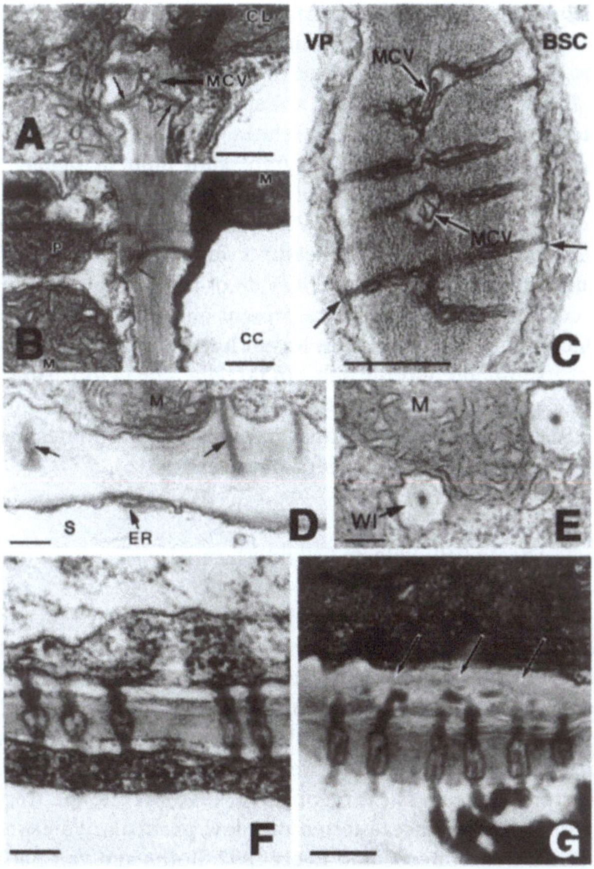

Fig. 5A–E. Plasmodesmata at bundle-sheath cell and companion cell interfaces with contiguous cells. **A** *Moricandia arvensis.* Plasmodesmata between bundle-sheath (*left*) and phloem parenchyma cells (*right*). *Unlabeled arrows* point to desmotubules, which are clearly discernible on both side of the interface. *CL* Chloroplast; *MCV* median cavity. *Bar* 0.25 μm (Beebe and Evert 1992). **B** *Moricandia arvensis.* Longitudinal view of a plasmodesma between bundle-sheath (*left*) and companion (*CC, right*) cells. Desmotubule (*unlabeled arrow*) typically is clearly discernible on the bundle-sheath side of the interface, but is not evident on the companion-cell side, where the plasmodesmal channel appears to be filled with a dense, amorphous substance. *M* Mitochondrion; *P* peroxisome. *Bar* 0.2 μm (Beebe and Evert 1992). **C** *Populus deltoides.* Longitudinal views of plasmodesmata in slightly thickened portion of cell wall between vascular parenchyma (*VP*) and bundle-sheath (*BSC*) cells. *Unlabeled arrows* point to sites of contact between endoplasmic reticulum and desmotubules of plasmodesmata. *MCV* median cavity. *Bar* 0.25 μm (Russin and Evert 1985). **D** *Sonchus oleraceus.* Cell wall between transfer cell (*above*) and sieve element (*S, below*), showing portions of plasmodesmata (*unlabeled arrows*) in wall thickening (*left*) and wall ingrowths (*right*). *ER* Endoplasmic reticulum; *M* mitochondrion. *Bar* 0.2 μm. **E** *Sonchus oleraceus.* Portion of a transfer cell, showing wall ingrowths (*WI*) in transverse section with a plasmodesma traversing the center of each ingrowth. *M* Mitochondrion. *Bar* 0.1 μm (Fisher 1991). **F, G** *Zea mays.* Plasmodesmata at the interface between bundle-sheath (*above*) and vascular parenchyma (*below*) cells in minor veins from the tips of wild type (**F**) and *sxd1* (**G**) leaves. Plasmodesmata at the bundle-sheath/vascular parenchyma interface in minor veins of the wild-type leaf clearly extend between the two cell types. In contrast, plasmodesmata at the same interface in minor veins of the *sxd1* lamina tip are greatly distorted and covered by cell wall material. *Bars* 0.2μm. (Russin et al. 1996)

MC/BSC interfaces. Like the BSC/MC plasmodesmata, these plasmodesmata typically have extensive median cavities, within which the desmotubule has a decidedly membranous appearance (Fig. 5A, C; Russin and Evert 1985; Warmbrodt and VanDerWoude 1990; Beebe and Evert 1992). There does not appear to be a difference in plasmodesmal structure across this interface that correlates with a species being either a C_3 or C_4 dicot, or functioning as an apoplasmic or symplasmic phloem loader.

In grass leaf blades the inner tangential wall of BSCs, regardless whether chlorenchymatous BS or MS, are in contact with VPC. In virtually all studied leaves, the plasmodesmata show different structural features on opposite sides of the cell wall. Sphincters occur at the orifices on the BSC side of the wall, but not on the VPC side (Fig. 5F). Neck constrictions typically are present on both sides of the interface. The desmotubule once again has an electron lucent lumen except in the areas where it is constricted by the sphincter and the suberin lamella. The cytoplasmic sleeve is more or less electron lucent. Electron dense, spoke-like structures radiate outward from the desmotubule to the plasmalemma in transverse views of plasmodesmata at this interface in the sugarcane leaf (Robinson-Beers and Evert 1991b).

2.6 The Bundle-Sheath Cell–Phloem Cell Interface

This interface is especially critical, because it represents the physical and physiological interface between the photosynthetic mesophyll tissues and the vascular system. Moreover, this cell-to-cell interface represents a likely site of unloading of assimilates from the symplasm into the apoplast in both dicot and monocot apoplasmic phloem loaders (Geiger 1976; Giaquinta 1983), while for symplasmic loaders this is the interface across which plasmodesmata act to prevent diffusive outflow and loss of raffinose oligosaccharide transport sugars (Turgeon 1991, 1996). While the frequency of plasmodesmata across these interfaces is often very low, particularly in species that translocate sucrose (Fisher 1990; Beebe and Evert 1992; Botha and van Bel 1992), plasmodesmata present at this interface show considerable variation in structure.

2.6.1 The Bundle-Sheath Cell–Companion Cell Interface

The degree of plasmodesmal structural complexity at the BSC/CC interface correlates strongly with differences in transport sugar composition (Turgeon et al. 1993), and hence, phloem-loading strategies (van Bel et al. 1992, 1994; van Bel 1993; Flora and Madore 1996; Grusak et al. 1996).

Species with transfer cells (TC; Gunning et al. 1968, 1974; Gunning and Pate 1969), e.g., *Pisum* (Wimmers and Turgeon 1991), *Sonchus* (Fisher 1991), *Cymbalaria* and *Linaria* (Turgeon et al. 1993), are representative of plants specialized for carrier-mediated uptake of sucrose across the plasmalemma of phloem cells (Grusak et al. 1996). In spite of the classification of TC-containing minor veins as "closed" (Gamalei 1989), TC are not symplasmically isolated from surrounding cells. *Sonchus* (Fisher 1991), *Linaria* (Turgeon et al. 1993), and *Vicia* (Gunning et al. 1974; Bonnemain et al. 1991) all have plasmodesmal connections with BSC, albeit at very low frequencies. Plasmodesmata at the BSC/TC interface, as well as across other TC interfaces, are unbranched. *Sonchus*, in particular, is unusual in having plasmodesmata that both traverse the cell-wall ingrowths themselves and occur in other thickened portions of the cell wall (Fig. 5D, E; Fisher 1991). In certain species, e.g., *Linaria* (see Fig. 4 in Turgeon et al. 1993), the plasmodesmal channels within the TC portion of the BSC/TC interface appear to be very narrow, lacking a discernible cytoplasmic sleeve and desmotubule.

Plants with ordinary CC, which lack wall ingrowths but are otherwise ultrastructurally similar to TC, are also species that are believed to be apoplasmic phloem loaders (Grusak et al. 1996). Ordinary CC have been among the most intensely studied parenchymatous elements of leaf minor vein phloem. Within the apoplasmic loading dicots, plasmodesmata between BSC and CC are typically branched, with the greater number of branches occurring on the BSC side of the interface. As for the BSC/TC interface, the number of plasmodesmata at the BSC/CC interface is usually very low, among the lowest of plasmodesmal frequencies that have been measured in minor veins (Evert and Mierzwa 1986; Fisher 1990; Beebe and Evert 1992). In some species, the plasmodesmal channel in the CC portion of the wall appears to have a markedly different structure compared to the continuation of the plasmodesma in the BSC wall, and is similar to the plasmodesmal channels within the BSC/TC interface in *Linaria* (Turgeon et al. 1993). The best example to date of this unique plasmodesmal structure has been observed in *Moricandia arvensis* (Beebe and Evert 1992). The plasmodesmal channels within the CC wall are modified such that the diameter of the channel is reduced along its entire length, the desmotubule is typically not distinct, and the channel is filled with a dense, amorphous material (Fig. 5B). What is striking is that the continuation of the very same plasmodesma in the BSC wall has a typical plasmodesmal structure: there is a visible desmotubule, surrounded by an electron lucent (open) cytoplasmic sleeve, in a channel having a diameter of normal dimension (Fig. 5B). In the region of the middle lamella, a median cavity, within which the desmotubule has a membranous appearance, interconnects the branches of small plasmodesmal aggregates (Beebe and Evert 1992).

Given that plasmodesmata at the BSC/CC and BSC/TC interfaces are not believed to play a direct role in apoplasmic phloem loading, what functional significance can be attributed to the their presence? Might they be non-functional remnants of connections that were once physiologically functional? Recent studies by Gagnon and Beebe (1996a, b) on leaf development and aspects of minor vein differentiation associated with the sink-to-source transition (Turgeon 1989) showed that plasmodesmata across the BSC/CC interface undergo secondary modification associated with the differentiation of the SE/CC complex. There did not appear to be a strict correlation between plasmodesmal differentiation at the BSC/CC interface and a given region's importing or non-importing status, as structurally mature, specialized plasmodesmata were found within all regions of the sampled leaves (Gagnon and Beebe 1996b).

Is there evidence to support symplasmic continuity via plasmodesmata at BSC interfaces? Analysis of intercellular dye-coupling following ionophoretic microinjection of various fluorescent dyes into cells within source leaves of both dicots (*Ipomea*: Madore et al. 1986, *Coleus*: Fisher 1988, *Cucurbita*: (D.U. Beebe and R. Turgeon, unpubl. results; Turgeon and Hepler 1989) and monocots (*Commelina*: Erwee et al. 1985; van Kesteren et al. 1988, *Hordeum*: Botha and Cross 1997) has demonstrated that there is symplasmic continuity that permits diffusive solute movement between the mesophyll and the long-distance pathway of the phloem. While the results of these studies do not reveal information about the specifics of the phloem-loading mechanisms in these species, they do establish that plasmodesmata along the MC›BSC›VPC›phloem cell pathway are open and may function in assimilate movement to the phloem.

In recent years, considerable attention has been focused on the possibility that certain species may have an entirely symplasmic phloem-loading pathway (for recent reviews, see van Bel 1989, 1992, 1993; Turgeon and Beebe 1991; Grusak et al. 1996; Turgeon 1996). A consensus has developed that such a mechanism is feasible for species translocating raffinose and stachyose in addition to sucrose. A recent report indicates that certain CAM species that translocate fructans may also use a symplasmic loading mechanism (Wang and Nobel 1998). The model described by Turgeon (1991) has been well received, particularly as the model's predictions have been supported by studies that have identified the site of raffinose oligosaccharide synthesis as the IC (Holthaus and Schmitz 1991; Beebe and Turgeon 1992; Turgeon and Gowan 1992), confirmed the distribution of the raffinose oligosaccharides within the source leaf (Haritatos et al. 1996), described the unusual structure and frequency of plasmodesmata across the BSC/IC interface (Fisher 1986; Turgeon et al. 1993; Beebe et al. 1996; Haritatos et al. 1996; Volk et al. 1996; Beebe and McEvoy 1997), and verified the correlation between the kinds of transport sugars and minor vein anatomy (van Bel et al. 1992; Turgeon et al. 1993; van Bel et al. 1994; Flora and Madore 1996).

Intermediary cells are specialized for the synthesis of the raffinose oligosaccharides (Holthaus and Schmitz 1991; Beebe and Turgeon 1992; Haritatos et al. 1996; Turgeon 1996). Although a large number of species are known to have ICs (Gamalei 1989), only three species have been studied in detail to date: *Coleus* (Fisher 1986), and two members of the squash family, *Cucurbita* and *Cucumis* (Turgeon et al. 1975; Schmitz et al. 1987; Beebe et al. 1996; Volk et al. 1996). Intermediary cells are large, densely cytoplasmic, contain a large number of small vacuoles, and are always found in a consistent spatial arrangement relative to the other cells of the minor vein. Unlike CC and TC, which have well-developed chloroplasts that are often observed to contain starch grains (Beebe and Evert 1992), the plastids of ICs are very poorly developed (lacking well-defined thylakoid membranes systems) and typically devoid of starch grains. A striking feature of these cells is the presence of a great many plasmodesmata at the BSC/IC interface (Turgeon et al. 1975; Fisher 1986; Schmitz et al. 1987; Gamalei 1989; Haritatos et al. 1996; Volk et al. 1996), where loading presumably occurs. The plasmodesmata at this interface are very different structurally from plasmodesmata seen at other cell-to-cell interfaces in the leaf (Fig. 6). They are distributed in large aggregates or fields (Fig. 6B) where the wall is thickened and they are branched on both sides, more so on the IC side (Figs. 6A, C). Computer generated three-dimensional reconstructions from serial ultrathin sections of BSC/IC plasmodesmal aggregates in *Cucumis* leaves have revealed an astounding complexity of interconnections between plasmodesmal channels on either side of the interface (G. Volk, pers. comm.). The IC channels are longer and narrower than those in the BSC wall and the desmotubule is commonly difficult to discern in the IC channels (Fig. 6). Interestingly, a median cavity is often well developed in the region of the middle lamella (Fig. 6C, unlabelled arrow) and, in a pattern similar to that seen in plasmodesmata across the *Moricandia* BSC/CC interface (Beebe and Evert 1992), the plasmodesmal channel within the BSC wall has a typical structure with an open cytoplasmic sleeve and a clearly visible appressed ER strand (Fig. 6). The symplasmic phloem-loading model (Fig 1B; Turgeon 1991, 1996)

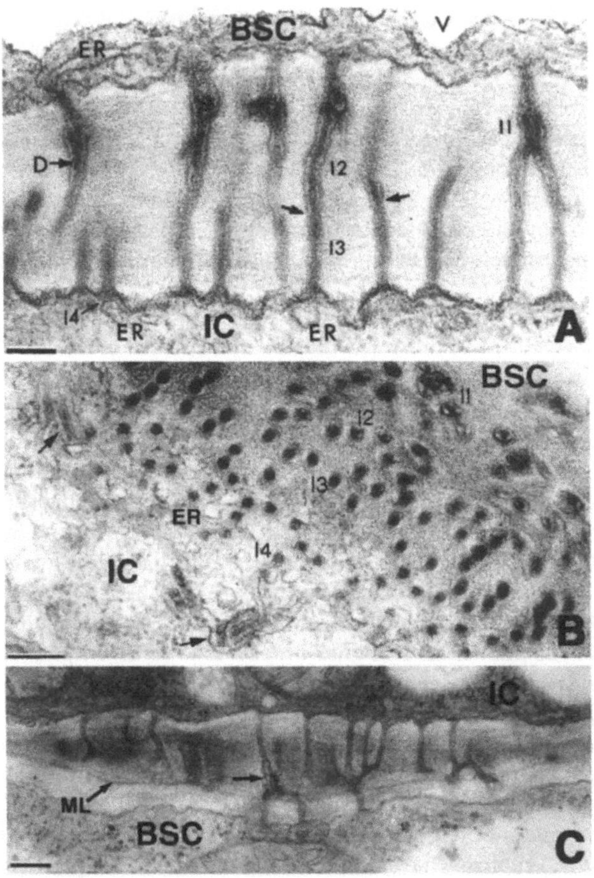

Fig. 6A–C. Plasmodesmata between bundle-sheath cells (*BSC*) and intermediary cells (*IC*). **A, B** *Coleus blumei.* **A** Longitudinal view of plasmodesmata between bundle-sheath (*BSC*) and intermediary cells (*IC*). Levels *11–14* correspond to similar levels 11–14 in **B**. *Unlabeled arrows* point to beginning of extended neck region. *ER* Endoplasmic reticulum. *Bar* 0.1 μm. **B** Glancing section through wall between bundle-sheath cell (*BSC*) and intermediary cell (*IC*), showing numerous plasmodesmata in transverse view. Plasmodesmata at locations *11–14* correspond to similar levels 11–14 in **A**. *Unlabeled arrows* point to endoplasmic reticulum-associated tubules. *ER* Endoplasmic reticulum. *Bar* 0.2 μm (Fisher 1986). **C** *Cucurbita pepo.* Longitudinal view of plasmodesmata between intermediary (*IC*) and bundle-sheath (*BSC*) cells from a minor vein in a fully expanded cotyledon. Note the highly branched plasmodesmal channels, asymmetric thickening and increased staining density of the IC wall, and the enlarged plasmodesmal channels near the region of the middle lamella (*ML*) in this wall. As in *Coleus*, there appears to be an extended, narrow neck region in the plasmodesmal channels on the IC side of the interface. *Bar* 0.2 μm. (D. U. Beebe, unpubl. results)

posits that the size exclusion limit (SEL) of plasmodesmata at the IC interface is reduced, thereby greatly inhibiting the diffusive loss of raffinose oligosaccharides. This prediction has been confirmed by Haritatos and coworkers (1996) from their detailed analyses of sugar concentrations in individual cell and tissue types, plasmodesmal frequency, and calculations of potential sugar fluxes through plasmodesmata across the

BSC/IC interface. They showed that the high concentrations of raffinose and stachyose measured within the SE-IC complexes supported the prediction that plasmodesmata at the BSC/IC interface have an SEL that is sufficiently small to eliminate diffusive loss of the raffinose oligosaccharides (Haritatos et al. 1996).

In monocot leaves, BSC (regardless whether chlorenchymatous BS or MS) have few if any plasmodesmal connections to CC (Chonan et al. 1985; van Bel et al. 1988; Botha 1992; Botha and van Bel 1992; Evert and Russin 1993; Evert et al. 1996b). The highest plasmodesmal frequency reported for monocots is in the sugarcane leaf at the MS-CC interface (Robinson-Beers and Evert 1991a). Illustrations and descriptions of the BSC-CC connections are lacking.

2.6.2 The Bundle-Sheath Cell–Sieve Element Interface

The typical spatial arrangement of phloem cells in a minor vein results in little direct contact between SE and BSC (Fig. 7A, B). Sieve elements are commonly found in an interior position within the vein and are typically isolated from the surrounding BSC, making an association between these two cell types very infrequent. Even if some degree of contact is present, there do not appear to be any plasmodesmata between these two cell types. During an analysis of plasmodesmal frequency in *Moricandia* (Beebe and Evert 1992), in which over 2600 plasmodesmata were counted along a total interface length of 20 700 μm within 15 higher-order minor veins, only 11 μm of cell-to-cell contact were measured between SE and BSC, and no plasmodesmata were observed across this length of interface.

In *Commelina* and several C_3 and C_4 grasses, there are two types of sieve tube member that can be distinguished on the basis of their structure, position, and time of maturation (van Bel et al. 1988; Evert and Russin 1991; Evert et al. 1996a, b). They are called thin-walled and thick-walled sieve-tube members (Fig. 7C). Thin-walled sieve-tube members (TnWST) typically have thinner walls, characteristic plastids, pore-plasmodesmal connections with CC (Fig. 7C, D), and reach maturity earlier than thick-walled sieve-tube members (TkWST). In the species studied to date, protophloem (in those

Fig. 7. A, B. Portions of minor veins and surrounding bundle-sheath cells in leaves of *Moricandia arvensis* (**A**) and *Beta vulgaris* (**B**). Note that the majority of sieve elements (*S*) have no contact with cells other than vascular cells, and rarely contact bundle-sheath cells (*BS*). *CC* Companion cell; *MB* multivesicular body; *PP* and *PhP* phloem parenchyma cell; *VP* vascular parenchyma cell. **A** *Bar* 5 μm (Beebe and Evert 1992). **B** *Bar* 4.3 μm (Evert and Mierzwa 1986). **C** *Zea mays*. Transverse section of a portion of a maize leaf minor vein showing two sieve tubes. Thick-walled sieve tube (*TkWST*) has numerous pore-plasmodesmal connections with three adjacent vascular parenchyma cells (*VP*). Thin-walled sieve tube (*TnWST*) has connections with a companion cell (*CC*). *V* Vessel member. *Bar* 0.71 μm (Evert et al. 1978). **D** *Hordeum vulgare*. Longitudinal section of pore-plasmodesmal connections between thin-walled sieve tube and a companion cell (*CC*) in a leaf minor vein. Branched plasmodesmata occur in thickened regions of cell wall on the companion cell (*CC*) side, callose-lined pores occur in slightly thickened areas of cell wall on the sieve-tube side. *Bar* 0.16 μm (Evert and Mierzwa 1989). **E** *Hordeum vulgare*. Detail of pore-plasmodesmal connection between thick-walled sieve tube (*lower left*) and vascular parenchyma cell (*upper right*) from a leaf minor vein. *Bar* 0.16 μm. (Evert and Mierzwa 1989)

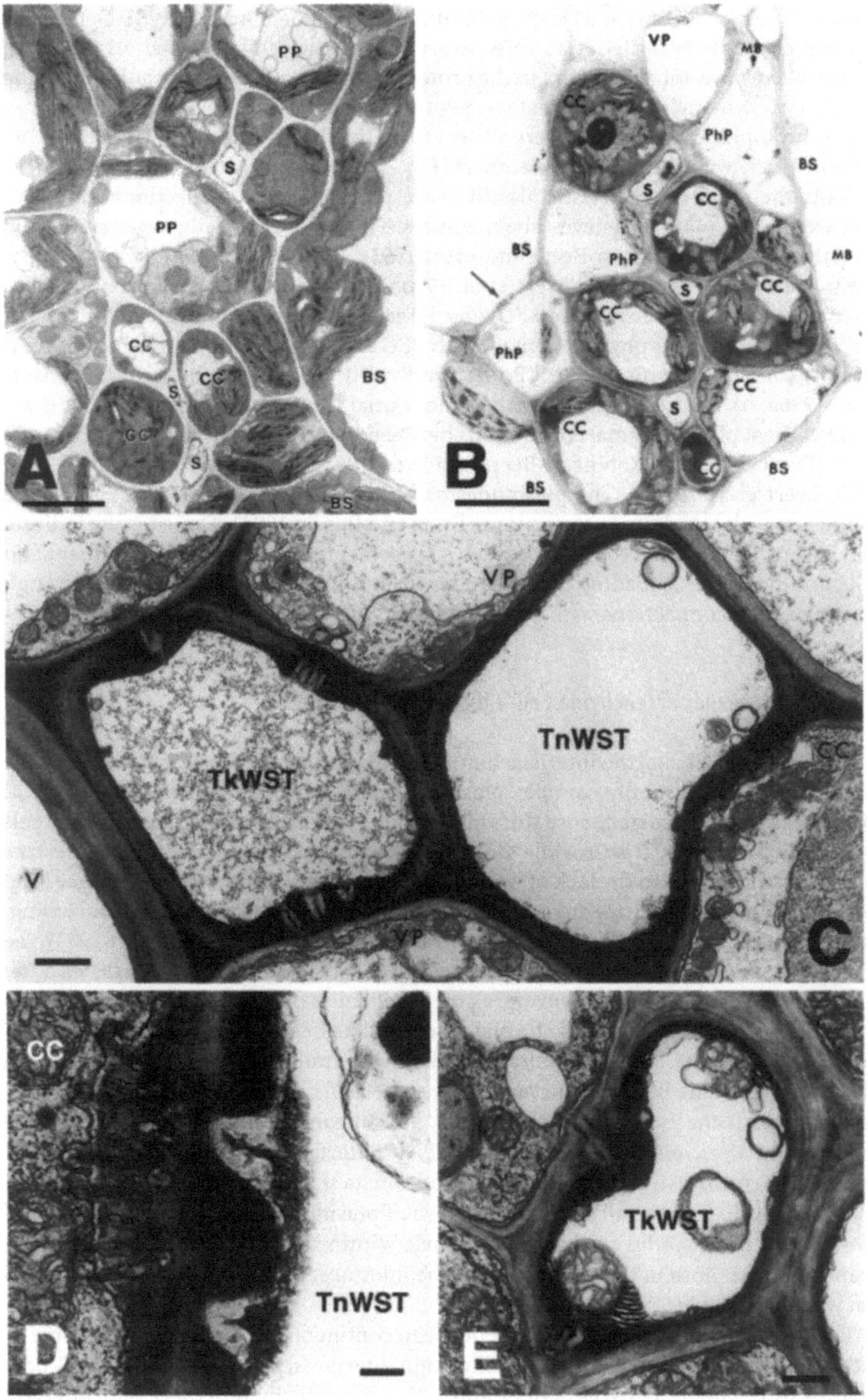

PP
S
PP
CC
CC
S
CC
S
BS
BS
A
VP
MB
CC
PhP
BS
S
CC
BS
PhP
MB
S
CC
CC
PhP
CC
S
CC
CC
BS
B
BS
VP
TnWST
TkWST
CC
V
VP
C
CC
TnWST
D
TkWST
E

classes of vein that contain it) and much of the metaphloem have TnWST as the conducting elements (van Bel et al. 1988; Evert and Russin 1991; Evert et al. 1996a, b). Thin-walled sieve-tube members in the protophloem are often directly adjacent to the BSC. Very few, if any, plasmodesmata were present in at the BSC/SE interface in studies where protophloem elements were observable (Evert and Russin 1993; Evert et al. 1996b). Thick-walled sieve-tube members (Fig. 7C, E), in contrast, typically have thicker walls, their own characteristic plastids, pore-plasmodesmal connections with VPC, and are the last maturing sieve-tube members. TkWST are virtually never connected directly with CC (Robinson-Beers and Evert 1991a, b; Botha 1992; Botha and van Bel 1992; Evert and Russin 1993; Evert et al. 1996b). In addition, TkWST occur in close proximity to the vessel members. Owing largely to their position near the xylem, TkWST have no direct physical contact with BSC and, therefore, are not in direct symplasmic connection with the BSC (Botha and Evert 1988; Evert and Russin 1993; Evert et al. 1996a, b). Studies centering on plasmodesmal frequency have shown that there are few direct plasmodesmal connections between bundle-sheath cells and either type of SE (Evert et al. 1977; Robinson-Beers and Evert 1991a; Botha 1992; Evert and Russin 1993; Evert et al. 1996b). The plasmodesmata that do interconnect BSC/TnWST are simple pore-plasmodesma connections (Evert and Russin 1993). These results indicate that except for possible connections between protophloem sieve-tube members and BSC, any symplasmic loading of sieve tubes from the BSC must be mediated through plasmodesmal connections with other cell types.

2.7 The Vascular Parenchyma Cell–Phloem Cell Interface

Within leaves of dicots, the interface between vascular parenchyma and phloem cells often contains structurally complex plasmodesmata, particularly across the VPC/CC interface. Plasmodesmata across this interface in *Populus* have long narrow channels within the thickened CC wall, while the channels within the VP portion of the interface are much shorter due to the lack of wall thickening on the VP side of the interface (Fig. 8A, B). The median cavities are extensive and the desmotubules form anastomosing, membranous complexes (Fig. 8B). In *Moricandia arvensis* (Beebe and Evert 1992), the VP/CC plasmodesmata are highly branched on the VP side of the interface, with the VP plasmodesmata channels converging on a median cavity. A slight neck constriction is usually apparent surrounding the plasmodesmal orifice. In contrast, the plasmodesmata in CC wall are typically unbranched and are similar to the CC portions of the plasmodesma at the BSC/CC interface in the same leaf, i.e., the desmotubule is not discernible, nor is the cytoplasmic sleeve (Fig. 5B). In contrast to the complexity of the interfaces in *Moricandia* and *Populus*, the VP/CC interface in *Solanum* (McCauley and Evert 1989) consists of rather simple plasmodesmata that occur in a slightly thickened area of the wall. The overall structural pattern of plasmodesmata is similar to those of *Moricandia* and *Populus* in that the channels within the VPC wall are more highly branched than those in the CC portion of the interface, although the median cavity is not well developed.

The distribution of plasmodesmata between contiguous VP cells can yield some of the highest frequency values among minor vein interfaces, particularly where it is possible to differentiate between xylem and phloem parenchyma (Evert and Mierzwa

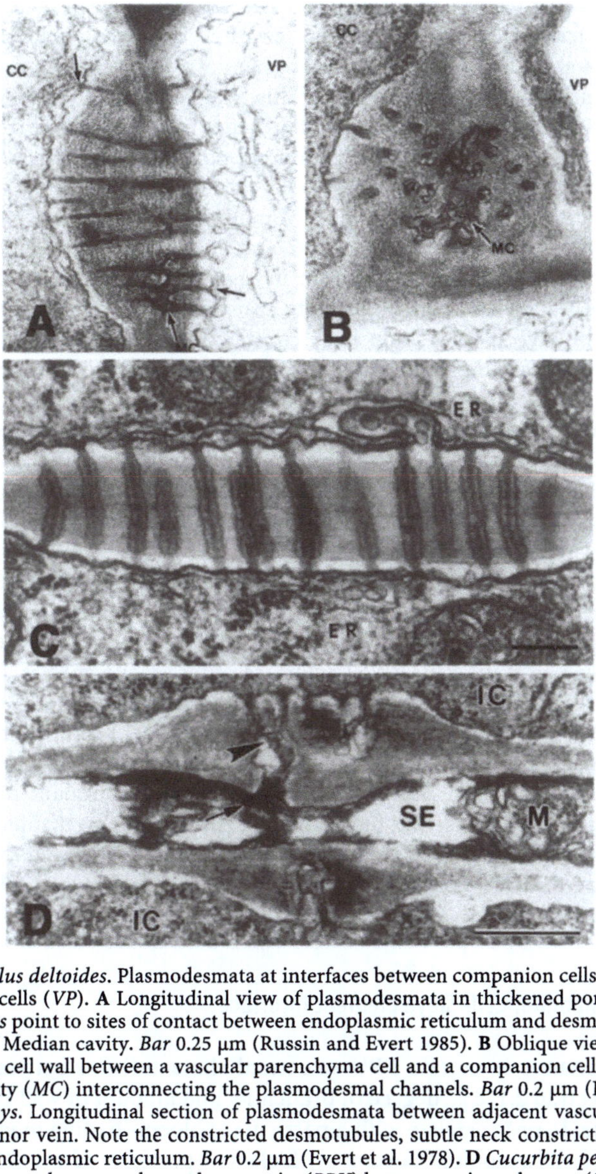

Fig. 8. A, B *Populus deltoides*. Plasmodesmata at interfaces between companion cells (*CC*) and vascular parenchyma cells (*VP*). **A** Longitudinal view of plasmodesmata in thickened portion of cell wall. *Unlabeled arrows* point to sites of contact between endoplasmic reticulum and desmotubules of plasmodesmata. *MC* Median cavity. *Bar* 0.25 μm (Russin and Evert 1985). **B** Oblique view of plasmodesmal aggregate in cell wall between a vascular parenchyma cell and a companion cell. Note the extensive median cavity (*MC*) interconnecting the plasmodesmal channels. *Bar* 0.2 μm (Russin and Evert 1985). **C** *Zea mays*. Longitudinal section of plasmodesmata between adjacent vascular parenchyma cells in a leaf minor vein. Note the constricted desmotubules, subtle neck constrictions, and lack of sphincters. *ER* Endoplasmic reticulum. *Bar* 0.2 μm (Evert et al. 1978). **D** *Cucurbita pepo*. Longitudinal section through complex pore-plasmodesma units (*PPU*) between a sieve element (*SE*) and two contiguous intermediary cells (*IC*) within a minor vein. Note the fusion (*arrowhead*) of the two IC desmotubules within the median cavity to form a larger, dilated central structure in the SE portion of the interface. *Unlabeled arrow* indicates parietal array of smooth endoplasmic reticulum associated with the PPU. *M* Mitochondrion. *Bar* 0.5 μm. (Volk et al. 1996)

1986; McCauley and Evert 1989; Beebe and Evert 1992). For example, an analysis of plasmodesmal frequencies in *Moricandia* leaves revealed that phloem parenchyma cells have more abundant plasmodesmal connections with adjacent cells than xylem parenchyma cells (Beebe and Evert 1992). This relatively large number of plasmodesmata at the VP/VP interface may reflect the presence of a symplasmic route that would permit a more even distribution of incoming photoassimilates around the region of the SE-CC complex.

Vascular parenchyma cells in the veins of monocot leaves can have plasmodesmal connections with other phloem cells: VP, CC, and TnWST and TkWST. Between VP/VP (Fig. 8C) and VP/CC plasmodesmata occur in aggregates in at least slightly thickened portions of cell wall. These plasmodesmata are structurally similar on both sides of the interface. All have at least slight neck constrictions, desmotubules that are more electron dense than at other interfaces, and cytoplasmic sleeves that appear electron dense and mottled; sphincters are absent (Robinson-Beers and Evert 1991b).

As in the dicots, interfaces between adjacent VP cells in monocots typically have very large plasmodesmal frequencies (Botha and Evert 1988; van Bel et al. 1988; Evert et al. 1996b). The plasmodesmal frequencies at the VP/CC interface are typically much lower than those found at the VP/VP interface. In addition, the VP/CC frequency is quite variable when compared among species.

The interface between VP and TkWST in all studied monocot species have typical pore-plasmodesma units (PPU) with small pores on the sieve-element side of the wall and branched plasmodesmata on the VP cell side (Fig. 7C, E). As with the VP/CC interface, the frequencies at the VP/TkWST interface differ among species. In *Bromus* and *Hordeum*, two C_3 grasses, there are very few plasmodesmata at the VP/TkWST interface (Botha 1992; Evert et al. 1996b; Botha and Cross 1997). In contrast, the other grass species examined (Robinson-Beers and Evert 1991a; Botha 1992; Evert and Russin 1993) and *Commelina* (van Bel et al. 1988) have more abundant plasmodesmata between VP and TkWST. Vascular parenchyma cells can also be symplasmically connected to TnWST. In contrast to the data from *Themeda* (Botha and Evert 1988), the VP/ST interfaces typically contain few plasmodesmata (van Bel et al. 1988; Robinson-Beers and Evert 1991a; Botha 1992; Evert and Russin 1993).

2.8 The Companion Cell–Sieve Element Interface

There are differences in certain PPU structural characteristics between the species identified in Section 2.1, e.g., degree of branching and amount of wall thickening, but these variations do not correlate with a particular phloem-loading mechanism. Pore-plasmodesma units commonly occur in thickened portions of the CC/SE interface (Figs. 7C, 8D). Most commonly, the plasmodesmal channels are branched, sometimes extensively. Complex branched, plasmodesmal groupings are observed between TC and adjacent SE (Gunning et al. 1968; Fisher 1991; Wimmers and Turgeon 1991). Similar elaborate PPUs are found in both *Cananga* (Fisher 1990) and *Populus* (Russin and Evert 1985). In other species (*Beta*, Evert and Mierzwa 1986), both simple and complex branching may occur. Again, there does not appear to be a pattern in the extent of plasmodesmal branching that can be associated with any particular species or a likely phloem-loading mechanism.

The pore of the PPU is typically the same diameter or slightly larger than the diameter of the plasmodesmal channels in the same PPU. Exceptions to this general characteristic can be found in the PPU of *Cucurbita* (Volk et al. 1996) and *Ipomea* (Madore et al. 1986), where the pore is quite large. In *Cucurbita*, the pore is up to four times the diameter of the plasmodesmal channels in the PPU (Fig. 8D).

In light of the structural complexity observed in some species at the CC interfaces with cells other than SE, e.g., *Cucurbita* (Volk et al. 1996), *Coleus* (Fisher 1986), and *Moricandia* (Beebe and Evert 1992), it is interesting to note that plasmodesmal channels of the PPU often lack the same structural modifications. Plasmodesmal channels in PPU commonly have a distinct cytoplasmic sleeve, whereas the plasmodesmal channels at IC and CC interfaces with surrounding BSC and VPC have narrow channels. Plasmodesmal channels across the IC/IC and CC/CC interfaces are apparently open, as judged by the presence of a distinct cytoplasmic sleeve (Fisher 1986; Beebe and Evert 1992; Volk et al. 1996) and by dye-coupling studies (D.U. Beebe and R. Turgeon, unpubl. results; Turgeon and Hepler 1989)

In the monocots, connections between companion cells with thin-walled sieve-tube members consist of small pores on the SE side of the wall and numerous branched plasmodesmata on the parenchymatous-cell side (Fig. 7D; Walsh and Evert 1975; Botha and Evert 1988; Robinson-Beers and Evert 1991b). The plasmodesmata occur in more or less thickened portions of the wall and are associated with neck constrictions (Evert and Mierzwa 1989; Evert and Russin 1993). Sphincters are lacking, and the desmotubules appear electron-dense for their entire length (Robinson-Beers and Evert 1991b).

In the monocot species that have been examined in detail, PPUs between CC and SE are abundant. However, the SE/CC complexes of most monocot species are almost completely isolated from all other minor vein cell types (Evert et al. 1977; van Bel et al. 1988; Robinson-Beers and Evert 1991b; Botha 1992; Evert and Russin 1993). One exception is *Themeda* (Botha and Evert 1988), where the SE/CC are in symplasmic continuity with other phloem cells.

Is the ER continuous through the PPU? While Esau and Thorsch (1985) stated that there is no continuity of ER from CC to SE, studies of *Nicotiana* (Ding et al. 1993) and *Cucurbita* (Volk et al. 1996) have presented convincing evidence for intercellular ER continuity (Fig. 8D). The continuity and substructure of the ER from CC to SE via the PPU remains unclear for the monocots. Hence, the exact nature of a common PPU substructure remains to be determined. We believe, however, that the PPU is fundamentally a highly modified plasmodesmal structure, and as the ER of adjoining SE is continuous through the sieve-plate pores, it is most likely also continuous from CC to SE via the PPU.

2.9 Plasmodesmal Modifications Associated with the Sink-to-Source Transition

In many species, e.g., *Beta vulgaris* (Fellows and Geiger 1974; Schmalstig and Geiger 1987), *Cucumis melo* and *Cucurbita pepo* (Turgeon and Webb 1976; Beebe et al. 1996; Volk et al. 1996; Beebe and McEvoy 1997), *Nicotiana tabacum* (Ding et al. 1988), *Populus deltoides* (Isebrands and Larson 1973), and *Zea mays* (Evert et al. 1996a), the structural maturation of minor vein phloem in expanding leaves is closely correlated with the initiation of photoassimilate export. The period during which a developing leaf makes the transition from an importing to an exporting status is termed the sink-to-source transition (Turgeon 1989). Structural and physiological changes associated with this transition are initiated in the apical region of the expanding leaf and progress basipetally. If modifications of existing primary plasmodesmata and/or the formation of secondary plasmodesmata occur in minor vein phloem prior to or concurrently with the sink-to-source transition, it is likely that these plasmodesmata may play a role in phloem loading.

Several recent studies have directly addressed this question. Gagnon and Beebe (1996b) analyzed ultrastructural changes that occurred in three higher-order minor vein classes during minor vein CC differentiation in expanding leaves of *Moricandia arvensis*. The differentiation of CC plasmodesmata (Fig. 9A–F) was linked to the maturation of the SE-CC complex, and not to the region's overall importing or nonimporting status. There was not a strict correlation between the termination of importing activity and the differentiation of plasmodesmata at CC interfaces, particularly as fully differentiated SE-CC complexes were observed in both importing and non-importing tissue. This observation supports the hypothesis that plasmodesmata at the CC interface with surrounding cells do not play a primary role in apoplasmic phloem loading, but function in a more secondary role: they are not actively involved in sucrose uptake and may simply act to prevent significant diffusive loss of loaded sucrose, although some leakage of sucrose via these channels could occur (Gagnon and Beebe 1996b).

In earlier studies on phloem unloading in the developing tobacco leaf (Ding et al. 1988), Ding and coworkers showed that, unlike in the cucurbits, there was a profound decrease (approx. 50%) in the number of plasmodesmata between the SE-CC complex and the surrounding cells. They also found that there was a significant decrease in plasmodesmal frequencies at all interfaces between the SE-CC complex and the mesophyll associated with a region's transition from an importing to a nonimporting status. They suggested that this decline could help explain why mature leaves do not import photoassimilate (Turgeon 1986, 1987). It may be, as for the *M. arvensis* leaf, that an increase in symplasmic isolation of the SE-CC complex is a necessary prerequisite for the establishment of phloem-loading capacity within species that have an apoplasmic phloem-loading mechanism.

In contrast to the *Moricandia arvensis* and *Nicotiana* results, Volk and coworkers (1996) showed that the increased plasmodesmal frequency and the development of structurally complex plasmodesmata across the BSC/IC and IC/IC interfaces in expanding leaves of *Cucumis* and *Cucurbita* (Fig. 9G, H) was tightly linked to the establishment of phloem-loading capacity and the initiation of photoassimilate export. Moreover, analysis of IC and BSC expansion and changes in plasmodesmal frequencies at the BSC/IC interface in importing, transition, and mature *Cucumis* leaf tissue revealed that many of the abundant plasmodesmata are true secondary plasmodesmata, i.e., formed de novo across the preexisting cell-to-cell interface. The formation of these modified plasmodesmata prior to the sink-to-source transition, initiation of phloem loading, and assimilate export supports the hypothesis that plasmodesmata at the BSC/IC interface provide a symplasmic channel for the movement of sugars from the mesophyll into the IC (Turgeon 1996; Volk et al. 1996).

Interest in the involvement of plasmodesmata in sink-to-source transition in grass leaves has centered on the MC/BSC interface in sugarcane (Robinson-Beers et al. 1990;

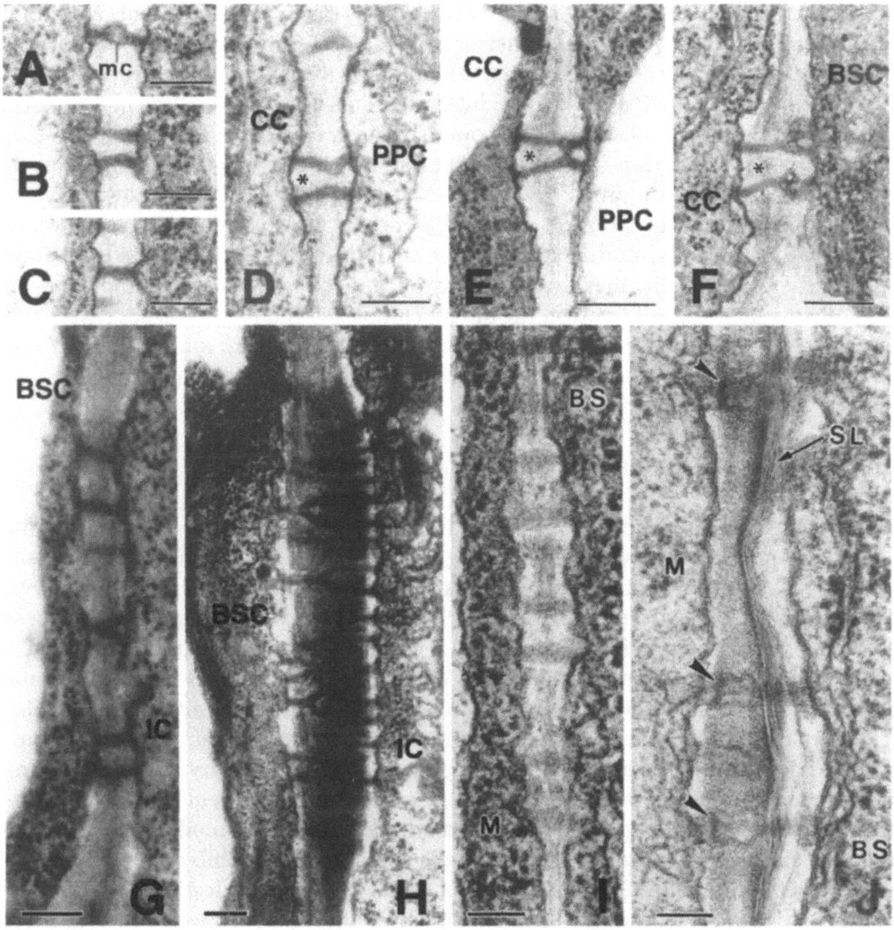

Fig. 9A–J. Plasmodesmal development during the sink-to-source transition. **A–F** *Moricandia arvensis* Specialized plasmodesmata between companion cell (*CC*) and phloem parenchyma cell (*PPC*) (**A–E**) and between CC and bundle-sheath cell (*BSC*) (**F**) interfaces at different stages of development in expanding *Moricandia* leaves. *Asterisks* indicate areas of asymmetric thickening on the CC side of the interface. *MC* Median cavity. *Bars* 0.25 µm (Gagnon and Beebe 1996b). **G, H** *Cucurbita pepo* Plasmodesmata at a bundle-sheath (*BSC*) and intermediary cell (*IC*) interface from importing (**G**) and non-importing (**H**) regions in an expanding squash leaf. *Bars* 0.2 µm (Volk et al. 1996). **I, J** *Zea mays* Portions of mesophyll (*M*) and bundle-sheath cell (*BS*) interfaces from small vascular bundles in developing (**I**) and expanded (**J**) *Zea* leaves. *Arrowheads* indicate sphincters; *SL* suberin lamella. *Bars* 0.1 µm. (Evert et al. 1996a)

Robinson-Beers and Evert 1991c) and maize (Evert et al. 1996a). This interface is of interest largely because its plasmodesmata undergo the most extensive structural changes during transition (Fig. 9I, J), and because of the interest in metabolite transport between these two cell types in C_4 grasses (e.g., Stitt and Heldt 1985).

In both maize and sugarcane, continuous suberin lamellae were present in the cell walls and the vascular tissues were found to be mature in advance of import cessation. Sphincters associated with plasmodesmata at the MC/BSC interface in both leaves were the last structures to mature (Fig. 9J). In maize, the sphincters appeared (simultaneously at the MC/MC and MC/BSC interfaces) in a region of the leaf that was still importing (Evert et al. 1996a). In sugarcane, sphincters appeared first on the MC side in a region that was in transition, then on the BSC side in a region that was nonimporting (Robinson-Beers and Evert 1991c). In both studies, changes observed in plasmodesmal structure at the MC/BSC interface were not directly related to cessation of phloem unloading. Plasmodesmal frequency analysis of vascular tissues in expanding maize leaves showed that there was a paucity of connections between protophloem SE and other cell types, and between the SE-CC complex and other cell types (Evert and Russin 1993). A decrease in symplastic continuity between SE and other cells was therefore not likely to be a factor involved in the cessation of phloem loading.

3 The Role of Plasmodesmata in Phloem Loading

In the apoplasmic phloem-loading model (Fig. 1A), plasmodesmata are present at the SE-CC complex interface with surrounding cells, but at very low frequencies. If the sucrose is taken up from the apoplast across the plasmalemma (Sauer and Stolz 1994; Stadler et al. 1995; Truernit and Sauer 1995; Rentsch and Frommer 1996), do these plasmodesmata function in phloem loading? The answer may be that their function is indirect, in that they appear to undergo structural modifications that occlude the intercellular diffusive pathway (Beebe and Evert 1992; Gagnon and Beebe 1996b). These modifications occur during differentiation of the minor vein phloem and appear to be related to the establishment of phloem-loading capacity in the expanding leaf (Gagnon and Beebe 1996b). If symplasmic isolation is a prerequisite for the initiation of assimilate loading, what function do the remaining plasmodesmata at the CC interface fulfill? They may simply provide the necessary intercellular pathway needed to maintain a minimal, but essential, level of cell-to-cell communication. Any diffusive loss of loaded sucrose from the SE-CC complex via these plasmodesmata would most likely be negligible in comparison with the movement of sucrose into the SE pathway. An alternate explanation may be that plasmodesmata across this interface become completely non-functional once a leaf makes the transition to export (Turgeon 1989), and therefore simply represent a vestige of former intercellular connections. At present, however, there is no experimental evidence for a symplasmic discontinuity or restriction across this interface, so the functional implication of this unusual plasmodesmal structure remains unknown.

Plasmodesmata across the BSC/IC interface in the symplasmic loading model (Fig. 1B) play a critical yet converse role. They act to ensure diffusive inflow of sucrose and other metabolically important molecules into the CC, but at the same time, they are the "filter" that prevents diffusive loss of newly synthesized raffinose oligosaccharides. Moreover, unlike plasmodesmal frequencies at CC and TC interfaces with contiguous cells in apoplasmic loading species, plasmodesmata are present in great abundance across the BSC/IC interface, at frequencies that are among the highest observed in minor veins (Fisher 1986, 1990; Haritatos et al. 1996). The filtering action at the BSC/IC interface is most likely related to the physical narrowing of plasmodesmal channels in the IC wall, as well as the increased length of these channels due to asymmetric wall thickening (Fig.6). Modeling of intercellular diffusive movement of different sugars through plasmodesmata indicated that the diffusive permeability of raffinose is only half that of sucrose, at the dimensions of a typical plasmodesmal microchannel (radius 1.5 nm; Turgeon 1992). With decreasing pore radius, the discrimination between sucrose and raffinose increases. With a reduction in pore size comes a reduction in overall diffusive flow, which may be compensated by the increased abundance of plasmodesmata at the BSC/IC interface (Fisher 1986; Haritatos et al. 1996).

Current data from studies of plasmodesmal ultrastructure, development, and frequency correlated with physiological and iontophoretic results indicate that phloem loading in both C_3 (*Hordeum*) and C_4 (*Zea*) grass leaves is apoplasmic (Heyser 1980; Evert 1986; Evert and Russin 1993). There are considerable numbers of plasmodesmata in the walls between contiguous MC and between MC and BSC (Tucker 1990). In addition, the continuous, impermeable suberin lamellae in the outer tangential and radial walls of the BSC (Hattersley and Browning 1981; Evert et al. 1985) act as effective barriers to apoplasmic solute movement (Heyser 1980; Evert et al. 1985). Therefore, the major transport pathway of primary assimilate movement between MC and BSC must be symplasmic (Heyser 1980; Evert 1986).

The BSC/VP interface has a high plasmodesmal frequency, and, in addition, the inner tangential wall of the BSC is suberized only in the region of the plasmodesmata (Evert et al. 1978). These characteristics suggest that transport of sucrose from BSC to the VP cells in minor veins could be either apoplasmic or symplasmic (Evert et al. 1978).

Independent studies of plasmodesmal frequency have been completed for two different barley cultivars (*H. vulgare* cv. Morex: Evert et al. 1996b and *H. vulgare* cv. Dyan: Botha and Cross 1997). Plasmodesmal counts that were obtained for the veins most closely associated with the loading process, the minor veins, support the conclusion that the greatest degree of symplasmic continuity occurs along the MC→BS→MS→VP→VP pathway. This pathway could be described as having high potential symplasmic conductivity. All the other combinations comprise less than 2% of the total plasmodesmata, indicating a very low potential symplasmic conductivity along alternate pathways.

Correlative microinjection experiments were performed on barley leaves to determine the possibility of symplasmic transport from MC to SE in minor veins (Farrar et al. 1992; Botha and Cross 1997). As might be predicted from plasmodesmal frequency results, Lucifer Yellow was unambiguously observed in other MC and also in parenchymatous BSC when the dye was injected into MC (Farrar et al. 1992). Unfortunately, difficulties in identifying specific cells in the vascular tissue at the light microscope level and with the low intensity of the fluorescence signal led the authors to conflicting conclusions regarding symplasmic transport into SE. Farrar et al. (1992) concluded that a symplasmic pathway was readily available for dye movement from MS to VP cells, and possibly into SE. In contrast, Botha and Cross (1997) concluded that phloem loading in the barley leaf is likely to be apoplasmic due to the relatively weak fluorescence signal obtained from the vascular tissue (in conjunction with the relatively low abundance of plasmodesmata). Regardless of their conclusions, the authors acknowledge that there was at least minimal symplasmic movement of the Lucifer Yellow into the phloem.

Linking the plasmodesmal frequency and dye coupling data leads to the conclusion that just a few plasmodesmata might form a potential symplasmic pathway across a given interface. As stated in one paper (Evert et al. 1996b), "... it cannot be precluded that the presence of even a few plasmodesmata between the pertinent cell types may have provided sufficient symplasmic continuity for the movement of the Lucifer Yellow."

3.1 Analysis of Plasmodesmal Frequencies in Studies of Phloem Loading

An assumption often implicit in the consideration of the role of plasmodesmata in phloem-loading pathways is that the presence of plasmodesmata, as measured by their distribution and frequency along a given cell-to-cell series, may be taken as prima-facie evidence for their direct participation in sugar movement. That is, where plasmodesmata exist, there is a priori an intercellular route for photosynthate transport to the phloem. This has yet to be demonstrated unequivocally. The presence of plasmodesmata between adjoining cells does not necessarily indicate that these cells are symplasmically connected (Erwee and Goodwin 1985; Palevitz and Hepler 1985). As pointed out previously by several workers (van Bel et al. 1988; Madore and Lucas 1989), the use of plasmodesmata frequencies as a basis for projecting the pathways of sugar movement rests on the assumption that plasmodesmata along the pathway can be equated to simple pores and are all functionally similar. Given the apparent structural complexity of plasmodesmata (cf. Ding et al. 1992b; Badelt et al. 1994; White et al. 1994; Overall and Blackman 1996) and the structural variation from interface to interface along the pathway(s), these inferences may oversimplify any plasmodesmata-related mechanisms that regulate or affect cell-to-cell traffic of photoassimilates (see also van Bel and Oparka 1995).

4 The *sxd1* Mutant of *Zea mays*: New Evidence for the Role of Plasmodesmata in Apoplasmic Phloem Loading

The maize sucrose export defective1 (*sxd1*) mutant presents a unique opportunity to study the effects of plasmodesmal interruption on photosynthate transport (Russin et al. 1996; Neuffer et al. 1997). Whole-leaf autoradiography of *sxd1* leaves showed that fixed carbon is not exported from the tip region, while labeled photosynthate was exported from the basal regions of similar-aged leaves (Russin et al. 1996). Export characteristics were correlated with structural modifications to plasmodesmata at a specific interface. Ultrastructural examination of the *sxd1* leaf blade showed that only plasmodesmata at the BSC-VP interface of minor veins in the tip region appeared contorted, discontinuous, and completely covered by a layer of wall material through which they did not extend (Fig. 5G). These results demonstrate a strong correlation between structural features and export competence in the *sxd1* leaves.

Evidence from the *sxd1* study indicates that disruption of plasmodesmal connections and, hence, symplasmic continuity at the BSC-VP interface in *sxd1* plants appears to block the loading of photosynthates. Therefore, plasmodesmata at the BSC/VP interface apparently are critical in the transfer of photosynthate from BSC to vascular tissue. It is likely that photosynthates follow a symplasmic route from the MC to the VP. The VP cells unload photosynthates into the apoplast, from which they are loaded into the SE-CC complex, a step hypothesized by van Bel et al. (1988) in *Commelina.*

Vascular parenchyma cells in the maize leaf are not symplasmically isolated, however, from TkWST or from other VP cells (Evert et al. 1977; Evert and Russin 1993). Although the VP cells in the minor veins of *sxd1* leaves were severely plasmolyzed, plasmodesmata at all other interfaces except at the BSC/VP interface appeared structurally normal. These results suggest that the plasmodesmata at the BSC-VP interface are in some way unique and that they may play a pivotal role in a variety of crucial processes (Ding et al. 1992a; Leisner et al. 1992; Lucas et al. 1993a).

5 Concluding Remarks

We have explored the variation in plasmodesmal structure, frequency, and possible functions in a relatively broad cross-section of dicots and monocots, including both apoplasmic and symplasmic phloem loaders and C_3 and C_4 species. A central theme that emerges from this examination is that there are plasmodesmata across virtually every interface from the mesophyll to the SE-CC complex. The abundance of plasmodesmata at some interfaces can be quite low, but apparent symplasmic continuity, as inferred from the presence of plasmodesmata, exists all along the pathway. Given the demonstrated changes in abundance, distribution, and substructure of plasmodesmata across the various mesophyll-phloem pathway interfaces during leaf expansion, it would seem that plants possess the necessary level of control to regulate fully the existence of plasmodesmata. Hence, it would appear unlikely that plasmodesmata across an interface are non-functional remnants of earlier symplasmic continuity, as has been suggested for CC plasmodesmata in apoplasmic loading species. We are left with the conclusion that the plasmodesmata are in place to maintain an intercellular communication network, which might, as has been suggested by Lucas and coworkers (1996), function to control photosynthesis in the mesophyll and the loading and export of photoassimilates in the minor vein phloem. For this system to work efficiently, small signaling molecules are hypothesized to move between the CC and the mesophyll. Jorgensen et al. (1998) have recently proposed that plasmodesmata are the pathway for intercellular movement of small RNA molecules that function in activation and/or suppression of gene activity. It almost goes without saying that plasmodesmata are now perceived as vital components in an endogenous mechanism of fine control over cell and tissue development.

At the moment, however, we remain unable to demonstrate the actual nature or mechanism of symplasmic communication, nor can we show unequivocally that plasmodesmata are the channels for this intercellular communication network. Frankly, there has been little true advancement in our understanding of the roles of plasmodesmata, despite an apparent general perception in the plasmodesmal biology community that there is a wealth of information that permits us to understand details of plasmodesmal function in macromolecular trafficking, assimilate movement, and phloem loading. We certainly know more about specific ultrastructural differences along the phloem-loading pathway and have a better and clearer vision of the substructure of plasmodesmata. There is a growing body of information on the differentiation of plasmodesmata and the postcytokinetic formation of secondary plasmodesmata. Aspects of plasmodesmal development are tightly linked to the establishment of phloem-loading capacity and the initiation of phloem export. We now have a testable model for an alternate mode of phloem loading, the symplasmic model of Turgeon (1991, 1992, 1996), and there is accumulating evidence in its support. We also know that in many species there exists a potential symplasmic pathway for assimilate movement, if we assume that the movement of fluorescent probes truly can be equated to a similar behavior by photoassimilates.

In spite of these advances, many questions remain to be addressed. We must continue to explore the substructure of plasmodesmata and make advances in the understanding of the functional significance of plasmodesmata component proteins. Several plasmodesmata proteins have been tentatively identified (see Chap. 9), but their characterization remains a critically important goal. We must experimentally define the functional basis for the structural variations seen across the different cell-to-cell interfaces in the phloem-loading pathway. Of further interest is the impact of various environmental factors on plasmodesmata structure, function, and phloem loading; and finally, we return to the central question: how can we unequivocally demonstrate the nature of plasmodesmal function in phloem loading? In the 20 years since Brian Gunning's review, we have made significant and important contributions to our understanding of plasmodesmal structure/function relationships in phloem loading. It continues to be a significant challenge to further the characterization of plasmodesmata in this important aspect of source leaf physiology.

References

Badelt K, White RG, Overall RL, Vesk M (1994) Ultrastructural specializations of the cell wall sleeve around plasmodesmata. Am J Bot 81:1422–1427

Beebe DU, Evert RF (1990) The morphology and anatomy of the leaf of *Moricandia arvensis* (L.) DC. (Brassicaceae). Bot Gaz 151:184–203

Beebe DU, Evert RF (1992) Photoassimilate pathway(s) and phloem loading in the leaf of *Moricandia arvensis* (L.) DC. (Brassicaceae). Int J Plant Sci 153:61–77

Beebe DU, McEvoy SM (1997) Structural and biochemical characterization of plasmodesmata from minor veins of *Cucurbita pepo* cotyledons. Plant Physiol 114:265

Beebe DU, Turgeon R (1992) Localization of galactinol, raffinose, and stachyose synthesis in *Cucurbita pepo* leaves. Planta 188:354–361

Beebe DU, McEvoy SM, Ben Achour Y (1996) Molecular and structural characterization of plasmodesmata in cotyledons and expanding leaves of *Cucurbita pepo* L. 3rd Int Worksh on Basic and Applied Research in Plasmodesmal Biology, March 1996, Zichron-Yakov, Israel, pp 4–7

Bonnemain JL, Bourquin S, Renault S, Offler C, Fisher DG (1991) Transfer cells: structure and physiology. In: Bonnemain J-L, Delrot S, Lucas WJ, Dainty J (eds) Recent advances in phloem transport and assimilate compartmentation. Ouest Editions, Nantes, pp 74–83

Botha CEJ (1992) Plasmodesmatal distribution, structure and frequency in relation to assimilation in C_3 and C_4 grasses in southern Africa. Planta 187:348–358

Botha CEJ, Cross RHM (1997) Plasmodesmatal frequency in relation to short-distance transport and phloem loading in leaves of barley (*Hordeum vulgare*). Phloem is not loaded directly from the symplast. Physiol Plant 99:355–362

Botha CEJ, Evert RF (1986) Free-space marker studies on the leaves of *Saccharum officinarum* and *Bromus unioloides*. S Afr J Bot 52:335–342

Botha CEJ, Evert RF (1988) Plasmodesmatal distribution and frequency in vascular bundles and contiguous tissues of the leaf of *Themeda trianda*. Planta 173:433–441

Botha CEJ, van Bel AJE (1992) Quantification of symplastic continuity as visualised by plasmodesmograms: diagnostic value for phloem-loading pathways. Planta 187:359–366

Botha CEJ, Hartley BJ, Cross RHM (1993) The ultrastructure and computer-enhanced digital image analysis of plasmodesmata at the Kranz mesophyll-bundle sheath interface of *Themeda triandra* var. *imberis* (retz) A. Camus in conventionally fixed leaf blades. Ann Bot 72:255–261
Bouché-Pillon S, Fleurat-Lessard P, Fromont J-C, Serrano R, Bonnemain J-L (1994) Immunolocalization of the plasma membrane H^+-ATPase in minor veins of *Vicia faba* in relation to phloem loading. Plant Physiol 105:691–697
Bush DR (1992) The proton-sucrose symport. Photosynth Res 32:155–165
Chonan N, Kawahara H, Matsuda T (1985) Ultrastructure of elliptical and diffuse bundles in the vegetative nodes of rice. Jpn J Crop Sci 54:393–402
Colbert JT, Evert RF (1982) Leaf vasculature in sugarcane (*Saccharum officinarum* L.). Planta 156: 136–151
Dannenhoffer JM, Ebert Jr. W, Evert RF (1990) Leaf vasculature in barley, *Hordeum vulgare* (Poaceae). Am J Bot 77:636–652
Dannenhoffer JM, Evert RF (1994) Development of the vascular system in the leaf of barley (*Hordeum vulgare* L.). Int J Plant Sci 155:143–157
Delrot S (1989) Loading of photoassimilates. In: Baker DA, Milburn JA (eds) Transport of photoassimilates. Longman, Essex, pp 167–205
DeWitt ND, Sussman MR (1995) Immunocytological localization of an epitope-tagged plasma membrane proton pump (H^+-ATPase) in phloem companion cells. Plant Cell 7:2053–2067
Ding B, Lucas WJ (1996) Secondary plasmodesmata: biogenesis, special functions, and evolution. In: Smallwood M, Knox P, Bowles D (eds) Membranes: specialized functions in plants. BIOS Scientific, Oxford, pp 489–506
Ding B, Parthasarathy MV, Niklas K, Turgeon R (1988) A morphometric analysis of the phloem-unloading pathway in developing tobacco leaves. Planta 176:307–318
Ding B, Haudenshield J, Hull R, Wolf S, Beachy RN, Lucas WJ (1992a) Secondary plasmodesmata are specific sites of localization of the tobacco mosaic virus movement protein in transgenic tobacco plants. Plant Cell 4:915–928
Ding B, Turgeon R, Parthasarathy MV (1992b) Substructure of freeze-substituted plasmodesmata. Protoplasma 169:28–41
Ding B, Haudenshield JS, Willmitzer L, Lucas WJ (1993) Correlation between arrested secondary plasmodesmal development and onset of accelerated senescence in yeast acid invertase transgenic tobacco plants. Plant J 4:179–189
Ehlers K, Kollmann R (1996) Formation of branched plasmodesmata in regenerating *Solanum nigrum* protoplasts. Planta 199:126–138
Ehlers K, Schulz M, Kollmann R (1996) Subcellular localization of ubiquitin in plant protoplasts and the function of ubiquitin in selective degradation of outer-wall plasmodesmata in regenerating protoplasts. Planta 199:139–151
Erwee MG, Goodwin PB (1985) Symplast domains in extrastelar tissues of *Egeria densa* Planch. Planta 163:9–19
Erwee MG, Goodwin PB, van Bel AJE (1985) Cell-cell communication in the leaves of *Commelina cyanea* and other plants. Plant Cell Environ 8:173–178
Esau K (1977) Anatomy of seed plants. Wiley, New York
Esau K, Thorsch J (1985) Sieve plate pores and plasmodesmata, the communication channels of the symplast: ultrastructural aspects and developmental relations. Am J Bot 72:1641–1653
Evert RF (1986) Phloem loading in maize. In: Shannon JC, Knievel DP, Boyer CD (eds) Regulation of carbon and nitrogen reduction and utilization in maize. American Society of Plant Physiology, Rockville, pp 67–81
Evert RF, Mierzwa RJ (1986) Pathway(s) of assimilate movement from mesophyll cells to sieve tubes in the *Beta vulgaris* leaf. In: Cronshaw J, Giaquinta RT, Lucas WJ (eds) Phloem transport. Alan Liss, New York, pp 419–432
Evert RF, Mierzwa RJ (1989) The cell wall-plasmalemma interface in sieve tubes of barley. Planta 177: 24–34
Evert RF, Russin WA (1991) Aspects of sieve element structure in *Zea mays*. In: Bonnemain J-L, Delrot S, Lucas WJ, Dainty J (eds) Recent advances in phloem transport and assimilate compartmentation. Ouest Editions, Nantes, pp 47–57
Evert RF, Russin WA (1993) Structurally, phloem unloading in the maize leaf cannot be symplastic. Am J Bot 80:1310–1317
Evert RF, Eschrich W, Heyser W (1977) Distribution and structure of the plasmodesmata in mesophyll and bundle-sheath cells of *Zea mays* L. Planta 136:77–89
Evert RF, Eschrich W, Heyser W (1978) Leaf structure in relation to solute transport and phloem loading in *Zea mays* L. Planta 138:279–294
Evert RF, Botha CEJ, Mierzwa RJ (1985) Free-space marker studies on the leaf of *Zea mays* L. Protoplasma 126:62–73

Evert RF, Russin WA, Bosabalidis AM (1996a) Anatomical and ultrastructural changes associated with sink-to-source transition in developing maize leaves. Int J Plant Sci 157:247–261
Evert RF, Russin WA, Botha CEJ (1996b) Distribution and frequency of plasmodesmata in relation to photoassimilate pathways and phloem loading in the barley leaf. Planta 198:572–579
Farrar J, van der Schoot C, Drent P, van Bel AJE (1992) Symplastic transport of Lucifer Yellow in mature leaf blades of barley: potential mesophyll-sieve-tube transfer. New Phytol 120:191–196
Fellows RJ, Geiger DR (1974) Structural and physiological changes in sugar beet leaves during sink to source conversion. Plant Physiol 54:877–885
Fisher DG (1986) Ultrastructure, plasmodesmatal frequency, and solute concentration in green areas of variegated *Coleus blumei* Benth. leaves. Planta 169:141–152
Fisher DG (1988) Movement of lucifer yellow in leaves of *Coleus blumei* Benth. Plant Cell Environ 11: 639–644
Fisher DG (1990) Distribution of plasmodesmata in leaves. A comparison of *Cananga odorata* with other species using different measures of plasmodesmatal frequency. In: Robards AW, Jongsma H, Lucas WJ, Pitts J, Spray D (eds) Parallels in cell to cell junctions in plants and animals, vol H46. Springer, Berlin Heidelberg New York, pp 199–221
Fisher DG (1991) Plasmodesmatal frequency and other structural aspects of assimilate collection and phloem loading in leaves of *Sonchus oleraceus* (Asteraceae), a species with minor vein transfer cells. Am J Bot 78:1549–1559
Fisher DG, Evert RF (1982a) Studies on the leaf of *Amaranthus retroflexus* (Amaranthaceae): morphology and anatomy. Am J Bot 69:1133–1147
Fisher DG, Evert RF (1982b) Studies on the leaf of *Amaranthus retroflexus* (Amaranthaceae): ultrastructure, plasmodesmatal frequency, and solute concentration in relation to phloem loading. Planta 155:377–387
Flora LL, Madore MA (1996) Significance of minor-vein anatomy to carbohydrate transport. Planta 198:171–178
Fritz E, Evert RF, Nasse H (1989) Loading and transport of assimilates in different maize leaf bundles. Digital image analysis of 14C microautoradiographs. Planta 178:1–9
Gagnon M-J, Beebe DU (1996a) Establishment of a plastochron index and analysis of the sink-to-source transition in leaves of *Moricandia arvensis* (L.) DC. (Brassicaceae). Int J Plant Sci 157:262–268
Gagnon M-J, Beebe DU (1996b) Minor vein differentiation and the development of specialized plasmodesmata between companion cells and contiguous cells in expanding leaves of *Moricandia arvensis* (L.) DC. (Brassicaceae). Int J Plant Sci 157:658–697
Gamalei Y (1989) Structure and function of leaf minor veins in trees and herbs. A taxonomic review. Trees 3:96–110
Gamalei YV, Pakhomova MV (1982) Distribution of plasmodesmata and parenchyma transport of assimilates in the leaves of several dicots. Sov Plant Physiol 28:649–661
Geiger DR (1975) Phloem loading. In: Zimmermann MH, Milburn JA (eds) Encyclopedia of plant physiology. Springer, Berlin Heidelberg New York, pp 395–431
Geiger DR (1976) Phloem loading in source leaves. In: Wardlaw IF, Passioura JB (eds) Transport and transfer processes in plants. Academic Press, New York, pp 167–183
Geiger DR, Giaquinta RT, Sovonick SA, Fellows RJ (1973) Solute distribution in sugar beet leaves in relation to phloem loading and translocation. Plant Physiol 52:585–589
Giaquinta RT (1983) Phloem loading of sucrose. Annu Rev Plant Physiol 34:347–387
Grusak M, Beebe DU, Turgeon R (1996) Phloem loading. In: Zamski E, Schaffer A (eds) Photoassimilate distribution in plants and crops: source-sink relationships. Dekker, New York, pp 209–227
Gunning BES (1976) The role of plasmodesmata in short distance transport to and from the phloem. In: Gunning BES, Robards AW (eds) Intercellular communication in plants: studies on plasmodesmata. Springer, Berlin Heidelberg New York, pp 203–227
Gunning BES, Pate JS (1969) "Transfer cells" plant cells with wall ingrowths, specialized in relation to short distance transport of solutes – their occurrence, structure and development. Protoplasma 68:107–133
Gunning BES, Pate JS, Briarty LG (1968) Specialized "transfer cells" in minor veins of leaves and their possible significance in phloem translocation. J Cell Biol 37:C7–C12
Gunning BES, Pate JS, Minchin FR, Marks I (1974) Quantitative aspects of transfer cell structure in relation to vein loading in leaves and solute transport in legume nodules. Symp Soc Exp Biol 28: 87–126
Haritatos E, Keller F, Turgeon R (1996) Raffinose oligosaccharide concentrations measured in individual cell and tissue types in *Cucumis melo* L. leaves: implications for phloem loading. Planta 198:614–622
Hattersley PW, Browning AJ (1981) Occurrence of the suberized lamella in leaves of grasses of different photosynthetic types. I. In parenchymatous bundle sheaths and PCR (Kranz) sheaths. Protoplasma 109:371–401

Heyser W (1980) Phloem loading in the maize leaf. Ber Dtsch Bot Ges 93:221–228
Hickey LJ (1979) A revised classification of the architecture of dicotyledonous leaves. In: Metcalf CR, Chalk L (eds) Anatomy of the dicotyledons, vol 1. Oxford, New York, pp 25–39
Holthaus U, Schmitz K (1991) Distribution and immunolocalization of stachyose synthase in *Cucumis melo* L. Planta 185:479–486
Isebrands JG, Larson PR (1973) Anatomical changes during leaf ontogeny in *Populus deltoides*. Am J Bot 60:199–208
Jorgensen RA, Atkinson RG, Forster RLS, Lucas WJ (1998) An RNA-based information superhighway in plants. Science 279:1486–1487
Kollmann R, Glockmann C (1991) Studies on graft unions. III. On the mechanism of secondary formation of plasmodesmata at the graft interface. Protoplasma 165:71–85
Kühn C, Franceschi VR, Schulz A, Lemoine R, Frommer WB (1997) Macromolecular trafficking indicated by localization and turnover of sucrose transporters in enucleate sieve elements. Science 275:1298–1300
Kuo J, O'Brien TP, Canny MJ (1974) Pit-field distribution, plasmodesmatal frequency, and assimilate flux in the mestome sheath of wheat leaves. Planta 121:97–118
Lee HJ, Ashley DA, Brown RH (1980) Sucrose concentration gradients in wheat leaves. Crop Sci 20:95–98
Leisner SM, Turgeon R, Howell SH (1992) Long distance movement of cauliflower mosaic virus in infected turnip plants. Mol Plant-Microbe Inter 5:41–47
Lucas WJ, Ding B, van der Schoot C (1993a) Plasmodesmata and the supracellular nature of plants. New Phytol 125:435–476
Lucas WJ, Olesinski A, Hull RJ, Haudenshield JS, Deom CM, Beachy RN, Wolf S (1993b) Influence of the tobacco mosaic virus 30-kDa movement protein on carbon metabolism and photosynthate partitioning in transgenic plants. Planta 190:88–96
Lucas WJ, Balachandran S, Park J, Wolf S (1996) Plasmodesmal companion cell-mesophyll communication in the control over carbon metabolism and phloem transport: insights gained from viral movement proteins. J Exp Bot 47:1119–1128
Madore MA, Lucas WJ (1989) Transport of photoassimilates between leaf cells. In: Baker DA, Milburn JA (eds) Transport of photoassimilates. Longman, Essex, pp 49–78
Madore MA, Oross JW, Lucas WJ (1986) Symplastic transport in *Ipomoea tricolor* source leaves. Demonstration of functional symplastic connections from mesophyll to minor veins by a novel dye-tracer method. Plant Physiol 82:432–442
McCauley MM, Evert RF (1988a) Morphology and vasculature of the leaf of potato (*Solanum tuberosum*). Am J Bot 75:377–390
McCauley MM, Evert RF (1988b) The anatomy of the leaf of potato, *Solanum tuberosum* L. 'Russet Burbank'. Bot Gaz 149:179–195
McCauley MM, Evert RF (1989) Minor veins of the potato (*Solanum tuberosum* L.) leaf: ultrastructure and plasmodesmatal frequency. Bot Gaz 150:351–368
Monzer J (1990) Secondary formation of plasmodesmata in cultured cells – structural and functional aspects. In: Robards AW, Jongsma H, Lucas WJ, Pitts J, Spray D (eds) Parallels in cell to cell junctions in plants and animals, vol H46. Springer, Berlin Heidelberg New York, pp 185–197
Monzer J (1991) Ultrastructure of secondary plasmodesmata formation in regenerating *Solanum nigrum* protoplast cultures. Protoplasma 165:86–95
Nelson RS, van Bel AJE (1998) The mystery of virus trafficking into, through and out of vascular tissue. Prog Bot 59:476–533
Neuffer MG, Coe EH, Wessler SR (1997) Mutants of maize. Cold Spring Harbor Laboratory, New York
Olesen P (1979) The neck constriction in plasmodesmata. Evidence for a peripheral sphincter-like structure revealed by fixation with tannic acid. Planta 144:349–358
Olesen P, Robards AW (1990) The neck region of plasmodesmata: general architecture and some functional aspects. In: Robards AW, Lucas WJ, Pitts JD, Jongsma HJ, Spray DC (eds) Parallels in cell to cell junctions in plants and animals, vol H 46. Springer, Berlin Heidelberg New York, pp 145–170
Overall RL, Blackman LM (1996) A model of the macromolecular structure of plasmodesmata. Trends Plant Sci 1:307–311
Palevitz BA, Hepler PK (1985) Changes in dye coupling of stomatal cells of *Allium* and *Commelina* demonstrated by microinjection of Lucifer Yellow. Planta 164:473–479
Pristupa NA (1964) Redistribution of radioactive assimilates in the leaf tissues of cereals. Sov Plant Physiol 11:31–36
Rentsch D, Frommer WB (1996) Molecular approaches towards an understanding of loading and unloading of assimilates in higher plants. J Exp Bot 47:1199–1204
Riesmeier JW, Hirner B, Frommer WB (1993) Potato sucrose transporter expression in minor veins indicates a role in phloem loading. Plant Cell 5:1591–1598
Riesmeier JW, Willmitzer L, Frommer WB (1994) Evidence for an essential role of the sucrose transporter in phloem loading and assimilate partitioning. EMBO J 13:1–7

Robinson-Beers K, Evert RF (1991a) Ultrastructure of and plasmodesmatal frequency in mature leaves of sugarcane. Planta 184:291–306
Robinson-Beers K, Evert RF (1991b) Fine structure of plasmodesmata in mature leaves of sugarcane. Planta 184:307–318
Robinson-Beers K, Evert RF (1991c) Structure and development of plasmodesmata at the mesophyll/bundle sheath cell interface of sugarcane leaves. In: Bonnemain J-L, Delrot S, Lucas WJ, Dainty J (eds) Recent advances in phloem transport and assimilate compartmentation. Ouest Editions, Nantes, pp 116–122
Robinson-Beers K, Sharkey TD, Evert RF (1990) Import of 14C-photosynthate by developing leaves of sugarcane. Bot Acta 103:424–429
Russell SH, Evert RF (1985) Leaf vasculature in *Zea mays* L. Planta 164:448–458
Russin WA, Evert RF (1984) Studies on the leaf of *Populus deltoides* (Salicaceae): morphology and anatomy. Amer J Bot 71:1398–1415
Russin WA, Evert RF (1985a) Studies on the leaf of *Populus deltoides* (Salicaceae): quantitative aspects, and solute concentrations of the sieve-tube members. Amer J Bot 72:487–500
Russin WA, Evert RF (1985b) Studies on the leaf of *Populus deltoides* (Salicaceae): ultrastructure, plasmodesmatal frequency and solute concentrations. Am J Bot 72:1232–1247
Russin WA, Evert RF, Vanderveer PJ, Sharkey TD, Briggs SP (1996) Modification of a specific class of plasmodesmata and loss of sucrose export ability in the sucrose export defective1 maize mutant. Plant Cell 8:645–658
Sauer N, Stolz J (1994) SUC1 and SUC2: two sucrose transporters from *Arabidopsis thaliana*: expression and characterization in baker's yeast and identification of the histidine tagged protein. Plant J 6:67–77
Schmalstig JG, Geiger DR (1987) Phloem unloading in developing leaves of sugar beet. II. Termination of phloem unloading. Plant Physiol 83:49–52
Schmitz K, Cuypers B, Moll M (1987) Pathway of assimilate transfer between mesophyll cells and minor veins in leaves of *Cucumis melo* L. Planta 171:19–29
Stadler R, Sauer N (1996) The *Arabidopsis thaliana* AtSUC2 gene is specifically expressed in companion cells. Bot Acta 109:299–306
Stadler R, Brandner J, Schulz A, Gahrtz M, Sauer N (1995) Phloem loading by the PmSUC2 sucrose carrier from *Plantago major* occurs into companion cells. Plant Cell 7:1545–1554
Stitt M, Heldt HW (1985) Generation and maintenance of concentration gradients between the mesophyll and bundle sheath in maize leaves. Biochim Biophys Acta 808:400–414
Truernit E, Sauer N (1995) The promoter of *Arabidopsis thaliana* SUC2 sucrose-H^+ symporter gene directs expression of β-glucuronidase to the phloem: evidence for phloem loading and unloading by SUC2. Planta 196:564–570
Tucker EB (1990) Calcium-loaded 1, 2-bis(2-aminophenoxy)ethane-N,N,N',N'-tetraacetic acid blocks cell-to-cell diffusion of carboxyfluorescein in staminal hairs of Setcreasea purpurea. Planta 182: 34–38
Turgeon R (1986) The import-export transition in dicotyledonous leaves. In: Cronshaw J, Lucas WJ, Giaquinta RT (eds) Phloem transport. Alan Liss, New York, pp 285–291
Turgeon R (1987) Phloem unloading in tobacco sink leaves: insensitivity to anoxia indicates a symplastic pathway. Planta 171:73–81
Turgeon R (1989) The sink-source transition in leaves. Annu Rev Plant Physiol Plant Mol Biol 40: 119–138
Turgeon R (1991) Symplastic phloem loading and the sink-source transition in leaves: a model. In: Bonnemain J-L, Delrot S, Lucas WJ, Dainty J (eds) Recent advances in phloem transport and assimilate compartmentation. Ouest Editions, Nantes, pp 18–22
Turgeon R (1992) Symplastic phloem loading: evaluation of a current model. 2nd Int Worksh on Basic and Applied Research in Plasmodesmatal Biology, Sept 1992, Oosterbeck, The Netherlands, pp 111–114
Turgeon R (1996) Phloem loading and plasmodesmata. Trends Plant Sci 1:418–423
Turgeon R, Beebe DU (1991) The evidence for symplastic phloem loading. Plant Physiol 96:349–354
Turgeon R, Gowan E (1990) Phloem loading in *Coleus blumei* in the absence of carrier-mediated uptake of export sugar from the apoplast. Plant Physiol 94:1244–1249
Turgeon R, Gowan E (1992) Sugar synthesis and phloem loading in *Coleus blumei* leaves. Planta 187: 388–394
Turgeon R, Hepler PK (1989) Symplastic continuity between mesophyll and companion cells in minor veins of mature *Cucurbita pepo* L. leaves. Planta 179:24–31
Turgeon R, Webb JA (1976) Leaf development and phloem transport in *Cucurbita pepo*: Maturation of the minor veins. Planta 129:265–269
Turgeon R, Webb JA, Evert RF (1975) Ultrastructure of minor veins in *Cucurbita pepo* leaves. Protoplasma 83:217–232

Turgeon R, Beebe DU, Gowan E (1993) The intermediary cell: minor-vein anatomy and raffinose oligosaccharide synthesis in the Scrophulariaceae. Planta 191:446–456

van Bel AJE (1989) The challenge of symplastic phloem loading. Bot Acta 102:183–185

van Bel AJE (1992) Pathways and mechanisms of phloem loading. In: Pollock CJ, Farrar JF, Gordon AJ (eds) Carbon partitioning within and between organisms. Bios, Oxford, pp 53–70

van Bel AJE (1993) Strategies of phloem loading. Annu Rev Plant Physiol Plant Mol Biol 44:253–281

van Bel AJE, Koops AJ (1985) Uptake of [14C]sucrose in isolated minor-vein networks of *Commelina benghalensis* L. Planta 164:362–369

van Bel AJE, Oparka KJ (1995) On the validity of plasmodesmograms. Bot Acta 108:174–182

van Bel AJE, van Kesteren WJP, Papenhuijzen C (1988) Ultrastructural indications for coexistence of symplastic and apoplastic phloem loading in *Commelina benghalensis* leaves. Planta 176:159–172

van Bel AJE, Gamalei YV, Ammerlaan A, Bik LPM (1992) Dissimilar phloem loading in leaves with symplasmic or apoplasmic minor-vein configurations. Planta 186:518–525

van Bel AJE, Ammerlaan A, van Dijk AA (1994) A three-step screening procedure to identify the mode of phloem loading in intact leaves. Evidence for symplasmic and apoplasmic loading associated with the type of companion cell. Planta 192:32–39

van Kesteren WJP, van der Schoot C, van Bel AJE (1988) Symplastic transfer of fluorescent dyes from mesophyll to sieve tubes in stripped leaf tissue and partly isolated minor veins of *Commelina benghalensis*. Plant Physiol 88:667–670

Volk GM, Turgeon R, Beebe DU (1996) Secondary plasmodesmata formation in the minor vein phloem of *Cucumis melo* L. and *Cucurbita pepo* L. Planta 199:425–432

Walsh MA, Evert RF (1975) Ultrastructure of metaphloem sieve elements in *Zea mays*. Protoplasma 83:365–388

Wang N, Nobel PS (1998) Phloem transport of fructans in the crassulacean acid metabolism species *Agave deserti*. Plant Physiol 116:709–714

Warmbrodt RD, VanDerWoude WJ (1990) Leaf of *Spinacia oleracea* (spinach): ultrastructure, and plasmodesmatal distribution and frequency, in relation to sieve-tube loading. Am J Bot 189: 1361–1377

White RG, Badelt K, Overall RL, Vesk M (1994) Actin associated with plasmodesmata. Protoplasma 180:169–184

Wimmers LE, Turgeon R (1991) Transfer cells and solute uptake in minor veins of *Pisum sativum* leaves. Planta 186:2–12

CHAPTER 16

P-Protein Trafficking Through Plasmodesmata

G. A. THOMPSON

Department of Plant Sciences, 303 Forbes Building,
University of Arizona, Tucson, Arizona 85721 USA

1 Introduction

Phloem tissue in flowering plants has a distinctive cellular composition in which the conducting elements are a complex of two unique cell types, sieve elements (SE) and companion cells (CC), that are developmentally related and functionally dependent upon one another. The sieve element-companion cell complex (SE-CC) of angiosperms originates from cambial derivatives as the result of an unequal cell division forming the undifferentiated SE and CC (reviewed in Schulz 1998). The differentiating SE undergoes selective autophagy of organelles resulting in apparent loss of the cell's ability for transcription and translation. Mature SEs are characteristically enucleate and lack ribosomes. Reorganization of the endomembrane system results from degeneration of the tonoplast and dictyosomes and changes in the endoplasmic reticulum (ER) that are distinct enough, according to some authors, to merit the special term sieve element reticulum or SER (Sjölund and Shih 1983). SER, plastids, and mitochondria are retained in a parietal position in close contact with the plasma membrane of the mature SE (Evert 1990).

In contrast to SEs, CCs are specialized parenchyma cells with structural characteristics typical of metabolically active cells. They have a prominent nucleus and dense cytoplasm due to large numbers of ribosomes, mitochondria, rough ER, and plastids (Evert 1990). CCs are capable of limited cell division in some species and are thought to be the source of metabolites and macromolecules required to maintain the enucleate SE. During early stages of differentiation, the SE-CC complex of transport phloem becomes virtually symplasmically isolated from the surrounding phloem parenchyma (van Bel and van Rijen 1994; Kempers et al. 1998). Specialized plasmodesmata form between the CC and SE during differentiation, allowing interactions to occur between the two cell types (reviewed in van Bel and Kempers 1997). The interdependence of SEs and CCs is illustrated by their common origin, numerous symplasmic connections, and coordinated cell death (Behnke 1975; Evert 1990). Recent work suggests that this interdependence extends to the exchange of proteins essential for the function of SEs.

Another significant feature of the phloem of most angiosperms is the appearance of distinct proteinaceous structures that accumulate in differentiating and mature SEs. The term P-protein or phloem-protein was introduced to describe proteinaceous filaments, tubules, and aggregations observed in SEs with light and transmission electron microscopy (Cronshaw and Esau 1967; Esau and Cronshaw 1967). The use of this general term is appropriate since P-protein also occurs in CCs and rarely in phloem parenchyma cells (Cronshaw and Esau 1967; Clark et al. 1997; Dannenhoffer et al. 1997). Other terms such as sieve-tube element proteins (STEPs) (Sakuth et al. 1993), phloem-specific proteins (Cronshaw and Sabnis 1990) or phloem-associated proteins (Ishiwatari et al. 1996) encompass all proteins that are characteristic of either SEs or phloem tissue. Observations that P-protein subunits move within the assimilate stream confuse previous distinctions between soluble and polymerized phloem-associated proteins (Tiedemann and Carstens-Behrens 1994). Since the nucleus and ribosomes degenerate during SE maturation, P-protein synthesis must occur either in immature SEs prior to organelle degeneration or in CCs from which they would then be transported, presumably via pore/plasmodesma units (PPUs), into SEs. In this chapter we will examine P-protein accumulation in the developing SE-CC complex from structural, biochemical, and molecular perspectives to evaluate the recent evidence that indicates P-proteins are synthesized in CCs, and traffic into SEs through PPUs.

2 P-Protein Distribution, Morphology and Development

Numerous ultrastructural studies have shown that P-protein is a common feature of SEs in most (if not all) dicotyledonous plants (Cronshaw 1975) and many monocotyledonous plants (Behnke 1981). Its appearance in all dicotyledonous species that have been examined indicates that P-protein is an integral component of SEs in this class of angiosperms. Among the monocots, P-protein has been observed in *Avena* and *Secale* (O'Brien and Thimann 1967), but is conspicuously absent in other grasses such as *Zea mays* (Walsh and Evert 1975; Tóth and Sjölund 1994), *Hordeum vulgare* (Evert et al. 1971; Tóth and Sjölund 1994) and *Triticum aestivum* (Kuo et al. 1972). Monocots other than grasses in which P-protein has been reported include *Narcissus* (Tóth and Sjölund 1994), *Iris* (Tóth and Sjölund 1994), *Dioscorea* (Behnke and Dörr 1967), *Elodea* (Currier and Shih 1968), *Tradescantia* (Heyser 1971), *Musa* (Sabnis and Sabnis 1995), *Tamus* (Sabnis and Sabnis 1995), and several palms (Parthasarathy 1974).

Deposition of P-proteins into distinctive structures is a dynamic process that occurs during SE differentiation (reviewed in Cronshaw 1975; Cronshaw and Sabnis 1990; Sabnis and Sabnis 1995). Ultrastructural examination of differentiating and mature SEs shows various morphological forms of P-protein including granular, filamentous or fibrillar, tubular, and crystalline, depending upon the species and stage of SE differentiation. Several of these forms can be present either during differentiation or within mature SEs of a single species. Accumulation of P-protein during differentiation of the SE-CC complex has been examined in detail in a number of species, providing a general picture of P-protein deposition in SEs. Although P-proteins are a prominent feature of SEs and are widely distributed throughout angiosperms, their function remains elusive. Yet the distinct structural changes that occur during vascular development suggest a significant physiological role for P-proteins.

P-protein is initially observed in the cytoplasm of immature nucleate SEs as small aggregates of fine fibrils or tubules that are intermixed with ribosomes, ER, and dictyosomes. The close structural association of P-protein with ribosomes and ER led to a widely held view that P-proteins are synthesized in the immature SEs prior to the period of selective autophagy (e.g. Northcote and Wooding 1965; Cronshaw and Esau 1968a; Zee 1969). However, it is notable that at this stage of development SEs and CCs have visibly differentiated and are connected by PPUs typical of the SE-CC complex (Northcote and Wooding 1965). Gagnon and Beebe (1996), in their study of minor-vein plasmodesmata of *Moricandia*, observed CCs in a more advanced stage of differentiation than their associated SEs. They suggested that plasmodesmata development was closely linked to maturation of the CC rather than the SE-CC complex.

During early stages of SE differentiation, P-protein fibrils or tubules coalesce to form discrete aggregates called P-protein bodies. While P-protein bodies are often closely associated with stacked ER, they are not membrane-bound (Cronshaw 1975). Enlarging protein bodies are located adjacent to the plasma membrane and are often composed of aggregated tubules ranging in diameter from 15–25 nm. The number and morphology of P-protein bodies that form within an individual SE is variable among different species. Many species form a single or few large P-protein bodies that could result from the coalescence of smaller bodies. The single P-protein body typical of tobacco SEs is ellipsoidal in the longitudinal direction measuring approximately 15 × 6 μm and roughly circular in transection (Cronshaw and Esau 1967). Unlike the compact aggregates of tobacco, P-protein bodies of *Ricinus communis* lack the tubular form of P-protein and form loose aggregates of either 15 or 17.5-nm fibrils that fuse, forming a single, large fibrillar body (Cronshaw 1975). Various types of P-protein bodies can coexist, as illustrated by the large P-protein bodies of *Acer pseudoplatanus* that are composed of either closely packed 18–24-nm fibres, loosely organized 9–10-nm fibrils, or a combination of both (Northcote and Wooding 1965). In contrast to SEs containing single bodies, individual SEs of *Cucurbita maxima* have approximately 70–100 protein bodies, each ~20 μm in diameter, organized in longitudinal rows (Cronshaw and Esau 1968a). These include two types of spherical P-protein bodies: an abundant, large body composed of fine 18-nm fibrils and a rarer, smaller body composed of groups of more condensed ~27.5-nm tubules.

Also notable are the crystalline P-protein bodies typical of the SEs of many legumes in the Fabaceae (Zee 1969; Wergin and Newcomb 1970; Palevitz and Newcomb 1971; Wergin et al. 1975; Lawton 1978). Crystalline P-protein bodies in mature SEs have two general morphologies: they are typically spindle-shaped in longitudinal section and square in cross-section and, depending upon the genus, either possess or lack sinuous tails (Lawton 1978). Formation of crystalline P-protein bodies in differentiating SEs of *Glycine max* proceeds through a tubular intermediate (Wergin and Newcomb 1970). Small bundles of 25–75 densely packed 13.5-nm tubules consolidate into one or two masses that undergo further compaction, creating the crystalline P-protein body. Typical of many papilionaceous legumes, tubular intermediates in *Desmodium canadense* appear later in development following the initial formation of small, thin crystalline inclusions in the young SE (Palevitz and Newcomb 1971). As the crystal body increases in size and tails form, 16-nm tubules contiguous with the enlarging crystalline P-protein body also accumulate. Four different forms of P-protein bodies have been observed in *Vicia faba* including an amorphous body composed of fine filamentous material, the typical crystalline body, large compact bodies composed of 14-nm tubules, and sparse aggregates of 35-nm fibrils (Zee 1969).

P-protein bodies disperse in maturing SEs as the dictyosomes disappear, the tonoplast and nuclear membranes degenerate, ribosomes are lost, and the endoplasmic reticulum is reorganized. In this phase of selective autophagy, the large P-protein bodies of tobacco expand and disaggregate as individual tubules separate in the degenerating cytoplasm (Cronshaw and Esau 1967; Cronshaw 1975). As the P-protein tubules disperse, they reorganize into smaller 15-nm striated fibrils. Similarly, in maturing SEs of *Acer pseudoplatanus* the compact 18-24-nm fibres in P-protein bodies disperse, forming smaller 9–10-nm striated fibrils (Northcote and Wooding 1965). In contrast, the 15-nm striated fibrils and 17.5-nm non-striated fibrils found in the loose P-protein bodies of *Ricinus communis* do not appear to reorganize and can be found in mature SEs (Cronshaw 1975). In members of the Cucurbitaceae, the P-protein bodies of the fascicular SEs typically disperse, forming fine 6-nm filaments, whereas the P-protein bodies of the extrafascicular SEs remain as aggregates (Cronshaw and Esau 1968b). Some variation also exists in the dispersal of crystalline P-protein bodies of the papilionaceous legumes (Lawton 1978). However, the persistence of at least some crystalline P-protein bodies within the mature SE is often a common feature. According to Palevitz and Newcomb (1971), some of the crystalline P-protein bodies undergo a limited form of dispersal where they expand and separate into fine filaments, but still remain as discrete bodies. Tubules of P-protein were also observed in the mature SE, some of which appeared to disperse, forming striated filaments similar to those of tobacco. In *Phaseolus vulgaris* (LaFlèche 1966) and *Glycine max* (Wergin and Newcomb 1970), the crystalline P-protein bodies were reported to completely disperse, forming masses of fine fibrils. In contrast, the crystalline P-protein bodies did not disperse in *Glycine max* tissues prepared by freeze substitution methods, although 10-nm fibrils were observed within mature SEs (Fisher 1975). Non-dispersive crystalline P-protein bodies were also present in SEs of *Phaseolus multiflorus* roots (Lawton 1978).

The most common structural form of P-protein in translocating SEs is filamentous (Cronshaw and Sabnis 1990). Although there has been disagreement as to the distribution of P-protein within the mature SE, studies where surge effects were minimized indicate a parietal distribution of P-protein filaments in close association with the plasma membrane and SER (Evert et al. 1973; Fisher 1975; Deshpande 1984; Knoblauch and van Bel 1998). Interactions with membranes of the mature SE are thought to immobilize P-protein filaments, preventing their movement in the assimilate stream (Cronshaw 1975; Smith et al. 1987). Ultrastructural studies also indicate that the sieve plate pores in mature functioning SEs are open and only become obstructed with P-protein as a rapid response to vascular wounding (Cronshaw 1975). The primary function most often suggested for P-proteins is to seal wounded SEs by accumulating at the sieve plate to form slime plugs, thereby occluding sieve-plate pores (Eschrich 1975). Such a mechanism is thought to reduce the loss of assimilates and maintain turgor pressure in disrupted SEs. Although slime plugs are a characteristic structural feature of disrupted SEs, their effectiveness in sealing SEs is questionable (Barclay 1982; Kempers et al. 1993; Kempers and Van Bel 1997). To date, the function of P-proteins in intact or disrupted SEs has not been convincingly established.

These ultrastructural studies point to the complexity of P-protein structures and indicate that dynamic structural changes occur during differentiation of the SE-CC complex. Variations observed among species may relate to differences in vascular morphology or transport physiology. However, despite the apparent variability, P-protein deposition follows common themes. At the earliest stages of SE-CC complex differentiation, P-protein filaments form and coalesce into P-protein bodies within the immature SE. The bodies enlarge before the cytoplasm degenerates and disperse during selective autophagy, forming a filamentous network that becomes an integral component of translocating cells. It is possible that evolutionary pressure allows some variation in the overall architecture of these structures, which is reflected in the biochemical characteristics of the subunits. Divergence of polypeptides that compose P-protein structures has yet to be determined because of the limited number of species that have been examined. Studying the genes encoding P-proteins from a wide range of species will shed light on whether the diversity observed is a result of their structural divergence or functional convergence.

Chapter 16

P-Protein Trafficking Through Plasmodesmata

G. A. Thompson

Department of Plant Sciences, 303 Forbes Building,
University of Arizona, Tucson, Arizona 85721 USA

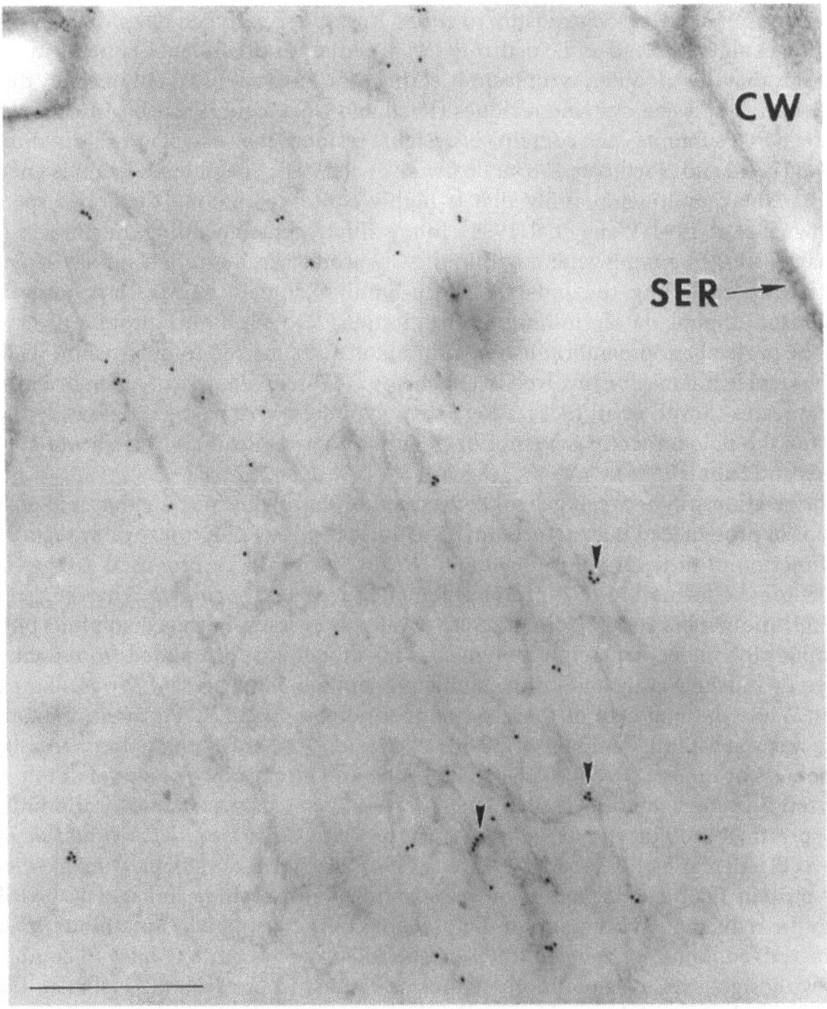

Fig 1. PP1 immunolocalizes to dispersed P-protein filaments within differentiated SEs of *Arabidopsis thaliana*. A mixture of two monoclonal antibodies raised against an 89-kDa phloem filament protein from *Streptanthus tortuosus* tissue cultures showed specific labelling of P-protein filaments (*arrowheads*) visualized with 10-nm colloidal gold. The label was absent from the sieve element reticulum (*SER*) and cell walls (*CW*). *Bar* 0.5 µm. (Tóth et al. 1994)

clearly shows that PP1 is the primary structural protein involved in the formation of P-protein bodies, filaments, and slime plugs and can be justifiably called the phloem filament protein.

PP2 is a 48-kDa dimeric lectin (haemagglutinin) that specifically binds poly(β-1,4-*N*-acetylglucosamine) (Allen 1979; Sabnis and Hart 1978). In contrast to purified PP1 that forms filaments upon oxidation, purified PP2 remains soluble or precipitates as amorphous material upon exposure to either atmospheric oxygen or oxidizing agents (Kleinig et al. 1975; Read and Northcote 1983b). Analysis of phloem filaments in vitro indicates that this globular, cytoplasmic lectin cross-links with PP1 polymers by disulphide bonds between cysteine residues (Read and Northcote 1983a, b). In *Cucurbita*, identical PP2 subunits each contain six cysteine residues that also play a role in dimerization (Read and Northcote 1983a; Bostwick et al. 1994). The phloem lectin is encoded by a small multigene family that is highly conserved among *Cucurbita* species (Bostwick et al. 1994; Wang et al. 1994). Interestingly, genes encoding the phloem lectin have diverged among genera within the Cucurbitaceae, but retain similar structural features including the putative chitin-binding domain (A.M. Clark and G.A. Thompson unpubl. data). Immunolocalization of PP2 with both P-protein filaments and the parietal endomembrane system in mature SEs has led to the hypothesis that the phloem lectin may be involved in anchoring P-protein filaments to membrane glycoconjugates (Smith et al. 1987). Alternative hypotheses regarding a defensive function for the phloem lectin have not been substantiated (Read and Northcote 1983a; Sabnis and Sabnis 1995).

The relationship between gel-sol conversion of P-proteins and microscopic observations of proteinaceous structures in SEs is unclear, generating controversy regarding the function of these proteins. Read and Northcote (1983a, b) proposed a structural model for the formation of P-protein filaments in SEs of *Cucurbita*. They suggested that PP1 monomers and PP2 dimers were covalently cross-linked via disulphide bonds forming high molecular weight polymers. This model was formulated from data obtained by biochemically analyzing soluble phloem filaments present in vascular exudates. While the majority of these proteins were complexed as P-protein filaments, there was a substantial amount of PP1 (~40%) and PP2 (~20%) that remained as free monomers or dimers (Read and Northcote 1983a). The existence of a pool of unpolymerized P-protein subunits is further supported by the apparent translocation of genus-specific P-proteins in intergeneric grafts between *Cucurbita ficifolia* and *Cucumis sativus* (Tiedemann and Carstens-Behrens 1994). Alosi et al. (1988) questioned whether P-protein filament formation or stabilization by disulphide linkages is possible when the reducing environment of the phloem sap is considered. Glutathione, a well-characterized donor of reduced thiols, is abundant in vascular exudates in combination with high levels of glutathione reductase activity in *Cucurbita* (Alosi et al. 1988) and glutaredoxin activity in *Ricinus* (Szederkényi et al. 1997). Recently, cDNAs have been isolated that encode glutaredoxin from *Ricinus communis* (Szederkényi et al. 1997) and cytoplasmic thioredoxin h from rice (*Oryza sativa*) (Ishiwatari et al. 1995). Further characterization of thiol-regulated systems in SEs may reveal their role in controlling P-protein polymerization.

4 P-Proteins Are Synthesized in Companion Cells and Accumulate in Sieve Elements

In the Cucurbitaceae, P-protein profiles from vascular exudates are remarkably consistent, regardless of the organ or developmental stage examined. Sodium dodecyl sulphate-polyacrylamide gel electrophoresis and isoelectric focusing showed no qualitative differences in P-protein patterns isolated from the following contrasting developmental stages: 5-day old seedlings and mature plants; young petioles compared with the oldest stem internodes; between dark-grown and light-grown seedlings; and from plants in either a vegetative or flowering state (Sabnis and Hart 1976, 1978, 1979; Smith et al. 1987). Indeed, *PP1* and *PP2* mRNA can be detected in all organs containing vascular tissue, including fruits and flowers (Clark et al. 1997). In contrast to this consistency at the whole-plant level, examining P-protein synthesis at the cellular level reveals an intricate and highly coordinated pattern of accumulation that parallels the complexity of vascular development.

Analysis of the temporal and spatial expression of PP1, PP2, and their mRNAs in *Cucurbita* seedlings illustrates the high level of developmental regulation that occurs during P-protein synthesis and deposition. In pumpkin hypocotyls, the genes encoding PP1 and PP2 are developmentally expressed: mRNA and protein are initially detected 3 and 4 days after germination, respectively, and accumulate in parallel, with maximum levels of expression correlating with the developmental maturation of the organ at about 10 to 12 days after germination (Sham and Northcote 1987; Clark et al. 1997; Dannenhoffer et al. 1997). Immunoprecipitation of in vitro translation products showed that the accumulation of translatable *PP2* mRNA was directly correlated with the increased concentration of the phloem lectin (Sham and Northcote 1987). Following the period of peak accumulation, steady-state levels of *PP1* and *PP2* mRNA decrease, while protein concentrations remain at high levels (Clark et al. 1997). *PP1* and *PP2* mRNAs accumulated in similar developmental patterns, but the average ratio between 6 and 24 days after germination was 26 *PP2* mRNA molecules for each *PP1* mRNA. Comparative analysis of the *PP1* and *PP2* promoters in transgenic plants supports the idea that *PP1* genes are less transcriptionally active (A.A. Carneiro and G.A. Thompson, unpubl. data). Although considerable differences occur in the steady-state levels of *PP1* and *PP2* mRNAs, similar concentrations of PP1 and PP2 accumulated at all stages of development with a stoichiometry ranging from three to five molecules of PP2 for each molecule of PP1. Pulse-labelling experiments further demonstrated the stability of PP1 and PP2, consistent with their designation as structural proteins (Dannenhoffer et al. 1997).

Accumulation of *PP1* and *PP2* mRNA is tightly linked to vascular differentiation in elongating hypocotyls of *Cucurbita* seedlings (Clark et al. 1997; Dannenhoffer et al. 1997). During the first 2 days of seedling growth, when *PP1* and *PP2* mRNAs are not detectable, the vascular tissue of the hypocotyl consists of procambial bundles containing few differentiated protoxylem and protophloem elements. *PP1* and *PP2* mRNAs were initially detected in hypocotyls 3 days after germination when SE-CC complexes are actively differentiating in the internal and external metaphloem and in the bundle-associated extrafascicular phloem. Accumulation of *PP1* and *PP2* mRNAs in hypocotyls increase substantially by 6 days after germination when the vascular bundles contain abundant metaphloem elements and the first secondary elements are differentiating from the vascular cambium.

Cellular specificity of P-protein gene expression and protein accumulation was determined in differentiating SE-CC complexes during vascular development in *Cucurbita* hypocotyls and stems (Clark et al. 1997; Dannenhoffer et al. 1997). *PP1* and *PP2* mRNAs were identified in the internal and external bundle phloem and bundle-associated extrafascicular phloem in 3-day-old hypocotyls by in situ hybridization of digoxigenin-labelled riboprobes observed by light microscopy (Fig. 2A). Transcripts were localized exclusively in CCs of differentiating SE-CC complexes (Fig. 2B, C). In contrast, P-protein mRNAs were never detected in immature metaphloem SEs characterized by intact nuclei and large central vacuoles (Fig. 2D), or in protophloem SEs lacking CCs. These results provide convincing evidence that the genes encoding PP1 and PP2 are transcribed and their mRNAs accumulate specifically in CCs of differentiating SE-CC complexes.

In contrast to CC-specific mRNA localization, the phloem lectin was immunocytochemically localized to both SEs and CCs (Smith et al. 1987; Dannenhoffer et al. 1997). Interestingly, there is a fundamental difference between P-protein deposition in the bundle and extrafascicular phloem of *Cucurbita* (Fig. 3A–G). Both CCs and SEs were heavily labelled when the extrafascicular phloem was viewed by light microscopy (Fig. 3E). High resolution immunogold labelling of PP2 in developing *Cucurbita moschata* stems showed PP2 accumulated in the persistent protein bodies of the extrafascicular SEs (Fig. 3F, G). In contrast, PP2 was conspicuously absent from either the filamentous P-protein bodies or the cytoplasm of immature SEs in the bundle phloem (Fig. 3C, D). Dense labelling of PP2 was observed in CCs of differentiating SE-CC complexes, suggesting a temporal and spatial difference between the accumulation of PP1 and PP2 in immature SE-CC complexes of the bundle phloem (Fig. 3B). PP2 was detected at sieve plates and in a parietal position within mature SEs. High-resolution immunolocalization of PP2, visualized with colloidal gold, showed the gold particles in close association with P-protein filaments and the SER that lines the plasma membrane of differentiated SE-CC complexes (Smith et al. 1987).

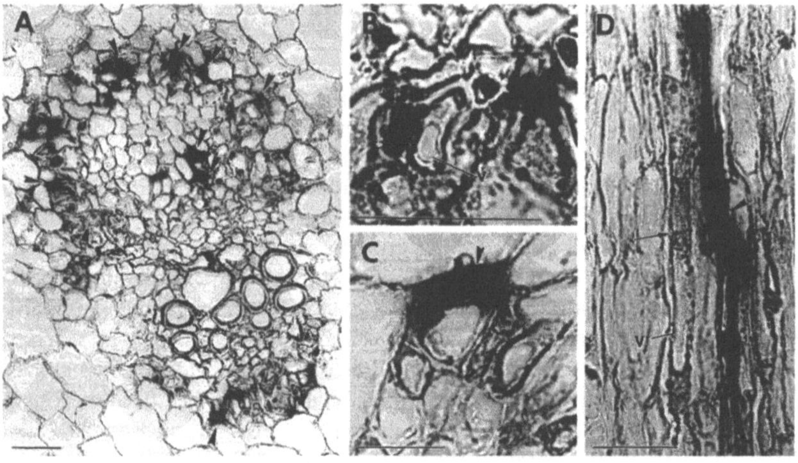

Fig 2A–D. Accumulation of *PP1* and *PP2* mRNA is companion cell-specific. In situ hybridization of *PP1* and *PP2* mRNA with digoxigenin-labelled antisense riboprobes in hypocotyls of *Cucurbita maxima* seedlings 3 days after germination. **A** *PP2* mRNA (*black cells*) was detected within the internal and external bundle metaphloem and the bundle-associated extrafascicular phloem. **B** Transverse section of two SE-CC complexes showing *PP2* mRNA within the CCs (*arrowheads*) but not immature SEs. Both SEs have P-protein bodies; one has a degenerating nucleus and the other an intact vacuole (*v*). **C** Transverse section of a SE-CC complex of the extrafascicular phloem showing *PP1* mRNA (*arrowhead*) within CCs. **D** Longitudinal section showing labelled CCs (*arrowhead*) and immature SEs. *Arrow* indicates developing sieve plate. *Bars* 50 μm (After Dannenhoffer et al. 1997)

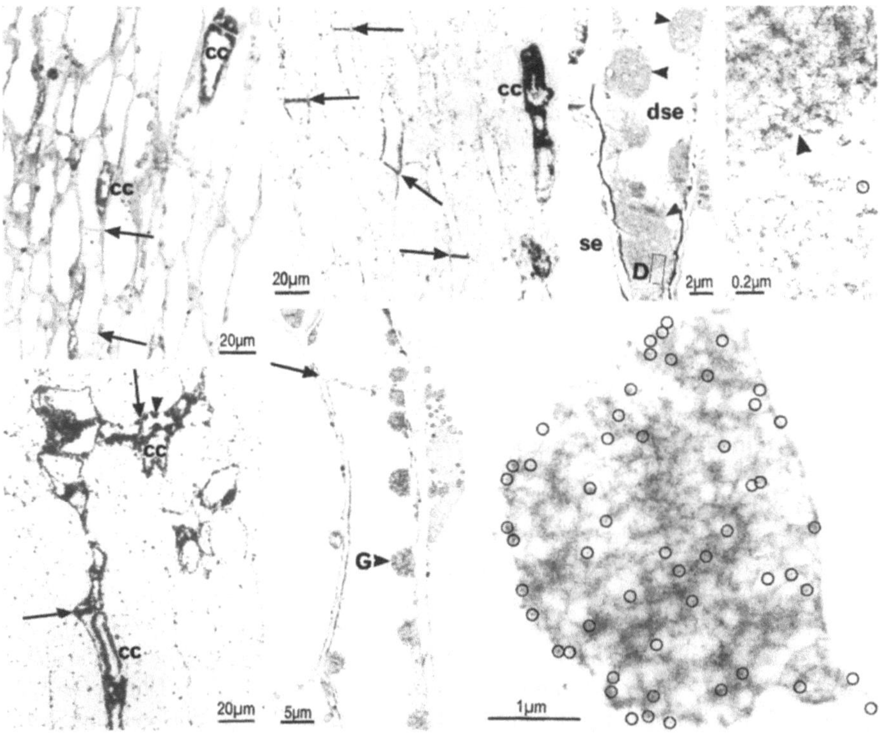

Fig 3A–G. PP2 differentially accumulates in SEs of the bundle and extrafascicular phloem of *Cucurbita moschata* stems. Bundle phloem: **A** Light micrograph of bundle phloem stained with crystal violet. Sieve plates (*arrows*) and companion cells (*cc*) are shown. **B** Light micrograph of bundle phloem immunolabelled with PP2 antiserum visualized by silver-enhanced colloidal gold. Sieve plates (*arrows*) and parietal regions of mature SEs are lightly labelled, and companion cells (*cc*) of developing SE-CC complexes are densely labelled. **C** Electron micrograph of a mature sieve element (*se*) with dispersed P-protein filaments and developing sieve element (*dse*) with P-protein bodies (*arrowheads*). **D** Higher magnification of the developing SE in **C**. Immunolabelling with 5-nm colloidal gold was not observed in either the cytoplasm or P-protein body (*arrowhead*) with the exception of a small amount of background (*circle*). Extrafascicular phloem: **E** Light micrograph of sieve plates (*arrows*) and persistent P-protein bodies (*arrowhead*) of extrafascicular SEs and their companion cells (*cc*) densely labelled by PP2 antiserum. **F** Electron micrograph of mature extrafascicular SEs with sieve plate (*arrow*) and persistent P-protein bodies (*arrowhead*). The *arrowhead labelled G* indicates the P-protein body shown in panel **G**. **G** Colloidal gold particles (*circles*) labelled the filaments of the persistent P-protein body. (Dannenhoffer et al. 1997)

Although P-protein synthesis occurs in companion cells of developing SE-CC complexes, it is not clear if synthesis continues in differentiated SE-CC complexes. Nuske and Eschrich (1976) isolated labelled basic proteins with molecular weights corresponding to PP1 and PP2 after applying ^{14}C-labelled amino acids to the cotyledons of *Cucurbita maxima* seedlings. Microautoradiography of vascular bundles in hypocotyls showed that labelled proteins were concentrated in CCs of differentiated internal and external metaphloem, suggesting that P-protein synthesis does occur in CCs of the mature phloem. In *Cucurbita maxima* seedlings, maximum levels of *PP1* and *PP2* mRNA occur at approximately 10–12 days after germination when the hypocotyls have fully extended (Clark et al. 1997). At this developmental stage, the primary bundle phloem is mature and the secondary phloem is differentiating from the vascular cambium. In addition, numerous bundle-associated and cortical extrafascicular SE-CC complexes have differentiated. *PP1* and *PP2* mRNAs localize specifically to CCs of differentiated SE-CC complexes in *Cucurbita maxima* seedlings at 10–12 days after germination (Bostwick et al. 1992; unpubl. results). Localization patterns for *PP1* and *PP2* mRNAs are identical: *PP1* and *PP2* digoxigenin-labelled antisense riboprobes hybridize to mRNA within the phloem of hypocotyl tissues in the fascicular, bundle-associated extrafascicular, and cortical extrafascicular phloem tissue. While P-protein synthesis is most active during rapid vascular development, these proteins also appear to be synthesized in CCs of differentiated SE-CC complexes.

5 Protein Movement Through Pore/Plasmodesma Units

Cucurbita P-proteins are synthesized in CCs and accumulate in SEs in a developmentally regulated manner. Although we have yet to understand the mechanisms involved in regulating their cell-to-cell movement, trafficking almost certainly involves the numerous symplasmic connections that occur between SEs and CCs. During the ontogeny of the SE-CC complex, multiple-branched plasmodesmata at the CC side of the wall are joined to a small pore on the SE side by a large central cavity in the middle lamella (e.g. Deshpande 1975; Esau and Thorsch 1985; Ding et al. 1993; Northcote 1995; Fig. 4). PPUs are the primary conduit for interactions that occur between the SE and CC (reviewed in van Bel and Kempers 1997). These interactions clearly include the movement of proteins between the two cell types for long-term maintenance of the enucleate SE and possibly long-distance signalling via the phloem. Analysis of phloem sap from rice, wheat, and castor bean revealed 100-200 distinct polypeptides that are present in SEs (Fisher et al. 1992; Nakamura et al. 1993; Sakuth et al. 1993). Fisher et al. (1992) presented a model, based on observations of ^{35}S-labelled phloem-mobile proteins at several points along the source-to-sink pathway in wheat, suggesting that proteins selectively move between CCs and mature SEs. These observations implicate PPUs as the pathway for symplasmic movement of macromolecules within the phloem. Since the majority of these phloem-mobile proteins range in size from M_r 10–70 kDa, the size limitation that is imposed by PPUs is an important regulatory consideration for protein movement between CCs and SEs.

Molecular exclusion limits ranging from 0.75–1.0 kDa for plasmodesmata in mesophyll cells were initially established by microinjecting cells with dyes and fluorescently labelled dextrans (Terry and Robards 1987; Wolf et al. 1989; Robards and Lucas 1990; see Chaps. 5 and 6 for further references). In contrast, PPUs in the transport phloem appear to allow passive transport of larger molecules than plasmodesmata in the mesophyll. Microinjection of low molecular weight dyes (Lucifer Yellow CH, M_r 457, and fluorescein-Gly$_6$, M_r 749) into CCs of the extrafascicular phloem of *Cucurbita maxima* resulted in the rapid movement of the fluorescent probes into the adjacent SE and the translocation stream (Kempers et al. 1993). Fluorescein isothiocyanate (FITC)-labelled insulin A (M_r 2921) injected into CCs also moved into SEs, suggesting a minimum molecular exclusion limit of 3 kDa for PPUs in *Cucurbita* extrafascicular phloem. Recent-

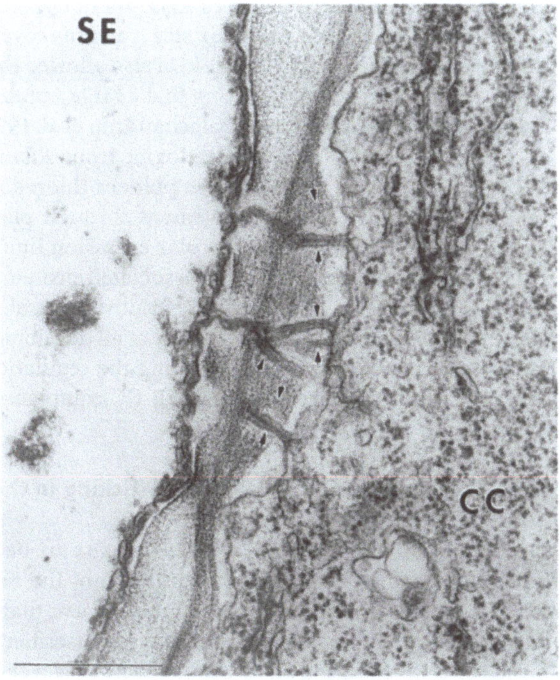

Fig 4. Pore/plasmodesma units (PPUs) form symplasmic connections between the sieve element and companion cell in *Cucurbita pepo*. Multiple branches of the plasmodesmata (*arrowheads*) are located on the CC side of the cell wall. *Bar* 0.5 µm. (Courtesy of Alexander Schulz, Royal Veterinary and Agricultural University, Copenhagen, Denmark)

ly, microinjection of fluorescently labelled dextrans into the SEs of the fascicular phloem of *Vicia faba* indicated that the molecular exclusion limit of PPUs is at least 10 kDa and could be as high as 25 kDa (Kempers and van Bel 1997). Fluorescently labelled 3, 4.4, and 10-kDa dextrans microinjected into SEs of the fascicular phloem in *Vicia faba* stems moved into adjacent CCs (Kempers and van Bel 1997). FITC-labelled chymotrypsinogen A (25 kDa) also moved into CCs from the injected SEs in a pattern similar to the 10-kDa dextran, while movement of a 40-kDa dextran was limited to the injected SE. The increased molecular exclusion limit for PPUs could reflect the unique nature of mature SEs. Recent studies have demonstrated the importance of actin filaments extending through plasmodesmata in the regulation of plasmodesmal transport (White et al. 1994; Heinlein et al. 1995; McLean et al. 1995; Ding et al. 1996; also see Chap. 3 and 9). Microtubules and actin filaments are commonly found in immature SEs but, with rare exceptions, do not remain in mature SEs following selective autophagy (Evert 1990; Schulz 1998). The absence of cytoskeletal components in mature SEs could be one mechanism to account for the increased molecular exclusion limits observed for PPUs.

Evidence is also emerging that phloem-associated proteins, including P-proteins, injected into mesophyll cells can increase the molecular exclusion limit of plasmodesmata, allowing their intercellular movement. Microinjection of FITC-labelled PP2 in mesophyll cells of *Cucurbita maxima* cotyledons showed movement of the labelled phloem lectin that extended 20–30 cells from the injection site (Balachandran et al. 1997). In coinjection studies, very low concentrations (0.025–0.04 mg ml^{-1} protein) of

PP2 also directed movement of 20-kDa, but not 40-kDa, FITC dextrans. Microinjection of mesophyll cells with each of six size fractions covering a range of *C. maxima* vascular exudate proteins from 10–200-kDa also allowed the movement of 10 and 20, but not 40-kDa, FITC dextrans, suggesting that a large number of phloem-associated proteins interact with plasmodesmata (Balachandran et al. 1997). Indeed, two phloem-associated proteins, cystatin and glutaredoxin, from *Ricinus communis* vascular exudates (Balachandran et al. 1997) and the phloem thioredoxin h from rice (Ishiwatari et al. 1998) mediate intercellular movement through plasmodesmata in mesophyll cells. This ability to increase the molecular exclusion limit is probably more important for regulating trafficking of phloem-associated proteins in immature SE-CC complexes when PPUs could be under more restrictive control, and for trafficking high molecular weight phloem-associated proteins at all developmental stages. Clearly, many questions remain to be answered regarding the regulation of protein movement through PPUs in differentiating and mature SE-CC complexes.

6 Perspectives on P-Protein Trafficking in the Developing SE-CC Complex

P-protein trafficking through PPUs appears to be a temporally regulated process linked to specific phases in differentiation of the SE-CC complex. The available evidence indicates that genes encoding the phloem filament protein PP1 and the phloem lectin PP2 in *Cucurbita* are coordinately transcribed and their mRNAs are translated in CCs of differentiating SE-CC complexes. However, differences between PP1 and PP2 in their size and pattern of accumulation during SE differentiation raise the possibility of alternate mechanisms regulating their intercellular movement. Such mechanisms could involve additional proteins.

PP1, synthesized in CCs, rapidly accumulates in filamentous P-protein bodies within immature (nucleate) SEs of the bundle phloem. Even if the increased molecular exclusion limit of PPUs is considered, cell-to-cell movement of this 96-kDa protein must involve modification of either the protein, the PPU, or both. The results of microinjection studies in cucurbit cotyledons suggest that PP1 can modify the molecular exclusion limit of mesophyll plasmodesmata (Balachandran et al. 1997). It is also likely that PP1 is transported through PPUs in an unfolded state soon after its synthesis. Several observations suggest that *PP1* mRNA is efficiently translated in CCs. However, the P-protein bodies that characterize immature SEs are not a typical feature of CCs. Molecular chaperones identified in vascular exudates (Schobert et al. 1995) could be involved in posttranslational (un)folding of PP1 monomers and in mediating their transport into SEs. Recently, the chaperone cyclophilin was identified as an abundant phloem-associated protein in *Ricinus* and *Cucurbita* (Schobert et al. 1998). Schobert and co-workers demonstrated cell-to-cell movement of fluorescently-labelled recombinant cyclophilin, and its colocalization with P-protein filaments in SEs of *Ricinus* and *Cucurbita* (Christian Schobert, University of California, Davis, pers. comm.). They speculated that proteins trafficking through plasmodesmata in an unfolded state undergo facilitated folding via interaction with cyclophilin. Heterocomplexes containing chaperones have important roles in folding proteins such as P pili subunits in *E. coli* (Bullit et al. 1996), tubulin (Tian et al. 1996), and actin (Gao et al. 1992) into native configurations required for the formation of polymeric structures. The role of chaperones in P-protein trafficking and filament formation has yet to be demonstrated, but offers interesting new approaches to investigate the dynamics of P-protein movement and deposition.

In contrast to the rapid accumulation of the phloem filament protein in immature SEs, PP2 is retained within CCs until late stages of SE differentiation in the bundle phloem. PP2 probably moves into SEs during or soon after the period of selective autophagy, cross-linking with dispersed PP1 filaments. Movement into the SE prior to this developmental stage could result in the formation of persistent (non-dispersive) P-protein bodies typical of those observed in the extrafascicular phloem of cucurbits (Dannenhoffer et al. 1997). Thus, trafficking of this cytoplasmic lectin might involve an additional mechanism preventing its premature accumulation in differentiating SEs of the bundle phloem. The developmental stage at which the trafficking occurs could also influence the types of interactions that occur between PPUs and the phloem lectin. PP2 appears to directly modify the molecular exclusion limit of mesophyll plasmodesmata to mediate its own transport (Balachandran et al. 1997). However, the increased molecular exclusion limit attributed to PPUs in differentiated SE-CC complexes might allow unregulated movement of the 24.5-kDa PP2 subunits. Microinjection studies of chymotrypsinogen A (M_r 25000) demonstrated diffusional exchange of this protein between SEs and CCs (Kempers and van Bel 1997). Even larger molecular exclusion limits for PPUs, speculated by Kempers and van Bel (1997), might also allow unrestricted movement of the 48-kDa PP2 dimer. They stated that molecular exclusion limits based on the distribution patterns of fluorescently labelled dextrans must be corrected to account for the tight packing of globular proteins. The Stokes radius of a PP2 dimer (2.9 nm; Anantharam et al. 1986) is within the experimentally demonstrated range for PPUs: movement of 10-kDa dextran conjugates (2.3 nm) from injected SEs into CCs was permitted, whereas 40-kDa dextran conjugates (4.4 nm) were excluded (Kempers and van Bel 1997). At this point, too little is known about the regulation of P-protein movement through PPUs to arrive at anything but speculative conclusions.

Several mechanisms probably exist for trafficking phloem-associated proteins from CCs to SEs or for their removal from the assimilate stream. An interesting new report suggests that plant mRNA might also traffic between CCs and SEs. Although transcribed in CCs, the sucrose transporter SUT1 mRNA was localized in close association with plasmodesmata in enucleate SEs of *Solanum tuberosum* (Kühn et al. 1997). This intriguing observation raises questions regarding mechanisms for plant RNA trafficking through PPUs and the significance of such trafficking into a cell that no longer has the capacity for translation. Clearly, many questions remain to be addressed concerning intercellular movement of phloem-associated macromolecules between these two very unusual cell types. So far, relatively few phloem-associated proteins have been identified, and only a handful of genes have been isolated encoding these proteins. However, this situation is rapidly changing, and we have entered an exciting period that promises to reveal the significance of complex regulatory and structural interactions that occur at the interface of the SE-CC complex.

Acknowledgements. Thanks are expressed to Dr. Anna M. Clark for her assistance with preparation of this manuscript, including the figures. The preparation of this chapter was supported in part by grant No. IBN-9727626 from the National Science Foundation Integrative Plant Biology Program, USA.

References

Allen AK (1979) A lectin from the exudate of the fruit of the vegetable marrow (*Cucurbita pepo)* that has a specificity for a β-1,4-linked *N*-acetylglucosamine oligosaccharide. Biochem J 183:133–137

Alosi MC, Melroy DL, Park RB (1988) The regulation of gelation of phloem exudate from *Cucurbita* fruit by dilution, glutathione, and glutathione reductase. Plant Physiol 86:1089–1094

Anantharam V, Patanjali SR, Swamy MJ, Sanadi AR, Goldstein IJ, Surolia A (1986) Isolation, macromolecular properties, and combining site of chito-oligosaccharide-specific lectin from the exudate of ridge gourd (*Luffa acutangula*). J Biol Chem 261:14621–14627

Balachandran S, Xiang Y, Schobert C, Thompson GA, Lucas WJ (1997) Phloem sap proteins from *Cucurbita maxima* and *Ricinus communis* have the capacity to traffic cell-to-cell through plasmodesmata. Proc Natl Acad Sci USA 94:14150–14155

Barclay GF (1982) Slime plugs do not inhibit surge flow in sieve tubes. Can J Bot 60:1281–1284

Behnke H-D (1975) Companion cells and transfer cells. In: Aronoff S, Dainty J, Gorham PR, Srivastava LM, Swanson CA (eds) Phloem transport. NATO ADV Study Inst Ser, Plenum Press, New York, pp 153–175

Behnke H-D (1981) Siebelement-Plastiden, Phloem-Protein und Evolution der Blütenpflanzen: II. Monokotyledonen. Ber Dtsch Bot Ges 94:647–662

Behnke H-D, Dörr I (1967) Zur Herkunft und Struktur der Plasmafilamente in Assimilatleitbahnen. Planta 74:18–44

Beyenbach J, Weber C, Kleinig H (1974) Sieve-tube proteins from *Cucurbita maxima*. Planta 119: 113–124

Blyth A (1958) Origin of primary extraxylary stem fibers in dicotyledons. Univ Calif Berkeley Publ Bot 30:145–232

Bostwick DE, Dannenhoffer JM, Skaggs MI, Lister RM, Larkins BA, Thompson GA (1992) Pumpkin phloem lectin genes are specifically expressed in companion cells. Plant Cell 4:1539–1548

Bostwick DE, Skaggs MI, Thompson GA (1994) Organization and characterization of *Cucurbita* phloem lectin genes. Plant Mol Biol 26:887–897

Bullit E, Jones CH, Striker R, Soto G, Jacob-Dubuisson F, Pinkner J, Wick MJ, Makowski L, Hultgren SJ (1996) Development of pilus organelle subassemblies in vitro depends on chaperone uncapping of a beta zipper. Proc Natl Acad Sci USA 93:12890–12895

Clark AM, Jacobsen KR, Dannenhoffer JM, Skaggs MI, Thompson GA (1997) Molecular characterization of a phloem-specific gene encoding the filament protein, Phloem Protein 1 (PP1), from *Cucurbita maxima*. Plant J 12:49–61

Crafts AS (1932) Phloem anatomy, exudation and transport of organic nutrients in cucurbits. Plant Physiol 7:183–225

Cronshaw J (1975) P-proteins. In: Aronoff S, Dainty J, Gorham PR, Srivastava LM, Swanson CA (eds) Phloem transport. NATO ADV Study Inst Ser, Plenum Press, New York, pp 79–147

Cronshaw J, Esau K (1967) Tubular and fibrillar components of mature and differentiating sieve elements. J Cell Biol 34:801–815

Cronshaw J, Esau K (1968a) P-Protein in the phloem of *Cucurbita*. I. The development of P-protein bodies. J Cell Biol 38:25–39

Cronshaw J, Esau K (1968b) P-Protein in the phloem of *Cucurbita*. II. The P-protein of mature sieve elements. J Cell Biol 38:292–303

Cronshaw J, Sabnis DD (1990) Phloem proteins. In: Behnke H-D, Sjolund RD (eds) Sieve elements, comparative structure, induction and development. Springer, Berlin Heidelberg New York, pp 257–283

Currier HB, Shih CY (1968) Sieve tubes and callose in *Elodea* leaves. Am J Bot 55:145–152

Dannenhoffer JM, Schulz A, Skaggs MI, Bostwick DE, Thompson GA (1997) Expression of the phloem lectin is developmentally linked to vascular differentiation in cucurbits. Planta 201:405–414

Deshpande BP (1975) Differentiation of the sieve plate of *Cucurbita*: a further view. Ann Bot 39: 1015–1022

Deshpande BP (1984) Distribution of P-protein in mature sieve elements of *Cucurbita maxima* seedlings subjected to prolonged darkness. Ann Bot 53:237–247

Ding B, Haudenshield JS, Willmitzer L, Lucas WJ (1993) Correlation between arrested secondary plasmodesmal development and onset of accelerated leaf senescence in yeast acid invertase transgenic tobacco plants. Plant J 4:179–189

Ding B, Kwon M-O, Warnberg L (1996) Evidence that actin filaments are involved in controlling the permeability of plasmodesmata in tobacco mesophyll. Plant J 10:157–164

Esau K, Cronshaw J (1967) Tubular components in cells of healthy and tobacco mosaic virus-infected *Nicotiana*. Virology 33:26–35

Esau K, Thorsch J (1985) Sieve plate pores and plasmodesmata, the communication channels of the symplast: ultrastructural aspects and developmental relations. Am J Bot 72:1641–1653
Eschrich W (1975) Sealing systems in phloem. In: Zimmerman MH, Milburn JA (eds) Transport in plants I. phloem transport. Springer, Berlin Heidelberg New York, pp 39–56
Eschrich W, Evert RF, Heyser W (1971) Proteins of the sieve-tube exudate of *Cucurbita maxima*. Planta 100:208–221
Evert RF (1990) Dicotyledons. In: Behnke H-D, Sjolund RD (eds) Sieve elements, comparative structure, induction and development. Springer, Berlin Heidelberg New York, pp 103–137
Evert RF, Eschrich W, Eichhorn SE (1971) Sieve-plate pores in leaf veins of *Hordeum vulgare*. Planta 100:262–267
Evert RF, Eschrich W, Eichhorn SE (1973) P-protein distribution in mature sieve elements of *Cucurbita maxima*. Planta 109:193–210
Fisher DB (1975) Structure of functional soybean sieve elements. Plant Physiol 56:555–569
Fisher DB, Wu Y, Ku MSB (1992) Turnover of soluble proteins in the wheat sieve tube. Plant Physiol 100:1433–1441
Gagnon M-J, Beebe DU (1996) Minor vein differentiation and the development of specialized plasmodesmata between companion cells and contiguous cells in expanding leaves of *Moricandia arvensis* (L.) DC. (Brassicaceae). Int J Plant Sci 157:685–697
Gao Y, Thomas JO, Chow RL, Lee G-H, Cowan NJ (1992) A cytoplasmic chaperone that catalyzes β-actin folding. Cell 69:1043–1050
Heinlein M, Epel BL, Padgett HS, Beachy RN (1995) Interaction of tobamovirus movement proteins with the plant cytoskeleton. Science 270:1983–1985
Heyser W (1971) Phloemdifferenzierung bei *Tradescantia albiflora*. Cytobiology 4:186–197
Ishiwatari Y, Honda C, Kawashima I, Nakumura S, Hirano H, Mori S, Fujiwara T, Hayashi H, Chino M (1995) Thioredoxin h is one of the major proteins in rice phloem sap. Planta 195:456–463
Ishiwatari Y, Fujiwara T, McFarland KC, Nemoto K, Hayashi H, Chino M, Lucas WJ (1998). Rice phloem thioredoxin h has the capacity to mediate its own cell-to-cell transport through plasmodesmata. Planta 205:12–22
Kempers R, van Bel AJE (1997) Symplasmic connections between sieve elements and companion cell in the stem phloem of *Vicia faba* L. have a molecular exclusion limit of at least 10 kDa. Planta 201: 195–201
Kempers R, Prior DAM, van Bel AJE, Oparka KJ (1993) Plasmodesmata between sieve element and companion cell of extrafascicular stem phloem of *Cucurbita maxima* permit passage of 3-kDa fluorescent probes. Plant J 4:567–575
Kempers R, Ammerlaan A, van Bel AJE (1998) Symplasmic constriction and ultrastructural features of the sieve element/companion cell complex in the transport phloem of apoplasmically and symplasmically phloem-loading species. Plant Physiol 116:271–278
Kleinig H, Dörr I, Kollmann R (1971) Vinblastine-induced precipitation of phloem proteins in vitro. Protoplasma 73:293–302
Kleinig M, Thones J, Dörr I, Kollmann R (1975) Filament formation in vitro of sieve tube protein from *Cucurbita maxima* and *Cucurbita pepo*. Planta 127:163–170
Knoblauch M, van Bel AJE (1998) Sieve tubes in action. Plant Cell 10:35–50.
Kollmann R, Dörr I, Kleinig H (1970) Protein filaments – structural components of the phloem exudate. I. Observations with *Cucurbita* and *Nicotiana*. Planta 95:86–94
Kühn C, Franceschi VR, Schulz A, Lemoine R, Frommer WB (1997) Macromolecular trafficking indicated by localization and turnover of sucrose transporters in enucleate sieve elements. Science 275:1298–1300
Kuo J, O'Brien TP, Zee SY (1972) The transverse veins of the wheat leaf. Aust J Biol Sci 25:721–737
LaFlèche D (1966) Ultrastructure et cytochimie des inclusions flagellées des cellules criblées de *Phaseolus vulgaris* L. J Microsc (Paris) 5:493–510
Lawton DM (1978) P-protein crystals do not disperse in uninjured sieve elements in roots of runner bean (*Phaseolus multiflorus*) fixed with glutaraldehyde. Ann Bot 42:353–361
McLean BG, Zupan J, Zambryski PC (1995) Tobacco mosaic virus movement protein associates with the cytoskeleton in tobacco cells. Plant Cell 7:2101–2114
Nakamura S, Hayashi H, Mori S, Chino M (1993) Protein phosphorylation in the sieve tubes of rice plants. Plant Cell Physiol 34:927–933
Northcote DH (1995) Aspects of vascular tissue differentiation in plants: parameters that may be used to monitor the process. Int J Plant Sci 156:245–256
Northcote DH, Wooding FBP (1965) Development of sieve tubes in *Acer pseudoplatanus*. Proc R Soc Lond Ser B 163:524–537
Nuske J, Eschrich W (1976) Synthesis of P-protein in mature phloem of *Cucurbita maxima*. Planta 132:109–118

O'Brien TP, Thimann KV (1967) Observations on the fine structure of the oat coleoptile. III. Correlated light and electron microscopy of the vascular tissues. Protoplasma 63:443–478
Palevitz BA, Newcomb EH (1971) The ultrastructure and development of tubular and crystalline P-protein in the sieve elements of certain papilionaceous legumes. Protoplasma 72:399–426
Parthasarathy MV (1974) Ultrastructure of phloem in palms. I. Immature sieve elements and parenchymatic elements. Protoplasma 79:59–92
Read SM, Northcote DH (1983a) Subunit structure and interactions of the phloem proteins of *Cucurbita maxima* (pumpkin). Eur J Biochem 134:561–569
Read SM, Northcote DH (1983b) Chemical and immunological similarities between the phloem proteins of three genera of Cucurbitaceae. Planta 158:119–127
Robards AW, Lucas WJ (1990) Plasmodesmata. Annu Rev Plant Physiol Plant Mol Biol 41:369–419
Sabnis DD, Hart JW (1976) A comparative analysis of phloem exudate proteins from *Cucumis melo, Cucumis sativus* and *Cucurbita maxima* by polyacrylamide gel electrophoresis and isolelectric focusing. Planta 130:211–218
Sabnis DD, Hart JW (1978) The isolation and some properties of a lectin (haemagglutinin) from *Cucurbita* phloem exudate. Planta 142:97–101
Sabnis DD, Hart JW (1979) Heterogeneity in phloem protein complements from different species: consequences to hypotheses concerned with P-protein functions. Planta 145:459–466
Sabnis DD, Sabnis HM (1995) Phloem proteins: Structure, biochemistry and function. In: Iqbal M (ed) The cambial derivatives. Gebruder Borntraeger, Berlin, pp 271–292
Sakuth T, Schobert C, Pecsvaradi A, Eichholz A, Komor E, Orlich G (1993) Specific proteins in the sieve-tube exudate of *Ricinus communis* L. seedlings: separation, characterization and in-vivo labelling. Planta 191:207–213
Schobert C, Grobmann P, Gottschalk M, Komor E, Pecsvaradi P, zur Nieden U (1995) Sieve-tube exudate form *Ricinus communis* L. seedlings contains ubiquitin and chaperones. Planta 196:205–210
Schobert C, Baker L, Szederkenyi J, Grossmann P, Komor E, Hayashi H, Chino M, Lucas WJ (1998) Identification of immunologically related proteins in sieve-tube exudate collected from monocotyledonous and dicotyledonous plants. Planta 206:245–252
Schulz A (1998) Phloem. Structure related to function. Prog Bot 59:429–475
Sham M-H, Northcote DH (1987) Transcription and translation of phloem protein (PP2) during phloem differentiation in *Cucurbita maxima.* Planta 170:392–399
Sjölund RD, Shih CY (1983) Freeze-fracture analysis of phloem structure in plant tissue cultures I. The sieve element reticulum. J Ultrastruct Res 82:111–121
Smith LM, Sabnis DD, Johnson RPC (1987) Immunochemical localisation of phloem lectin from *Cucurbita maxima* using peroxidase and colloidal-gold labels. Planta 170:461–470
Szederkényi J, Komor E, Schobert C (1997) Cloning of the cDNA for glutaredoxin, an abundant sieve-tube exudate protein from *Ricinus communis* L. and characterization of the glutathione-dependent thiol-reduction system in sieve tubes. Planta 202:349–356
Terry BR, Robards AW (1987) Hydrodynamic radius alone governs the mobility of molecules through plasmodesmata. Planta 171:145–157
Tian G, Huang Y, Rommelaere H, Vandekerckhove J, Ampe C, Cowan NJ (1996) Pathway leading to correctly folded b-tubulin. Cell 86:287–296
Tiedemann R, Carstens-Behrens U (1994) Influence of grafting on the phloem protein patterns in Cucurbitaceae. I. Additional phloem exudate proteins in *Cucumis sativus* grafted on two *Cucurbita* species. J Plant Physiol 143, 189–194
Tóth KF, Sjölund RD (1994) Monoclonal antibodies against phloem P-protein from plant tissue cultures. II. Taxonomic distribution of cross-reactivity. Am J Bot 81:1378–1383
Tóth KF, Wang Q, Sjölund RD (1994) Monoclonal antibodies against phloem P-protein from plant tissue cultures. I. Microscopy and biochemical analysis. Am J Bot 81:1370–1377
van Bel AJE, Kempers R (1997) The pore/plasmodesm unit; key element in the interplay between sieve element and companion cell. Prog Bot 58:278–291
van Bel AJE, van Rijen HVM (1994) Microelectrode-recorded development of the symplasmic autonomy of the sieve element/companion cell complex in the stem phloem of *Lupinus luteus* L. Planta 192:165–175
Walker TS (1972) The purification and some properties of a protein causing gelling in phloem sieve tube exudate from *Cucurbita pepo.* Biochim Biophys Acta 257:433–444
Walsh MA, Evert RF (1975) Ultrastructure of metaphloem sieve elements in *Zea mays.* Protoplasma 83:365–388
Wang MB, Boulter D, Gatehouse JA (1994) Characterization and sequencing of a cDNA clone encoding the phloem protein PP2 of *Cucurbita pepo.* Plant Mol Biol 24:159–170
Weber C, Franke WW, Kartenbeck J (1974) Structure and biochemistry of phloem proteins isolated from *Cucurbita maxima.* Exp Cell Res 87:79–106

Wergin WP, Newcomb EH (1970) Formation and dispersal of crystalline P-protein in sieve elements of soybean (*Glycine max* L.). Protoplasma 71:365–388
Wergin WP, Palevitz BA, Newcomb EH (1975) Structure and development of P-protein in phloem parenchyma and companion cells of legumes. Tissue Cell 7:227
White RG, Badelt K, Overall RL, Vesk M (1994) Actin associated with plasmodesmata. Protoplasma 180:169–184
Wolf S, Deom CM, Beachy RN, Lucas WJ (1989) Movement protein of tobacco mosaic virus modifies plasmodesmatal size exclusion limit. Science 246:377–379
Zee S-Y (1969) Fine structure of the differentiating sieve elements of *Vicia faba*. Aust J Bot 17:441–456

Chapter 17

Plasmodesmata and Long-Distance Virus Movement

P. M. Derrick · R. S. Nelson

Samuel Roberts Noble Foundation, 2510 Sam Noble Parkway, P.O. Box 2180, Ardmore, Oklahoma 73402, U.S.A.

1 Introduction

Long-distance movement (or systemic movement) of viruses is a term used to refer to the transport of viruses between organs of infected plants via the vascular bundles. For the vast majority of viruses, long-distance movement occurs through sieve elements (SEs). During this process, invading viruses must therefore successfully spread from initially infected cells, enter veins, move through the sieve tubes to distal organs and exit the vasculature in order to propagate in the systemically infected tissues. Insect-transmitted viruses may be deposited directly into SEs by their vectors and so enter the phloem passively in the natural situation. During movement, viruses must circumvent interactions with host factors that may result in plant resistance to movement. As viruses are obligate non-mobile parasites, however, their replication and movement within plants requires interaction with host proteins and subcellular structures such as plasmodesmata. Thus for a successful systemic invasion of the host, viruses must maintain or enhance interactions with host factors that are required for movement through plasmodesmata, while avoiding interactions with host factors that prevent it.

The papers of Holmes (1930), Samuel (1934) and Bennett (1940) are landmarks describing long-distance movement of tobacco mosaic virus (TMV) within plants and providing evidence suggesting that the rapid spread of viruses between organs of plants occurs through the vasculature. During the following decades, the inaccessibility, high turgor pressure, and loss of structural integrity on wounding of the SE/companion cell (CC) complex has greatly hampered progress in studying long-distance movement of viruses. This notwithstanding, the availability of electron microscopy enabled a greater understanding of virus invasion of the vasculature and the distribution of different viruses in different cell types. These observations were enhanced further by use of antibodies to purified viral proteins and, later, products of cloned viral genes. In situ hybridization of labelled nucleic acid probes has also proved particularly suitable for following long-distance virus movement (e.g. Melcher 1989; Leisner et al. 1992; Simon et al. 1992) as has the introduction of marker protein genes (e.g. the gene encoding β-glucuronidase) into viral genomes for expression of marker proteins fusions with viral proteins or free marker proteins (e.g. Dolja et al. 1994). The more recent introduction of native and modified *Aequoria victoria* green fluorescent proteins (GFPs) as markers, particularly in combination with the high resolution of confocal scanning laser microscopy (CLSM), now promises substantial advances for analyzing movement of viruses within vascular tissue (reviewed in Oparka et al. 1996, 1997b; Chap. 6). CLSM and GFPs permit near real-time analysis of viral movement in intact plant organs, thereby overcoming the problems imposed by the small, high-pressured and wound-sensitive SEs and CCs located deeply embedded within vascular tissue.

Studies on virus cell-to-cell movement have yielded results of interest to those studying host macromolecular cell-to-cell movement (for review see Mezitt and Lucas 1996). It is likely that viruses have utilized existing host vascular transport systems to systemically infect their hosts. Therefore studies on the long-distance movement of viruses, although still in their infancy, will most likely provide researchers with clues to how plants regulate their activities on a supracellular basis (e.g. Jorgensen et al. 1998). In this chapter, we summarize the progress and potential in this expanding area of research.

In regard to terminology, we will use the terms long-distance movement and vascular-dependent accumulation when discussing virus movement. It is our opinion that the term long-distance movement should represent only the movement of virus in the phloem transport system. The term vascular-dependent accumulation represents the measurable accumulation of virus in sink tissue after vascular movement by the virus. Thus, one can observe long-distance movement without vascular-dependent accumulation if the virus moves through the phloem SEs but does not accumulate or replicate to measurable levels in the sink tissue. Defining these terms now will help to limit any confusion that may appear as researchers begin to dissect the transport pathway.

The number of reviews on virus movement published in recent years reflects the high levels of interest and research activity in this field (Carrington et al. 1996; Gilbertson and Lucas 1996; Deom et al. 1997; Nelson and van Bel 1998 in addition to Chap. 6 of this Vol.). Many recent reviews concerning the structure and function of plasmodesmata and vascular tissue in a broader sense have also been published (Lucas et al. 1993a, b; van Bel 1993; Lucas 1995; van Bel and Oparka 1995; Grusak et al. 1996; Mezitt and Lucas 1996; Overall and Blackman 1996; Turgeon 1996; Ding 1997; Ghoshroy et al. 1997; McLean et al. 1997; Sjölund 1997; van Bel and Kempers 1997; Schulz 1998).

2 Vascular-Dependent Virus Accumulation

Vascular-dependent virus accumulation is affected by various factors at every step in the infection process (e.g. during replication, cell-to-cell movement in inoculated leaves, or vascular movement) and its measurement represents the culmination of this process. In this section, we will review literature describing the rates of systemic virus movement and compare these results with those reporting the rates of virus accumulation in inoculated cells and the rates of virus cell-to-cell movement.

Symptoms, most often reflecting virus accumulation, appear on systemically infected leaves at varying times, depending on the virus, the host, the age of the host and the leaves inoculated, and the environment (see Matthews 1991 for review). For many viruses inoculated onto vegetative herbaceous hosts, however, the timing of systemic symptom appearance will be from 5–10 days postinoculation (e.g. Oxelfelt 1970; Roossinck and Palukaitis 1990; Heaton et al. 1991; Xiong et al. 1993; Bransom et al. 1995; Brugidou et al. 1995; Wang et al. 1996).

The appearance of symptoms is a late phase in vascular-dependent virus accumulation, and is the product of not only long-distance movement of the virus but, as previously mentioned, the replication of virus in the invaded tissue. A better indication of the rate of movement of virus is obtained through leaf-detachment experiments. In these studies, inoculated leaves are detached from the plants at various times postinoculation. Thus, the point at which adequate amounts of virus have spread through the vasculature of the leaf petiole to establish a systemic infection will be identified. The earliest time observed for systemic spread in such studies is 24–30 h postinoculation for cucumber mosaic virus (CMV; Gal-On et al. 1994). Although interpreting results from these studies must take into account the possibility that some viruses may move through the vasculature without establishing a systemic infection, these studies represent the best estimate of movement rates. Thus they should be compared with the reported rate of progeny virus accumulation in inoculated cells and the rate of virus cell-to-cell movement.

Rates of progeny virus accumulation in inoculated cells have been determined for TMV. By 40 min postinoculation, host translation and viral replicase-associated proteins disassemble the virus and encapsidated progeny virus begins to appear in tobacco protoplasts (Wu et al. 1994; Wu and Shaw 1997). Whether this rate reflects what occurs in cells within intact leaves has yet to be determined, however, considering that encapsidation is a late event in TMV replication and TMV does not need to encapsidate to move from cell to cell, it may be a conservative estimate of the time required for the appearance of progeny competent for cell-to-cell movement. If one considers that most cells lie within six cells of vascular tissue (see Nelson and van Bel 1998 for review), the virus should reach vascular tissue within 6 h unless other factors limit virus spread. The finding that viruses can move through phloem at non-limiting rates of centimetres per hour (see Bennett 1956 for review) suggests that other factors, such as the rate of cell-to-cell spread in the inoculated leaves, do limit the rate of vascular-dependent virus accumulation.

Researchers studying TMV and tobacco rattle virus (TRV) determined that virus spread from the inoculated cells did not occur for 4 h postinoculation (Fry and Matthews 1963; Derrick et al. 1992). Interestingly, the subsequent movement of virus between cells took only 1-2 h. Derrick et al. (1992) speculated that viral factors necessary for moving virus cell-to-cell or establishing infection in the second cell were present in the second cell earlier than in the first cell during the respective infections. One such factor may be the so-called movement protein (MP) of some viruses. The 30-kDa MP of TMV has been shown to increase the size exclusion limit (SEL) of plasmodesmata several mesophyll cells from a microinjection site (Waigmann et al. 1994). Indeed, the 30-kDa, 3a, 35-kDa, and BL1 proteins of TMV, CMV, red clover necrotic mosaic virus (RCNMV), and bean dwarf mosaic virus (BDMV), respectively, have been shown to move at least one cell from the microinjected cell (Fujiwara et al. 1993; Noueiry et al. 1994; B. Ding et al. 1995; Waigmann and Zambryski 1995). Itaya et al. (1997) have determined that the 3a protein moves between cells after biolistic gene bombardment of a leaf with a plasmid containing the 3a gene; a procedure that minimizes potential artifacts caused by pressure injection and the lack of in planta post-translational modifications of a bacterially-produced protein. However, the ability of the MP from TMV to exert distant effects through its own cell-to-cell movement recently has been questioned (Padgett et al. 1996; Oparka et al. 1997a). Future work will determine whether the presence of these proteins in advance of the translation and replication cycle of the virus accounts for the differing rate of spread between the first and second infected cells and between cells infected later.

A delay of a few hours is still not sufficient to fully account for the difference in time between the exit of virus from an inoculated leaf and the estimate one can make of when this should occur based on measured rates of cell-to-cell movement in non-vascular tissue. Thus, other sites controlling systemic movement are likely to be present in inoculated leaves. The following sections will discuss evidence for regulation points within the vascular tissue of the inoculated leaves for both virus and photosynthate transport.

2.1 Entry of Viruses into Veins

Viruses that are not limited to the vasculature will move through mesophyll cells (MCs) in a leaf on inoculation and eventually encounter a vein. Veins can be defined as major or minor based on their structure and location, branching pattern and function (Esau 1977; Hickey 1979; Grusak et al. 1996; Nelson and van Bel 1998). For dicotyledons, major veins are often defined as those enclosed in parenchyma tissue with few chloroplasts and which form the rib rising above the leaf surface (Esau 1977; Grusak et al. 1996), have branched usually no more than twice [i.e. they are first (midrib) through third order veins; Hickey 1979, Mauseth 1988], and function in long-distance transport of water, inorganic nutrients, photoassimilates and other organic matter (Grusak et al. 1996). These veins may also be involved in photoassimilate unloading (Turgeon 1986; Roberts et al. 1997). Minor veins are generally embedded in mesophyll cells (Esau 1977), are the result of three or more branches from the first-order vein (Hickey 1979) and function in photoassimilate loading in mature leaves (Grusak et al. 1996; Nelson and van Bel 1998). Cell numbers can vary in minor veins, but they often contain fewer than 20 cells (B. Ding et al. 1988; X.S. Ding et al. 1998). The smallest classes of veins usually form areoles in dicotyledonous leaves. Areoles are defined as the smallest area of mesophyll cells bounded by intersecting veins (Esau 1977; Mauseth 1988). In some instances, branches of minor veins end blindly in the areole (Esau 1977; Mauseth 1988; Horner et al. 1994; Nelson and van Bel 1998). It is important to realize that unbranched veins can change size, shape and function throughout their length. Therefore the vein system should be viewed as a continuum, changing its function according to position (Grusak et al. 1996).

In monocotyledons, the difference between major and minor veins is not so apparent, although smaller and larger veins do alternate in leaves with parallel veins (Esau 1977). Parallel veins converge and join at the apex of the leaf (Esau 1977). The parallel veins are interconnected by transverse veins (also referred to as commissural bundles) (Esau 1977) which often have only a single SE and are not thought to actively participate in phloem loading (see review by Nelson and van Bel 1998). Considerable evidence exists that minor veins from monocotyledonous plants function in phloem loading (for review see Nelson and van Bel 1998).

Whether viruses enter veins through termini of minor veins, branch points of veins, or through bundle-sheath cells (BSCs) encasing minor or major veins remains to be determined (Ding et al. 1998; Table 1). In fact, whether minor veins, major veins or both serve as sites of infection leading to vascular-dependent infections remains to be determined (Nelson and van Bel 1998). Some evidence does exist that minor veins are sites for virus colonization and movement. Leisner et al. (1992) observed a pattern of lesions on inoculated leaves suggesting that cauliflower mosaic virus (CaMV) invades the vascular system through minor veins. Also minor veins certainly make up the majority of the vascular network in leaves and thus would be the first veins that a virus would contact during spread (Grusak et al. 1996; Ding et al. 1998).

Within veins there is also controversy about the route of virus movement between the vascular cells. Ding et al. (1998) determined that certain legumes (*Phaseolus vulgaris* and *Pisum sativum*) were systemically infected before virus was detected in the CCs or transfer cells (TCs) of the inoculated leaf. They hypothesized that virus is either infecting SEs directly from vascular parenchyma cells (VPCs) within the vein or that

Table 1. Potential control points for vascular-dependent accumulation of virus in the host

Site	Evidence/comment	Sample references
MC:MC	+[a]/Virus deficiencies or host resistance may limit cell-to-cell spread	Mise and Ahlquist (1995); Callaway et al. (1996)
Vein termini or branch points of veins	?/	
Major veins	?/	
Minor veins, MC:BSC	+/–/Evidence both for and against this as a site of regulation	B. Ding et al. (1992a); S. W. Ding et al. (1995); Thompson and García-Arenal (1998)
Minor veins, BSC:VPC	+/Considerable evidence indicating that this is a unique site of regulation	D'Arcy and de Zoeten (1979); Shepardson et al. (1980); Goodrick et al. (1991); B. Ding et al. (1992a); Sanger et al. (1994); X. S. Ding et al. (1995); van den Heuvel et al. (1995); Wintermantel et al. (1997); Thompson and García-Arenal (1998)
Minor veins, BSC or VPC:SE/CC complex	+ and –/(+) For species with smooth walled CCs or TCs and (–) for species with ICs	McCauley and Evert (1989); X. S. Ding et al. (1995); X. S. Ding et al. (1998); Thompson and García-Arenal (1998)
CC:SE	+/Some evidence that short-distance movement of luteovirus is regulated (possibly via 17-kDa MP)	Derrick and Barker (1997); Schmitz et al. (1997)
SE:SE	+/Some evidence that virus in transit is regulated	Golinowski et al. (1987); Wilson and Jones (1992); Arce-Johnson et al. (1997)
External phloem: Internal phloem	+/Evidence accumulating that some host/virus determinants control virus spread between these tissues in the transport phloem	Derrick and Barker (1992, 1997); Andrianifahanana et al. (1997)
All interfaces during leaf development	+/ Much evidence that leaf development regulates vascular-dependent accumulation	Solberg and Bald (1962); Lucy et al. (1996); Más and Pallás (1996); Wang et al. (1996); Roberts et al. (1997)

[a] + evidence for; – evidence against; ? no evidence either way

one of the alternative routes of virus invasion (i.e. through vein termini, branch points or major veins) was responsible for the systemic infection. Although the remainder of this section will discuss virus entry into veins from MCs through BSCs or mestome sheath cells (MSCs) surrounding the vein, results from future research will determine whether this is the primary or only route taken by virus during the vascular-dependent accumulation of virus in the host.

Only a few studies have examined the infection of BSCs or MSCs from MCs by viruses. To summarize, in studies with TMV in tobacco there is no evidence that the MP modifies the plasmodesmata between the MCs and BSCs differently than those between MCs nor is there evidence that access to or replication within the BSCs by wild-type virus is limited (B. Ding et al. 1992a; X. S. Ding et al. 1995). Recently, the spread of a CMV-tomato aspermy virus (TAV) reassorted virus, containing RNAs 1 and 2 of CMV and RNA 3 of TAV, in cotyledons of *Cucumis sativus* was studied (Thompson and García-Arenal 1998). These researchers did not state the specific percentage of infected MCs (i.e. they simply said most MCs were infected). However, the results suggest a slight delay in CMV or the reassortant virus accumulation in BSCs compared with MCs. More experiments are required with other viruses and hosts to further determine the traits of this boundary. Also studies comparing the entrance of virus into BSCs surrounding major and minor veins should be undertaken to determine if the vein type affects virus accumulation in these cells. Indeed, Thompson and García-Arenal (1998) observed more BSCs from minor veins than from major veins infected with virus.

Passage of virus through the boundary between the encircling sheath cells and the internal vein cells (i.e. VPCs and CCs) is also an understudied area. However, unlike the findings noted above for entrance into sheath cells, the studies published to date examining this boundary clearly indicate it to be a significant barrier to virus movement and accumulation. Interestingly, this boundary also regulates photoassimilate transport. It has been shown that (1) the SELs of plasmodesmata between BSCs and VPCs in veins of tobacco are not modified by the MP of TMV, unlike those between MCs and between MCs and BSCs (Ding et al. 1992a); (2) the percentage of internal vein cells infected by two strains of TMV or a CMV-TAV reassortant is limited compared with the percentage of BSCs infected (X. S. Ding et al. 1995; Thompson and García-Arenal 1998); (3) the percentage of VPCs infected by two strains of TMV are markedly different indicating that a viral factor (i.e. the 126- or 183-kDa proteins) limits virus accumulation in these cells, most likely by limiting virus transport into these cells from the BSCs (X.S. Ding et al. 1995); (4) the phloem-associated viruses are limited in their ability to accumulate in BSCs compared with the internal vein cells (e.g. D'Arcy and de Zoeten 1979; Shephardson et al. 1980; Sanger et al. 1994; van den Heuvel et al. 1995); (5) host factors regulate the ability of virus and photosynthate to accumulate in, or flux through, internal vein cells from BSCs (Goodrick et al. 1991; Russin et al. 1996); and (6) the expression of a viral transgene, encoding the 2a protein of CMV, protects tobacco plants from virus challenge by preventing virus entry into VPCs and CCs from BSCs in minor veins (Wintermantel et al. 1997).

The studies described above indicate that viral and host factors different from those required for virus movement between the MCs control the flux of molecules into the internal vein cells (see Ding 1997; Ghoshroy et al. 1997 for reviews on virus and host macromolecular cell-to-cell movement). These results generally indicate that the plasmodesmata between the BSCs and internal vein cells are different from those between MCs.

For dicotyledons, it is interesting that many researchers have reported either subtle or no differences in plasmodesmata structure between BSCs and VPCs and those between MCs or MCs and BSCs in mature exporting leaves (Turgeon et al. 1975; Fisher and Evert 1982; Evert and Mierzwa 1986; Fisher 1986; Madore et al. 1986; Ding et al. 1988; McCauley and Evert 1989; Wimmers and Turgeon 1991; Beebe and Evert 1992). This is in sharp contrast to the unique branched, deltoid structure generally observed for plasmodesmata between the CCs and SEs (e.g. Beebe and Evert 1992; Volk et al.

1996). Symplasmic isolation between the BSCs and VPCs is somewhat unexpected as the osmotic potentials within BSCs and VPCs generally are similar and much less negative than those in the SE/CC complex (e.g. Geiger et al. 1973; Fisher and Evert 1982; Beebe and Evert 1992). Also, dye-coupling experiments with Lucifer Yellow have indicated symplasmic continuity between these cells (Madore et al. 1986; Ding et al. 1992a). Thus, studies on virus transport have helped to define symplasmic domains for photosynthate accumulation that were not recognized through structural and physiological studies until recently (e.g. Ding et al. 1993; Russin et al. 1996). It is also interesting to note that BSCs and VPCs generally originate from different source tissue (for review, see Ding and Lucas 1996).

In monocotyledons, structural differences in plasmodesmata connecting MSCs and VPCs versus those at other boundaries are more easily detectable (e.g. van Bel et al. 1988; Robinson-Beers and Evert 1991). There have been no detailed studies describing the development of virus infections within specific vascular cells over time for monocotyledons. It would be interesting to determine whether structural differences observed in plasmodesmata between vascular cells in monocots correlate with different infection patterns by viruses in those cells.

Another parameter that affects virus movement into and within veins is the changing structure of vein and vein-associated cells and their plasmodesmata as the leaf develops from a sink to a source. Several studies have noted changes in plasmodesmata structure and number within developing veins (e.g. Ding et al. 1988, 1992b; Volk et al. 1996). Some work analyzing the development of virus infections in maturing leaves has been reported (Al-Kaff and Covey 1996; Lucy et al. 1996; Más and Pallás 1996; Wang et al. 1996). The geminiviruses, BDMV and maize streak virus (MSV), have been determined to escape their vascular association in older leaves of *Phaseolus vulgaris* and *Nicotiana benthamiana* and *Zea mays*, respectively (Lucy et al. 1996; Wang et al. 1996). Inoculation of tobacco plants with cherry leaf roll virus yielded results indicating that the virus distribution in the infected plants, as well as the susceptibility of the plants to systemic infection, decreased with the development of the inoculated leaves at the time of virus inoculation (Más and Pallás 1996). It will be interesting to determine whether this change in susceptibility of the tobacco plant to systemic infection correlates with an inability of the virus to pass through particular plasmodesmata within the inoculated leaf.

While the plasmodesmata between the BSCs and VPCs were a somewhat unexpected regulated interface for virus and photoassimilate transport, the determination that the interface between the SE/CC complex, especially in species with smooth-walled CCs or TCs, and other cells is regulated for virus transport could have been expected for the following reasons. In most instances, there are few plasmodesmata connecting the SE/CC complex to other cells in species with smooth-walled CCs or TCs (see Grusak et al. 1996; Schulz 1998 for review). Considering the rigid control these cells must be under in order to maintain osmotic potentials greatly different from their neighbouring cells, it is a wonder that viruses, which almost certainly rely on symplasmic transport routes, can invade these cells at all. Ding et al. (1998) have indeed questioned whether CCs or TCs are part of the direct pathway for vascular invasion. It is possible that some of the infections observed in the CCs of mature leaves are due to the movement of virus out of SEs containing the pathogen due to the infection and spread of virus into cells upstream of them in the photoassimilate flow.

An exception to the limited connectivity of the SE/CC complex with other cells is found in those plants that contain intermediary cells (ICs). ICs have many plasmodesmata connecting them to BSCs (Grusak et al. 1996; Schulz 1998). These numerous connections should seemingly help viruses, which most likely require symplasmic transport to invade these cells and thereby gain entrance to the SEs. In fact, Thompson and García-Arenal (1998) observed numerous ICs infected with CMV in cotyledons from *Cucumis sativus* inoculated 6 days previously. However, there is no indication that plants with widely different plasmodesmata numbers between the SE/CC complex and other cells (i.e. symplasmic versus apoplasmic phloem loaders) become systemically infected with virus over widely different time periods (Ding et al. 1998). Thus, as mentioned previously, studies should be conducted that unambiguously determine whether CCs in inoculated leaves are in the direct pathway for virus long-distance movement.

2.2 Entry of Viruses into SEs and Virus Transport in the Phloem

Plant viruses must pass into the SEs to initiate systemic invasion of the host. Although the timing of exit of virus from inoculated leaves has been approximated for a number of viruses by leaf-detachment experiments (see Sect. 2), and systemic movement of viruses generally appears to follow sink-source fluxes (Leisner and Turgeon 1993; Roberts et al. 1997), very little is understood of how virus entry into SEs is achieved. It is, however, assumed that viruses enter SEs through the plasmodesmata linking SEs and CCs. Certainly, the plasmodesmata between the SEs and CCs allow host RNAs and proteins of significant size to flux between these cells (Bostwick et al. 1992; Clark et al. 1997; Dannenhoffer et al. 1997; Kühn et al. 1997; Ishiwatari et al. 1998; Chap. 15), a system that may be appropriated by viruses for their transport. It is also generally held that the plasmodesmata between SEs and CCs are highly specialized and the term pore-plasmodesm unit (PPU) has been coined to describe this special connection (reviewed in van Bel and Kempers 1997). The PPU between the SEs and CCs in mature leaves are branched (i.e. deltoid-shaped) in the cell wall next to the CCs. They often differ from other plasmodesmata in their fine structures, orientation of branching, and physiology. For example, in *Moricandia arvensis* the PPUs between the SEs and CCs are not occluded and have distinct desmotubules, whereas the plasmodesmata between phloem parenchyma cells (PPCs) or BSCs and CCs are often occluded with a dense, amorphous material and they lack desmotubules on the CC side of the wall when observed under the electron microscope (Beebe and Evert 1992; see Chap. 15). The deltoid shape of the PPU is conserved among all plants, suggesting an important function for this structure in plant survival. The SEL of the PPU of transport phloem is greater than that observed for plasmodesmata between other cells and the PPU does not close after dye injection unlike the plasmodesmata between other cells (reviewed by van Bel and Kempers 1997). These results indicate that the PPU are different from plasmodesmata connecting other cells, and this would suggest that different host factors are required for virus transport into the SEs compared with other cells.

A large, unresolved problem underpinning long-distance movement of viruses is the nature of the infectious material that enters SEs and that which is transported in SEs. Viruses may enter and move within SEs as nucleoprotein complexes or as virus particles. For many viruses, coat proteins (CPs) are required for efficient long-distance movement (see Table 2 and Nelson and van Bel 1998 for detailed lists), but for these viruses, it is unclear whether the CP is required for formation of virus particles or for another, as yet unidentified, function. For TMV, there is a preponderance of evidence supporting the viewpoint that it moves over long distances as virus particles. For example, the *flavum* mutant of TMV has a temperature-sensitive CP and accumulates systemically only at temperatures at which the mutant also forms virus particles (Oxelfelt 1975). Similarly, a mutant in which the origin of assembly in the RNA sequence was mutated to disrupt encapsidation of the MP open reading frame (ORF), but not its translation or the function of the protein, moved poorly in tobacco (Saito et al. 1990). For TRV, strains which lack RNA-2 (containing the CP ORF) invade hosts via slow cell-to-cell movement rather than phloem-mediated movement (Sänger 1969). For RCNMV, it has also been suggested that intact virions are necessary for long-distance transport (Vaewhongs and Lommel 1995). Even for TMV, however, others have published results suggesting that the role of the CP is to interact with the viral RNA, but not to form a virion (Dawson and Schlegel 1973; Dorokhov et al. 1984; Gera et al. 1995). X.S. Ding et al. (1996) have shown that the TMV CP is not necessary for the virus to enter VPCs. The CP appears to be important for entry of TMV into CCs, but in one instance an encapsidation incompetent virus expressing a defective CP was observed in a CC (X. S. Ding et al. 1996). For this mutant virus, it is possible that the CP functions apart from capsid formation to allow entry of the virus into the SEs.

Much evidence indicates a requirement for the presence of the CP for efficient long-distance movement of many other viruses, such as tobacco etch virus (TEV), cowpea chlorotic mottle virus (CCMV), turnip crinkle virus (TCV), and beet western yellows virus (BWYV). However, its effect in these instances may be more on cell-to-cell movement of the virus (for review see Nelson and van Bel 1998; also see Rodríguez-Cerezo et al. 1997; Rojas et al. 1997 regarding potyvirus CPs). For TEV and CCMV, it is becoming clear that a portion of the CP is required for their cell-to-cell movement, while the entire CP is required for their vascular-dependent accumulation (Dolja et al. 1994, 1995; Schneider et al. 1997).

Recently, it has been determined that the CP of potato virus X (PVX), which is necessary for cell-to-cell movement of this virus, localizes to the plasmodesmata between all cells of *N. benthamiana* except those between CCs and PPs or SEs (Santa Cruz et al. 1998). The authors noted that these results may indicate that PVX accesses the long-distance transport pathway in a form different from the one assumed for transport through the other cellular boundaries, or that the large SEL of the plasmodesmata between the CC and SE permit freer transport of virions.

It should be borne in mind that no evidence has yet been published demonstrating that virus enters SEs as intact virus particles. In recent studies with CMV, it was determined that encapsidated virus was not observed in the plasmodesmata between the CC and SE of minor veins within inoculated leaves of *N. clevelandii*; however, both encapsidated virus and MP were observed in SEs (Blackman et al. 1998). Therefore, it is possible that some viruses enter SEs as non-virion nucleoprotein complexes and assemble within SEs prior to translocation. It is also possible that, as is the case for cell-to-cell movement elsewhere in the plant, different viruses use different mechanisms to achieve long-distance movement. Encapsidation is not an absolute prerequisite for vascular-dependent accumulation of viruses such as barley stripe mosaic virus (BSMV, Petty and Jackson 1990), various geminiviruses (e.g. Ingham et al. 1995; Padidam et al. 1995), pea enation virus (Demler et al. 1996), the umbraviruses, which do not form virus particles (Falk et al. 1979; Reddy et al. 1985; Murant et al. 1996), and viroids, which do not encode proteins and do not require a helper virus (Palukaitis 1987).

Other viral proteins also are important for vascular-dependent accumulation (Table 2). As noted in Section 2.1, one or both of the TMV 126-/183-kDa proteins appears to regulate movement of this virus between BSCs and VPCs (X. S. Ding et al. 1995). These proteins may also regulate vascular-dependent virus accumulation in other ways. It has been determined that a TMV mutant altered in its 126-/183-kDa protein sequence can replicate and invade VPCs and CCs within minor veins normally, but does not accumulate systemically (X. S. Ding and R. S. Nelson, unpublished results). As speculated for the TMV CP, the site where these proteins regulate virus movement also may be at the plasmodesmata between the CC and SE. The helper component-proteinase (HC-Pro) of TEV, a protein known to be essential for vascular-dependent accumulation of this virus, accumulates in what appears to be both VPCs and CCs (Cronin et al. 1995). These authors speculated that the protein may be necessary to allow entry or exit from SEs of minor veins. If it functions in entry, a possible site of effect is the plasmodesmata between the CC and SE.

Table 2. Viral factors that affect vascular-dependent accumulation of the virus in the host

Viral factor	Virus	Known or potential site of effect	Comment	Reference
CP	TMV, TRV, RCNMV	For TMV, possibly at entrance to the SE/CC complex or within the SEs	1. Only a sampling of viruses that show little or no effect on local cell-to-cell spread are listed (see Nelson and van Bel 1998 for a detailed discussion of the literature) 2. For some viruses, the CP is not required for vascular-dependent accumulation (Nelson and van Bel 1998 for review)	Siegel et al. (1962); Sänger (1969); Takamatsu et al. (1987); Dawson et al. (1988); Hamilton and Baulcombe (1989); Vaewhongs and Lommel (1995); X.S. Ding et al. (1996)
HC-Pro	TEV	Possibly for entrance into or exit from SEs		Cronin et al. (1995)
VPg domain of NIa	TEV	Possibly for entrance into or exit from SEs	Mutants overcome specific resistance genes. Mutant sequences do not elicit a general resistance response in host	Schaad et al. (1997)
126/183-kDa proteins	TMV	Entrance of virus into VPCs and CCs or possibly for entrance of virus into SEs, passage through sieve tubes, or exit from SEs	For some mutants, effect may be due to a decreased expression of other viral products or an induction of a host resistance	Watanabe et al. (1987); Lewandowski and Dawson (1993); X.S. Ding et al. (1995); X. S. Ding and R. S. Nelson (unpubl.)
129/186-kDa proteins	SHMV	?		Deom et al. (1994); Deom et al. (1997)

Table 2. Continued

Viral factor	Virus	Known or potential site of effect	Comment	Reference
1a	CMV	?		Roossinck and Palukaitis (1990); Lakshman and Gonsalves (1985); Gal-On et al. (1994)
1a	BMV	?		De Jong and Ahlquist (1995)
1	CCMV	?		Wyatt and Kuhn (1980)
αa	BSMV	?		Weiland and Edwards (1994, 1996)
2	BMV	?	Effect may be due to decreased cell-to-cell spread	Traynor et al. (1991)
2	CMV	Possibly host resistance to cell-to-cell spread		Edwards et al. (1983)
2b	CMV	?		S.W. Ding et al. (1995)
p19	TBSV	?	?	Scholthof et al. (1995)
gene VI	CaMV	?	Light influences the effect of the viral factor on vascular-dependent accumulation	Wintermantel et al. (1993)
RNA 3	BNYVV	?	May still be a protein effect through truncated N protein	Lauber et al. (1998)
5′ leader of RNAγ	BSMV	?	May affect vascular-dependent accumulation indirectly since the translation of the γa protein, involved in virus replication, is downregulated by mutating this sequence	Petty et al. (1990)
MP	TMV	a) entry or exit to vasculature b) transport phloem	Difficult to prove an effect on vascular entry/exit of long- distance movement due to the known function of these proteins on cell-to-cell movement	a) Hilf and Dawson (1993); Fenczik et al. (1995) b) Arce-Johnson et al. (1997)

In many of these studies, the boundaries for virus spread were identified through the analysis of mutant viruses expressing altered proteins. It is possible that these altered proteins are triggering a host defence response at a specific subcellular location. For example, it has been determined that a host-resistance gene to a fungal pathogen encodes a pathogenesis-related protein, PRms, that localizes to plasmodesmata in maize radicles (Murillo et al. 1997). It is possible that other induced host-resistance genes may function at the plasmodesmata between SEs and CCs (or within plasmodesmata at other specific cellular interfaces such as those between BSCs and VPCs). In regard to virus infections, Schaad and Carrington (1996) found that a recessive and multigenic resistance in *N. tabacum* inhibited vascular-dependent accumulation of TEV. Virus was able to accumulate in VPCs and possibly CCs in the inoculated leaves. It was concluded that the resistance functioned somewhere at or beyond the SE/CC interface to prevent the systemic accumulation of virus. These researchers recently determined that coinfection with two TEVs which do or do not accumulate, respectively, in systemically infected leaf tissue, does not inhibit the vascular-dependent accumulation of the mobile strain (Schaad et al. 1997), nor does the non-mobile strain elicit a visible resistance response. This suggests that a host defence reponse is not limiting vascular-dependent accumulation of the non-mobile virus in this system. For other mutant viruses it will be important to determine whether their lack of vascular-dependent accumulation is due to a host defense response or to a failure of a viral protein to function to allow virus movement.

Once virus is in the SEs, what viral and host factors affect transport, and do plasmodesmata play a role in modulating this transport? Through electron microscopy, it has been shown that virus particles can accumulate in SEs (e.g. Esau 1968; Esau and Hoefert 1971); but these studies did not determine when virus first appeared relative to the development of infection. Therefore, it is impossible to determine whether these particles represent a movement form or accumulation of virus long after movement has occurred. It is also possible that virus produced in immature nucleated SEs may remain in mature, enucleated SEs. These particles may or may not be a part of the package of infectious material that causes systemic infection in plants. It is therefore extremely important to know the developmental stage of the host tissue and cells with which one works if meaningful results on vascular spread of virus are to be obtained (Séron and Haenni 1996; Nelson and van Bel 1998).

Little work has been published on the accumulation of virus in stem SEs (i.e. transport phloem SEs) during a period when systemic symptoms are just appearing. De Zoeten and Gaard (1983) inoculated fully expanded leaves of *P. sativum* with pea enation mosaic virus and then analyzed stem sections directly above the node of the inoculated leaf. By 4 days postinoculation, vesicles were apparent in mature SEs. The appearance of double-stranded RNA coincided with the appearance of the vesicles. Although the origin of the dsRNA was not determined in this study, this laboratory had previously shown that RNA-dependent RNA polymerase and virus-specific dsRNA were present in vesicles of infected cells (Powell and de Zoeten 1977; Powell et al. 1977 and references therein). It is interesting that RNA-2 of this virus can cause a systemic infection without the presence of RNA-1, which encodes the CP for this virus (reviewed in Demler et al. 1996). Therefore it may be that this virus needs replication-associated proteins and not the CP for long-distance movement.

There is growing support in the literature for the involvement of viral MPs during virus vascular movement apart from their role in cell-to-cell movement. Arce-Johnson et al. (1997) determined that TMV MP, expressed in vascular tissue of a stem intergraft, was necessary to complement the movement between rootstock and scion of a TMV mutant lacking an MP gene (MP-). The timeframe in which this observation was made was long, thereby potentially allowing virus to move cell-to-cell outside the SEs in the transgenic intergraft to reach the susceptible scion. This may have confounded the interpretation of the results from this study. Indeed, Gera et al. (1995) observed that an MP- mutant of TMV moves rapidly through a stem intergraft not expressing the MP gene thus questioning the need for the MP in vascular movement. However, it has recently been shown that the phloem-specific expression of the TMV MP affects the carbon metabolism and partitioning in transgenic potatoes (Almon et al. 1997). This indicates that the TMV MP can function in this region to alter plant metabolism and thus it may function in this region to support virus movement. Further work will be necessary to determine the true involvement of the MP of TMV during long-distance virus movement.

Another potential regulatory site within the host for virus movement is between the external and internal phloem in the stem tissue. Multiple plant species, including important crop species, have internal and external phloem (*e.g. Capsicum annuum, Solanum tuberosum, Nicotiana tabacum, Cucurbita pepo;* Hayward 1938). Evidence exists that photoassimilate transport out of source leaves occurs in the abaxial phloem (connected to external phloem) and not in the adaxial phloem (connected to internal phloem; Bonnemain 1969a, b; Turgeon and Webb 1976). Recently, Andrianifahanana et al. (1997) determined that vascular-dependent accumulation of pepper mottle virus (PepMoV), a potyvirus, occurred in a defined pattern within the stem of *Capsicum annuum*. Through tissue immunoblots, virus was shown to move first through the external phloem towards the roots. Then at or somewhere between the cotyledonary node and the roots, the virus entered the internal phloem, through which it moved to the shoot apex. Derrick and Barker (1992, 1997) have also observed a dichotomy of infection of stem phloem tissue in naturally resistant or transgenic *Solanum tuberosum* infected with potato leafroll virus (PLRV). In their studies, PLRV accumulation was decreased in the external phloem bundles in the stems of the resistant plants compared with susceptible plants. Although there are some inconsistencies between these reports [e.g. why do the internal phloem bundles of transgenic *Solanum tuberosum* become significantly infected if the external phloem bundles, the putative source of virus in the model proposed by Andrianifahanana et al. (1997), contain little virus?], they both indicate the potential regulation of virus transport in the stem. The plasmodesmata in the stem connections between the external and internal phloem may have a role in this regulated movement.

Host factors that regulate vascular-dependent virus accumulation have not been isolated and characterized, although many are known and their genetics studied (Holmes 1955; Kuhn et al. 1981; Lei and Agrios 1986; Barker 1987; Golinowski et al. 1987; Dufour et al. 1989; Law et al. 1989; Derrick and Barker 1992; Simon et al. 1992; Wilson and Jones 1992; Leisner et al. 1993; Murphy and Kyle 1995; Schaad and Carrington 1996; Tang and Leisner 1997). A minority of these reports have characterized where the resistance to virus transport occurs. Besides the previously mentioned work of Schaad and Carrington (1996; Sect. 2.2) and Derrick and Barker (1992, 1997; this Sect.), Golinowski et al. (1987) and Wilson and Jones (1992) characterized host resistances to PLRV in *Solanum tuberosum*. Golinowski et al. (1987) determined that excessive callose formed in the sieve tubes of the resistant cultivar, Apta, compared with the susceptible cultivar, Osa, when both were challenged with PLRV. The ER in SEs were dilated and the desmotubules in plasmodesmata from petioles were occluded with densely staining fibrillar material. The timing of appearance of these symptoms suggest that these alterations in the plasmodesmata are a direct, as opposed to an indirect, response to the virus, but this must be proven through further studies.

Plant development can also affect the symplasmic isolation of the stem phloem. Van Bel and van Rijen (1994) determined that the SE/CC complex in transport phloem was increasingly electrically isolated from other cells, such as the VPCs, as they matured. It is interesting that van Lent and Verduin (1987) detected CCMV CP in phloem, BSCs, and cortex cells of systemically infected petiolules of *V. unquiculata* by 24-h posttransfer of tissue from 10 to 25°C. The low temperature inhibits virus replication greatly but not vascular movement. In another study, non-vascular stem and petiole cells of tobacco became infected by TMV, within a time period restricting the source of viral inoculum in these tissues to the phloem (Nelson et al. 1993; Derrick et al. 1997). How could virus be moving out of the SE/CC complex in mature petiolule, petiole, or stem tissue that is electrically isolated and therefore possibly symplasmically isolated from other cells? Kempers et al. (1998; Chap. 14 for more on this) recently published a report indicating that a constricted, but consistent, number of plasmodesmata connecting the SE/CC complex with VPCs existed in transport phloem of symplasmically and apoplasmically loading plant species. Patrick and Offler (1996) demonstrated that when the prevailing source/sink ratio favours net assimilate storage in stems, unloading of photoassimilate may be symplasmic. Perhaps viruses utilize this symplasmic transport system to invade non-vascular tissue associated with the transport phloem.

Recent work by Knoblauch and van Bel (1998) may light the way for real time analysis of virus transport through SEs. Using CLSM and phloem-mobile fluorochromes, these researchers have observed flow of material through SEs in planta. Researchers may be able to use such a system to follow the initial long-distance movement of fluorescently labeled virus proteins or RNA through these elements.

2.3 Exit of Viruses from the Phloem

It has long been known that viruses invade leaves according to their status as a source or sink and therefore that the age of an uninoculated leaf at the time of infection profoundly affects disease symptom development (e.g. Holmes 1934; Zech 1952; Reid and Matthews 1966; Nilsson-Tillgren et al. 1969; Ferguson and Matthews 1993). These studies did not detail how viruses exit phloem during vascular-dependent invasion of roots, stems, and leaves. However, some detailed analyses of viral egress from phloem have been attempted. Hatta and Matthews (1974) published results of a time-course study of turnip yellow mosaic virus (TYMV) egress from phloem cells in veins of young (4 cm long) Chinese cabbage leaf tissue. The veins were described as 5th- or 6th-order veins containing 30 to 60 cells in cross-section, and so were not the smallest veins in these leaves; the smallest veins contained approximately 8 cells. Unfortunately, no indication was given as to whether these were the smallest veins from which virus invaded leaves. They did determine that virus first appeared in cells adjacent to the SE/CC complex and VPCs and not xylem-associated cells. Hoefert et al. (1988) conducted similar studies with lettuce infectious yellows virus in systemically infected leaves of *Lactuca sativa* over time. They identified virus in TCs and VPCs at 5 days postinoculation. The developmental stage of the leaves and the vein orders analyzed were not described in detail.

More recently, X. S. Ding et al. (1995) determined that in the minor veins (containing 9–12 cells) of young tobacco leaves systemically invaded by TMV, the percentage of VPCs and CCs infected was 60 and 50%, respectively. In mature, inoculated leaves of these plants, 90% of the VPCs and 10–30% of the CCs were infected. The greater abundance of CCs infected and the infection of VPCs in the systemically infected leaves was attributed to the greater abundance of plasmodesmata between SEs and other cell types in the immature cells of minor veins in the young leaves studied. Unfortunately, virus exit from veins in these leaves was not followed over time, and thus this immature tissue may have been infected by virus moving cell to cell from an adjacent larger vein (i.e. an indirect infection of the minor vein cells). It is interesting, though, that in studies with CaMV, the minor veins of systemic leaves infected late in leaf development appeared to be infected without infection of adjacent mesophyll cells (Al-Kaff and Covey 1996). Invasion of minor veins without infection of adjacent mesophyll cells also appears to be the situation for BDMV on systemically infected leaves of *P. vulgaris* (Sudarshana et al. 1998). Thus, perhaps direct infection of minor veins by vascular movement of the virus is dependent on the virus and/or the age of the leaf at the time of invasion (see next paragraph, however).

The most informative study of virus emergence from the phloem published so far is that of Roberts et al. (1997), who studied long-distance movement of a GFP-expressing PVX in *Nicotiana benthamiana* plants. By CLSM, GFP, representing the location of the virus, was observed to emerge exclusively from relatively large veins (third order and larger), as did the phloem tracer dye, carboxyfluorescein diacetate. This pattern of virus exit was observed regardless of the age of the systemically infected leaf. This report illustrates the enormous value of GFP as a reporter molecule for future studies of long-distance virus movement. Also it illustrates the importance of monitoring the developmental stage of the tissue under study when making conclusions on experimental observations.

The phloem-associated viruses, notably the luteoviruses, closteroviruses, and geminiviruses are able to move along SEs and from SEs to CCs and VPCs (e.g. D'Arcy and de Zoeten 1979; Shepardson et al. 1980; Horns and Jeske 1991; Sanger et al. 1994; van den Heuvel et al. 1995; Lucy et al. 1996; Wang et al. 1996). In some instances, the ability of phloem-associated viruses to accumulate in mesophyll protoplasts has been demonstrated (Kubo and Takanami 1979; Barnett et al. 1981;Barker and Harrison 1982; Veidt et al. 1992; Sanger et al. 1994; Schmitz et al. 1997), so phloem association of these pathogens appears to be due to a lack of movement rather than lack of replication outside the phloem. Indeed, some phloem-associated viruses do replicate in non-vascular tissue after the tissue has reached a specific stage of development (e.g. Lucy et al. 1996; Wang et al. 1996). Interestingly, phloem limitation of bean golden mosaic virus, a geminivirus, is lost if plants are coinfected with sunnhemp mosaic virus, a tobamovirus (Carr and Kim 1983). Similarly, coinfection of potatoes with PLRV and any of a number of helper viruses, such as potato virus Y (PVY), resulted in greater accumulation of PLRV in infected plants (Barker 1987). Furthermore, as a greater proportion of mesophyll protoplasts isolated from PVY / PLRV coinfected potato plants contained PLRV, it was concluded that the helper viruses supplied a factor which enabled some degree of release from phloem limitation. It was assumed that the helper virus MP in some way aided movement of the phloem-limited virus out of the phloem in these cases. More recently, a 17-kDa protein encoded by PLRV has been described which possesses biochemical properties typical of other viral movement proteins and which is localized around the PPUs (Tacke et al. 1993; Schmitz et al. 1997). A phosphorylated form of the protein is present in SEs, phosphorylation being mediated by a membrane-associated protein kinase (Sokolova et al. 1997). Schmitz et al. (1997) expressed a full-length clone of PLRV in transgenic *Solanum tuberosum* from behind a 35S promoter, and observed the location of the 17-kDa protein throughout the plant by immunocytochemistry. Although both mesophyll and vascular cells were infected due to the expression of the virus RNA from the 35S promoter, the 17-kDa protein was observed only in PPUs. This localization of 17-kDa protein to specific cell types was not observed in transgenic plants expressing only the 17-kDa protein. Cooper and Dodds (1995) also observed a difference in the subcellular location of TMV and CMV MP in infected non-transformed and non-infected MP-expressing transgenic plants. Schmitz et al. (1997) noted that perhaps other viral or virally induced host factors are responsible for the sequestering of the 17-kDa protein to the PPUs during virus infection. Regardless of this, it appears that this MP functions specifically for the transport of virus between SEs and CCs.

3 Viral Factors Involved in Vascular-Dependent Accumulation

Viral proteins and potentially some viral RNAs regulate the vascular-dependent accumulation of viruses in hosts beyond a simple involvement in virus accumulation in initially inoculated cells or protoplasts (Table 2). Some of these factors decrease systemic accumulation indirectly by decreasing local spread of the virus or by inducing a host-resistance response (e.g. Edwards et al. 1983; Petty et al. 1990). However, some have what appear to be more direct effects on vascular-dependent accumulation (noted in the previous sections). Perhaps the most intriguing aspect of this compilation is that virus movement may be linked to non-structural viral proteins other than those classically defined as movement proteins (Table 2). Some of these proteins are known to affect virus replication (e.g. 126/183 kDa protein; see Buck 1996 for a review of viral replicases). Identifying interactions between viral factors or viral and host factors that directly, and only, affect vascular spread will be the next great hurdle in this research.

4 Concluding Remarks

Many of the regulation points for vascular-dependent accumulation of plant viruses discussed in the previous sections are summarized in Table 1. It is possible that the plasmodesmata between the cells at each location play a crucial role in this regulation. In no instance has a host factor associated with the plasmodesmata been identified that regulates this movement (note, however, that actin is a likely participant; White et al. 1994; B. Ding et al. 1996). Multiple viral factors have been identified that affect vascular-dependent virus accumulation (Table 2). MPs associate with plasmodesmata, but their function in virus spread into, through, and out of veins remains to be determined. Those viral proteins that do not associate with plasmodesmata themselves may affect virus transport by associating with cellular or viral factors that then become associated with plasmodesmata, or by conducting a specialized function that occurs only in the vascular tissue. The ability to label viral proteins and visualize their location (e.g. through viral protein-GFP fusions and CLSM) will lead to a more complete understanding of their ability to move during viral infections and their associations with specific host factors. Results obtained by CLSM showing colocalization of specific host and viral factors will lead to molecular, genetic and biochemical studies to determine the specific interactions between these factors.

The current research is helping to define the specific host-pathogen interactions in vascular tissue that previously were characterized only by visual description (e.g. barley stripe mosaic was most likely named due to symptoms mediated by vascular movement). Considering that the vast majority of crop yield and quality declines are due to vascular-dependent accumulation of the pathogen, research studying the influence of plasmodesmata within the vascular tissue on virus spread will be of great interest to plant pathologists and agronomists. Also for cell and developmental biologists, studies of virus vascular transport will most likely serve as a paradigm for host macromolecule vascular transport, just as studies on viral cell-to-cell movement have served as a paradigm for host macromolecule trafficking between cells.

To indicate how quickly research is moving in this area, some recent exciting findings need to be mentioned. It has recently become clear that some viral factors known to affect vascular-dependent accumulation of viruses in the host also function to turn on or stabilize virus or transgene RNA and protein accumulation. Specifically, HC-Pro of the potyviruses can supress post-transcriptional gene silencing of GFP or GUS and their associated transcripts in mature tissue of transgenic plants (Anandalakshmi et al. 1998; Kasschau and Carrington 1998; Brigneti et al. 1998). The 2b protein of CMV prevents the initiation of post-transcriptional gene silencing (Brigneti et al. 1998). Thus the lack of phloem-dependent accumulation of viruses mutated in the HC-Pro or 2b ORFs is likely due to an inability of the mutant viruses to prevent virus induced gene silencing in the systemically infected tissue. It will be interesting to see if other viral proteins, such as the 126-/183-kDa protein of TMV, function in a similar fashion.

Acknowledgements. The authors wish to thank Drs. Biao Ding and William Schneider for comments on portions of the manuscript. We thank Allyson Wilkins for text and table preparation.

References

Al-Kaff NS, Covey SN (1996) Unusual accumulation of cauliflower mosaic virus in local lesions, dark green leaf tissue, and roots of infected plants. Mol Plant-Microbe Interact 9:357–363

Almon E, Horowitz M, Wang H-L, Lucas WJ, Zamski E, Wolf S (1997) Phloem-specific expression of the tobacco mosaic virus movement protein alters carbon metabolism and partitioning in transgenic potato plants. Plant Physiol 115:1599–1607

Anandalakshmi R, Pruss GJ, Ge X, Marathe R, Mallory AC, Smith TH, Vance VB (1998) A viral suppressor of gene silencing in plants. Proc Natl Acad Sci 95:13079–13084

Andrianifahanana M, Lovins K, Dute R, Sikora E, Murphy JF (1997) Pathway for phloem-dependent movement of pepper mottle potyvirus in the stem of *Capsicum annuum*. Phytopathology 87: 892–898

Arce-Johnson P, Reimann-Philipp U, Padgett HS, Rivera-Bustamente R, Beachy RN (1997) Requirement of the movement protein for long distance spread of tobacco mosaic virus in grafted plants. Mol Plant-Microbe Interact 10:691–699

Barker H (1987) Invasion of non-phloem tissue in *Nicotiana clevelandii* by potato leafroll luteovirus is enhanced in plants also infected with potato Y potyvirus. J Gen Virol 68:1223–1227

Barker H, Harrison BD (1982) Infection of potato mesophyll protoplasts with five plant viruses. Plant Cell Rep 1:247–249

Barnett A, Hammond J, Lister RM (1981) Limited infection of cereal leaf protoplasts by barley yellow dwarf virus. J Gen Virol 57:397–401

Beebe DU, Evert RF (1992) Photoassimilate pathway(s) and phloem loading in the leaf of *Moricandia arvensis* (L.) DC. (Brassicaceae). Int J Plant Sci 153:61–77
Bennett CW (1940) Relation of food translocation to movement of virus of tobacco mosaic. J Agric Res 60:361–389
Bennett CW (1956) Biological relations of plant viruses. Annu Rev Plant Physiol 7:143–170
Blackman LM, Boevink P, Santa Cruz S, Palukaitis P, Oparka KJ (1998) The movement protein of cucumber mosaic virus traffics into sieve elements in minor veins of *Nicotiana clevelandii*. Plant Cell 10:525–537
Bonnemain J (1969a) Transport du ^{14}C assimile a partir des feuilles de tomate en voie de croissance et vers celles-ci. CR Acad Sci 269:1660–1663
Bonnemain J (1969b) Le phloème interne et le phloème inclus des cidotylédonnes, leur histogénèse et leur physiologie. Rev Gen Bot 76:5–36
Bostwick DE, Dannenhoffer JM, Skaggs MI, Lister RM, Larkins BA, Thompson GA (1992) Pumpkin phloem lectin genes are specifically expressed in companion cells. Plant Cell 4:1539–1548
Bransom KL, Weiland JJ, Tsai CH, Dreher TW (1995) Coding density of the turnip yellow mosaic virus genome: roles of the overlapping coat protein and p206-readthrough coding regions. Virology 206:403–412
Brigneti G, Voinnet O, Li WX, Ji LH, Ding SW, Baulcombe DC (1998) Viral pathogenicity determinants are suppressors of transgene silencing in *Nicotiana benthamiana*. EMBO J 17:6739–6746
Brugidou C, Holt C, Yassi MNA, Zhang S, Beachy R, Fauquet C (1995) Synthesis of an infectious full-length cDNA clone of rice yellow mottle virus and mutagenesis of the coat protein. Virology 206:108–115
Buck KW (1996) Comparison of the replication of positive-stranded RNA viruses of plants and animals. Adv Virus Res 47:159–251
Callaway A, Liu W, Andrianov V, Stenzler L, Zhao J, Wettlaufer S, Jayakumar P, Howell SH (1996) Characterization of cauliflower mosaic virus (CaMV) resistance in virus-resistant ecotypes of *Arabidopsis*. Mol Plant-Microbe Interact 9:810–818
Carr RJ, Kim KS (1983) Evidence that bean golden mosaic virus invades non-phloem tissue in double infections with tobacco mosaic virus. J Gen Virol 64:2489–2492
Carrington JC, Kasschau KD, Mahajan SK, Schaad MC (1996) Cell-to-cell and long-distance transport of viruses in plants. Plant Cell 8:1669–1681
Clark AM, Jacobsen KR, Bostwick DE, Dannenhoffer JM, Staggs MI, Thompson GA (1997) Molecular characterization of a phloem-specific gene encoding the filament protein, phloem protein 1 (PP1), from *Cucurbita maxima*. Plant J 12:49–61
Cooper B, Dodds JA (1995) Differences in the subcellular localization of tobacco mosaic virus and cucumber mosaic virus movement proteins in infected and transgenic plants. J Gen Virol 76:3217–3221
Cronin S, Verchot J, Haldeman-Cahill R, Schaad MC, Carrington JC (1995) Long-distance movement factor: a transport function of the potyvirus helper component proteinase. Plant Cell 7:549–559
D'Arcy CJ, de Zoeten GA (1979) Beet western yellows virus in phloem tissue of *Thlaspi arvense*. Phytopathology 69:1194–1198
Dannenhoffer JM, Schulz A, Skaggs MI, Bostwick DE, Thompson GA (1997) Expression of the phloem lectin is developmentally linked to vascular differentiation in cucurbits. Planta 201:405–414
Dawson WO, Schlegel DE (1973) Differential temperature treatment of plants greatly enhances multiplication rates. Virology 53:476–478
Dawson WO, Bubrick P, Grantham GI (1988) Modifications of the tobacco mosaic virus coat protein gene affecting replication, movement, and symptomology. Phytopathology 78:783–789
De Jong W, Ahlquist P (1995) Host-specific alterations in viral RNA accumulation and infection spread in a brome mosaic virus isolate with an expanded host range. J Virol 69:1485–1492
Demler SA, de Zoeten GA, Adam G, Harris KF (1996) Pea enation mosaic enamovirus: Properties and aphid transmission. In: Harrison BD, Murant AF (eds,) The plant viruses, vol 5: Polyhedral virions and bipartite RNA genomes. Plenum Press, New York, pp 303–344
Deom CM, He XZ, Beachy RN, Weissinger AK (1994) Influence of heterologous tobamovirus movement protein and chimeric-movement protein genes on cell-to-cell and long-distance movement. Virology 205:198–209
Deom CM, Quan S, He XZ (1997) Replicase proteins as determinants of phloem-dependent long-distance movement of tobamoviruses in tobacco. Protoplasma 199:1–8
Derrick PM, Barker H (1992) The restricted distribution of potato leafroll luteovirus antigen in potato plants with transgenic resistance resembles that in clones with one type of host gene-mediated resistance. Ann Appl Biol 120:451–457
Derrick PM, Barker H (1997) Short and long distance spread of potato leafroll luteovirus: effects of host genes and transgenes conferring resistance to virus accumulation in potato. J Gen Virol 78:243–251

Derrick PM, Barker H, Oparka KJ (1992) Increase in plasmodesmatal permeability during cell-to-cell spread of tobacco rattle virus from individually inoculated cells. Plant Cell 4:1405–1412
Derrick PM, Carter SA, Nelson RS (1997) Mutation of the 126-/183-kDa proteins of tobacco mosaic tobamovirus: the relationship of phloem-dependent accumulation with viral protein accumulation. Mol Plant-Microbe Interact 10:589–596
de Zoeten GA, Gaard G (1983) Mechanisms underlying systemic invasion of pea plants by pea enation mosaic virus. Intervirology 19:85–94
Ding B (1997) Cell-to-cell transport of macromolecules through plasmodesmata: a novel signalling pathway in plants. Trends Cell Biol 7:5–9
Ding B, Lucas WJ (1996) Secondary plasmodesmata: biogenesis, special functions and evolution. In: Smallwood M, Knox P, Bowles D (eds) Membranes: specialized functions in plants. BIOS Scientific Publishers, Oxford, pp 489–506
Ding B, Parthasarathy MV, Niklas K, Turgeon R (1988) A morphometric analysis of the phloem-unloading pathway in developing tobacco leaves. Planta 176:307–318
Ding B, Haudenshield JS, Hull RJ, Wolf S, Beachy RN, Lucas WJ (1992a) Secondary plasmodesmata are specific sites of localization of the tobacco mosaic virus movement protein in transgenic tobacco plants. Plant Cell 4:915–928
Ding B, Turgeon R, Parthasarathy MV (1992b) Substructure of freeze-substituted plasmodesmata. Protoplasma 169:28–41
Ding B, Haudenshield JS, Willmitzer L, Lucas WJ (1993) Correlation between arrested secondary plasmodesmal development and onset of accelerated leaf senescence in yeast acid invertase transgenic tobacco plants. Plant J 4:179–189
Ding B, Li Q, Nguyen L, Palukaitis P, Lucas WJ (1995) Cucumber mosaic virus 3a protein potentiates cell-to-cell trafficking of CMV RNA in tobacco plants. Virology 207:345–353
Ding B, Kwon M-O, Warnberg L (1996) Evidence that actin filaments are involved in controlling the permeability of plasmodesmata in tobacco mesophyll. Plant J 10:157–164
Ding SW, Li WX, Symons RH (1995) A novel naturally occurring hybrid gene encoded by a plant RNA virus facilitates long distance virus movement. EMBO J 14:5762–5772
Ding XS, Shintaku MH, Arnold SA, Nelson RS (1995) Accumulation of mild and severe strains of tobacco mosaic virus in minor veins of tobacco. Mol Plant-Microbe Interact 8:32–40
Ding XS, Shintaku MH, Carter SA, Nelson RS (1996) Invasion of minor veins of tobacco leaves inoculated with tobacco mosaic virus mutants defective in phloem-dependent movement. Proc Natl Acad Sci 93:11 155–11 160
Ding XS, Carter SA, Nelson RS (1998) Tobamovirus and potyvirus accumulation in minor veins of inoculated leaves from representatives of the Solanaceae and Fabaceae. Plant Physiol 116: 125–136
Dolja VV, Haldeman R, Robertson NL, Dougherty WG, Carrington JC (1994) Distinct functions of capsid protein in assembly and movement of tobacco etch potyvirus in plants. EMBO J 13: 1482–1491
Dolja VV, Haldeman-Cahill R, Montgomery AE, Vandenbosch KA, Carrington JC (1995) Capsid protein determinants involved in cell-to-cell and long distance movement of tobacco etch potyvirus. Virology 206:1007–1016
Dorokhov YL, Alexandrova NM, Miroshnichenko NA, Atabekov JG (1984) Stimulation by aurintricarboxylic acid of tobacco mosaic virus-specific RNA synthesis and production of informosome-like infection-specific ribonucleoprotein. Virology 135:395–405
Dufour O, Palloix A, Selassie KG, Pochard E, Marchoux G (1989) The distribution of cucumber mosaic virus in resistant and susceptible plants of pepper. Can J Bot 67:655–660
Edwards MC, Gonsalves D, Provvidenti R (1983) Genetic analysis of cucumber mosaic virus in relation to host resistance: location of determinants for pathogenicity to certain legumes and *Lactuca saligno*. Phytopathology 73:269–273
Esau K (1968) Viruses in plant hosts. Form, distribution, and pathologic effects. The University of Wisconsin Press, Madison, Wisconsin
Esau K (1977) Anatomy of Seed Plants. In: Esau K (ed) John Wiley, New York, pp 321–332
Esau K, Hoefert LL (1971) Cytology of beet yellows virus infection in *Tetragonia* III. Conformations of virus in infected cells. Protoplasma 73:51–65
Evert RF, Mierzwa R (1986) Pathway(s) of assimilate movement from mesophyll cells to sieve tubes in the *Beta vulgaris* leaf. In: Cronshaw J, Lucas WJ, Giaquinta RT (eds) Phloem transport. Alan Liss, New York, pp 419–432
Falk B, Morris T, Duffus J (1979) Unstable infectivity and sedimentable ds-RNA associated with lettuce speckles mottle virus. Virology 96:239–248
Fenczik CA, Padgett HS, Holt CA, Casper SJ, Beachy RN (1995) Mutational analysis of the movement protein of odontoglossum ringspot virus to identify a host-range determinant. Mol Plant-Microbe Interact 8:666–673

Ferguson A, Matthews REF (1993) Mosaic disease induced by turnip yellow mosaic *Tymovirus*. Biochimie 75:555–559
Fisher DG (1986) Ultrastructure, plasmodesmatal frequency and solute concentration in green areas of variegated *Coleus blumei* Benth. leaves. Planta 169:141–152
Fisher DG, Evert RF (1982) Studies on the leaf of *Amaranthus retroflexus* (Amaranthaceae): ultrastructure, plasmodesmatal frequency, and solute concentration in relation to phloem loading. Planta 155:377–387
Fry PR, Matthews REF (1963) Timing of some early events following inoculation with tobacco mosaic virus. Virology 19:461–469
Fujiwara T, Giesman-Cookmeyer D, Ding B, Lommel SA, Lucas WJ (1993) Cell-to-cell trafficking of macromolecules through plasmodesmata potentiated by the red clover necrotic mosaic virus movement protein. Plant Cell 5:1783–1794
Gal-On A, Kaplan I, Roossinck MJ, Palukaitis P (1994) The kinetics of infection of zucchini squash by cucumber mosaic virus indicate a function for RNA 1 in virus movement. Virology 205:280–289
Geiger DR, Giaquinta RT, Sovonick SA, Fellows RJ (1973) Solute distribution in sugar beet leaves in relation to phloem loading and translocation. Plant Physiol 52:585–589
Gera A, Deom CM, Donson J, Shaw JJ, Lewandowski DJ, Dawson WO (1995) Tobacco mosaic tobamovirus does not require concomitant synthesis of movement protein during vascular transport. Mol Plant-Microbe Interact 8:784–787
Ghoshroy S, Lartey R, Sheng J, Citovsky V (1997) Transport of proteins and nucleic acids through plasmodesmata. Annu Rev Plant Physiol Plant Mol Biol 48:27–50
Gilbertson RL, Lucas WJ (1996) How do viruses traffic on the 'vascular highway'? Trends Plant Sci 1: 260–267
Golinowski W, Tomenius K, Oxelfelt P (1987) Ultrastructural studies on potato phloem cells infected with potato leaf roll virus-comparison of two potato varieties. Acta Agric Scand 37:3–19
Goodrick BJ, Kuhn CW, Hussey RS (1991) Restricted systemic movement of cowpea chlorotic mottle virus in soybean with nonnecrotic resistance. Phytopathology 81:1426–1431
Grusak MA, Beebe DU, Turgeon R (1996) Phloem loading. In: Zamski E, Schaffer AA (eds) Photoassimilate distribution in plants and crops. Marcel Dekker, New York, pp 209–227
Hamilton WDO, Baulcombe DC (1989) Infectious RNA produced by in vitro transcription of a full-length tobacco rattle virus RNA-1 cDNA. J Gen Virol 70:963–968
Hatta T, Matthews REF (1974) The sequence of early cytological changes in Chinese cabbage leaf cells following systemic infection with turnip yellow mosaic virus. Virology 59:383–396
Hayward HE (1938) The structure of economic plants. Macmillan, New York
Heaton LA, Lee TC, Wei N, Morris TJ (1991) Point mutations in the turnip crinkle virus capsid protein affect the symptoms expressed by *Nicotiana benthamiana*. Virology 183:143–150
Hickey LJ (1979) A revised classification of the architecture of dicotyledonous leaves. In: Metcalfe CR, Chalk L (eds) Anatomy of the dicotyledons. Oxford University Press, New York, pp 25–39
Hilf ME, Dawson WO (1993) The tobamovirus capsid protein functions as a host-specific determinant of long-distance movement. Virology 193:106–114
Hoefert LL, Pinto RL, Fail GL (1988) Ultrastructural effects of lettuce infectious yellows virus in *Lactuca sativa*L. J Ultrastruct Mol Struct Res 98:243–253
Holmes FO (1930) Local and systemic increase of tobacco mosaic virus. Am J Bot 17:789–805
Holmes FO (1934) A masked strain of tobacco mosaic virus. Phytopathology 24:845–873
Holmes FO (1955) Additive resistances to specific viral diseases in plants. Ann Appl Biol 42:129–139
Horner HT, Lersten NR, Wirth CL (1994) Quantitative survey of sieve tube distribution in foliar terminal veins of ten dicot species. Am J Bot 81:1267–1274
Horns T, Jeske H (1991) Localization of abutilon mosaic virus (AbMV) DNA within leaf tissue by in situ hybridization. Virology 181:580–588
Ingham DJ, Pascal E, Lazarowitz SG (1995) Both bipartite geminivirus movement proteins define viral host range, but only BL1 determines viral pathogenicity. Virology 207:191–204
Ishiwatari Y, Fujiwara T, McFarland KC, Nemoto K, Hayashi H, Chino M, Lucas WJ (1998) Rice phloem thioredoxin h has the capacity to mediate its own cell-to-cell transport. Planta 205:12–22
Itaya A, Hickman H, Bao Y, Nelson R, Ding B (1997) Cell-to-cell trafficking of cucumber mosaic virus movement protein: green fluorescent protein fusion produced by biolistic gene bombardment in tobacco. Plant J 12:1223–1230
Jorgensen RA, Atkinson RG, Forster RLS, Lucas WJ (1998) A RNA-based superhighway in plants. Science 279:1486–1487
Kasschau KD, Carrington JC (1998) A counterdefensive strategy of plant viruses: suppression of post-transcriptional gene silencing. Cell 95:461–470
Kempers R, Ammerlaan A, van Bel AJE (1998) Symplasmic constriction and ultrastructural features of the sieve element/companion cell complex in the transport phloem of apoplasmically and symplasmically phloem-loading species. Plant Physiol 116:271–278

Knoblauch M, van Bel AJE (1998) Sieve tubes in action. Plant Cell 10:35–50
Kubo S, Takanami Y (1979) Infection of tobacco mesophyll protoplasts with tobacco necrotic dwarf virus, a phloem-limited virus. J Gen Virol 42:387–398
Kühn C, Francheschi VR, Schulz A, Lemoine R, Frommer WB (1997) Macromolecular trafficking indicated by localization and turnover of sucrose transporters in enucleate sieve elements. Science 275:1298–1300
Kuhn CW, Wyatt SD, Brantley BB (1981) Genetic control of symptoms, movement, and virus accumulation in cowpea plants infected with cowpea chlorotic mottle virus. Phytopathology 71:1310–1315
Lakshman DK, Gonsalves D (1985) Genetic analyses of two large-lesion isolates of cucumber mosaic virus. Phytopathology 75:758–762
Lauber E, Guilley H, Tamada T, Richards KE, Jonard G (1998) Vascular movement of beet necrotic yellow vein virus in *Beta macrocarpa* is probably dependent on an RNA 3 sequence domain rather than a gene product. J Gen Virol 79:385–393
Law MD, Moyer JW, Payne GA (1989) Effect of host resistance on pathogenesis of maize dwarf mosaic virus. Phytopathology 79:757–761
Lei JD, Agrios GN (1986) Mechanisms of resistance in corn to maize dwarf mosaic virus. Phytopathology 76:1034–1040
Leisner SM, Turgeon R (1993) Movement of virus and photoassimilate in the phloem: a comparative analysis. BioEssays 15:741–748
Leisner SM, Turgeon R, Howell SH (1992) Long-distance movement of cauliflower mosaic virus in infected turnip plants. Mol Plant-Microbe Interact 5:41–47
Leisner SM, Turgeon R, Howell SH (1993) Effects of host plant development and genetic determinants on the long-distance movement of cauliflower mosaic virus in *Arabidopsis*. Plant Cell 5:191–202
Lewandowski DJ, Dawson WO (1993) A single amino acid change in tobacco mosaic virus replicase prevents symptom production. Mol Plant-Microbe Interact 6:157–160
Lucas WJ (1995) Plasmodesmata: intercellular channels for macromolecular transport in plants. Curr Opin Cell Biol 7:673–680
Lucas WJ, Ding B, van der Schoot C (1993a) Plasmodesmata and the supracellular nature of plants. New Phytol 125:435–476
Lucas WJ, Olesincki A, Hull RJ, Haudenshield JS, Deom CM, Beachy RN, Wolf S (1993b) Influence of the tobacco mosaic virus 30-kDa movement protein on carbon metabolism and photosynthate partitioning in transgenic tobacco plants. Planta 190:88–96
Lucy AP, Boulton MI, Davies JW, Maule AJ (1996) Tissue specificity of *Zea mays* infection by maize streak virus. Mol Plant-Microbe Interact 9:22–31
Madore MA, Oross JW, Lucas WJ (1986) Symplastic transport in *Ipomea tricolor* source leaves. Plant Physiol 82:432–442
Más P, Pallás V (1996) Long-distance movement of cherry leaf roll virus in infected tobacco plants. J Gen Virol 77:531–540
Matthews REF (1991) Plant virology. Academic Press, San Diego
Mauseth JD (1988) Plant anatomy, Benjamin/Cummings, Menlo Park, California
McCauley MM, Evert RF (1989) Minor veins of the potato (*Solanum tuberosum* L.) leaf: ultrastructure and plasmodesmatal frequency. Bot Gaz 150:351–368
McLean BG, Hempel FD, Zambryski PC (1997) Plant intercellular communications via plasmodesmata. Plant Cell 9:1043–1054
Melcher U (1989) Symptoms of cauliflower mosaic virus infection in *Arabidopsis thaliana* and turnip. Bot Gaz 150:139–147
Mezitt LA, Lucas WJ (1996) Plasmodesmal cell-to-cell transport of proteins and nucleic acids. Plant Mol Biol 32:251–273
Mise K, Ahlquist P (1995) Host-specificity restriction by bromovirus cell-to-cell movement protein occurs after initial cell-to-cell spread of infection in nonhost plants. Virology 206:276–286
Murant A, Roberts I, Goold R (1996) Carrot mottle-a persistent aphid-borne virus with unusual properties and particles. J Gen Virol 4:329–341
Murillo I, Cavallarin L, San Segundo B (1997) The maize pathogenesis-related PRms protein localizes to plasmodesmata in maize radicles. Plant Cell 9:145–156
Murphy JF, Kyle MM (1995) Alleviation of restricted systemic spread of pepper mottle potyvirus in *Capsicum annuum* cv. Avelar by coinfection with a cucumovirus. Phytopathology 85:561–566
Nelson RS, van Bel AJE (1998) The mystery of virus trafficking into, through and out of the vascular tissue. Prog Bot 59:476–533
Nelson RS, Li G, Hodgson RAJ, Beachy RN, Shintaku MH (1993) Impeded phloem-dependent accumulation of the masked strain of tobacco mosaic virus. Mol Plant-Microbe Interact 6:45–54
Nilsson-Tillgren T, Kolehmainen-Sevrus L, von Wettstein D (1969) Studies on the biosynthesis of TMV I. A system approaching a synchronized virus synthesis in a tobacco leaf. Mol Gen Genet 104:124–141

Noueiry AO, Lucas WJ, Gilbertson RL (1994) Two proteins of a plant DNA virus coordinate nuclear and plasmodesmal transport. Cell 76:925–932
Oparka KJ, Boevink P, Santa Cruz S (1996) Studying the movement of plant viruses using green fluorescent protein. Trends Plant Sci 1:412–418
Oparka KJ, Prior DAM, Santa Cruz S, Padgett HS, Beachy RN (1997a) Gating of epidermal plasmodesmata is restricted to the leading edge of expanding infection sites of tobacco mosaic virus (TMV). Plant J 12:781–789
Oparka KJ, Roberts AG, Santa Cruz S, Boevink P, Prior DAM, Smallcombe A (1997b) Using GFP to study virus invasion and spread in plant tissues. Nature 388:401–402
Overall RL, Blackman LM (1996) A model of the macromolecular structure of plasmodesmata. Trends Plant Sci 1:307–311
Oxelfelt P (1970) Development of systemic tobacco mosaic virus infection. I. Initiation of infection and time course of virus multiplication. Phytopathol Z 69:202–211
Oxelfelt P (1975) Development of systemic tobacco mosaic virus infection. IV. Synthesis of viral RNA and intact virus and systemic movement of two strains as influenced by temperature. Phytopathol Z 83:66–76
Padgett HS, Epel BL, Kahn TW, Heinlein M, Watanabe Y, Beachy RN (1996) Distribution of tobamovirus movement protein in infected cells and implications for cell-to-cell spread of infection. Plant J 10:1079–1088
Padidam M, Beachy RN, Fauquet CM (1995) Tomato leaf curl geminivirus from India has a bipartite genome and coat protein is not essential for infectivity. J Gen Virol 76:25–35
Palukaitis P (1987) Potato spindle tuber viroid: Investigation of the long-distance, intra-plant transport route. Virology 158:239–241
Patrick JW, Offler CE (1996) Post-sieve element transport of photoassimilates in sink regions. J Exp Bot 47:1165–1177
Petty ITD, Jackson AO (1990) Mutational analysis of barley stripe mosaic virus RNA b. Virology 179: 712–718
Petty ITD, Edwards MC, Jackson AO (1990) Systemic movement of an RNA plant virus determined by a point substitution in a 5′ leader sequence. Proc Natl Acad Sci 87:8894–8897
Powell CA, de Zoeten GA (1977) Replication of pea enation mosaic virus RNA in isolated pea nuclei. Proc Natl Acad Sci 74:2919–2922
Powell CA, de Zoeten GA, Gaard G (1977) The localization of pea enation mosaic virus-induced RNA-dependent RNA polymerase in infected peas. Virology 78:135–143
Reddy DVR, Murant AF, Raschké JH, Mayo MA (1985) Properties and partial purification of infective material from plants containing groundnut rosette virus. Ann Appl Biol 107:65–78
Reid MS, Matthews REF (1966) On the origin of the mosaic induced by turnip yellow mosaic virus. Virology 28:563–570
Roberts A, Santa Cruz S, Roberts IM, Prior DAM, Turgeon R, Oparka K (1997) Phloem-unloading in sink leaves of *Nicotiana benthamiana* is symplastic and regulated by class III veins: comparison of fluorescent solute with fluorescent virus. Plant Cell 9:1381–1396
Robinson-Beers K, Evert RF (1991) Fine structure of plasmodesmata in mature leaves of sugarcane. Planta 184:307–318
Rodríguez-Cerezo E, Findlay K, Shaw JG, Lomonossoff GP, Qiu SG, Linstead P, Shanks M, Risco C (1997) The coat and cylindrical inclusion proteins of a potyvirus are associated with connections between plant cells. Virology 236:296–306
Rojas MR, Zerbini FM, Allison RF, Gilbertson RL, Lucas WJ (1997) Capsid protein and helper component-proteinase function as potyvirus cell-to-cell movement proteins. Virology 237:283–295
Roossinck MJ, Palukaitis P (1990) Rapid induction and severity of symptoms in zucchini squash (*Cucurbita pepo*) map to RNA 1 of cucumber mosaic virus. Mol Plant-Microbe Interact 3:188–192
Russin WA, Evert RF, Vanderveer PJ, Sharkey TD, Briggs SP (1996) Modification of a specific class of plasmodesmata and loss of sucrose export ability in the sucrose export defectivel maize mutant. Plant Cell 8:645–658
Saito T, Yamanaka K, Okada Y (1990) Long-distance movement and viral assembly of tobacco mosaic virus mutants. Virology 176:329–336
Samuel G (1934) The movement of tobacco mosaic virus within the plant. Ann Appl Biol 21:90–111
Sänger HL (1969) Functions of the two particles of tobacco rattle virus. J Virol 3:304–312
Sanger M, Passmore B, Falk BW, Bruening G, Ding B, Lucas WJ (1994) Symptom severity of beet western yellows virus strain ST9 is conferred by the ST9-associated RNA and is not associated with virus release from the phloem. Virology 200:48–55
Santa Cruz S, Roberts AG, Prior DAM, Chapman S, Oparka KJ (1998) Cell-to-cell and phloem-mediated transport of potato virus X: the role of virions. Plant Cell 10:495–510
Schaad MC, Carrington JC (1996) Suppression of long-distance movement of tobacco etch virus in a nonsusceptible host. J Virol 70:2556–2561

Schaad MC, Lellis AD, Carrington JC (1997) VPg of tobacco etch potyvirus is a host genotype-specific determinant for long-distance movement. J Virol 71:8624–8631
Schmitz J, Stussi-Garaud C, Tacke E, Prüfer D, Rohde W, Rohfritsch O (1997) In situ localization of the putative movement protein (pr17) from potato leafroll luteovirus (PLRV) in infected and transgenic potato plants. Virology 235:311–322
Schneider WL, Greene AE, Allison RF (1997) The carboxy-terminal two-thirds of the cowpea chlorotic mottle bromovirus capsid protein is incapable of virion formation yet supports systemic movement. J Virol 71:4862–4865
Scholthof HB, Scholthof K-BG, Kikkert M, Jackson AO (1995) Tomato bushy stunt virus spread is regulated by two nested genes that function in cell-to-cell movement and host-dependent systemic invasion. Virology 213:425–438
Schulz A (1998) Phloem. Structure related to function. Prog Bot 59:429–475
Séron K, Haenni A-L (1996) Vascular movement of plant viruses. Mol Plant-Microbe Interact 9: 435–442
Shepardson S, Esau K, McCrum R (1980) Ultrastructure of potato leaf phloem infected with potato leafroll virus. Virology 105:379–392
Siegel A, Zaitlin M, Sehgal OP (1962) The isolation of defective tobacco mosaic virus strains. Proc Natl Acad Sci 48:1845–1851
Simon AE, Li XH, Lew JE, Stange R, Zhang C, Polacco M, Carpenter CD (1992) Susceptibility and resistance of *Arabidopsis thaliana* to turnip crinkle virus. Mol Plant-Microbe Interact 5:496–503
Sjölund RD (1997) The phloem sieve element: a river runs through it. Plant Cell 9:1137–1146
Sokolova M, Prüfer D, Tacke E, Rohde W (1997) The potato leafroll virus 17K movement protein is phosphorylated by a membrane-associated protein kinase from potato with biochemical features of protein kinase C. FEBS Lett 400:201–205
Solberg RA, Bald JG (1962) Virus invasion and multiplication during leaf histogenesis. Virology 17: 359–361
Sudarshana MR, Wang HL, Lucas WJ, Gilbertson RL (1998) Dynamics of bean dwarf mosaic geminivirus cell-to-cell and long-distance movement in *Phaseolus vulgaris* revealed, using the green fluorescent protein. Mol Plant-Microbe Interact 11:277–291
Tacke E, Schmitz J, Prüfer D, Rohde W (1993) Mutational analysis of the nucleic acid-binding 17-KDa phosphoprotein of potato leafroll luteovirus identifies an amphipathic α-helix as the domain for protein/protein interactions. Virology 197:294–282
Takamatsu N, Ishikawa M, Meshi T, Okada Y (1987) Expression of bacterial chloramphenicol acetyltransferase gene in tobacco plants mediated by TMV-RNA. EMBO J 6:307–311
Tang W, Leisner SM (1997) Cauliflower mosaic virus isolate NY8153 breaks resistance in *Arabidopsis* ecotype En-2. Phytopathology 87:792–798
Thompson JR, García-Arenal F (1998) The bundle sheath-phloem interface of *Cucumis sativus* is a boundary to systemic infection by tomato aspermy virus. Mol Plant-Microbe Interact 11:109–114
Traynor P, Young BM, Ahlquist P (1991) Deletion analysis of brome mosaic virus 2a protein: effects on RNA replication and systemic spread. J Virol 65:2807–2815
Turgeon R (1986) The import-export transition in dicotyledonous leaves. In: Cronshaw J, Lucas WJ, Giaquinta RT (eds) Phloem transport. Alan R Liss, New York, pp 285–291
Turgeon R (1996) Phloem loading and plasmodesmata. Trends Plant Sci 1:418–423
Turgeon R, Webb JA, Evert RF (1975) Ultrastructure of minor veins of *Curcurbita pepo* leaves. Protoplasma 83:217–232
Turgeon R, Webb JA (1976) Leaf development and phloem transport in *Cucurbita pepo*: maturation of the minor veins. Planta 129:265–269
Vaewhongs AA, Lommel SA (1995) Virion formation is required for the long-distance movement of red clover necrotic mosaic virus in movement protein transgenic plants. Virology 212:607–613
van Bel AJE (1993) The transport phloem. Specifics of its functioning. Prog Bot 54:134–150
van Bel AJE, Kempers R (1997) The pore/plasmodesm unit; key element in the interplay between sieve element and companion cell. Prog Bot 58:278–291
van Bel AJE, Oparka KJ (1995) On the validity of plasmodesmograms. Bot Acta 108:174–182
van Bel AJE, van Rijen HVM (1994) Microelectrode-recorded development of the symplasmic autonomy of the sieve element/companion cell complex in the stem phloem of *Lupinus luteus* L. Planta 192:165–175
van Bel AJE, van Kesteren WJP, Papenhuijzen C (1988) Ultrastructural indications for coexistence of symplastic and apoplastic phloem loading in *Commelina benghalensis* leaves. Planta 176:159–172
van den Heuvel JFJM, de Blank CM, Peters D, van Lent JWM (1995) Localization of potato leafroll virus in leaves of secondarily infected potato plants. Eur J Plant Pathol 101:567–571
van Lent JWM, Verduin BJM (1987) Detection of viral antigen in semi-thin sections of plant tissue by immunogold-silver staining and light microscopy. Neth J Pathol 93:261–272

Veidt I, Bouzaoubaa SE, Leiser RM, Ziegler-Graff V, Guilley H, Richards K, Jonard G (1992) Synthesis of full-length transcripts of beet western yellows virus RNA: messenger properties and biological activity in protoplasts. Virology 186:192–200

Volk GM, Turgeon R, Beebe DU (1996) Secondary plasmodesmata formation in the minor-vein phloem of *Cucumis melo* L and *Cucurbita pepo* L. Planta 199:425–432

Waigmann E, Zambryski P (1995) Tobacco mosaic virus movement protein-mediated protein transport between trichome cells. Plant Cell 7:2069–2079

Waigmann E, Lucas WJ, Citovsky V, Zambryski P (1994) Direct functional assay for tobacco mosaic virus cell-to-cell movement protein and identification of a domain involved in increasing plasmodesmal permeability. Proc Natl Acad Sci 91:1433–1437

Wang HL, Gilbertson RL, Lucas WJ (1996) Spatial and temporal distribution of bean dwarf mosaic geminivirus in *Phaseolus vulgaris* and *Nicotiana benthamiana*. Phytopathology 86:1204–1214

Watanabe Y, Morita N, Nishiguchi M, Okada Y (1987) Attenuated strains of tobacco mosaic virus reduced synthesis of a viral protein with a cell-to-cell movement function. J Mol Biol 194:699–704

Weiland JJ, Edwards MC (1994) Evidence that the aa gene of barley stripe mosaic virus encodes determinants of pathogenicity to oat (*Avena sativa*). Virology 201:116–126

Weiland JJ, Edwards MC (1996) A single nucleotide substitution in the aa gene confers oat pathogenicity to barley stripe mosaic virus strain ND18. Mol Plant-Microbe Interact 9:62–67

White RG, Badelt K, Overall RL, Vesk M (1994) Actin associated with plasmodesmata. Protoplasma 180:169–184

Wilson CR, Jones RAC (1992) Resistance to phloem transport of potato leafroll virus in potato plants. J Gen Virol 73:3219–3224

Wimmers LE, Turgeon R (1991) Transfer cells and solute uptake in minor veins of *Pisum sativum* leaves. Planta 186:2–12

Wintermantel WM, Anderson EJ, Schoelz JE (1993) Identification of domains within gene VI of cauliflower mosaic virus that influence systemic infection of *Nicotiana bigelovii* in a light-dependent manner. Virology 196:789–798

Wintermantel WM, Banerjee N, Oliver JC, Paolillo DJ, Zaitlin M (1997) Cucumber mosaic virus is restricted from entering minor veins in transgenic tobacco exhibiting replicase-mediated resistance. Virology 231:248–257

Wu X, Shaw JG (1997) Evidence that a viral replicase protein is involved in the disassembly of tobacco mosaic virus particles in vivo. Virology 239:426–434

Wu X, Xu Z, Shaw JG (1994) Uncoating of tobacco mosaic virus RNA in protoplasts. Virology 200: 256–262

Wyatt SD, Kuhn CW (1980) Derivation of a new strain of cowpea chlorotic mottle virus from resistant cowpeas. J Gen Virol 49:289–296

Xiong Z, Kim KH, Giesman-Cookmeyer D, Lommel SA (1993) The roles of the red clover necrotic mosaic virus capsid and cell-to-cell movement proteins in systemic infection. Virology 192:27–32

Zech H (1952) Untersuchungen über den infektionsvorgang und die wanderung des tabakmosaikvirus im pflanzenkörper. Planta 40:461–514

Subject Index

References to illustrations are given in **boldface** type.

MIX
Papier aus verantwortungsvollen Quellen
Paper from responsible sources
FSC® C105338

If you have any concerns about our products,
you can contact us on
ProductSafety@springernature.com

In case Publisher is established outside the EU,
the EU authorized representative is:
Springer Nature Customer Service Center GmbH
Europaplatz 3, 69115 Heidelberg, Germany

Printed by Libri Plureos GmbH
in Hamburg, Germany